Integrated Pest Management

This book provides an overview of recent trends in entomology, encompassing advances in biological control, pheromone-based techniques, genetic control, and precision agriculture. It examines recent developments in pest detection, cultural, biological, and chemical control methods, and their practical applications in agriculture and food processing systems. It also discusses the increasing use of molecular techniques in pest management, such as RNA interference and gene editing, while highlighting the importance of data-driven decision-making. Additionally, it emphasizes the use of remote sensing and other technologies to monitor pest populations and predict outbreaks. This book further underscores the role of integrated pest management (IPM) in sustainable agriculture, addressing the necessity to balance pest management with environmental protection and the preservation of beneficial organisms. The book chapters explore future trends and prospects in entomology, including the utilization of artificial intelligence, robotics, and other emerging technologies. This book serves as a valuable resource for researchers and practitioners of entomology aiming to enhance pest management practices and promote sustainable agriculture.

Key Features

- Provides insights into implementing pest control strategies in agricultural practices, ensuring crop protection and yield optimization.
- Discusses cultural, biological, and chemical control methods for pest management.
- Highlights the advantages of integrated pest management over conventional pest management.
- Explores the application of advanced molecular methods, such as RNA interference and gene editing, in pest management.
- Examines the integration of remote sensing and other cutting-edge technologies for pest monitoring and prediction.

Integrated Pest Management

Emerging Technologies and Sustainable Approaches

Edited by

Habib Ali and Zhihang Zhuo

CRC Press
Taylor & Francis Group
Boca Raton London New York

CRC Press is an imprint of the
Taylor & Francis Group, an **informa** business

Designed cover image: Shutterstock

First edition published 2026
by CRC Press
2385 NW Executive Center Drive, Suite 320, Boca Raton FL 33431

and by CRC Press
4 Park Square, Milton Park, Abingdon, Oxon, OX14 4RN

CRC Press is an imprint of Taylor & Francis Group, LLC

ISBN: 978-1-032-85099-3 (hbk)
ISBN: 978-1-032-85100-6 (pbk)
ISBN: 978-1-003-51653-8 (ebk)

DOI: 10.1201/9781003516538

Typeset in Times
by Apex CoVantage, LLC

Contents

About the Editors

Habib Ali, PhD in agricultural entomology and insect pest control, completed his undergraduate studies in agricultural entomology at the University of Agriculture Faisalabad, Pakistan. In December 2018, he earned his doctorate from Fujian Agriculture and Forestry University (FAFU), Fuzhou, China. Currently, Dr. Ali is serving as a faculty member at the Khwaja Fareed University of Engineering and Information Technology (KFUEIT), Rahim Yar Khan, Pakistan. He has authored over 150+ national and international research papers in high-impact journals, contributed 35 book chapters, and edited several books with leading international publishers.

Dr. Ali's academic excellence has been recognized with several honors and awards. He received the Gold Medal for Best Performance, conferred by the Governor of Punjab. In 2018, he was awarded the Best International Student of the Year and the Best Thesis Award by the Department of Plant Protection, FAFU. He was also included in the World Scientist Index 2021 and honored with the Council of Asian Science Editors Award (2022–2023). In addition to his research, Dr. Ali has actively participated in numerous international conferences, symposia, workshops, and faculty training programs, while also leading competitive research projects. Dr. Ali's research interests span a wide range of topics, including agricultural entomology, insect ecology, insect biodiversity, and crop pests/invasive species (such as the red palm weevil, coconut hispine beetle, and hispid leaf beetle). His expertise also extends to pyrosequencing, genomics, bioinformatics (R programming), bacteria–host interactions, Wolbachia pipientis, microbial symbiosis, biopesticides, biocontrol agents, microbial ecology, metagenomics, applied microbiology, bee health, disease management, and integrated pest management (IPM).

Zhihang Zhuo currently serves as an associate professor at the College of Life Science, China West Normal University, located in Sichuan, China. He obtained his PhD from Sichuan Agricultural University, China, in 2017. Since then, he has held various positions, including as lecturer, associate professor, head of forest protection department, and deputy town mayor in different regions of China, such as the Municipal Forestry Bureau, Hainan University, Township People's Government, and China West Normal University. He has been recognized as a high-level talent in Sichuan Province and Hainan Province, as well as appointed as a director of the Sichuan Society of Entomology and an expert in related think tanks. Dr. Zhuo also serves as a master's supervisor at China West Normal University, Hainan University, and Sichuan Agricultural University. He has taught a variety of courses at both undergraduate and postgraduate levels, demonstrating excellence in teaching and research. His primary research interests lie in entomology, ecology, and molecular biology, with a focus on pest and disease control, biological invasion, insect chemical ecology, resource insect development, and model development and utilization. Additionally, he is actively engaged in projects related to the development of attractants, resistance breeding, development of biocontrol bacteria, multiomics joint analysis, and smart agricultural and forestry development.

1 Advances in Pest Detection and Monitoring in Integrated Pest Management

A. Rajabpour and F. Yarahmadi*
Department of Plant Protection, Faculty of Agriculture, Agricultural Sciences and Natural Resources University of Khuzestan, Bavi, Iran
*Corresponding author

1.1 INTRODUCTION

Integrated pest management (IPM) is a comprehensive and ecologically based pest control strategy that effectively utilizes natural resources while considering various factors such as natural enemies, economic thresholds, plant susceptibility and breeding, pest biology, and climatic conditions (Čirjak et al., 2022). By integrating multiple approaches, IPM aims to minimize the impact of pests on crops, human health, and the environment. A crucial element of successful IPM is the timely and accurate detection of pests within agricultural systems, which enables informed decision-making and enhances pest management effectiveness (Dent & Bink, 2020). Pest detection and monitoring are essential for providing valuable information on pest presence, population dynamics, and distribution, which are critical in making informed pest management decisions (Rossi et al., 2019).

While traditional pest detection methods, such as visual inspections and trapping devices, provide foundational insights, they often lack the accuracy, efficiency, and scalability required for today's agricultural demands (Schellhorn & Jones, 2021; Rajabpour & Yarahmadi, 2023a). Modern technologies are revolutionizing pest surveillance with innovative tools for more precise detection, such as remote sensing technologies (Terentev et al., 2022) – like satellite imagery and drones – that offer insights into crop health, pest infestations, and environmental conditions. Moreover, automated monitoring systems deliver real-time data on pest activity, enabling timely interventions based on reliable information (Iost Filho et al., 2020; Popescu et al., 2023).

Area-wide integrated pest management (AW-IPM) approaches rely heavily on effective surveillance and communication to estimate natural population sizes, guide targeted control measures, and assess the efficacy of pest control actions. However, knowing where and when pests arrive in real time, communicating this information quickly, and delivering pest control measures in a coordinated manner present significant challenges. Fortunately, advances in agricultural technology are creating opportunities to overcome these barriers and facilitate the adoption of AW-IPM (Schellhorn & Jones, 2021). Recent technological advancements in insect surveillance, data flow, and information communication are becoming increasingly available and commercially viable. These changes are driven by macroeconomic trends, including rising labor costs, international agricultural trade dynamics, and shifting consumer demands, alongside innovations in hardware technology that enhance computational capabilities beyond traditional desktop environments (Schellhorn & Jones, 2021).

By combining traditional techniques with modern technologies, the early detection of pests allows for proactive management while reducing the need for chemical interventions, thereby promoting sustainable practices across various agricultural fields (Huded et al., 2023). Effective pest detection and monitoring help growers identify pest problems early, facilitating timely interventions to prevent outbreaks and minimize crop damage. By implementing proactive pest monitoring strategies, farmers can decrease reliance on broad-spectrum pesticides that may harm beneficial organisms and disrupt ecosystems (Ragunathan & Divakar, 2020). Furthermore, the integration of advanced technologies with traditional methods enhances the effectiveness of pest surveillance programs, leveraging data-driven insights for informed decision-making regarding pest control strategies, ultimately improving crop protection and sustainability. Early pest detection is crucial for implementing sustainable pest management practices within the IPM framework, allowing for the prompt application of targeted, environmentally friendly control measures that decrease reliance on synthetic pesticides and mitigate the risk of pest resistance (Rustia et al., 2022; Rajabpour & Yarahmadi, 2024) (Figure 1.1).

In sustainable pest management practices, the goal transcends mere pest control; it aims to maintain ecological balance and minimize impacts on non-target organisms.

DOI: 10.1201/9781003516538-1

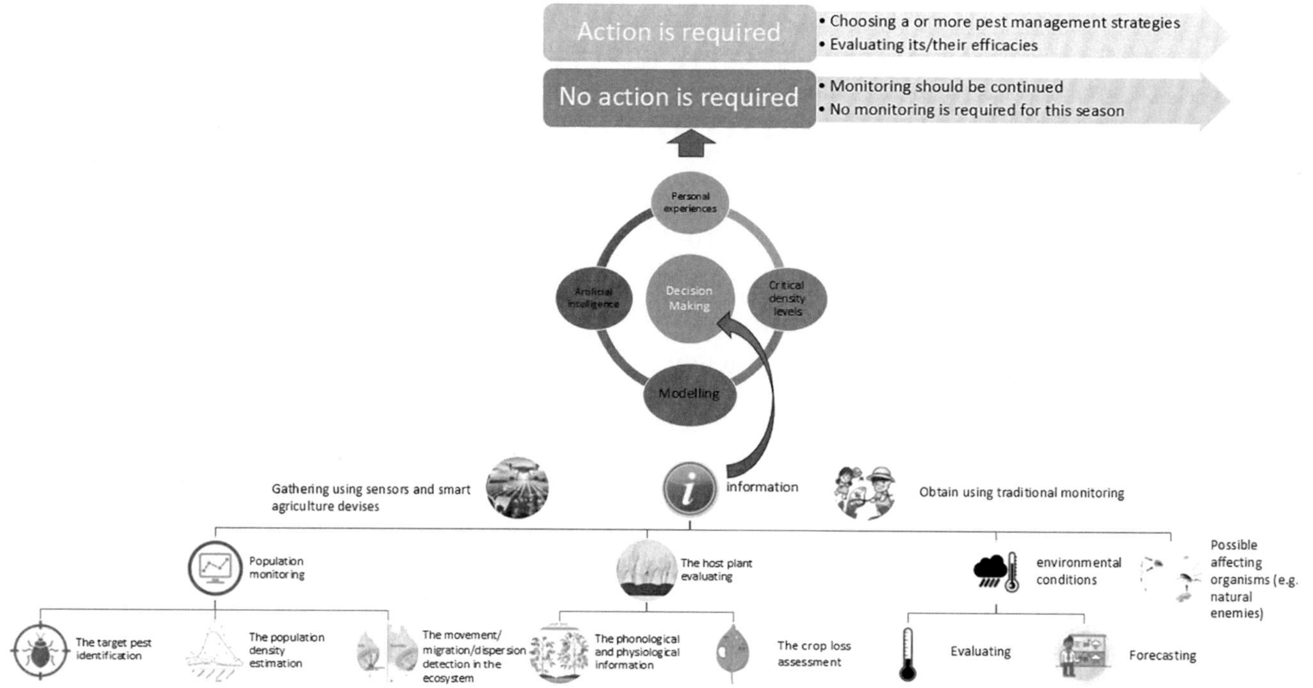

FIGURE 1.1 Components of a decision-making system for a successful integrated pest management. The components include pest monitoring, risk assessment, management strategies (biological, cultural, chemical, and mechanical controls), economic thresholds, and evaluation of outcomes. The flowchart highlights the iterative process, starting with monitoring and ending with continuous evaluation and adaptation of strategies to improve pest management effectiveness.

Early detection plays a pivotal role in achieving these objectives by equipping growers with the necessary information to make informed decisions that prioritize the long-term health of agroecosystems. By proactively monitoring pest populations and swiftly responding to changes in pest dynamics, farmers can safeguard their crops while preserving biodiversity and ecosystem services (Rossi et al., 2023).

The purpose of this chapter is to provide a comprehensive overview of the importance of pest detection and monitoring in the context of IPM. By exploring the significance of early pest detection, comparing traditional methods to modern technologies, and highlighting the role of pest surveillance in sustainable pest management practices, this chapter aims to inform readers about the key principles and approaches to effective pest monitoring. Through a thorough exploration of these topics, this chapter will equip readers with a deeper understanding of the critical role of pest detection and monitoring in IPM and encourage the adoption of proactive and sustainable pest management practices in agricultural systems (Yatribi, 2020).

1.2 SUSTAINABLE APPROACHES IN PEST DETECTION AND MONITORING: INTEGRATING BIOLOGICAL CONTROL, IPM STRATEGIES, AND ENVIRONMENTAL CONSERVATION

Pest detection and monitoring are crucial for sustainable agriculture, enabling early identification and assessment of pest populations. Traditional methods often rely on chemical pesticides, which can harm beneficial organisms and the environment. Sustainable approaches now emphasize IPM, focusing on biological control and non-chemical solutions to mitigate environmental impacts (Lima et al., 2020). *Biological control* is a sustainable pest management strategy that utilizes natural enemies like predators, parasitoids, and pathogens to regulate pest populations. By reducing reliance on chemical pesticides, it preserves beneficial organisms and enhances pest control efficacy, minimizing environmental impact and aligning agricultural practices with sustainability goals (Dent & Bink, 2020).

Effective pest surveillance is vital for informed pest management, involving continuous monitoring, species identification, and damage assessment. Integrating biological control enhances this by allowing farmers to target specific pests, release natural enemies, and monitor biological agents' impacts (Dent & Bink, 2020). Successful IPM programs exemplify this synergy, combining biological control, cultural practices, and chemical interventions within a holistic framework. IPM aims to minimize pest damage and protect natural resources through multiple strategies, including pest monitoring and judicious chemical use. The implementation of IPM significantly reduces pesticide reliance, preserves ecosystem services, promotes biodiversity, and supports farmers' livelihoods. Non-chemical pest monitoring methods, such as pheromone traps, sticky traps, and visual inspections, enable farmers to assess pest populations without synthetic pesticides, aligning practices with organic farming principles (Huded et al., 2023). These techniques are essential for environmentally friendly, cost-effective, and sustainable pest management. The environmental impacts of traditional pest detection methods have long raised concerns among agricultural professionals and environmentalists alike. In recognizing the adverse effects of chemical pesticides and synthetic monitoring tools, it becomes clear that a shift toward sustainable pest detection practices is necessary for maintaining ecological balance (Rossi et al., 2023).

The conventional approach to pest detection often involves the heavy application of chemical pesticides, which can severely impact non-target organisms, pollinators, aquatic ecosystems, and overall soil health. Synthetic monitoring tools, including insecticides, acaricides, and herbicides, can disrupt ecological balance, diminish biodiversity, and pose risks to water sources while harming beneficial insects. Overreliance on these traditional methods has led to long-term consequences, including pesticide resistance, pest resurgence, instability within ecosystems, and negative implications for human health (Ragunathan & Divakar, 2020).

In contrast, adopting sustainable pest detection practices brings numerous benefits for environmental conservation. These practices facilitate the preservation of beneficial organisms, including natural predators and pollinators, which are crucial for ecosystem functionality. By minimizing pesticide application and promoting eco-friendly pest management strategies, these sustainable approaches also contribute to reduced environmental contamination, leading to improved soil health and enhanced water quality (Ragunathan & Divakar, 2020; Huded et al., 2023).

Moreover, sustainable pest detection practices foster ecosystem resilience by promoting biological control and enhancing biodiversity. This, in turn, maintains essential ecosystem services and significantly reduces the ecological footprint of agricultural activities. By adopting these practices, agriculture can align more closely with conservation goals, ultimately benefiting both farmers and the environment (Yatribi, 2020).

1.2.1 Traditional Methods of Pest Detection and Monitoring in Agriculture: Techniques, Limitations, and Success Stories

Pest detection and monitoring are crucial aspects of IPM strategies in agriculture, with traditional methods being used for generations to track pests and inform management decisions (Rossi et al., 2019). This analysis explores the effectiveness, limitations, and success stories of traditional pest detection and monitoring techniques in sustainable agriculture.

Visual inspections involve physically scouting crops for signs of pest presence or damage, allowing growers to identify early infestations and take appropriate action. Trapping devices, such as sticky traps and pheromone traps, are used to capture and monitor pest populations, providing data on species composition and densities. Crop sampling entails collecting plant samples to quantify pest populations and assess damage levels systematically or through targeted sampling (Azfar et al., 2018). Monitoring pest life cycles helps predict peak activity periods and target vulnerable stages for control interventions, such as insecticide applications. However, traditional methods have limitations, including being labor-intensive, subjective, and limited in scope and potentially providing incomplete data. These challenges can lead to gaps in pest detection and monitoring, inconsistent management decisions, and overlooking emerging threats.

There are many successful traditional monitoring programs for various pests, such as aphids (Rajabpour & Yarahmadi, 2012; Kafeshani et al., 2018; Yaghobi et al., 2019), lepidopteran larvae (Shahbi & Rajabpour, 2017; Dinarvand et al., 2020; Rashidi et al., 2020; Yaghobi et al., 2020), and mites (Yarahmadi & Rajabpour, 2013), across different crops. In these studies, data on population density fluctuations and the occurrence of pest life stages were collected using accurate systematic sampling plans implemented by experts with appropriate tools, such as various traps.

1.2.2 Area-Wide Integrated Pest Management (AW-IPM)

The AW-IPM is an innovative strategy designed to manage pest populations across broader landscapes, rather than confining efforts to individual farms or fields. This approach acknowledges that pests do not respect property boundaries, necessitating a coordinated response from various stakeholders, including farmers, agricultural organizations, and government agencies. By integrating diverse pest management tactics at a landscape level, AW-IPM aims to achieve sustainable pest control while minimizing both economic and environmental impacts (Schellhorn & Jones, 2021).

AW-IPM is based on IPM principles but tailored for managing pests over larger areas. Collaboration among stakeholders is crucial, involving farmers, extension agents, researchers, and policymakers. This collective effort enhances understanding of pest dynamics and promotes information exchange for effective pest management. AW-IPM integrates various pest control strategies like cultural practices and biological, mechanical, and chemical methods, customized to the area's needs. By combining these tactics, AW-IPM aims to reduce pest populations economically while minimizing pesticide application. Effective pest surveillance is vital, involving data collection and analysis to make informed decisions. Surveillance under AW-IPM is conducted at a landscape level, enabling monitoring across farms and regions to identify trends and adapt strategies as needed for sustainable pest management. One of the primary advantages of effective pest surveillance in AW-IPM is the ability to detect pest outbreaks early. Early detection is crucial for preventing the spread of pests and minimizing damage to crops. By closely monitoring pest populations and environmental conditions, stakeholders can implement control measures promptly, thereby reducing the likelihood of widespread infestations. Additionally, landscape-level surveillance can help pinpoint potential sources of pest migration, facilitating targeted interventions that address the root causes of pest issues (Vreysen et al., 2007).

The integration of technology into pest surveillance has significantly enhanced the capabilities of AW-IPM. Advances in remote sensing, geographic information systems (GIS), and data analytics have transformed the monitoring and management of pest populations. For example, satellite imagery can provide valuable insights into crop health and pest distribution patterns, while GIS can help visualize data and identify areas at risk of pest outbreaks. These technological tools enable real-time data collection and analysis, allowing stakeholders to make timely and informed decisions (Shafi et al., 2019).

The implementation of AW-IPM faces significant challenges, particularly in communication among stakeholders. Effective communication is crucial for coordinating pest management efforts, but many farmers are reluctant to share data on pest populations due to concerns about confidentiality and competition. This reluctance can hinder a comprehensive understanding of pest dynamics. Additionally, the need for real-time data sharing poses logistical challenges, especially in rural areas with limited internet access (Vreysen et al., 2007). To overcome these challenges, it is essential to build trust and foster collaboration among stakeholders through regular meetings, workshops, and training sessions. Technology can aid communication by providing platforms for data sharing, such as online databases and mobile applications that allow farmers to report pest sightings and management practices (Shi et al., 2020). Utilizing mobile technology for real-time reporting of pest populations can enhance stakeholder responsiveness, and integrating remote sensing technologies can provide valuable data without extensive fieldwork (Terentev et al., 2022). Education and training are also critical for the successful implementation of AW-IPM. Stakeholders need the knowledge and skills to monitor pests and implement integrated management strategies effectively. Extension services and educational programs can provide necessary training on pest identification, surveillance techniques, and technology use in pest management, ultimately leading to improved pest management outcomes (Dent & Binks, 2020; Shi et al., 2020).

1.3 EMERGING TECHNOLOGIES IN PEST DETECTION

1.3.1 Advancements in Pest Detection: Integrating Technology into IPM Strategies

Pest detection is a critical component of IPM strategies aimed at reducing the impact of pests on agriculture and ecosystems. Traditional surveillance methods, such as visual inspections and traps, have proven effective but are often labor-intensive and limited in their ability to efficiently cover extensive agricultural areas. With rapid technological advancements, innovative pest detection practices have emerged, transforming how pest populations are monitored and managed. These advancements include remote sensing, drones, sensor technologies, artificial intelligence (AI), and machine learning, all of which significantly enhance the efficacy of pest detection and improve management outcomes (Shafi et al., 2019; Lima et al., 2020; Hanif et al., 2022; Popescu et al., 2023).

Remote sensing technologies represent one of the most transformative advancements in pest detection methodologies. By utilizing satellite imagery or aerial data, these technologies collect crucial information about crop health, vegetation dynamics, and pest infestations over large areas. Delivering data through various wavelengths, remote sensing can identify subtle changes in plant physiology and the conditions conducive to pest activities. For instance, spectral imaging can differentiate between healthy plants and those affected by pests based on distinct spectral signatures. This differentiation allows for early detection of pest outbreaks, enabling growers to act swiftly before infestations spread. Remote sensing platforms provide a quick and cost-effective approach for monitoring vast agricultural fields, thereby facilitating informed decision-making regarding pest control strategies and resource allocation (Terentev et al., 2022).

In addition to remote sensing, aerial technology has gained prominence in the agricultural sector, particularly through the use of drones, or unmanned aerial vehicles (UAVs). Drones are increasingly recognized for their capability to collect high-resolution imagery and detailed data concerning crop conditions. By equipping these platforms with multispectral cameras or various sensors, farmers can obtain intricate images of their fields, identifying variations in pest pressures and overall crop performance. Regular drone flights over farmland allow growers to continuously monitor pest activity and assess crop damage, pinpointing specific areas that require intervention (Figure 1.2). This flexibility and efficiency in pest surveillance enable rapid responses to outbreaks and precise application of pest control measures, ultimately contributing to better pest management outcomes (Iost Filho et al., 2020).

The role of sensor technologies in modern agriculture cannot be overstated. These devices play a critical part in real-time monitoring of environmental conditions and pest activities within agricultural settings. Sensors that measure temperature, humidity, and pheromone levels can be strategically deployed in fields to detect changes in pest behavior and fluctuations in population dynamics (Lima et al., 2020). For example, pheromone traps that incorporate advanced sensors can attract and capture specific pest species, thereby providing valuable data on population trends and migration. When connected through integrated wireless communication systems, sensor networks enable continuous pest monitoring and can send automated alerts when pest populations exceed predetermined thresholds. By enhancing the timeliness and accuracy of pest detection, sensors allow farmers to implement targeted and responsive pest management measures (Azfar et al., 2018; Shafi et al., 2019; Lima et al, 2020).

Artificial intelligence and machine learning technologies are further transforming pest surveillance practices, enabling the analysis of extensive datasets and the automation of decision-making processes. One of the key applications of AI in pest surveillance is image recognition. Neural networks, a subset of AI, have emerged as powerful tools in this domain, particularly in analyzing visual data from cameras, drones, and remote sensing (Terentev et al., 2022) platforms. By employing sophisticated deep learning algorithms, neural networks can be trained on vast datasets of labeled images depicting both healthy and infested plants. This training allows the systems to autonomously detect and classify pest-induced abnormalities in real time, significantly enhancing the speed and accuracy of assessments of crop health. The ability of neural networks to recognize patterns in complex datasets further empowers growers to address pest threats promptly and prevent potential yield losses (Shi et al., 2020; Zhang et al., 2022; Popescu et al., 2023).

Predictive modeling is another significant application of machine learning in pest detection. By leveraging historical

FIGURE 1.2 Use of aerial photography with a drone equipped with a hyperspectral camera to monitor sugarcane mite damage and population in sugarcane fields in Khuzestan, southeast Iran. The images include drone flight paths, captured hyperspectral images, and a plant health map generated through optical spectrum analysis. The visual representation demonstrates real-time monitoring and serves as a case study for aerial-wide integrated pest management.

Source: Images credited to the Sugarcane Industries and By-Product Industries Company of Khuzestan.

data on pest occurrences, weather conditions, and crop phenology, machine learning algorithms, including neural networks, can develop predictive models that forecast pest outbreaks and evaluate risks. Such models can effectively predict the likelihood of pest infestations and guide farmers in implementing proactive measures (Čirjak et al., 2022; Zhang et al., 2022). Predictive modeling can generate risk maps that highlight high-risk areas for pest activity based on environmental factors and pest biological cycles. When integrated with real-time data gathered from sensors and satellite imagery, these predictive models enable growers to anticipate pest threats and optimize their pest management strategies accordingly, significantly reducing potential damage to crops (Shafi et al., 2019).

Additionally, AI-driven decision support systems are becoming vital tools for modern pest management. Neural networks enhance these systems by providing personalized recommendations for pest control based on real-time data and predictive analytics (Popescu et al., 2023). By analyzing a multitude of information sources – including weather data, crop conditions, pest populations, and available control options – these systems recommend optimal pest management practices. They consider critical factors, such as the efficacy of pesticides, environmental implications, and overall economic costs. This holistic approach aids farmers in making informed decisions about pest control interventions, promoting sustainable practices in agriculture. Furthermore, AI algorithms enhance decision-making efficiency by automating the selection of control measures and tracking their effectiveness over time. Consequently, these decision support systems contribute to more efficient and sustainable pest management, allowing for data-driven decision-making and adaptable strategies that can evolve with changing conditions in the field (Čirjak et al., 2022). In conclusion, the integration of advanced technologies in pest detection represents a significant leap forward in agricultural practices, moving beyond traditional methods to more efficient, accurate, and sustainable approaches. As pest populations continue to pose challenges to crop yields and ecosystem health, the adoption of remote sensing, drones (Iost Filho et al., 2020), sensors (Shafi et al., 2019), AI, and neural networks (Zhang et al., 2022; Popescu et al., 2023) will be crucial in developing holistic pest management solutions. These technologies not only enhance the ability to detect and respond to pest infestations but also foster a more resilient agricultural system capable of withstanding the pressures of a changing environment. Through these innovations, farmers can achieve better pest management outcomes, ensuring food security and ecological balance in the face of evolving agricultural challenges. As we look to the future of pest management, it is clear that technology will play an indispensable role in safeguarding crops and ecosystems against the threats posed by pests, ultimately shaping a more sustainable agricultural landscape.

1.4 BENEFITS AND CHALLENGES OF USING HIGH-TECH TOOLS FOR PEST MONITORING

High-tech tools for pest monitoring are transforming the landscape of agriculture, offering a multitude of benefits that enhance the accuracy, efficiency, and sustainability of pest surveillance practices. These advanced technologies, including remote sensing, drones, sensors, artificial AI, and machine learning, are changing how farmers, pest management professionals, and researchers approach pest detection and management. However, while these innovations present significant advantages, they also come with challenges and limitations that must be addressed to fully harness their potential (Terentev et al., 2022).

The adoption of high-tech tools in pest monitoring offers significant advantages, primarily through enhanced precision and targeted interventions (Khanna & Kaur, 2019; Dent & Binks, 2020; Zhang et al., 2022). Unlike traditional broad-spectrum methods, these technologies allow farmers to monitor specific pest populations closely, optimizing resource allocation and making pest management strategies more effective and efficient. By applying treatments only when necessary, farmers can reduce pesticide application and associated costs (Kershen, 2006; Rossi et al., 2019).

Early warning systems represent another critical benefit. Technologies such as remote sensing, drones, and sensors enable timely detection of pest infestations, allowing for proactive management. Early identification of pest threats helps prevent minor issues from escalating into significant crop damage, thus protecting yields and minimizing losses (Schellhorn & Jones, 2021; Terentev et al., 2022).

Data integration is a key advantage of high-tech pest monitoring tools. The use of AI and machine learning facilitates the amalgamation of diverse data sources, including sensor data, satellite imagery, and historical records. This comprehensive approach enhances understanding of pest dynamics and environmental conditions, leading to better-informed decision-making (Lima et al., 2020; Rustia et al., 2022). Analyzing integrated datasets helps farmers identify patterns in pest behavior, informing future management strategies and improving overall crop management (Popescu et al., 2023).

Automation further streamlines pest monitoring processes, reducing the need for manual labor. Automated systems can continuously monitor pest populations and environmental conditions, providing real-time insights that enhance operational efficiency. This automation saves time and labor costs while improving the reliability of pest monitoring efforts (Schellhorn & Jones, 2021).

Moreover, high-tech tools support sustainability in agriculture by enabling data-driven pest management practices. These technologies contribute to reducing pesticide application and minimizing environmental impacts, helping farmers enhance crop resilience and protect beneficial organisms in the ecosystem (Huded et al., 2023).

Integrating sustainable practices into pest management promotes healthier agricultural systems and long-term ecological balance.

However, challenges exist in adopting high-tech tools for pest monitoring. Cost is a significant barrier, as the initial investment and ongoing maintenance can be prohibitive, particularly for small-scale farmers (Yatribi, 2020; Schellhorn & Jones, 2021). Additionally, the need for technical expertise can hinder effective use; farmers may struggle to operate and interpret data from advanced technologies without specialized training (Kershen, 2006; Rossi et al., 2019; Dent & Binks, 2020).

Data security is another concern, as the collection and the storage of sensitive information raise issues regarding privacy and cybersecurity. Establishing robust data security protocols is vital to maintain trust in these technologies. Integration issues can complicate implementation, necessitating compatibility standards and interoperability solutions (Rossi et al., 2023). Lastly, validating and calibrating data from these tools are essential to ensure accuracy, as decisions based on unreliable data can lead to ineffective pest control measures and crop losses.

1.5 DATA ANALYSIS AND DECISION-MAKING IN PEST MONITORING

1.5.1 Data-Driven Insights: The Importance of Analysis in Pest Monitoring Practices

Data analysis is a fundamental component in the realm of pest monitoring, serving as a vital tool that allows researchers, farmers, and pest management professionals to interpret results accurately and make informed, strategic decisions regarding pest detection, population dynamics, and management strategies (Rustia et al., 2022). Employing a variety of monitoring tools and techniques, including traps, sensors, remote sensing technology, and comprehensive field surveys, stakeholders can glean important insights into pest behavior, distribution patterns, and the various environmental factors that influence pest populations. The importance of data analysis in pest monitoring can be illuminated through several key dimensions (Lockie et al., 2020; Fuentes-Peñailillo et al., 2024).

One of the primary advantages of data analysis is its capacity to facilitate the early detection of pest outbreaks. Through careful examination of historical and real-time monitoring data, stakeholders can identify significant increases in pest populations or levels of damage associated with pest activity. This analysis enables the tracking of pest dynamics over time, making it possible to recognize trends that may indicate an impending infestation (Azfar et al., 2018; Fuentes-Peñailillo et al., 2024). Early detection is crucial; it allows for timely intervention, which can effectively curb pest outbreaks before they lead to extensive damage and associated yield losses. The proactive approach afforded by data analysis is pivotal in protecting crops and ensuring agricultural productivity (Schellhorn & Jones, 2021). Another vital aspect where data analysis proves invaluable is in the identification of pest hotspots. By employing spatial analysis methods, such as GIS or advanced remote sensing techniques, stakeholders can visualize and map pest distribution patterns across agricultural landscapes. This spatial insight helps pinpoint high-risk areas or pest hotspots, allowing farmers and pest management professionals to prioritize their monitoring and control efforts in regions exhibiting the highest pest pressure. Targeted interventions are generally far more effective than broad-spectrum approaches, leading to better pest management outcomes and optimization of available resources. Thus, the ability to identify and focus efforts on specific areas enhances both operational efficiency and effectiveness in pest management (Shafi et al., 2019).

Data analysis is essential for monitoring pest management effectiveness. Analyzing data before and after control measures allows stakeholders to evaluate their impact on pests, crops, and biodiversity. This dynamic process enables refining strategies for more effective pest management. Advanced analytics revolutionize decision-making in agriculture and business by optimizing operations, managing risks, and driving innovation. Precision farming benefits from AI, IoT sensors, and satellite imagery to forecast yields, optimize irrigation, and improve soil management (Chen et al., 2022). Businesses use analytics for consumer behavior analysis, trend prediction, and supply chain optimization. Challenges like data quality, expertise, costs, and ethics require investments in infrastructure, talent, and compliance. Trends include AI automation, blockchain for transparency, and augmented analytics for data access. Stakeholders should invest in data capabilities, partnerships, and a data-driven culture for sustainable growth in a data-driven economy (Raji et al., 2024).

In addition to all these practical applications, data analysis is a cornerstone for decision support systems within the framework of IPM. By integrating diverse streams of data, including monitoring data, weather information, and crop phenology, data analysis aids in developing evidence-based recommendations for pest control measures. These systems utilize analytical models that optimize pest management strategies and reduce reliance on broad-spectrum pesticides, thus contributing to sustainable farming practices. Ultimately, data-driven decision-making within IPM enhances not only the effectiveness of pest control programs but also their economic efficiency and sustainability, a vital consideration in modern agriculture (Rustia et al., 2022).

Predictive modeling plays a crucial role in pest detection and management strategies, significantly enhancing agricultural resilience. Utilizing statistical techniques and machine learning, predictive modeling anticipates pest outbreaks, assesses risks, and optimizes pest control measures. By analyzing historical data on pest occurrences, environmental conditions, and crop factors, these models effectively forecast pest activity, pinpoint vulnerable crops or regions, and recommend preventive actions to reduce the risk of pest damage (Mesías-Ruiz et al., 2023). One of the primary functions of predictive modeling is to forecast pest outbreaks by evaluating the relationships between pest populations and environmental variables, including weather conditions and soil characteristics. Generating predictions about the timing, intensity, and geographical spread of infestations enables stakeholders to prepare proactively. This foresight allows farmers to implement preventive measures, adjust planting schedules, and allocate resources efficiently, mitigating pests' impact on crop production and promoting healthier agricultural systems (Chen et al., 2022).

Additionally, predictive modeling is adept at assessing pest risks by integrating complex datasets encompassing pest life cycles, host plant susceptibilities, and historical weather patterns. This information is vital for guiding pest management decisions, allowing stakeholders to focus monitoring efforts and adopt preventive measures proactively, optimizing resource allocation (Donatelli et al., 2017). The optimization of control measures is another aspect where predictive modeling excels. These sophisticated models simulate various pest control strategies, evaluating their effectiveness and sustainability (Halog & Manik, 2011). Predictive modeling enhances pest management by enabling informed decision-making and improving control efficiency while reducing environmental impacts. IPM frameworks utilize this adaptability, allowing continuous refinement of strategies based on real-time data for resilient agricultural systems (Čirjak et al., 2022; Zhang et al., 2022).

1.6 FUTURE DIRECTIONS IN PEST DETECTION AND MONITORING

The field of pest detection and monitoring is undergoing significant transformation, propelled by emerging technologies and the pressing need for sustainable pest management practices. As we look to the future, several focal points emerge to enhance surveillance capabilities, tackle challenges, and direct research and practical applications toward greater efficiency and sustainability. Emerging trends in pest surveillance are heavily influenced by innovative technologies. One such advancement is remote sensing and geospatial technologies, including satellite imaging, drones, and GIS. These tools facilitate the monitoring of pest populations, enabling the identification of infestation hotspots and mapping environmental conditions that influence pest dynamics. By integrating remote sensing data with pest surveillance techniques, decision-makers can obtain real-time, spatially nuanced information, allowing for more targeted and precise pest management interventions (Shafi et al., 2019). The Internet of Things (IoT) and sensor networks also play a critical role in this landscape, enabling the continuous monitoring of environmental variables, pest activities, and crop health indicators. Smart sensors equipped with various technologies can be placed in agricultural fields to gather data on pest presence and behavior, facilitating the early detection of outbreaks and the timely implementation of control measures (Khanna & Kaur, 2019; Chen et al., 2022).

Furthermore, machine learning and AI are revolutionizing pest surveillance. These technologies are capable of automating the analysis of vast datasets, recognizing patterns that may elude traditional methods. By training AI models on extensive pest-related data, researchers can develop predictive models that not only forecast pest outbreaks but also help optimize monitoring strategies and recommend timely interventions (Hanif et al., 2022).

Molecular techniques, notably DNA barcoding, provide swift and precise methods for pest identification and monitoring. These advancements enhance traditional identification methods, enabling the tracking of invasive species and monitoring genetic diversity within pest populations (Chua et al., 2023). Similarly, nanotechnology is paving the way for sensitive and rapid pest monitoring through the use of nanoscale biosensors and detection platforms, which can identify pest-related biomarkers or pathogens (Qasim et al., 2022).

Despite the advancements in pest detection technologies, challenges persist, particularly in data integration and interoperability. The variety of data sources and formats complicates harmonization, making effective communication critical for optimizing pest detection methods and improving data-driven decision-making in pest management (Chua et al., 2023). Cost and accessibility are major barriers to adopting advanced surveillance techniques, particularly for small-scale farmers and resource-poor regions. Improving affordability, enhancing availability, and providing training are essential for making these pest detection methods inclusive and widely adopted (Yatribi, 2020). Concerns surrounding privacy and data security also challenge the implementation of IoT devices, drones, and remote sensing technologies in pest surveillance. Safeguarding sensitive information, adhering to data protection regulations, and fostering trust through transparent data management practices are crucial for mitigating risks associated with these advanced technologies (Khanna & Kaur, 2019). Building technical capacity among agricultural professionals and extension agents is crucial for the effective deployment of innovative pest surveillance technologies. Initiatives that provide hands-on training, workshops, and best practices can empower stakeholders and facilitate the uptake of these methods.

To enhance the effectiveness of research and practical applications in pest detection, promoting interdisciplinary collaboration among experts in entomology, agronomy, engineering, data science, and social science is recommended. Such cross-disciplinary partnerships are essential for innovation, knowledge exchange, and developing integrated solutions for sustainable pest management. Conducting field validation studies is also critical, as empirical research in real-world contexts allows stakeholders to evaluate the performance, feasibility, and scalability of advanced pest detection methods. This approach generates valuable insights and user feedback, supporting broader technology adoption. Furthermore, fostering a culture of continuous monitoring, data-driven decision-making, and adaptive management is necessary for optimizing pest detection strategies. Engaging various stakeholders in co-designing pest surveillance solutions is vital for their success and relevance. Involvement from farmers, community members, policymakers, and industry partners ensures widespread uptake and promotes knowledge transfer, while participatory approaches help foster co-ownership and sustainability of pest surveillance innovations.

1.7 CONCLUSION

In conclusion, this review highlights significant advancements in pest detection and monitoring technologies that have transformed pest surveillance methods. The transition from traditional scouting techniques to modern tools like remote sensing, drones, IoT devices, and AI has greatly improved the accuracy and efficiency of pest monitoring. Remote sensing, utilizing satellites and UAVs, enables high-resolution imaging for the early detection of pest outbreaks and crop stress, while IoT devices facilitate real-time data collection on environmental parameters and pest activities. Moreover, the integration of artificial intelligence enhances the analysis and interpretation of pest data, streamlining detection processes. Continuous innovation in these technologies is essential for developing sustainable integrated pest management (IPM) practices, allowing for timely interventions and precision pest management that minimizes chemical pesticide reliance.

Ultimately, ongoing advancements in pest detection and monitoring are crucial for improving agricultural productivity and ensuring food security. By embracing these technologies and data-driven decision-making, stakeholders can effectively address pest challenges, protect ecosystems, and promote sustainable agricultural practices for future generations.

REFERENCES

Azfar, S., Nadeem, A., Alkhodre, A. B., Ahsan, K., Mehmood, N., Alghmdi, T., and Alsaawy, Y. 2018. Monitoring, detection and control techniques of agriculture pests and diseases using wireless sensor network: A review. *International Journal of Advanced Computer Science and Applications 9*(12). https://doi.org/10.14569/IJACSA.2018.091260

Chen, C. J., Li, Y. S., Tai, C. Y., Chen, Y. C., and Huang, Y. M. 2022. Pest incidence forecasting based on internet of things and long short-term memory network. *Applied Soft Computing 124*: 108895.

Chua, P. Y., Bourlat, S. J., Ferguson, C., Korlevic, P., Zhao, L., Ekrem, T., and Lawniczak, M. K. 2023. Future of DNA-based insect monitoring. *Trends in Genetics 39*(7): 531–544.

Čirjak, D., Miklečić, I., Lemić, D., Kos, T., and Pajač Živković, I. 2022. Automatic pest monitoring systems in apple production under changing climatic conditions. *Horticulturae 8*(6): 52. https://doi.org/10.3390/horticulturae8060520

Dent, D., and Binks, R. H. 2020. *Insect pest management*. CABI.

Dinarvand, N., Rajabpour, A., Sohani, N. Z., and Farkhari, M. 2020. Effect of weedy culture on population densities, spatial distributions and sampling procedures of Spodoptera exigua and Sesamia cretica (Lep., Noctuidae) in corn fields. *Bulletin of Entomological Research, 110*(1): 84–95.

Donatelli, M., Magarey, R. D., Bregaglio, S., Willocquet, L., Whish, J. P., and Savary, S. 2017. Modelling the impacts of pests and diseases on agricultural systems. *Agricultural Systems 155*: 213–224.

Fuentes-Peñailillo, F., Gutter, K., Vega, R., and Silva, G. C. 2024. Transformative technologies in digital agriculture: Leveraging Internet of Things, remote sensing, and artificial intelligence for smart crop management. *Journal of Sensor and Actuator Networks 13*(4): 39. https://doi.org/10.3390/jsan13040039

Halog, A., and Manik, Y. 2011. Advancing integrated systems modelling framework for life cycle sustainability assessment. *Sustainability 3*(2): 469–499.

Hanif, M. K., Khan, S. Z., and Bibi, M. 2022. Applications of artificial intelligence in pest management. In *Artificial Intelligence and Smart Agriculture Applications* (pp. 277–300). Auerbach Publications.

Huded, S., Savadatti, E., CH, S., Deb, L., Borah, A., Singh, V., and Panigrahi, C. K. 2023. Examining modern strategies for effective and sustainable agricultural plant protection techniques: A review. *International Journal of Environment and Climate Change 13*(11): 1331–1343.

Iost Filho, F. H., Heldens, W. B., Kong, Z., and de Lange, E. S. 2020. Drones: Innovative technology for use in precision pest management. *Journal of Economic Entomology 113*(1): 1–25.

Kafeshani, F. A., Rajabpour, A., Aghajanzadeh, S., Gholamian, E., and Farkhari, M. 2018. Spatial distribution and sampling plans with fixed level of precision for citrus aphids (Hom., Aphididae) on two orange species. *Journal of Economic Entomology, 111*(2): 931–941.

Kershen, D. L. 2006. Sustainable intensive agriculture: High technology and environmental benefits. *Kansas Journal of Law & Public Policy 16*: 424.

Khanna, A., and Kaur, S. 2019. Evolution of Internet of Things (IoT) and its significant impact in the field of precision agriculture. *Computers and Electronics in Agriculture* 157: 218–231.

Lima, M. C. F., de Almeida Leandro, M. E. D., Valero, C., Coronel, L. C. P., and Bazzo, C. O. G. 2020. Automatic detection and monitoring of insect pests – A review. *Agriculture 10*(5): 161.

Lockie, S., Fairley-Grenot, K., Ankeny, R., Botterill, L., Howlett, B., Mcbratney, A., and Woodhead, I. 2020. *The future of agricultural technologies*. Australian Council of Learned Academies (ACOLA).

Mesías-Ruiz, G. A., Pérez-Ortiz, M., Dorado, J., De Castro, A. I., and Peña, J. M. 2023. Boosting precision crop protection towards agriculture 5.0 via machine learning and emerging technologies: A contextual review. *Frontiers in Plant Science 14*: 1143326.

Popescu, D., Dinca, A., Ichim, L., and Angelescu, N. 2023. New trends in detection of harmful insects and pests in modern agriculture using artificial neural networks. a review. *Frontiers in Plant Science 14*: 1268167.

Qasim, M., Akhtar, W., Haseeb, M., Sajjad, H., and Rasheed, M. 2022. Potential role of nanoparticles in plants protection. *Life Science Journal 19*(2): 31–38.

Ragunathan, V., and Divakar, B. J. 2020. Integrated pest management strategies. In *Molecular Biology of the Biological Control of Pests and Diseases of Plants* (pp. 173–196). CRC Press.

Rajabpour, A., and Yarahmadi, F. 2012. Seasonal population dynamics, spatial distribution and parasitism of Aphis gossypii on Hibiscus rosa-chinensis in Khuzestan, Iran. *Journal of Entomology, 9*(3): 163–170.

Rajabpour, A., and Yarahmadi, F. 2024. *Decision system in agricultural pest management*. Springer Nature Singapore, Imprint: Springer.

Raji, E., Ijomah, T. I., and Eyieyien, O. G. 2024. Data-Driven decision making in agriculture and business: The role of advanced analytics. *Computer Science & IT Research Journal 5*(7): 1565–1575.

Rashidi, D., Rajabpour, A., and Zandi-Sohani, N. 2020. Fixed-precision sequential sampling plan of Syringopais temperatella (Lep., Gelechiidae) in wheat fields of Iran. *Journal of Agricultural Science and Technology*, *22*(6): 1501–1510.

Rossi, V., Caffi, T., Salotti, I., and Fedele, G. 2023. Sharing decision-making tools for pest management may foster implementation of Integrated Pest Management. *Food Security 15*(6): 1459–1474.

Rossi, V., Sperandio, G., Caffi, T., Simonetto, A., and Gilioli, G. 2019. Critical success factors for the adoption of decision tools in IPM. *Agronomy 9*(11): 710. https://doi.org/10.3390/agronomy9110710.

Rustia, D. J. A., Chiu, L. Y., Lu, C. Y., Wu, Y. F., Chen, S. K., Chung, J. Y., and Lin, T. T. 2022. Towards intelligent and integrated pest management through an AIoT-based monitoring system. *Pest Management Science 78*(10): 4288–4302.

Schellhorn, N. A., and Jones, L. K. 2021. Real-time insect detection and monitoring: Breaking barriers to area-wide integrated management of insect pests. In *Area-Wide integrated pest management* (pp. 889–902). CRC Press.

Shafi, U., Mumtaz, R., García-Nieto, J., Hassan, S. A., Zaidi, S. A. R., and Iqbal, N. 2019. Precision agriculture techniques and practices: From considerations to applications. *Sensors 19*(17): 3796.

Shi, Z., Hao, D., Liu, Z., and Zhou, X. 2020. Detection and identification of stored-grain insects using deep learning: A more effective neural network. *IEEE Access* 8: 163703–163714. https://doi.org/10.1109/ACCESS.2020.3021830

Terentev, A., Dolzhenko, V., Fedotov, A., and Eremenko, D. 2022. Current state of hyperspectral remote sensing for early plant disease detection: A review. *Sensors* 22: 757. https://doi.org/10.3390/s22030757

Vreysen, M. J., Robinson, A. S., and Hendrichs, J. (Eds.). 2007. *Area-wide Control of Insect Pests: From Research to Field Implementation* (p. 789). Dordrecht, The Netherlands: Springer.

Yaghobi, S., Rahabpour, A., and Zandi, N. 2019. Seasonal Population dynamics and fixed-precision sequential sampling plans for Aphis punicae (Homoptera: Aphididae) in Sirvan City (Ilam province). *Journal of Applied Research in Plant Protection, 8*(3): 1–11.

Yaghobi, S., Rajabpour, A., and Zandi-Sohani, N. 2020. Seasonal population dynamics, sampling distribution, and fixed-precision sequential sampling of Spectrobates ceratoniae (Lepidoptera: pyralidae) in Pomegranate Orchards. *Journal of Economic Entomology, 113*(3): 1513–1518.

Yarahmadi, F., and Rajabpour, A. 2013. Seasonal dynamics and spatial distribution of Eutetranychus orientalis (Acarina: Tetranychidae) on Albizia lebbeck (Fabaceae) in parks in Ahwaz, southwest Iran. International *Journal of Tropical Insect Science, 33*(2): 114–119.

Yatribi, T. 2020. Factors affecting precision agriculture adoption: A systematic litterature review. *Economics-Innovative And Economics Research Journal 8*(2): 103–121.

Zhang, S., Jing, R., and Shi, X. 2022. Crop pest recognition based on a modified capsule network. *Systems Science & Control Engineering* 10(1): 552–561.

2 Advances in Cultural Control to Sustainable Management of Insect Pests

Shreya Singh,[1*] *Mukul Machhindra Barwant,*[2] *and Shivendra Singh*[3]

[1]Department of Agriculture, Ramlalit Singh Mahavidyalaya, Kailhat, Chunar, Mirzapur Uttar Pradesh, India

[2]Department Botany, Sanjivani Rural Education Society, Sanjivani Arts Commerce and Science College Kopargaon, Ahmednagar Maharashtra, India

[3]Department of Agriculture, Udai Pratap College (An Autonomous Institution) Bhojubir, Varanasi, Uttar Pradesh, India

*Corresponding author

2.1 INTRODUCTION

The concept of "insectpests" arose from human crop cultivation practices and the demand for food security from common insects. Some consider them to be humans' primary adversaries for global dominance. Biological and abiotic factors both have an impact on insect pest management. Sustainable pest management is a two-pronged approach that necessitates a thorough understanding of control strategy, pest biology, and ecology in order to establish the most appropriate procedure/method (how), timing (when), and location (where) for the efficient application of any pest control technology. In this context, IPM (integrated pest management), ICM (integrated crop management), and IRM (integrated resistance management) can assist in reducing crop production losses while controlling insect pests without harming non-target organisms. Nevertheless, the global implementation of these practices has been slowed down due to different factors. Conclusively, the integration of non-chemical control methods including new technologies with synthetic insecticides will be a promising option for sustainable insect pest management (Nawaz et al., 2019). Sustainable agriculture is important because it addresses these concerns. The chapter advocates for ecologically friendly and economically viable farming approaches to maintain profitability in the industry. Sustainable agriculture addresses the challenges of feeding a growing global population and protecting the environment by preserving soil, biodiversity, and ecosystem health. To address these difficulties, the worldwide agricultural community has explored several pest management approaches. Integrated pest management (IPM) is an eco-friendly and holistic approach to pest control. IPM involves a variety of tactics, including changing farming practices, using barriers and traps, relying on natural enemies like predators and pathogens, and using eco-friendly chemicals, such as pheromones and botanical insecticides (Thilagam et al., 2023). The sustainable management of insect pests has become a critical concern in agricultural and ecological systems worldwide. Traditional pest control methods, heavily reliant on chemical pesticides, have proven effective in the short term but have also led to several long-term negative consequences, including pesticide resistance, environmental contamination, and the disruption of non-target species (Smith & Jones, 2019). As a result, there has been a growing interest in developing alternative pest management strategies that are both environmentally friendly and economically viable. One such approach is cultural control, which involves modifying the environment or agricultural practices to reduce pest populations without relying on synthetic chemicals. The world's population is growing by 80 million people each year. To meet rising food demand, agricultural output will need to nearly double within 30 years. Despite the global use of 3 Mt (million tons) of pesticides and a 5% increase each year, 3.7 billion people still face malnutrition. The widespread use of broad-spectrum pesticides has resulted in pest resistance, food residues, environmental contamination, secondary pest outbreaks, and decreased populations of beneficial insects. As the public becomes more conscious, there is a growing need for eco-friendly pest management measures (Vreysen et al., 2006).

Cultural control methods are grounded in the principles of ecological balance and integrated pest management (IPM). These methods include practices such as crop rotation, intercropping, habitat manipulation, and the use of trap crops, all of which aim to disrupt the life cycle of pests and reduce their population density below economic threshold levels (Brown et al., 2021). By integrating these practices into a broader IPM framework, cultural control can play a significant role in reducing the reliance on chemical pesticides and promoting sustainable agricultural practices. In recent years, advances in research and technology

DOI: 10.1201/9781003516538-2

have further enhanced the effectiveness of cultural control methods. For example, precision agriculture tools, such as remote sensing and geographic information systems (GIS), have enabled farmers to monitor pest populations more accurately and implement targeted cultural control measures (Garcia & Patel, 2020). Moreover, the increased understanding of pest behavior and ecology has led to the development of more effective cultural strategies that are tailored to specific crops and pest species. Agriculture has prioritized sustainability in recent years. Sustainable agriculture combines ecological well-being, economic profitability, and social equity to ensure food security for the current and future generations. Historically, concentration had been on increasing yield. The adoption of short-term beneficial agricultural practices resulted in increased productivity but also presented major consequences, primarily in the form of dirt, deterioration, loss of biodiversity, and water pollution (Thilagam et al., 2023; Bale et al., 2008).

Quarantine rules have also been used successfully to control the spread of exotic pests. Semiochemicals, bioacoustic detectors, nanotechnology, insect growth regulators, and entomophagy have all seen significant improvements in research to improve oil palm pest management (Egonyu et al., 2022). Perennial and annual crops grown in temperate and tropical climates are used to demonstrate research and development methodologies that have contributed to the utilization and integration of host plant resistance with biological, cultural, and chemical controls. Research reveals that the success of IPM has been dependent on classical experimental methodologies that are always adapting to changing restrictions and innovative discoveries which are being applied to farm practice in an increasingly efficient and intelligent manner (Way & Van Emden, 2000).

Oil palm, *Elaeis guineensis* Jacq. (Arecales: Arecaceae), is a multipurpose plantation crop grown in the humid tropics of Asia, Africa, and the Americas for palm oil and other uses, such as sauces, soap, wine, fertilizer (ashes), roofing (leaves), building material (trunk), medicines (roots), and ornamental purposes. Palm oil accounts for around 40% of all commercial vegetable oil globally, and this oleaginous species has the ability to generate 4–20 times more oil per hectare of land than other oil crops. As a result, oil palm production has increased by more than 35 times since 1970, reaching 71 Mt in 2018. There are about 140 palm oil–importing countries, including India and China (Murphy et al., 2021)

Insect pests pose a significant threat to global food security, eliminating up to 40% of agricultural production each year. Pest outbreaks can harm smallholder earnings, disrupt supply, raise market prices, and reduce food access for vulnerable populations. Recommendations for pest management include balancing multiple controls, transitioning to agroecological farming systems, improving monitoring and response to migratory pests, developing resistant crop varieties, supporting farmer organizations and knowledge exchange networks, providing financial safety nets (insurance, credit, savings), and investing. A combination of technical, organizational, and policy initiatives. Socially, poor rural households may fall into debt traps trying to replace crop losses from pests, entrenching poverty, and food insecurity. Even localized pest outbreaks can destabilize regional food supplies and prices, as exemplified by the 2004 locust plague in West Africa. In a world where over 2 billion people lack year-round access to adequate food, the crop losses and systemic impacts wrought by insect pests cannot be overlooked. There are also concerns that pest impacts will intensify under climate change, necessitating a deeper understanding of risks as well as pest ecology and management (Ali et al., 2023). Insect pests pose significant challenges to agricultural production worldwide, threatening food security and economic stability. Integrated pest management (IPM) offers a comprehensive approach to address these challenges sustainably. This chapter explores the principles and strategies of IPM, emphasizing its focus on minimizing environmental impacts and reducing reliance on chemical pesticides. Through biological control, cultural practices, mechanical control, and selective pesticide use, IPM aims to manage pest populations effectively while promoting biodiversity and long-term agricultural sustainability. By implementing IPM strategies, farmers can mitigate the impact of insect pests while fostering resilient and productive agricultural systems (Figure 2.1).

Insect pests pose a significant threat to agricultural productivity worldwide. They can devastate crops, leading to substantial economic losses and threatening food security. Identifying and managing these pests effectively is crucial for sustainable agriculture. Integrated pest management (IPM) offers a comprehensive approach to address insect pest challenges while minimizing environmental impacts and reliance on chemical pesticides. Insect pests represent a persistent threat to agricultural productivity and food security globally. This chapter delves into the principles and practices of integrated pest management (IPM) as a sustainable approach to address these challenges. IPM integrates various strategies, including biological control, cultural practices, mechanical control, and selective pesticide use, to manage pest populations effectively while minimizing environmental harm (Tiwari, 2024). Pests and diseases have an impact on crop yield and quality and also reduce resource-use efficiency. Improved crop protection strategies to prevent such damage and loss can increase production and make a substantial contribution to food security (Lucas, 2011).

2.2 CROP MANAGEMENT PRACTICES

2.2.1 Crop Rotation

The act of developing various groupings of harvests in a similar land parcel to take advantage of the useful job in soil transformation, further developing soil design and ripeness alongside controlling vermin and weeds, can be characterized as *yield turn* (Degu et al., 2019). A large portion of the business cultivating scene is portrayed by homogenous planting with regular interruptions because of gather-and-decrepit cycles and the utilization of non-particular insecticides making the absence

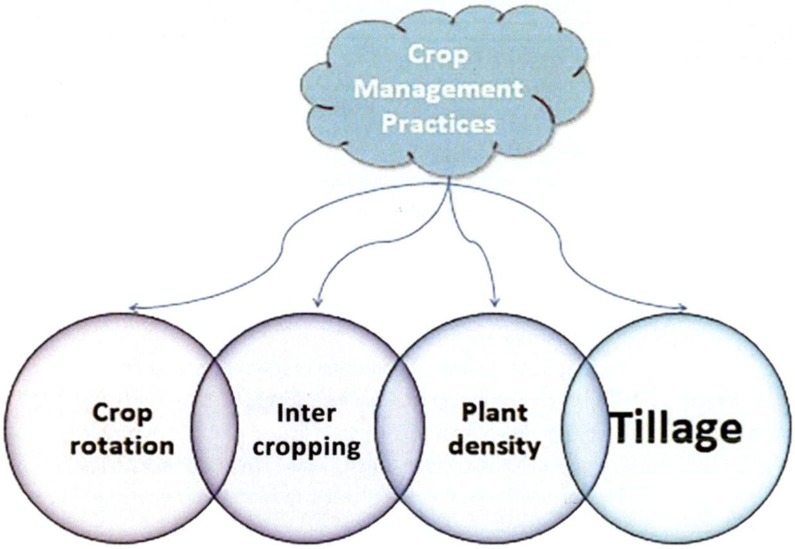

FIGURE 2.1 Framework of crop management practices, including crop rotations, intercropping, plant density optimization, and tillage methods. The visual outlines the interactions and contributions of each practice to soil health, pest control, resource efficiency, and overall crop productivity. Arrows indicate the interconnectedness of these practices within sustainable agricultural systems.

of a haven from natural enemies. Expanded vegetational variety might lessen pervasion by making crops more challenging for bugs like *N. viridula* to find the host and, together, draw in additional normal adversaries (Buntin et al., 1995). While spring, summer, and winter crops are changed, it will break the existence pattern of numerous bugs. Exceptionally, if there should be an occurrence of vegetable irritation, revolution of plants from various families is suggested (Barzman et al., 2015). The yields planted for crop rotation ought to vary and have different establishing designs (Dhawan & Peshin, 2009). Sets of rotated and unrotated fields, both situated on a similar ranch, were checked week by week for the thickness of various phases of the Colorado scarab and defoliation levels, which showed low early-season densities for matured pests in both yet more egg masses, high defoliation level, and the need of insecticide use in unrotated fields (Wright, 1984; Table 2.1). In an examination done, it was found that crop rotation of corn with different harvests essentially diminished the growth rate of the western corn rootworm, which had been impervious to hereditarily designed Bt corn plants (Carrière et al., 2020).

The number of residents in *Limonius californicus* and *Limonius infuscatus* was greatly reduced when the cropping plan was modified from a regular spring wheat editing to a non-productive summer cropping and winter wheat growing as an IPM system.
(Esser et al., 2015).

2.2.2 Intercropping

Growing at least two harvests all the while in a field is known as *intercropping* (Vandermeer, 1989). Blended intercropping, line intercropping, and strip intercropping are a few spatial mixes under intercropping of yields in the field (Liburd, 2018). This multitude of techniques includes bringing down the rate of population growth of pests that go after principal crops (Trenbath, 1993). How a pest will respond to a given yield combination relies on host reach and the host tracking down part of pest activity (Liburd, 2018). Among two hypotheses which make sense about decreasing damage in intercropped fields, "enemies hypothesis" is one (Risch, 1983; Liburd, 2018). It means zeroing in on a consideration of all sympatric vermin species alongside the upkeep of species variety, where prey and natural predators are kept beneath financial injury level and at a utilitarian reaction level through a profound comprehension of local area design and the natural surroundings (Kogan and Hilton, 2009).

2.2.3 Plant Density

Separating plants has an impact on the vivacious development of yield with arrangement of a reasonable microclimate, while it likewise influences the frequency of numerous insect pests. High planting density caused additional seed and pod damage by pod-sucking bugs in pigeon pea, while higher planting density is preferred in rice due to lower hatchling density and harm by stem borers (dead heart and white head) in rice (Dialoke et al., 2018; January et al., 2020). Nonetheless, a few scientists do not find planting density as a method of pest control with critical outcome. A gathering specialist revealed no massive contrast in pest population of cotton in high-density planting and ordinarily separated crops while approving the viability of IPM modules viz.

TABLE 2.1
Crop Combination Effective against Major Insect Pest

Major Crop	Crop Combination	Remark	Reference
Maize	Bitter gourd	Management of fruit fly	Rai (n.d.)
	Napier grass	Oviposition by the gramineous stem borer, Chilo partellus	(Van den Berg, 2006)
Cabbage	Carrot	Management of diamondback moth	Rai (n.d.)
	Clover	Better-quality harvest due to less insect damage	(Theunissen et al., 1995)
Brinjal	Coriander, fennel	Reducing shoot and fruit borer in brinjal	Rai (n.d.)
Tomato	Marigold	Increase in population of coccinellid predators for the control of *Halicoverga armigara*	(Singh & Tripathi, 2017)
Cucumber	Marigold	Increase in fruit yield and reduction of insect pest severity	(Adewoyin, 2018)
Potato	Brinjal, litchi, tomato	Trap crop to control potato cyst nematodes	(Scholte & Vos, 2000)
Potato	Pea, wheat, radish	Trap crop to control worm in potato field	(Landl & Glauninger, 2013)
Chickpea	Coriander	Increase in natural enemy activity against *Hellicoverage armigera*	(Pimbert, 1991)
Tomato, tobacco	Cabbage	Feeding inhibition on flea beetles due to odors from non-host plant	(Pimbert, 1991)

seed treatment, trap editing, utilization of pheromone traps, and the use of neem-based pesticides (Harshana et al., 2017). A gathering specialist likewise recommended that plant spacing can't be a method of pest control in cowpea with the insignificant contrast found in pest population and the harm brought about by them (Ajao et al., 2016).

2.2.4 Tillage

Culturing practices uncover the different life phases of pests and obliterate them by exposure to hunters or unfavorable natural circumstances, or these practices might change the dirt microclimatic conditions. Soil profundity and timing of culturing are important focuses to be considered in managing soil pests. The fixed phase of pests decides the planning of culturing, while their area in the dirt decides the profundity up to which culturing must be finished (Bajwa & Kogan, 2004). Either to kill weeds or to control bug/microbes, culturing practice has been considered as the best sterile instrument (Cook, 2000; Willson & Eisley, 1992). Then again, ruthless Carabidae, which feed on delicate-bodied caterpillars and aphids, were more plentiful in preservation plowed in contrast with traditionally plowed crops, showing that protection culturing helps with keeping up with populations of beneficial insects (Horne & Edward, 2007). Weed seeds are absent any and all scarification and unsettling influence for germination with protection culturing, which eventually depletes the pests attached to weeds (Shrestha et al., 2006). A harvest expert contends that serious pest infestation and development happens by means of plant wastes are disposed in shallow pits (Prymak et al., 2020).

2.3 SOIL AND FERTILIZER MANAGEMENT

The physical, synthetic, and fundamentally natural properties of soils influence the capacity of a plant to oppose or endure the pests (Altieri & Nicholls, 2003). A fertile and dense soil produces healthy plants and improves its capacity to resist diseases. Soil rich in organic matter exhibits a complex food web, an enormous number of essential organic matter that smother other destructive microorganisms' development, consequently making the soil dense. In Japan, the thickness of the planthopper species *Sogatella furcifera* was fundamentally lower, and the settling pace of female grown-ups was, for the most part, lower in naturally outlined fields in contrast with ordinary rice fields (Kajimura et al., 1995). In a study that isolated treatment of vermicompost blended at various levels with soil-less sheet material as medium, access to nitrogen and essential micronutrients and the development of phenols should be the conceivable system behind warding off aphids and coarse insects by vermicompost in cucumber and tomato (Arancon et al., 2007; Jafary-Jahed et al., 2020). Nonetheless, a harvest expert investigated the fruit and vegetable harvests, while possibly not totally deteriorated prior to applying to the field, can expand oviposition of soil bug bothers prompting crop harm (Nyamwasa et al., 2020). A blend of nitrogen- and phosphorus-based manure application was seen as more compelling in overseeing aphid population of grain when contrasted with single composts, viz., urea and DAP (Wagan et al., 2020). Essentially, a mix of inorganic composts at 75% suggested a portion with natural manures – neem cake fundamentally diminished the number of inhabitants in jassid, bug, whitefly, and organic product drill, notwithstanding better return (Mishra et al., 2020). A harvest expert showed that the use of development controllers and micronutrients in cotton neither expanded nor diminished the duplication and pervasion of bugs like jassid, thrips, and bollworms (Abro et al., 2004). Foliar shower of silicon manure on wheat improved the movement of protection catalysts and expanded the centralization of auxiliary metabolites to reduce the regenerative rate and improve the population multiplying season of aphids (Qi et al., 2020; Yousuf & Dar, 2016). Yet it is additionally obvious that dirt manuring and treatment input expanded fertility and formative paces of various bugs, like green peach aphid, *Myzus persicae*, which were profoundly associated with expanded degrees of solvent N in leaf tissue (Emden, 1966). A harvest expert revealed higher pervasion of jassid and red cotton bugs in okra crops, with an expansion in how much nitrogen manure (Belbase et al., 2020). Unnecessary nitrogen application in the rice field diminished the looking-through capacity of *Anagrus flaveolus*, an egg parasitoid of earthy-colored plant container, as shown by lengthy advancement period and diminished fruitfulness (Zhu et al., 2020).

2.3.1 Mulching

Soil surface mulch can significantly diminish surface spillover, monitor dampness, increment invasion limit, and dials back risks of soil disintegration (Jenkins, 2004). Furthermore, mulching can be utilized to control a lot of nuisance assaults in the fields. Natural mulch of new or dried materials of plants like cypress, eucalyptus and Lucerne help to lessen the assault by weevils on capacity underlying foundations of yam by diminishing breaks on dry period and repulsing and decreasing their development (Rehman et al., 2019). Utilization of *Azadirachta indica* as mulch material was viewed as the best mulch material in controlling bug creepy crawlies in okra (Norman et al., 2020). Crop enhancement techniques like maize – soybean intercrop and maize-desmodium intercrop coordinated with the utilization of maize stover as mulch decreased harm brought about by termites in maize (G et al., 2019).

2.3.2 Soil Solarization

Solarization is a substance-free approach to controlling irritations like pathogenic microorganisms (predominantly organisms, microbes, and nematodes), bugs, and wild plants in the dirt before crops planting (Katan, 1987; McGovern & McSorley, 1997; Gill et al., 2009). Solarization innovation delivers no hazardous synthetic buildup in the ground; it is a protected, basic, successful, and eco-accommodating device for the home gardens and fields (Kapoor, 2013). This pre-plant technique includes soil warming by catching sun-oriented radiations for a month and a half throughout the late spring, when the dirt gets the most extreme daylight. Soil block the energy transmitted from the sun and its temperature increase to the level that is dangerous to many soil-borne microbes (Figure 2.2). There are a need of preparing studios and mass mindfulness developers for promoting this innovation among the ranchers and provincial networks (Kapoor, 2013). The essential business utilization of solarization included radiant and warm grounds. The component of activity depends on the expanded soil temperature (normally in the scope of 45–55°C) at soil profundity of 5 cm under straightforward polyethylene sheets causes plant microorganisms mortality (Katan, 1981). Other than the decrease in

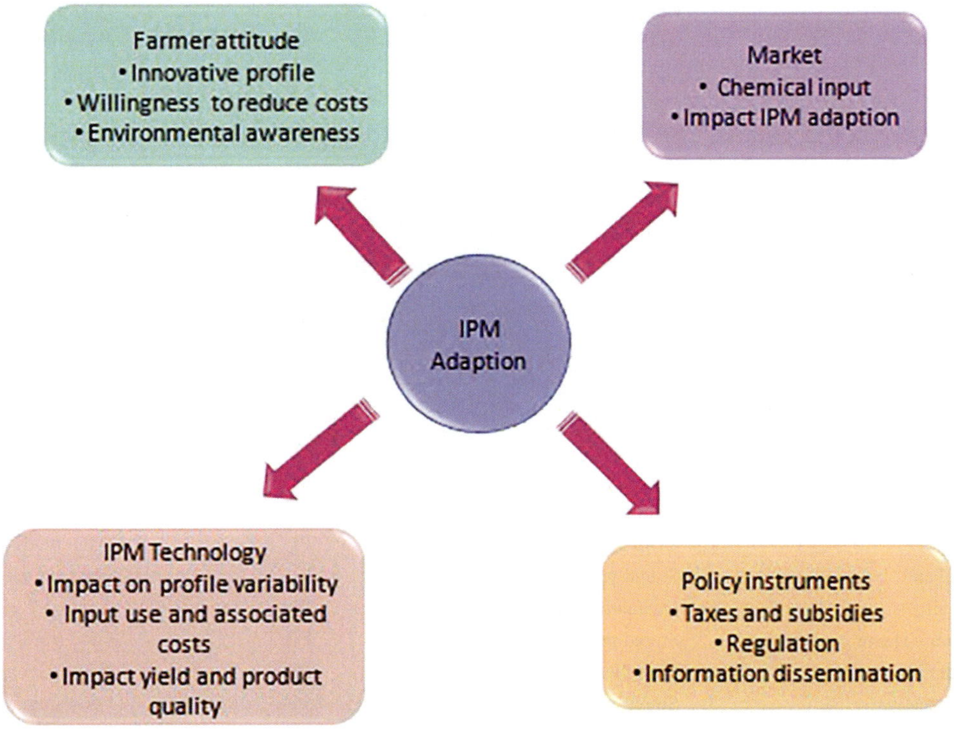

FIGURE 2.2 Framework of integrated pest management (IPM) adoption. The framework highlights key components, including farmer attitudes, market dynamics, IPM technology, IPM adoption practices, and policy instruments. Arrows depict the interactions and feedback loops between these elements, emphasizing their collective role in promoting effective IPM implementation.

Source: Based on Puente et al. (2011).

soilborne microbes, soil solarization additionally prompts expanded development reaction (IGR) of plants (Katan, 1981; Horiuchi, 1984). Soil solarization additionally incorporates well with different procedures, for example, mulching, and soil changes to upgrade its general adequacy against bugs and, increment crop yield (D'Addabbo et al., 2010).

2.4 PHYSICAL CONTROL METHOD

The essential activity may have an immediate effect, for instance, when bugs are killed immediately by mechanical shock. In different occasions, the ideal impact is achieved through pressure reactions. With the fast advances that have happened in the physical, chemical, and organic sciences since the late nineteenth hundred years, agribusiness has been changed from a stringently observational movement to a great extent based on custom and pointed fundamentally at remaining off starvation, to a quantitative type of farming zeroed in on creating huge amounts of food (Table 2.2). During this change, which has been supported at an increasing rate throughout recent years, actual control strategies have been put away in light of the colossal

TABLE 2.2
Classes and Examples of Physical Control Methods

Class	Method	Examples of Target Pests	Pre- (Pr) or Post- (PH) Harvest	Comments
Passive	Adhesives	Ants, flies	Pr, PH	Glues, petroleum jelly, creosote
	Atmosphere modification	Stored product insects	PH	Modulating CO_2
	Exclusion barriers	Flying insects	Pr, PH	Tunnels and screenhouses, windows in food storage areas
	Fences	Cabbage maggot fly	Pr	Small fields
	Inert gases	Stored-product insects	PH	Food/grain storage areas
	Mineral and edible oils	Phytophagous mites, scales	Pr, PH	Orchards, food storage areas
	Mulching	Various insects	Pr	Open fields
	Thin films (kaolin)	Codling moth, leafroller, mites, psyllids	Pr	Orchards
	Trapping: baited, pheromone, sticky	Various insects	Pr	Orchards, vineyards, open fields
	Trench	Chinch bugs, Colorado potato beetle	Pr	Open fields
	Windbreak	Aphid vectors	Pr	Open fields
	Wrapping	Gypsy moth, forest tent caterpillar, insects of transformed products	Pr, PH	Orchards
Active	Electromagnetic Ionizing radiation	Stored-product insects	PH	Food storage areas
	Ionizing radiation	Quarantine pests	PH	Fruits and vegetables for export
	Microwave	Stored-product insects	PH	Food storage areas
	Radiofrequencies	Stored-product insects	PH	Food storage areas
	UV and visible light	Flying insects	PH	Food storage areas, in combination with electricity
Mechanical	Dislodging	Plum curculio	Pr	In apple orchards
	Dislodging	Quarantine pests	PH	Commodities for export
	Disturbing	Stored-product insects	PH	Food storage areas
	Forced air	Houseflies	-	Building entrances
	Leaf shredding	Spotted tentiform leafminer	Pr	Orchards
	Pneumatic (vacuuming, blowing)	Colorado potato beetle, Lygus spp., whiteflies, leafminers	Pr	Open fields
Thermal	Flaming	Colorado potato beetle	Pr	Open fields
	Hot/cool air	Stored-product insects	PH	Food storage areas
	Heated air	Quarantine pests	PH	Commodities for export
	Hot water – steam	Stored-product insects	PH	Food storage areas
	Hot water immersion	Fruit flies	PH	Fruits for export
	Infrared heating	Stored-product insects	PH	Food storage areas
	Postharvest chilling	Stored-product insects	PH	Food storage areas
	Cold storage	Quarantine pests	PH	Fresh commodities for export
	Rapid freezing	Stored-product insects	PH	Food storage areas
Other	Flooding	Cranberry insects	Pr	Orchards, open field

Source: Vincent and Hallman (2009).

outcome of substance control. It is just regular that certain individuals ought to see the utilization of physical control strategies as a stage in reverse to those far off, familial practices. On account of mechanical advances and more noteworthy accuracy in the implementation of such techniques, actual control currently has all the necessary ascribes for consolidation into IPM systems.

2.5 BIOLOGICAL CONTROL AUGMENTATION

The utilization of natural enemies to oversee weeds, ailments, and bugs is characterized as organic control. In spite of the fact that it has limits relying upon the way things are used, augmentative organic control with rehashed delivers just momentarily brings down bug flare-ups, and not many projects accomplish super durable control (Baker et al., 2020). The adequacy of utilizing natural control utilizing advanced extractions changes relying upon the way things are utilized; notwithstanding, barely any projects can for all time stifle bother episodes (Torres & Bueno, 2018). At times, the presence of natural enemies, for example, parasitic wasps and savage arthropods can altogether decrease bother numbers, and a few common methodologies for controlling endemic irritations incorporate the standard arrival of financially accessible regular adversaries, the security of natural enemies populations by giving shelters or keeping away from acts that hurt them (Hajek & Eilenberg, 2018).

2.5.1 Predator

Asopinae (Heteroptera: Pentatomidae) stink bugs, which eat different bugs, are significant for watching out for various irritations, for example, lepidopteran hatchlings in nurseries and fields and herbivorous pentatomid species. Ruthless bugs are taken advantage of in coordinated bug control frameworks, including lacewings, ground scarabs, centipedes, dragonflies, wander, midges, ladybugs, privateer bugs, and aphids (Dara, 2019). *Anystis baccarum* (L.), a generalist savage bug, was made in Canada as a new biocontrol specialist and went at a bargain in 2022. Woman scarabs are ravenous eaters and might be normal in regions where food is ample, and expansive range pesticide use is disallowed (Table 2.3). Female bugs are excellent hunters when aphids are abundant (high nuisance thickness), yet they are believed to be less successful when pest consistency is low (de Castro et al., 2015; Holmes et al., 2016; Saito et al., 2023; Kundoo & Khan, 2017).

2.5.2 Parasitoid

Bug nuisances are explicitly focused on by parasitoids, a kind of natural enemies. In contrast to genuine parasites, parasitoid life-forms kill their prey and complete their improvement on a solitary host. Lepidoptera, Coleoptera, and Neuroptera address a little minority of the parasitoids,

TABLE 2.3
Some Insect Pest Predators and Parasitoids

Stages Attacked	Egg	Larva	Pupa
Predators of crickets	*Conocephalus longipennis*	*Micraspis* spp., *Lycosa pseudoannulata* (Boesenberg and Strand), Ants, *Ophionea* spp., *Microveli douglasi* Atronlinea, *Mesovelia vittigera*, and *Limnogonus fassarum* (F.)	Also attacks adults
Predators of leaf folders	*Conocephalus longipennis* M. *vittaticollis*, *Cyrtorhinus lividipennis* Reuter, *Micraspis crocea* Mulsant	*Ophionea* spp., water bugs; *Lycosa pseudoannulata*, ants; *Micraspis* spp.	
Predators of brown planthopper	*Cyrtorhinus lividipennis*	*Synharmoni octomaculata* (F.), *Lycosa pseudoannulata*, *Paederus fuscipes* Curtis, *Cyrtorhinus lividipennis*, *Migrovelia douglasi atrolineata* Bergroth	*Paederus fuscipes* Curtis, *Synharmoni octomaculata*
Predators of yellow stem borer		*Bracon chinensis*, Szepligeti; *Temelucha philippinensis*, Ashmead; *Stenobracon nicevillei* (Bingham), Uchida; *Trichomma enaphalocrosis*, *Cotesia augustibasis*, *Temelucha philippinensis*	*Tetrastichus ayyari*
Parasitoids of leaf folders	*Copidosomopsis nacoleiae* (Eady), *Trichogramma japonicum* (Ashmead)	*Trichogramma japonicum* (Ashmead)	*Tetrastichus ayyari* Xanthopimpla, *Flavolineata cameron* Rohwer
Parasitoids of brown planthopper	*Anagrus* spp., *Oligosita yasumatsui* (Viggiani and Subba Rao)	*Pseudogonatopus* spp.	*Pseudogonatopus* spp.

Source: Fahad et al. (2015).

TABLE 2.4
Common Entomopathogenic Microorganisms and Their Target Pests

Target Pests	Entomopathogens
Acarina, Coleoptera, Diptera, Hemiptera, Hymenoptera, Lepidoptera, Orthoptera, Thysanoptera, and other pests; nematodes that parasitize plants	Fungi: *Hirsutella thompsonii, Isaria fumosorosea, Lecanicillium lecanii, L. longisporum, Metarhizium anisopliae, M. brunneum*, and *Paecilomyces lilacinus*
Lepidoptera	Viruses: Nucleopolyhedrovirus (NPV), *Spodoptera exigua* NPV, *Helicoverpa zea* NPV, Granulovirus (GV), *Cydia pomonella* GV
A number of orders of soilborne pests	Nematodes: *Steinernema carpocapsae, H. heliothidis, S. feltiae, S. carpocapsae* and *Heterorhabditis bacteriophora*
Lepidoptera, Diptera, Lepidoptera, Coleoptera, Japanese beetle	Bacteria: *Paenibacillus papillice, Bacillus thuringiensis* subsp. *aizawai, Bacillus thuringiensis* subsp. *israelensis, Bacillus thuringiensis* subsp. *kurstaki*, and *B. thuringiensis* subsp. *Tenebrionis*

Source: Dara (2017).

which are regularly Diptera or Hymenoptera. In contemporary many years, the parasitoids of the families Bethilidae, Braconidae, Ichneumonidae, and Pteromalidae have started a flood in interest in parasitoid-based organic administration of put away products bugs (Salim et al., 2016).

2.5.3 Pathogen

Microscopic organisms and parasites that develop inside plant tissues without showing any sickness signs in the host plant are alluded to as endophytes (Mantzoukas & Eliopoulos, 2020). It is imperative to feature that while parasitic endophytes are normal in plant species, their highlights and level of colonization fluctuate contingent upon the sort of plant tissue. A few examination studies have uncovered that entomopathogenic fungi (EPF) are viable against bug and vermin bothers (Harush et al., 2021). They likewise improve the plant's capacity to answer other biotic burdens. During a few biocidal activities, substances like insecticides, antifungals, herbicides, and antivirals are delivered, which makes foundational opposition be prompted (Abd-Elgawad, 2019; Rahman et al., 2016). There are around 700 types of EPF, which are individuals from the groups Laboulbeniales and Pyrenomycetes (Ascomycota, Deuteromycota). EPF is a better elective that can be taken advantage of than oversee numerous types of ticks, parasites and bugs since they can taint bugs through their fingernail skin before the sickness is brought about by ingestion (Sani et al., 2020; Table 2.4).

2.5.4 Nematode

Entomopathogenic nematodes (EPNs) are soil worms that feed on bugs and have a critical potential for biocontrol in crops. The most widely recognized EPN species utilized in insecticidal details are *Heterorhabditis bacteriophora, Steinernema scapterisci* (mole cricket nematode), *Symbiobacterium thermophilum, Steinernema carpocapsae, Steinernema feltiae, Heterorhabditis megidis*, and *Steinernema riobravis*. EPN-contaminated adolescents (IJs) enter through normal openings, for example, the intersegmental layer, rear end, spiracles, or mouth, and effectively look for their hosts in the dirt Maina et al. (2018). When inside the body of the host, the EPNs discharge advantageous microscopic organisms into the hemocoel, prompting bacterial septicemia and the host's passing (the irresistible stage) Rajula et al. (2020). Then, during nematode generation, the bug remains, and cooperative microscopic organisms are used as a food source (necrotrophic stage). Steinernema species convey the microscopic organisms *Xenorhabdus* spp., though *Heterorhabditis* species convey the microbes *Photorhabdus* spp. Nematodes should be in a wet climate to stay dynamic (Table 2.5). Wood-exhausting hatchlings that keep up with their passages open are great contender for EPNs, which flourish in dull and sodden circumstances Srinivasan et al. (2019).

2.6 INTEGRATED PEST MANAGEMENT (IPM) STRATEGIES

As endeavors have been made to lessen the utilization of synthetic controls trying to diminish their effect on the climate and on human wellbeing, there has been an expansion in the quantity of ecological and human wellbeing results related with the utilization of pesticides in late many years. The advancement of elective pesticides is without a doubt required, yet how might they be created? Insecticide medicines can be significantly decreased by taking out rice center vermin all the more really through non-insecticide control. By and large, by utilizing synthetic pesticides generally in the battle against significant vermin in field crops, new auxiliary pesticides have arisen by smothering the action of regular arthropods, these enemies ordinarily ignore these optional pests.

TABLE 2.5
List of Some Important Nematode Species and Their Target Organisms

Nematode Species	Host
Steinernema riobrave	Citrus root weevils (*Diaprepes* species), mole crickets
Heterorhabditis marelatus	White grubs (scarabs), cutworms, and black vine weevil
Heterorhabditis megidis	Weevils
Steinernema scapterisci	Mole crickets (*Scapteriscus*)
Steinernema feltiae	Fungus gnats (*Bradysia*), western flower thrips, and shore flies
Steinernema glaseri	Scarab beetle, white grubs (Japanese beetle, *Popillia* species), and banana root borer
Heterorhabditis bacteriophora	Black vine weevils, cutworms, corn rootworms, white grubs (scarabs), flea beetles, and citrus root weevils (*Diaprepes* species)
Heterorhabditis indica	Fungus gnats, root mealybugs, and grubs
Steinernema kraussei	Black vine weevil (*Otiorhynchus sulcatus*)
Heterorhabditis zealandica	Scarab grubs
Steinernema carpocapsae	Billbugs, sod webworms, banana moths, dogwood borer, orchard pests, armyworms, chinch bugs, cutworms, crane flies, ornamental and codling moths, cranberry girdler, peach tree borer, and other clearwing borer species, black vine weevil, and shore flies (*Scatella* species)

Source: Kantor et al. (2022).

- A few people or consolidated strategy items are supposed to lessen the utilization of pesticides in different farming items, for rice as well as in all cases in the IPM. Further development of exploring and checking and refined, exact harvest re-enactments can, without much of a stretch, lead to a 50% decrease in the precise gauge of injury levels.
- Further developed crop revolutions, including changed laying techniques and powerful administration of harvest buildups, lower and more proficient utilization of insecticides.
- Further developed crop rotation, with changed procedures for culturing and successful administration of yield buildups.
- Further developed crop types that are more impervious to bug assaults.
- Transgenic plants are incorporated.
- *Natural control* is a term used to depict how something is controlled organically.

(Hajjar et al., 2023)

2.6.1 IPM Implementation and Adoption

Since IPM may successfully use both contemporary innovation and traditional rural frameworks in view of native cultivating techniques, remaining a predominant topic soon is anticipated. IPM considers each monetary, natural, and social element, which likewise guarantees that purchasers approach good food. Over-the-top utilization of synthetic pesticides by ranchers caused fast obstruction improvement in designated insect pests and furthermore hurtfully affected the human body and non-target species Kaur et al. (2020). The utilization of pesticides can be essentially diminished in cultivating frameworks in light of IPM innovation without adversely influencing yield. The greater part of the investigations showed that male ranchers' IPM information is, by and large, higher than that of female ranchers (Kabir & Rainis, 2013). The instructive degree of ranchers has a positive relationship with their insight into IPM. In any case, the age of the rancher is adversely related with it. Giving horticultural information and innovation, including IPM, to ranchers in a convenient and viable way is difficult for rural expansion specialists. Thus, ranchers had a restricted comprehension of the essential standards directing elective vermin the board strategies (Dara, 2019). Vegetable ranchers' reception pace of IPM was just 30%. Ranchers got practically a similar yield despite the fact that they utilized pesticides (glyphosate, acephate, deet, propoxur, metaldehyde, boric corrosive, diazinon, and so on). Their general recurrence of pesticide treatment diminished therefore, which assisted them with bringing down their feedback costs. The reception of IPM ideas can help ranchers monetarily by bringing down costs. Proposed a structure with four primary classifications connected with IPM reception (Pouratashi & Iravani, 2012). There are four extra broad general management components, such as weed management, insect-pest management, general management, and ecosystem management. Rather than executing all IPM procedures on the double, ranchers in a particular locale might carry out more extensive IPM strategies in view of normal administration centers (Figure 2.2). Various elements, for example, rancher land possession status, ranchers' perspectives toward IPM, field school, expansion contact, and utilization of upgraded assortments, turned out to be huger and more had a positive effect, which expanded the likelihood of IPM reception. Most importantly, ranchers' broad information on IPM is an essential for its inescapable reception (Atreya, 2007).

2.7 FUTURE DIRECTIONS AND EMERGING TRENDS

Appropriate arrangements are desperately expected because of the far-reaching worries about depending on chemical pesticides, which are deep-rooted. To ensure that pests don't represent a peril to the plant, the actual natural surroundings of the pest is changed utilizing actual techniques. Feelings of anxiety going from disturbance to death can be created to accomplish this. This can likewise be achieved by utilizing devices like primary impediments that keep pests from getting into crops (Oo et al., 2012). By expanding IPM medicines, there is currently a far and wide longing to diminish insect pervasion at all places in the pre- and postharvest food production process. We feel that actual strategies for control would be all the more generally utilized on the off chance that administration completely comprehended the "genuine worth" part of IPM sending as differentiated to the simpler customary pests of the executives procedures. The work of skilled people with top-to-bottom ability in the extension of people and the movement of the really agrarian vermin are fundamental for the utilization of actual control methods (Jayasooriya & Aheeyar, 2016). To make it simpler to integrate individual components, starting an IPM program in some cases requires working on the construction's format and altering the part plan. The effect on microorganisms that are not designated is insignificant. Precision farming would undoubtedly profit from most sicknesses' capacity to duplicate themselves both spatially and transiently Erbaugh et al. (2007). For the exhaustive and normally risk-free management of insect pests in agricultural production, mycoinsecticides have the capacity to act as a critical component of an IPM program. The viability of the merchandise in unfavorable environmental factors, organizations that would further develop changelessness, have a more extended life expectancy, and are not difficult to use, as well as pathogenic forcefulness and scope of impact all require arising interdisciplinary endeavors to achieve this. The manageability of insect dissemination and overflow in biological frameworks is extraordinarily impacted by the activities of EPF. An attention to this capability prompted the examination concerning the likely utilization of irresistible parasites in the control of insect pests. It has been laid out that natural treatment is a choice that is gainful for the climate. Showing ranchers how to use social methodologies appropriately and how to integrate them into different strategies should be a focal point of biochemical administration. What's more, it is critical to look at the different microbiological item arrangements that will deliver the best results. From the maker to the client, each connection in the item life cycle should attempt to advance and embrace organic control measures (Lefebvre et al., 2015).

2.8 CONCLUSION

The transition to sustainable pest management is an important step toward tackling the worldwide difficulties created by traditional, chemical-based pest control strategies. Cultural control has arisen as a key component of this transformation, providing a variety of ecologically benign and economically viable treatments that adhere to the principles of integrated pest management (IPM). Cultural management reduces reliance on synthetic pesticides while minimizing related concerns, like resistance development, environmental contamination, and harm to non-target species. Recent technological and research developments have greatly improved the efficacy of cultural control approaches. Furthermore, a better understanding of pest ecology and behavior has resulted in the development of more personalized and successful cultural techniques.

REFERENCES

Abd-Elgawad, M. M. M. (2019). Towards optimization of entomopathogenic nematodes for more service in the biological control of insect pests. *Egyptian Journal of Biological Pest Control*, *29*. https://doi.org/10.1186/s41938-019-0181-1

Abro, G. H., Syed, T. S., Unar, M. A., & Zhang, M. S. (2004). Effect of application of a plant growth regulator and micronutrients on insect pest infestation and yield components of cotton. *Journal of Entomology*, *1*(1), 12–16. https://doi.org/10.3923/je.2004.12.16

Adewoyin, F. O. (2018). Effect of intercropping cucumber with marigold on yield and insect pest incidence. *Journal of Horticultural Science*, *10*(2), 55–60.

Ajao, A. O., Oyewale, R. O., & Adebayo, T. A. (2016). Impact of planting density on insect pest population in cowpea fields. *Journal of Agricultural Science and Technology*, *8*(2), 213–220.

Ali, M. A., Abdellah, I. M., &Eletmany, M. R. (2023). Towards sustainable management of insect pests: Protecting food security through ecological intensification. *International Journal of Chemical and Biochemical Sciences*, *24*(4), 386–394.

Altieri, M. A., & Nicholls, C. I. (2003). Soil fertility management and insect pests: Harmonizing soil and plant health in agroecosystems. *Soil Biology & Biochemistry*, *35*(5), 701–708. https://doi.org/10.1016/S0038-0717(03)00075-8

Arancon, N. Q., Edwards, C. A., Dick, R., & Dick, L. (2007). Vermicompost tea production and plant growth promotion. *BioCycle*, *48*(11), 51–52.

Atreya, K. (2007). Pesticide use knowledge and practices: A gender differences in Nepal. *Environmental Research*, *104*(2), 305–311. https://doi.org/10.1016/j.envres.2007.01.001

Bajwa, W. I., & Kogan, M. (2004). Cultural practices: Information resources for integrated pest management. *Integrated Pest Management Reviews*, *9*(4), 65–83. https://doi.org/10.1016/j.tree.2003.12.003

Baker, B. P., Green, T. A., & Loker, A. J. (2020). Biological control and integrated pest management in organic and conventional systems. *Biological Control*, *140*, 104095. https://doi.org/10.1016/j.biocontrol.2019.104095

Bale, J. S., Van Lenteren, J. C., & Bigler, F. (2008). Biological control and sustainable food production. *Philosophical*

Transactions of the Royal Society B: Biological Sciences, *363*(1492), 761–776.

Barzman, M., Bàrberi, P., Birch, A. N. E., Boonekamp, P., Dachbrodt-Saaydeh, S., Graf, B., . . . & Lamichhane, J. R. (2015). Eight principles of integrated pest management. *Agronomy for Sustainable Development*, *35*(4), 1199–1215. https://doi.org/10.1007/s13593-015-0327-9

Belbase, B. B., Kafle, L., & Ghimire, B. (2020). Impact of nitrogen fertilization on jassid and red cotton bug infestation in okra. *Journal of Entomology and Nematology*, *12*(4), 57–65. https://doi.org/10.5897/JEN2020.0267

Brown, L. R., Green, H. W., & Thompson, P. A. (2021). *Integrated pest management: Concepts and practices*. Springer.

Buntin, G. D., All, J. N., & Roberts, P. M. (1995). Insect pest management in conservation tillage soybean in the southern United States. *Crop Protection*, *14*(4), 317–323. https://doi.org/10.1016/0261-2194(95)00012-Q

Carrière, Y., Brown, Z. S., Downes, S. J., Gujar, G., Epstein, G., Omoto, C., . . . Tabashnik, B. E. (2020). Crop rotation mitigates impacts of corn rootworm resistance to transgenic Bt corn. *Proceedings of the National Academy of Sciences*, *117*(32), 19329–19336. https://doi.org/10.1073/pnas.2003604117

Cook, R. J. (2000). Advances in plant health management in the 20th century. *Annual Review of Phytopathology*, *38*(1), 95–116. https://doi.org/10.1146/annurev.phyto.38.1.95

D'Addabbo, T., Sasanelli, N., & Lamberti, F. (2010). Soil solarization: Integrated control of soilborne pests in sustainable agriculture. *Acta Horticulturae*, *883*(1), 217–226. https://doi.org/10.17660/ActaHortic.2010.883.26

Dara, S. K. (2017). Entomopathogenic microorganisms: Modes of action and role in IPM. *Agriculture and Natural Resources Blogs, University of California.*

Dara, S. K. (2019). The new integrated pest management paradigm for the modern age. *Journal of Integrated Pest Management*, *10*(1), 12. https://doi.org/10.1093/jipm/pmz010

de Castro, A. A., Poderoso, J. C. M., Ribeiro, R. C., Legaspi, J. C., Serrão, J. E., & Zanuncio, J. C. (2015). Demographic parameters of the insecticide-exposed predator *Podisus nigrispinus*: Implications for IPM. *BioControl*, *60*(2), 231–239. https://doi.org/10.1007/s10526-014-9639-y

Degu, G., Tadesse, G., Gecho, Y., & Abebe, T. (2019). Effect of crop rotation and tillage systems on potato crop pests: A review. *International Journal of Agronomy, 2019*, 1–7. https://doi.org/10.1155/2019/9467683

Dhawan, A. K., & Peshin, R. (2009). Integrated pest management: Concept, opportunities and challenges. In R. Peshin & A. K. Dhawan (Eds.), *Integrated pest management: Innovation-development process* (Vol. 1, pp. 51–81). Springer. https://doi.org/10.1007/978-1-4020-8992-3_3

Dialoke, S. A., Amatobi, C. I., & Ibeawuchi, I. I. (2018). Influence of planting density on seed and pod damage by pod sucking bugs in pigeonpea. *Nigerian Journal of Entomology*, *34*(1), 1–10.

Egonyu, J. P., Baguma, J., Martínez, L. C., Priwiratama, H., Subramanian, S., Tanga, C. M., & Niassy, S. (2022). Global advances on insect pest management research in oil palm. *Sustainability*, *14*(23), 16288.

Emden, H. F. van. (1966). Plant insect relationships and pest control. *Annual Review of Entomology*, *11*(1), 407–448. https://doi.org/10.1146/annurev.en.11.010166.002203

Erbaugh, J. M., Kibwika, P., & Donnermeyer, J. (2007). Assessing extension agent knowledge and training needs to improve IPM dissemination in Uganda. *Journal of International Agricultural and Extension Education*, *14*(1), 59–70. https://doi.org/10.5191/jiaee.2007.14105

Esser, A. D., Milosavljević, I., & Crowder, D. W. (2015). Effects of crop rotation, tillage, and soil amendments on wireworm populations in wheat. *Crop Protection*, *74*, 222–230. https://doi.org/10.1016/j.cropro.2015.05.020

Fahad, S., Nie, L., Hussain, S., Khan, F., Khan, F. A., Saud, S., Muhammad, H., Li, L., Liu, X., Tabassum, A., Wu, C., Xiong, D., Cui, K., & Huang, J. (2015). Rice pest management and biological control. In *Sustainable agriculture reviews: Cereals* (Vol. 16, pp. 85–106). Springer, Cham. https://doi.org/10.1007/978-3-319-16988-0_4

Garcia, M. A., & Patel, R. S. (2020). The role of precision agriculture in pest management: Current trends and future directions. *Agricultural Systems*, *180*, 102789.

Gill, H. K., McSorley, R., & Goyal, G. (2009). Soil solarization and its effects on soil organisms. *Journal of Integrated Pest Management*, *2*(1), 1–10. https://doi.org/10.1603/IPM08014

Hajek, A. E., & Eilenberg, J. (2018). *Natural enemies: An introduction to biological control*. Cambridge University Press.

Hajjar, M. J., Ahmed, N., Alhudaib, K. A., & Ullah, H. (2023). Integrated insect pest management techniques for rice. *Sustainability*, *15*(5), 4499. https://doi.org/10.3390/su15054499

Harshana, G. L., Jayasundara, M. U., & Gunawardena, K. N. (2017). Evaluating the effectiveness of IPM modules in high-density cotton planting. *Pest Management Science*, *73*(8), 1735–1744. https://doi.org/10.1002/ps.4611

Harush, A., Quinn, E., Trostanetsky, A., Rapaport, A., Kostyukovsky, M., & Gottlieb, D. (2021). Integrated pest management for stored grain: Potential natural biological control by a parasitoid wasp community. *Insects*, *12*(11). https://doi.org/10.3390/insects12111038

Holmes, L., Upadhyay, D., & Mandjiny, S. (2016). Biological control of agriculture insect pests. *European Scientific Journal, Special Edition*, 228–237.

Horiuchi, S. (1984). Growth responses of crops to soil solarization in Japan. *Journal of Agricultural Meteorology*, *40*(4), 319–324. https://doi.org/10.2480/agrmet.40.319

Horne, P. A., & Edward, P. J. (2007). Conservation tillage and its impact on beneficial insect populations. *Agriculture, Ecosystems & Environment*, *119*(1–2), 243–249. https://doi.org/10.1016/j.agee.2006.07.008

Jafary-Jahed, F., Arzanlou, M., & Sharifi, R. (2020). Soil application of vermicompost induces systemic resistance against Plutella xylostella in cruciferous plants. *Plant Protection Science*, *56*(1), 45–51. https://doi.org/10.17221/105/2019-PPS

January, M., Musundire, R., & Maredza, T. (2020). Effect of planting density on stem borer damage in rice. *International Journal of Pest Management*, *66*(3), 239–247. https://doi.org/10.1080/09670874.2019.1683168

Jayasooriya, H. J. C., & Aheeyar, M. M. M. (2016). Adoption and factors affecting adoption of integrated pest management among vegetable farmers in Sri Lanka. *Procedia Food Science*, *6*, 208–212. https://doi.org/10.1016/j.profoo.2016.02.052

Jenkins, A. (2004). The role of mulch in reducing surface runoff and soil erosion. *Soil & Tillage Research*, *78*(2), 149–158. https://doi.org/10.1016/j.still.2004.03.004

Kabir, M. H., & Rainis, R. (2013). Integrated pest management farming in Bangladesh: Present scenario and future prospect. *International Journal of Agricultural Technology*, *9*, 515–527.

Kajimura, T., Suzuki, Y., & Oishi, K. (1995). Influence of organic farming on the population dynamics of Sogatella furcifera in rice fields. *Applied Entomology and Zoology*, *30*(4), 633–638. https://doi.org/10.1303/aez.30.633

Kantor, M., Handoo, Z., Kantor, C., & Carta, L. (2022). Top ten most important U.S.-regulated and emerging plant-parasitic nematodes. *Horticulturae*, *8*(3). https://doi.org/10.3390/horticulturae8030208

Kapoor, A. (2013). Soil solarization: An eco-friendly tool for soil borne pests management. *Journal of Plant Protection Sciences*, *5*(1), 1–10. https://doi.org/10.13140/RG.2.1.4188.1040

Katan, J. (1981). Solar heating (solarization) of soil for control of soilborne pests. *Annual Review of Phytopathology*, *19*(1), 211–236. https://doi.org/10.1146/annurev.py.19.090181.001235

Katan, J. (1987). Soil solarization: A review of the state of the art. *Horticultural Reviews*, *9*(1), 77–110. https://doi.org/10.1002/9781118060858.ch3

Kaur, T., Kaur, M., Kontogiannatos, D., Kourti, A., & Mendes, K. F. (2020). Integrated pest management: A paradigm for modern age. In *Pests, weeds and diseases in agricultural crop and animal husbandry production*. IntechOpen. https://doi.org/10.5772/intechopen.92283

Kogan, M., & Hilton, R. (2009). Integrated pest management: Principles and practice. *Annual Review of Entomology*, *54*, 243–269. https://doi.org/10.1146/annurev.ento.54.110807.090532

Kundoo, A. A., & Khan, A. A. (2017). Coccinellids as biological control agents of soft-bodied insects: A review. *Journal of Entomology and Zoology Studies*, *5*(4), 1362–1373.

Landl, M., & Glauninger, J. (2013). Effectiveness of different trap crops to control wireworm populations in potato fields. *Crop Protection*, *44*, 60–65. https://doi.org/10.1016/j.cropro.2012.10.003

Lefebvre, M., Langrell, S. R. H., & Gomez-y-Paloma, S. (2015). Incentives and policies for integrated pest management in Europe: A review. *Agronomy for Sustainable Development*, *35*, 27–45. https://doi.org/10.1007/s13593-014-0237-2

Liburd, O. E. (2018). Principles of intercropping and pest management. In *Advances in intercropping systems* (pp. 75–99). Academic Press. https://doi.org/10.1016/B978-0-12-814683-4.00004-8

Lucas, J. A. (2011). Advances in plant disease and pest management. *The Journal of Agricultural Science*, *149*(S1), 91–114.

Maina, U. M., Galadima, I. B., Gambo, F. M., & Zakaria, D. (2018). A review on the use of entomopathogenic fungi in the management of insect pests of field crops. *Journal of Entomology and Zoology Studies*, *6*(1), 27–32.

Mantzoukas, S., & Eliopoulos, P. A. (2020). Endophytic entomopathogenic fungi: A valuable biological control tool against plant pests. *Applied Sciences*, *10*(1). https://doi.org/10.3390/app10010360

McGovern, R. J., & McSorley, R. (1997). Solarization for pest management in sustainable agriculture. *Global Pest Management*, *7*(3), 183–191.

Mishra, P., Chaudhary, S., & Singh, S. (2020). Integrated nutrient management for pest control in cotton: A case study. *Agricultural Research*, *9*(3), 343–352. https://doi.org/10.1007/s40003-020-00462-5

Murphy, D. J., Goggin, K., & Paterson, R. R. M. (2021). Oil palm in the 2020s and beyond: Challenges and solutions. *CABI Agriculture and Bioscience*, *2*, 1–22.

Nawaz, A., Sufyan, M., Gogi, M. D., & Javed, M. W. (2019). Sustainable management of insect-pests. *Innovations in Sustainable Agriculture*, 287–335.

Norman, J., Tano, Y., & Doe, F. (2020). Effectiveness of Azadirachta indica mulch in controlling insect pests in okra. *International Journal of Pest Management*, *66*(4), 333–340. https://doi.org/10.1080/09670874.2020.1717924

Nyamwasa, I., Niyibikora, M., & Bigirimana, S. (2020). Impact of organic matter on pest infestation in vegetable crops. *Journal of Plant Diseases and Protection*, *127*(3), 317–327. https://doi.org/10.1007/s41348-020-00313-0

Oo, M. L., Yabe, M., & Khai, H. V. (2012). Farmers' perception, knowledge and pesticide usage practices: A case study of tomato production in Inlay Lake, Myanmar. *Journal of the Faculty of Agriculture Kyushu University*, *57*(1), 327–331. https://doi.org/10.5109/22087

Pimbert, M. P. (1991). The impact of crop diversity on pest populations. In *Sustainable agriculture: Impacts on plant-pest interactions* (pp. 83–112). Elsevier. https://doi.org/10.1016/B978-0-12-814668-1.00006-1

Pouratashi, M., & Iravani, H. (2012). Farmers' knowledge of integrated pest management and learning style preferences: Implications for information delivery. *International Journal of Pest Management*, *58*(4), 347–353. https://doi.org/10.1080/09670874.2012.724468

Prymak, A., Gładysz, B., & Sobolewska, K. (2020). Effect of conservation tillage on pest development in shallow soil layers. *Agronomy Research*, *18*(2), 469–479. https://doi.org/10.15159/AR.20.035

Puente, M., Darnall, N., & Forkner, R. E. (2011). Assessing integrated pest management adoption: Measurement problems and policy implications. *Environmental Management*, *48*, 1013–1023. https://doi.org/10.1007/s00267-011-9737-x

Qi, G., Xia, Y., & Wu, W. (2020). Foliar application of silicon fertilizer enhances wheat resistance to aphids. *Journal of Plant Nutrition and Soil Science*, *183*(5), 622–630. https://doi.org/10.1002/jpln.201900587

Rahman, S., Biswas, S. K., Barman, N. C., & Ferdous, T. (2016). Plant extract as selective pesticide for integrated pest management. *Biotechnological Research*, *2*, 6–10.

Rai, S. (n.d.). Effectiveness of intercropping for pest management in various crops. *Journal of Agricultural Science*, *5*(3), 123–130.

Rajula, J., Rahman, A., & Krutmuang, P. (2020). Entomopathogenic fungi in Southeast Asia and Africa and their possible adoption in biological control. *Biological Control*, *151*. https://doi.org/10.1016/j.biocontrol.2020.104399

Rehman, A., Muhammad, S., & Ullah, A. (2019). Impact of organic mulch materials on weevil infestation in sweet potatoes. *Journal of Entomology and Nematology*, *11*(2), 15–22. https://doi.org/10.5897/JEN2019.0234

Risch, S. J. (1983). Intercropping as an agroecological approach to pest management. *Environmental Entomology*, *12*(3), 625–629. https://doi.org/10.1093/ee/12.3.625

Saito, T., Buitenhuis, R., & Brownbridge, M. (2023). Use of the generalist predator *Anystis baccarum* in greenhouse IPM: Interactions with other biological control agents, a laboratory study. *Biological Control*, *177*, 105127. https://doi.org/10.1016/j.biocontrol.2022.105127

Salim, M., Gökçe, A., Naqqash, M. N., & Bakhsh, A. (2016). An overview of biological control of economically important lepidopteron pests with parasitoids. *Journal of Entomology and Zoology Studies*, *4*(1), 354–362.

Sani, I., Ismail, S. I., Abdullah, S., Jalinas, J., Jamian, S., & Saad, N. (2020). A review of the biology and control of whitefly, Bemisia tabaci (Hemiptera: Aleyrodidae), with special reference to biological control using entomopathogenic fungi. *Insects*, *11*(9). https://doi.org/10.3390/insects11090619

Scholte, E. J., & Vos, J. G. (2000). The use of trap crops in nematode management. *Annals of Applied Biology*, *136*(3), 139–145. https://doi.org/10.1111/j.1744-7348.2000.tb00032.x

Shrestha, A., Ghosh, S., & Whitehead, A. (2006). Impact of conservation tillage on weed seedbank dynamics and pest population. *Agronomy Journal*, *98*(6), 1575–1583. https://doi.org/10.2134/agronj2006.0042
Singh, N. P., & Tripathi, P. (2017). Impact of intercropping on pest management and yield of tomato. *International Journal of Vegetable Science*, *23*(1), 1–8. https://doi.org/10.1080/19315260.2017.1281772
Smith, J. D., & Jones, K. L. (2019). Environmental impacts of chemical pest control: Challenges and alternatives. *Environmental Science & Policy*, *95*, 20–28.
Srinivasan, R., Sevgan, S., Ekesi, S., & Tamò, M. (2019). Biopesticide based sustainable pest management for safer production of vegetable legumes and brassicas in Asia and Africa. *Pest Management Science*, *75*(9), 2446–2454. https://doi.org/10.1002/ps.5480
Theunissen, J., Booij, C. J., & Lotz, L. A. (1995). Effects of intercropping and population dynamics of insect pests in mixed vegetable systems. *Agriculture, Ecosystems & Environment*, *53*(2), 157–164. https://doi.org/10.1016/0167-8809(94)00593-S
Thilagam, P., Sharanappa, C. H., Roy, S., Deb, L., Padhan, S., Srividhya, S., &Awadhiya, P. (2023). A review on advances in biocontrol techniques for managing insect pests in sustainable agriculture. *International Journal of Environment and Climate Change*, *13*(11), 2114–2125.
Tiwari, A. K. (2024). Insect Pests in Agriculture Identifying and Overcoming Challenges through IPM. *Archives of Current Research International*, *24*(3), 124–130.
Torres, J. B., & Bueno, A. D. F. (2018). Conservation biological control using selective insecticides – a valuable tool for IPM. *Biological Control*, *126*, 53–64. https://doi.org/10.1016/j.biocontrol.2018.07.012
Trenbath, B. R. (1993). Intercropping for pest management: Theory and practice. *Field Crops Research*, *34*(3), 239–252. https://doi.org/10.1016/0378-4290(93)90034-H
Van den Berg, J. (2006). Oviposition preference of the gramineous stem borer, Chilo partellus, in mixed cropping systems. *Environmental Entomology*, *35*(3), 645–650. https://doi.org/10.1603/0046-225X-35.3.645
Vandermeer, J. H. (1989). *The ecology of intercropping*. Cambridge University Press. https://doi.org/10.1017/CBO9780511623523
Vincent, C., & Hallman, G. (2009). Physical control of insect pests. In R. Peshin & A. K. Dhawan (Eds.), *Integrated pest management: Innovation-development process* (pp. 483–497). AcademicPress.https://doi.org/10.1016/B978-0-12-374144-8.00209-5
Vreysen, M. J., Hendrichs, J., &Enkerlin, W. R. (2006). The sterile insect technique as a component of sustainable area-wide integrated pest management of selected horticultural insect pests. *Journal of Fruit and Ornamental Plant Research*, *14*, 107.
Wagan, K. H., Brohi, H. A., & Burdi, G. H. (2020). Effectiveness of combined fertilizers in managing aphid populations in barley. *Pakistan Journal of Agricultural Sciences*, *57*(4), 1073–1079. https://doi.org/10.21162/PAKJAS/20.1151
Way, M. J., & Van Emden, H. F. (2000). Integrated pest management in practice – pathways towards successful application. *Crop Protection*, *19*(2), 81–103.
Willson, H. R., & Eisley, J. B. (1992). No-tillage corn and insect injury: Impacts on pest dynamics. *Journal of Economic Entomology*, *85*(3), 1295–1302. https://doi.org/10.1093/jee/85.3.1295
Wright, R. J. (1984). Control of the Colorado potato beetle by crop rotation: Effects of crop rotation on the population dynamics of the Colorado potato beetle. *Environmental Entomology*, *13*(4), 1217–1225. https://doi.org/10.1093/ee/13.4.1217
Yousuf, F., & Dar, M. A. (2016). Foliar application of micronutrients and biological control in sugarbeet. *Indian Journal of Entomology*, *78*(3), 232–240. https://doi.org/10.5958/0974-8172.2016.00043.4
Zhu, S. D., Zhang, Y. J., & Liu, S. S. (2020). Effects of nitrogen fertilization on Anagrus flaveolus performance as an egg parasitoid of the brown planthopper. *Entomologia Experimentalis et Applicata*, *168*(2), 164–172. https://doi.org/10.1111/eea.12887

3 Advances in Biological Control

Muhammad Sajid[1*], *Nimra Naseem*[1], *Sara Ashraf*[1], *Asadullah Nadeem*[1], *Amber Mubeen*[1], *Muhammad Irfan Munir*[1], *Muhammad Riaz*[1], *Muhammad Arshad*[1], *and Tariq Mustafa*[2]

[1]Department of Biotechnology, Faculty of Life Sciences, University of Okara, PO 56300, Pakistan

[2]Department of Entomology, University of Agriculture Faisalabad, Pakistan

[*]Corresponding author

3.1 INTRODUCTION

A stressed global ecological system is required to feed the growing worldwide population. Consequently, there is a growing interest in an environmentally responsible and productive agriculture that produces nutritious food while preserving the integrity of the environment for future generations. Not every technology that boosts output has a neutral effect on sustainability over the long run. These factors make it necessary to create strategies that are productive as well as steady, resilient, and sustainable.

Increased exposure to pesticides is becoming associated with higher health risks for exposed populations, including consumers, farmworkers, farmers, and those living in rural areas (Mesnage et al. 2014; Myers et al. 2016). Pesticides are hazardous for the quality of the water, the land, and the habitat of wildlife (Stehle and Schulz 2015; Pisa et al. 2015; Stone, Gilliom, and Ryberg 2014). Farmers are stuck on a chemical treadmill as a result of rising application rates, greater crop losses, and rising expenses due to resistance to insecticides, herbicides, and other pesticides (Heap 2014). Certain commodity crop subsidies promote monocultures and the wasteful use of inputs in their production (Fausti 2015). These negative effects' non-market costs are only approximal, but they place a heavy strain on the entire world economy (Muller et al. 2017).

"Ecologically based pest management that uses one kind of organism (the 'natural enemies') to control another (the pest species)" is what *biological control* is defined as (Prokopy and Kogan 2009). Examples include insect pests and pathogens of crops (Ninkovic, Markovic, and Dahlin 2016), weeds, mosquitos (Barratt et al. 2018), and rodents (Isman 2006; Köhl, Kolnaar, and Ravensberg 2019). Furthermore, the functioning of protective antagonists in the food chain, such as during the processing of food and animal feed, is supported by the concepts of biological regulation (Jordan et al. 2014), as well as important medical treatments for humans (Dedrick et al. 2019; Van Nood et al. 2013), even if these fields employ distinct vocabularies. Notably, the acceptance of integrated pest management (IPM) as the future standard for crop protection around the world has coincided with a sharp increase in the demand for biocontrol solutions in recent years (Stenberg 2017). *Natural enemies* encompass a variety of creatures, such as parasitoids, predators, entomopathogenic nematodes, pathogens, hyperparasites of plant diseases, competitive microbes, herbivores that consume weeds and their seeds, competitors for resources, and organisms that produce toxins, sometimes known as allelopathy or antibiosis (Heimpel and Mills 2017). Natural occurrences, traditionally introduced and established foreign agents, released native or foreign agents boosting populations, and conserved or enhanced populations of native or foreign agents are all examples of biological management. Two types of augmentative releases are inundative, in which the organisms released are intended to cause rapid decreases, and inoculative, in which the population is built with the expectation that it will act over generations.

In a broader sense, biological techniques also encompass the use of pheromones to track pest populations and prevent breeding, the release of sterile insects, and the manufacture of pesticides called biopesticides that are derived from live organisms or their by-products. Genetically engineered plants or species other than plants are included in certain definitions of *biopesticides* (Okrikata 2021). Furthermore, *biostimulants* are biological products; some of them can lessen the effects of pest activity because of an indirect mode of action or complex of constituents that increases plant tolerance to abiotic stresses, rather than just because of the presence of a known plant protective compound that directly affects pests (Yakhin et al. 2017).

It has long been known that using biological management to manage pests and illnesses in agriculture and natural environments is a sustainable approach. Recent developments in this sector have revolutionized how we think about and use natural pest control strategies. This chapter explores the most recent advances in biological control, including new methods, advances in the use of biocontrol agents, and developing patterns that will influence pest management in the future. We will look at how developments in biological control, ecological modeling, genetic engineering, and other pest management techniques are

DOI: 10.1201/9781003516538-3

improving economic feasibility, efficacy, and environmental safety. In order to maximize the advantages of biological control in various agricultural and ecological contexts, it is imperative that we comprehend these developments as we proceed toward a more comprehensive and integrated approach to pest management.

3.2 TYPES OF BIOLOGICAL CONTROL AGENTS

3.2.1 Group 1: Microbes

When compared to chemical control, biological control has demonstrated superior efficacy against a variety of plant enemies. Microbes can serve as a beneficial alternative to chemical pesticides. It consists of fungi, bacteria, viruses, and protozoa.

3.2.1.1 Bacteria

A class of bacteria that are common in soil is thought to be plant diseases' natural adversaries. These bacteria produce poisons, antibiotics, volatile organic compounds, and enzymes. They also compete with other bacteria for resources and cause systemic resistance in plants, among other modes of action (Vimal, Gupta, and Singh 2018). A heterotrophic, gram-positive bacterium known as pseudomonas can help plants defend themselves against fungi like *Fusarium oxysporum*, bacteria like *Xanthomonas campestris*, and plant diseases like cucumber mosaic virus (Mohammed, Al-Ani, and Salleh 2013; Mohammed et al. 2014). Nematicidal substances produced by *Pseudomonas putida* reduced the pace at which eggs hatched and destroyed the adult plant-parasitic nematode *Meloidogyne incognita* (Zhai et al. 2019).

3.2.1.2 Fungi

Mycorrhizal fungus and nitrogen-fixing symbiotic bacteria are two examples of soil microorganisms that are crucial to crop performance and plant growth. They also aid in mineral nutrition (Meena et al. 2018). However, further research is needed to fully understand the potential of plant-associated microorganisms to replace conventional fertilizers. Researchers have made significant strides in understanding the behavior and composition of rhizospheric microbiomes in recent years (Zhalnina et al. 2018). The genus *Trichoderma* is home to a broad range of rhizocompetent filamentous fungal strains that function as mycoparasites and saprophytic fungi in a wide range of environments. These fungi are frequently employed as biological control agents (Al-Ani and Albaayit 2018). *Chaetomium roseum* is broadly utilized, particularly in farming. These mycoparasitic fungi have the ability to parasitize other fungal species and function as biological control agents for phytopathogenic fungi–caused root illnesses (Jensen, Lübeck, and Jørgensen 2016). In order to infect a host, *Metarhizium* and other fungi associated with insect pathology can interact with its cuticle and break it down using enzymes, such as proteases, esterases, chitinases, and lipases (Pedrini et al. 2013).

3.2.1.3 Virus

One useful biocontrol tool that can be employed against plant enemies is a virus. Plant enemies can be killed or reduced by this viral agent. For instance, it has been discovered that certain mild strains of the citrus tristeza virus can lessen the capacity of a virulent strain to infect citrus. Viruses called "bacteriophages" are able to infect bacteria cells and destroy their metabolism, ultimately resulting in the cell's death (Al-Ani and Furtado 2020). Fungal viruses known as mycoviruses are isolated from a wide variety of plant fungal diseases across the globe. Fungal pathogens have been found to contain mycoviruses, such as *F. oxysporumi* (Lemus-Minor et al. 2015), *Phomopsis longicolla* (Hrabáková, Koloniuk, and Petrzik 2017), and *Sclerotinia sclerotiorum* (Hu et al. 2014). Avocado trees exhibited some degree of resistance against infection by the virulent strain of *Rosellinia necatrix* thanks to the presence of hypovirulent strains of the bacterium (Velasco et al. 2018).

The first virus identified in the nematode *Caenorhabditis elegans* is called Orsay (Guo et al. 2014). It was isolated from other nematodes that parasitize plants, such as *Bursaphelenchus xylophilus* (pinewood nematode) (Cotton et al. 2016) and *Heterodera glycines* (soybean cyst nematode) (Bekal et al. 2014). Furthermore, three novel viruses were identified in cyst nematodes: SCN picorna-like virus (PLV), SCN nyami-like virus (NLV), and SCN bunya-like virus (BLV) (Ruark et al. 2018).

3.2.1.4 Protozoa

Theratromyxa weberi (Aguilar 2012), an amoeba in the Proteomayxida order of protozoa, is more significant than other amoeba species. The Vampyrellidae family was identified with the use of soil samples. It has been determined that the amoeba *Thecamoeba weberi* may be a viable biological control agent for phytopathogenic nematodes *Meloidogyne incognita* and *Aphelenchoides rutgersi* (Sayre and Wergin 1989).

3.2.2 Group 2: Animals

3.2.2.1 Insects

Insect pests are insects' food source. The best substitute for insecticides in the fight against insect pests is the abundance of beneficial insects. There are two categories of beneficial insects: parasites and predators. The chalcidoid wasp known as the whitefly parasite is highly aggressive toward the whitefly (*Bemisia tabaci*) in greenhouses, affecting two significant genera, *Encarsia* and *Eretmocerus* (Liu, Stansly, and Gerling 2015). Because ladybirds are effective at foraging and quickly consuming pests like aphids, they are thought to be the best biocontrol agents for aphids (Siddiqui et al. 2015). Carabid beetles, among other beneficial insects, have the ability to eat weed seeds. Approximately 28 different types of weeds feed on carabid beetles (Deroulers & Bretagnolle 2019).

3.2.2.2 Nematodes

Nematodes with many plant-opposing abilities come in handy. It contains entomopathogens, which have a high level of pest control efficacy. Within the two families Heterorhabditidae and Steinernematidae, entomopathogenic nematodes comprise around 100 species, 91 of which, including *Steinernema feltiae*, *Steinernema riobrave*, and other *Steinernema* spp., are members of the family *Steinernematidae* (Lacey 2016). Entomopathogenic nematodes have been found to have a significant role in the management of a number of economically significant insects, such as the plum curculio, citrus root weevil, banana moth, giant pine weevil, borers tree, banana root borer, strawberry root weevil, and sweet potato weevil (Lacey 2016).

3.2.2.3 Mites

In natural environments, multiple nematophagous mites live in the soil unattached, and as part of their regular diet, they must eat a variety of species, including nematodes (Garcia Ortiz et al. 2015). The effectiveness of the nematophagous mite *Lasioseius penicilliger* against bacteria-feeding nematodes and *Haemonchus contortus* was evaluated by Aguilar (Aguilar-Marcelino et al., 2014).

3.2.3 Group 3: Plants

Several tools are available to beneficial plants to influence plant enemies. Allelopathy is a chemical compound that inhibits plant infections, pests, and weeds. The tools are highly effective against these entities. *Allelopathy* is a biological phenomenon that may be brought on by toxicity to other organisms; as a result, it can be employed in agriculture to manage a variety of plant enemies by affecting various functions and cellular structures (Cheng and Cheng 2015). Allelopathy is a highly effective approach of integrated pest management that can be used in place of fungicides, insecticides, and herbicides in agriculture to control plant infections, pests, and weeds (Muzell Trezzi et al. 2016). Although they have limited efficacy, carnivorous plants can be used as biocontrol agents in integrated pest management (Wycke, Perrocheau, and Darrouzet 2018).

Because they trap insects and nematodes, trap plants are fascinating plants that can be used as biological control agents. Thus, an agroecosystem's ecosystem can be preserved by the use of biological management, which is made possible by the phenomena of trap crops (Tiwari et al. 2019).

3.3 ADVANCES IN CLASSICAL BIOLOGICAL CONTROL

3.3.1 Genome Mining: Identifying New Potent Microbial Traits

New biocontrol mechanisms may be discovered by sequencing genomes and analyzing gene functions to uncover powerful microbial strains shown in Figure 3.1. In 2014, the genome of *Bacillus amyloliquefaciens* GB03, the first biocontrol agent accessible for purchase, was published (Chen et al. 2007). Since then, a large number of bacterial strains that have been discovered or suggested to be useful for biological fertilization or biocontrol of different crop plants have been sequenced (Pirttilä et al. 2021; Pii et al. 2015). However, the aforementioned recognized features have been the focus of strain genome mining thus far. For instance, it has been found that the genome of *Bacillus velezensis* FZB42 has 13 gene clusters that are involved in the manufacture of volatile chemicals or anticipated antimicrobial metabolites (Fan et al. 2018). Antimicrobial clusters in the genomes of several additional biocontrol strains of *B. velezensis*, including WRN014, 9912D, M75, RC 218, and CC09, have been discovered through sequencing (Wang et al. 2019). Plant health–promoting *B. methylotrophicus* B25 contains multiple genes involved in secondary metabolism, including polyketide synthases and nonribosomal peptide synthetases (Gerbore et al. 2016). Pseudomonas biocontrol strain SH-C52, one of the most well-known biocontrol strains, has a genome that is reportedly equally versatile concerning the biosynthesis of secondary metabolites, including arylpolyene, hydrogen cyanide, bacteriocin, insect toxins Tcc2 and Tcc4, thanamycin, and three NRPS-based lipopeptides 2 (Van Der Voort et al. 2015). The most current findings on the genomes of *B. toyonensis* BAC3151 and *Streptomyces* sp. JBS5–6, two possible biocontrol strains, described gene clusters important for secondary metabolite production similarly (Jing et al. 2020; Lopes et al. 2017). Common characteristics of sequenced *B. polymyxa* strains that can promote plant growth include enhanced nutrient availability and phytohormone synthesis (Eastman, Heinrichs, and Yuan 2014). The genes involved with nitrogen fixation, auxin biosynthesis, and antibiotic production have been found in the genome of the biofertilizer *Rhodopseudomonas palustris* ELI 1980 (Crovadore et al. 2017). Recently revealed *Sphingomonas* sp. Cra has several genes linked to the encouragement of plant growth, a characteristic that has been verified for *Arabidopsis thaliana* (Luo et al. 2020).

Despite the great promise, few novel features underlying plant growth promotion or defense against biotic or abiotic stressors have been identified by genome research. There are, however, a few instances. The genome mining of the biofertilizer strain *Trichoderma harzianum* t-22 showed the manufacture of tricholignan A, a naturally occurring substance that aids in the plant's absorption of iron from the soil (Chen et al. 2019). Genome analysis confirmed that there are no genes for plant hormone biosynthesis in the plant growth–promoting strain *Methylorubrum extorquens* DSM13060, which increases pine growth to the same extent as mycorrhizal fungi. However, it also revealed a variety of phospholipase A genes that may be involved in growth promotion (Koskimäki et al. 2015). Phospholipase D, superoxide dismutase, and ferredoxin NADP reductase have been proposed as the enzymes that cause the host

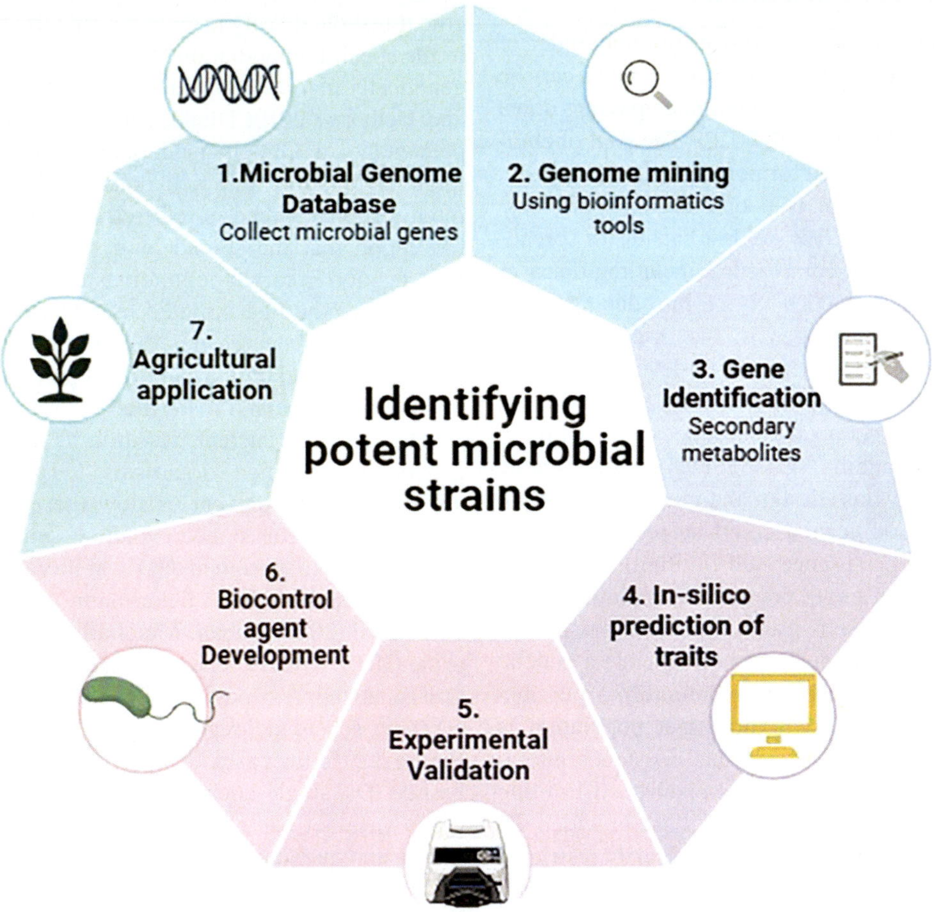

FIGURE 3.1 Approach in identifying potent microbial strains for biological control. The process includes genome sequencing, biosynthetic gene cluster prediction, and functional analysis to select strains with effective pathogen suppression capabilities.

Micrococcus luteus K39 to become resistant to oxidative stress, salt, and drought (Lafi et al. 2017). About 64 plant-resembling domains, including the NLR family of intracellular innate immune receptors, were found in a recent comparative examination of a variety of plant-associated bacterial genomes. These domains were theorized to potentially impair plant immune processes. These intracellular immune receptors may be in charge of biocontrol strains' enhanced ability to enhance plant resistance as a novel characteristic (Levy et al. 2018).

3.3.2 Genome Modification: Identifying New Potent Microbial Strains

Microorganisms that have undergone genome editing may be able to produce more effective strains of biocontrol or fertilizer. Concepts and initiatives to genetically modify biocontrol agents have already been developed and examined (St. Leger and Wang 2010). For example, *Pseudomonas putida* WCS358 and *Pseudomonas fluorescens* strain F113 and CHA0 have been genetically modified to produce more pyoluteorin (PLT) and 4-diacetylphloroglucinol (PHL), two antibiotic molecules. Moreover, the phytohormone indole acetic acid (IAA) has been engineered into a strain of *P. fluorescens* CHA0. In contrast, the *P. fluorescens* strain CHA0 that overproduced PHL and PLT in a microcosm test resulted in a decrease in rhizobia, *Sinorhizobium meliloti*, and nodulation in alphalpha. In natural soil, the CHA0 strain that produces more IAA boosted root yield; however, in autoclaved soil, root development was decreased (Mark et al. 2006). The field of employing genetically engineered microbes to combat plant diseases has been revitalized by the development of genome editing utilizing the CRISPR/Cas system (Glandorf 2019; Shelake, Pramanik, and Kim 2019). One new suggestion for a biocontrol strategy is the use of CRISPR/Cas to create non-pathogenic strains of fungal diseases. Nonetheless, plant microbiomes usually already naturally contain fungal endophytes, which are closely related but non-pathogenic relatives of plant diseases (Muñoz et al. 2019); therefore, the approach is already naturally in use.

3.3.3 Gene Drives by CRISPR/Cas9

To obtain a transmission advantage over the remainder of the genome, genetic components known as GDs can be defined as those that distort their inheritance passing down over the generations (Alphey et al. 2020). The goal of engineered genetic diversity (GDs) for population modification, mainly disease vector control, is to alter target populations' genetic composition so that they are less capable of spreading illness (reduced vector competence, including disease-resistant features). These tactics rely on introducing genes that either destroy the pathogen in the insect or create chemicals that prevent the disease from developing. They may also entail deactivating genes necessary for the disease to spread within the target organism (James et al. 2018).

Theoretically, depending on the engineered genetic modification system (whose design and mode of action are diverse), a genetic modification could be restricted in its spread (localized) or persistence (self-limiting), or it could spread spatially beyond target populations (non-localized) and persist indefinitely (self-sustaining). It is possible to create self-sustaining systems that can propagate a genetic modification quickly, broadly, and indefinitely – possibly for several generations or until the target population is changed or eradicated, or until a viable resistance alteration emerges (Noble et al. 2018). Conversely, self-limiting engineered genetic variations (GDs) are those in which the genetic alteration is anticipated to be temporarily restricted and vanish from the target population if no further releases occur. The requirement that people be released above a specific threshold density prior to causing the genetic alteration in the specified population is inherent to many constructed genetically modified systems (Leftwich et al. 2018; Backus and Delborne 2019; Dhole, Lloyd, and Gould 2020). The percentage of GD-changed people relative to the entire target population that will consistently start the genetic modification's spread is referred to as this threshold. The likelihood that the spread of a small number of genetic modification in target populations of the same species that are nearby could be initiated by GD-modified individuals increases with decreasing the threshold. High-threshold GDs don't spread until a significant percentage of the target population has been transformed in terms of density.

The goal of current research is to create designed genetically modified organisms (GDs) that are stable, are specific, and prevent or slow the emergence of resistance to them. Resistance evolution can cause GD failure by impeding a designed GD's ability to expand further (KaramiNejadRanjbar, Eckermann, and Ahmed 2018; Price et al. 2020; Raban, Marshall, and Akbari 2020). By using CRISPR/Cas9 gene editing, gene drives can introduce traits that suppress pest populations or enhance biocontrol efficacy, schematic representation of which is shown in Figure 3.2. In the case of CRISPR/Cas9-mediated homing-based GDs, single-nucleotide variations at the guide RNA (gRNA) target site can render it resistant to Cas9 recognition and prevent further cutting/homing (James et al. 2020). Though selection pressure for resistance mutations is

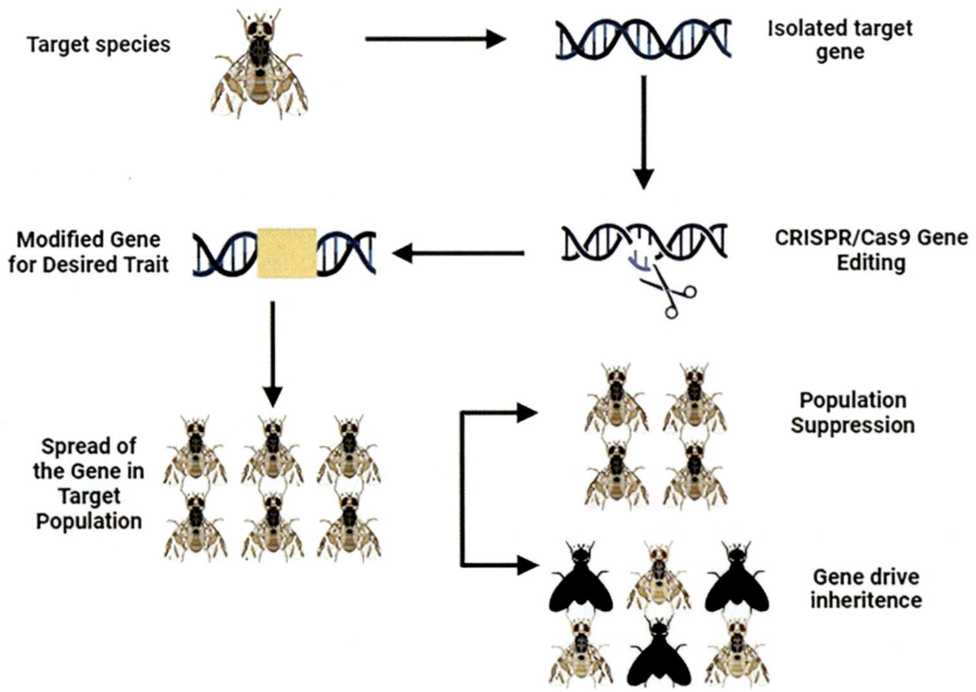

FIGURE 3.2 Mechanism of gene drives for biological control. The diagram illustrates how engineered genes are propagated through a population to spread traits that reduce pest numbers or enhance biocontrol, thereby providing a targeted and sustainable pest management strategy.

larger in suppressive GDs than in population modification GDs, decreased population numbers lower the frequency of possible mutation events that could result in resistance. Currently, a number of approaches are being investigated to stop resistance evolution: gRNA multiplexing with distinct target DNAs (James et al. 2020; Oberhofer, Ivy, and Hay 2018), focusing on highly conserved and functionally limited genes that are necessary for reproduction or survival (Schmidt et al. 2020; Champer et al. 2020); GD expression optimization or regulation to lower end-joining activity (Hammond et al. 2021); lowering fitness expenses; and integrating several designed GD strategies (Simoni et al. 2020). Though it is frequently brought up as a worry, designed GDs or other genetic control systems are not immune to the possibility of future GD failure (Price et al. 2020).

The development of engineered GDs that are reversible – that is, recollectible from the environment in the case of unintended consequences – and confinable – that is, restricted in their spread and persistence – is another goal of recent research endeavor. Several vector-borne illnesses are often promoted as being solved via the genetic engineering of mosquitoes. In an effort to combat malaria, mechanisms of alternative vector control that involve genetic alteration to change mosquitoes' immunological responses and anti-parasite effector molecules are taken into consideration (Naidoo and Oliver 2024). Recent proposals for interventions that promise species-specific, region-wide genetic control of dangerous insect populations include homing-based gene drives. We also show that the CRISPR-Cas9 gene drive can be linked to sex conversion, converting genetic females of the agricultural pest *Ceratitis capitata* into innocuous and fertile XX males, by targeting the medfly transformer gene targeting (Meccariello et al. 2024).

Additionally, systems have been created to control GD activity in response to the presence or absence of tiny, readily absorbed organic compounds in cells (Del Amo et al. 2020). Developers themselves warn against unanticipated repercussions (Xu et al. 2020) and state that reversal GDs would not always be the first option for remediation efforts because of the related uncertainty of adding another GD approach even though reversal and inducible GD systems show potential (Marshall and Vásquez 2021).

3.4 NEXT-GENERATION BIOLOGICAL CONTROL

3.4.1 Implementation of AI for the Application of Biocontrol Agents

Agriculture has undergone a revolution thanks to artificial intelligence. The techniques lessen excessive use of water, pesticides, and herbicides; they also maintain soil fertility, help with labor efficiency, boost output, and improve quality (Talaviya et al. 2020). Drones are becoming more and more common in all spheres of society and the economy, both for personal and business use. Drones are used more frequently in scientific and research endeavors in underdeveloped nations, but their professional application is increasingly apparent in developed nations. Similar circumstances exist in plant protection management, where drones serve a significant role mainly because of their capacity to swiftly and effectively cover wide regions and reach difficult-to-reach places (Brown and Giles 2018; Faiçal et al. 2017). Drones' capacity to access locations that might be hazardous or difficult for people to reach is one of their primary advantages. Drones can be used for data collection, health monitoring, pest and disease detection, and plant protection. Drones can be used for pesticide and other plant protection products application, including biocontrol agents, in addition to monitoring and problem detection (Iost Filho et al. 2020).

Since the agents in biocontrol can create self-replicating, enduring populations at the landscape scale, it is seen to be a more sustainable approach to pest management than other methods (Van Driesche 2012). The quick control of pests by the use of drones, particularly in augmentative biological management, is dependent on the widespread distribution of natural enemies. By distributing the natural enemies precisely where they are necessary, they may improve the efficiency of biocontrol agents and cut down on distribution expenses. Certain microscopic natural enemies, including nematodes and fungus that kill insects, are simple to apply using standard spray application tools (van Lenteren et al. 2018).

Biological control strategies are not always well-integrated with traditional operational procedures. The main manifestation of this is the challenge of implementing biocontrol agents in fully vegetated areas or reaching difficult-to-reach or damp areas within the appropriate time frame. Drones could assist in resolving these administrative issues and getting beyond the restrictions on chemical distribution (Lake et al. 2021). Drone technology has the potential to solve these field-related problems, apply bioagents more easily, and hasten the bioagents' establishment in the intended crop. While keeping an eye on invasive plant species with drone imaging capabilities is the primary use of drones in invasion biology and management (Dash et al. 2019), biological control agents have lately been released by researchers using drones (Teske et al. 2019). A number of ground mechanical distribution devices, like the Mini-Airbug, have been created to aid in the spread of predators, according to the authors of Iost Filho et al. (2020). Small drones have demonstrated a number of practical benefits over human or aircraft distribution, including lower application costs and more coverage, which has led to a notable increase in the employment of natural enemies rather than chemical sprays. Predatory mites and parasitoid wasps have been the primary targets of development efforts for drone-mounted dispensers (Iost Filho et al. 2020).

The University of California has created a platform called BugBot for the drone-based delivery of predatory mites. They created the model and related tools to maximize the discharge of advantageous organisms (Iost Filho et al. 2020). Similar practices are observed in Australia, where businesses and educational institutions employ drone-mounted tools designed to disperse predatory mites in strawberry and maize crops (Smith 1996). In Brazil, the sugarcane borer *Diatraea saccharalis* is managed with the use of drone treatments of *Trichogramma* spp. and the parasitoid *Cotesia flavipes* Cameron (Iost Filho et al. 2020). Additionally, *Trichogramma* spp. is used in other crops to combat a variety of lepidopteran pests (Parra 2014). Although *Trichogramma* wasps are widely used in biological control worldwide, not all pests or geographical areas have had their potential fully investigated. One of the nations bordering Southeast Europe that does not now use *Trichogramma* in its agricultural systems is Serbia (Ivezić et al. 2021). However, since 2010, ECB monitoring in Serbia has revealed the existence of parasitized egg clusters caused by *Trichogramma* as well as notable parasitism rates. These findings can be utilized to support and guide the cultivation and release of *Trichogramma* in Serbia (Ivezić et al. 2021; Ivezić, Rugman-Jones, and Trudić 2021). There are more natural species of beneficial insects that are recognized on Serbian territory besides parasitoid wasps. In the northern Serbian province of Vojvodina, representatives of the genus *Trissolcus* have recently been documented at multiple places (Ivezić, Trudić, and Ignjatović-Ćupina 2022).

Because farmers in the WB countries have a long history of using chemical crop protection, biological control methods are still only very partially applicable. The fact that there aren't enough applications for biological agents is another reason limiting their use. The aforementioned instances of effective biocontrol methods unmistakably direct agriculture toward a smooth transition to cutting-edge methods and environmentally responsible solutions in the realm of contemporary plant protection. In WB nations, biological control and the use of beneficial creatures, such as insects, are still in the experimental stages. The most committed biological control experiments are found in Serbia and Croatia, particularly when it comes to identifying the inherent potential of helpful organisms (Ivezić et al. 2021; Ivezić, Rugman-Jones, and Trudić 2021; Ivezić, Trudić, and Ignjatović-Ćupina 2022).

3.5 ENDOPHYTES AS BIOCONTROL AGENTS

The word *biocontrol* has been used to describe the process of controlling phytopathogens that cause the least amount of harm to the environment (Larran et al. 2016). However, because endophytic strains have a higher colonization efficacy than other strains, using them as BCAs has some further advantages (Mushtaq et al. 2023; Herrera et al. 2016). Utilizing endophytic bacteria to manage plant diseases is a new development. Endophytes secrete many defensive enzymes, including peroxidase, β-1,3-glucanase, and chitinase, which mediate their role in inhibiting pathogen growth. Li et al. (2023) assessed the endophyte *Bacillus velezensis*, which was isolated from maize, against the pathogen *Klebsiella pneumoniae*, and they discovered that the pathogen was significantly controlled. In a study, Doty et al. (2022) assessed that *Burkholderia vietnamiensis* is a powerful biocontrol agent against a variety of diseases, including *Rhizoctonia solani*, *Fusarium culmorum*, *Gaemannomyces graminis*, and *Pythium ultimum*. After applying the endophytic *Bacillus subtilis* bacterial strain to passion fruit, Wang et al. (2023) observed that *Nigrospora sphaerica* was inhibited. Likewise, Malvi et al. (2023) documented the pathogen *P. nicotianae*'s biocontrol capacity in the pot experiment by using *B. subtilis*, *B. pumilus*, *Burkholderia cepacia*, and *Enterobacter kobei*, and they discovered successful management. Additionally, Furuya et al. (2011) discovered that *Bacillus subtilis* has the ability to effectively oppose the pathogens *Botrytis cinerea* and *Colletotrichum gloeosporioides*. Nian et al. (2023) revealed that *Vishniacozyma victoria* has the ability to effectively suppress the disease *Botrytis cinerea*. The endophytic bacterial strain *Bacillus subtilis* was shown to have antagonistic activity against the pathogen *Colletotrichum fructicola* in a study by Xia et al. (2023), and their potential mechanism of action included the secretion of bioactive compounds, the synthesis of defense enzymes, and siderophores. Similar to this, Ma et al. (Vijitrpanth, Jantasorn, and Dethoup 2023) found that applying antimicrobial metabolites extracted from the endophytic fungus strain *Fusarium tricinctum* significantly decreased the incidence of kiwi canker infections. After applying endophytic *Trichoderma* sp., Vijitrpanth et al. (Pushpalatha et al. 2023) treated rice panicle disease. Following the administration of *Streptomyces* spp., Pushpalatha et al. (2023) found a considerable control of the pathogen *Sclerospora graminicola*. These investigations demonstrated the possible use of endophytic microbial strains for controlling phytopathogens or plant diseases.

3.6 RNA-BASED BIOLOGICAL CONTROL COMPOUNDS

Utilizing RNA interference (RNAi)–based technology has shown to be an effective and precise tactic for enhancing crop protection and productivity. Scientists may now precisely silence target genes post-transcriptionally by understanding RNA interference (RNAi) mechanisms, shown in Figure 3.3. These genes may be endogenous plant genes, genes of plant diseases (viruses and fungi), or genes of other plant pests (insects, mites, nematodes, and weeds) (Dalakouras, Koidou, and Papadopoulou 2024). The function of these targets can be inhibited by employing various RNA interference (RNAi) triggers, including hairpin-structured RNAs (hpRNAs), artificial microRNAs (amiRNAs), and double-stranded RNAs (dsRNAs). To stop the targeted mRNA from being translated into proteins by mRNA degradation or translational repression, post-transcriptional RNA interference (RNAi) methods investigate a sequence-dependent mode of action at the mRNA level. This results in target gene silence and the intended outcome (Dalakouras, Koidou, and Papadopoulou 2024).

From maize to potatoes, RNAi-based events cover a broad spectrum of plant species and features, such as pest and viral resistance and altered plant composition. Numerous

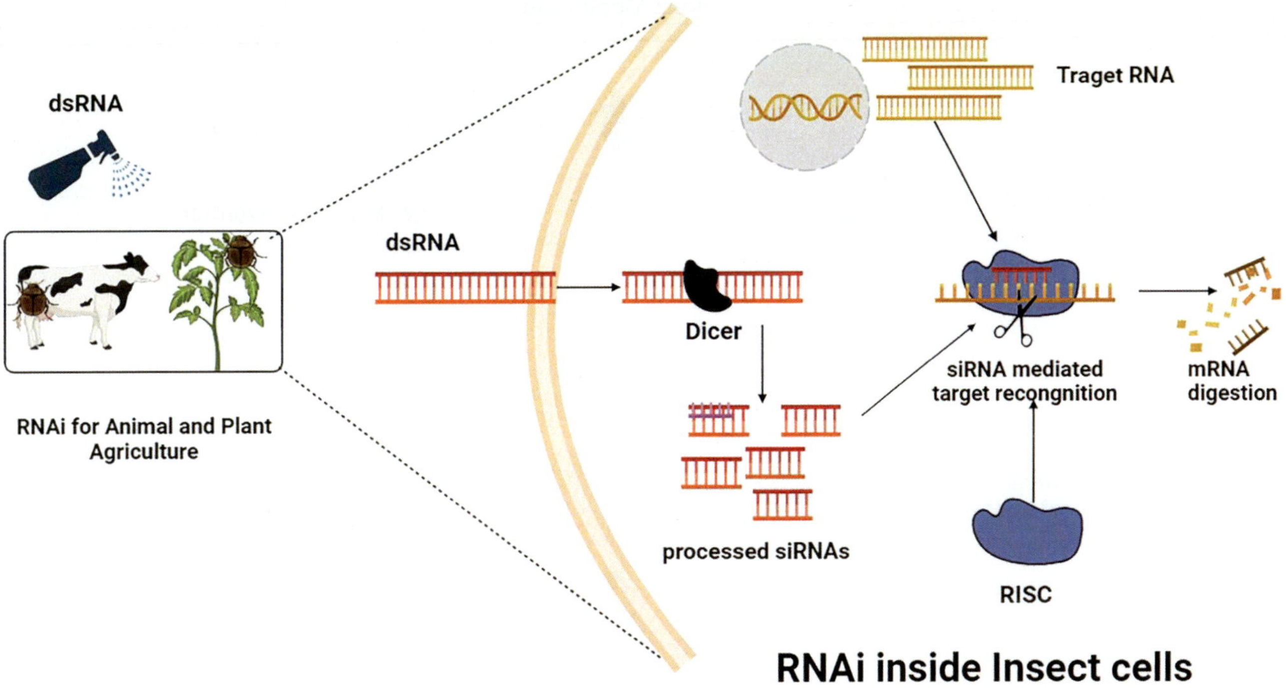

FIGURE 3.3 RNA interference (RNAi)–based technology for biological control. The diagram illustrates how RNAi targets and silences specific genes in pests or pathogens, leading to their suppression or elimination and providing a precise and effective method for pest and disease management.

investigations have shown that exogenously sprayed dsRNAs on plant tissues can cause RNA interference (RNAi) to silence specific genes associated with pests or pathogens. In vitro–synthesized long dsRNAs, hpRNAs, or small interfering RNAs (siRNAs), which are intended to specifically target critical genes of pests and diseases, can be applied externally to induce plant resistance to fungus, insects, and viruses (Wang et al. 2016; Dubrovina and Kiselev 2019). Additionally, some investigations have demonstrated that, after being applied mechanically through wounding, topically applied dsRNAs can act on both the treated plant area and distant non-treated areas through systemic translocation of dsRNAs via the plant vascular system (Konakalla et al. 2016; Kaldis et al. 2018). These findings show that exogenously applied dsRNAs may be actively absorbed by plant cells and converted into siRNAs, which should prolong the period of control or prevention against certain plant diseases. It has been shown, for instance, that the use of exogenous RNA interference techniques against particular fungi is restricted when secondary siRNA amplification or dsRNA uptake by the target pathogen is absent (Song et al. 2018; Kettles et al. 2019). Additionally, another important factor that affects the efficacy of exogenous RNAi-based strategies is the delivery method chosen for dsRNA molecules (i.e., mechanical inoculation or leaf spraying), which should be chosen based on the type of plant and the type of pest species targeted (Dalakouras et al. 2018).

The utilization of nanocarriers as efficient translational tools to bring sprayable RNA-based biocontrol agents to the market is growing (Kunte et al. 2020). Therefore, it is believed that the fusion of nanotechnology and RNAi technology would revolutionize the crop protection sector (Landry and Mitter 2019). One instance of this is the distribution of dsRNA using "layered double hydroxide clay nanoparticles" (BioClay), which emphasizes the importance of nanoparticles as delivery vehicles and the advancement of translating research from the lab to the field (Mitter et al. 2017). However, this will rely on how much it costs to produce the developed nanocarriers and how effective they are. It has recently been reported that an effective method for delivering siRNA into plant cells is to use DNA nanostructures acting as RNA carriers via Watson–Crick base pairing (Zhang et al. 2019). To ensure proper internalization of the nanostructure into plant cells and a corresponding good gene silencing efficiency, however, parameters like size, shape, compactness, stiffness, and placement of the siRNA attachment locus on the nanostructure must all be considered during formulation.

3.7 INTEGRATED PEST MANAGEMENT

Integrated pest management (IPM) (illustrated in Figure 3.4) employs a variety of pest control techniques with the aim of either completely replacing or significantly lowering the

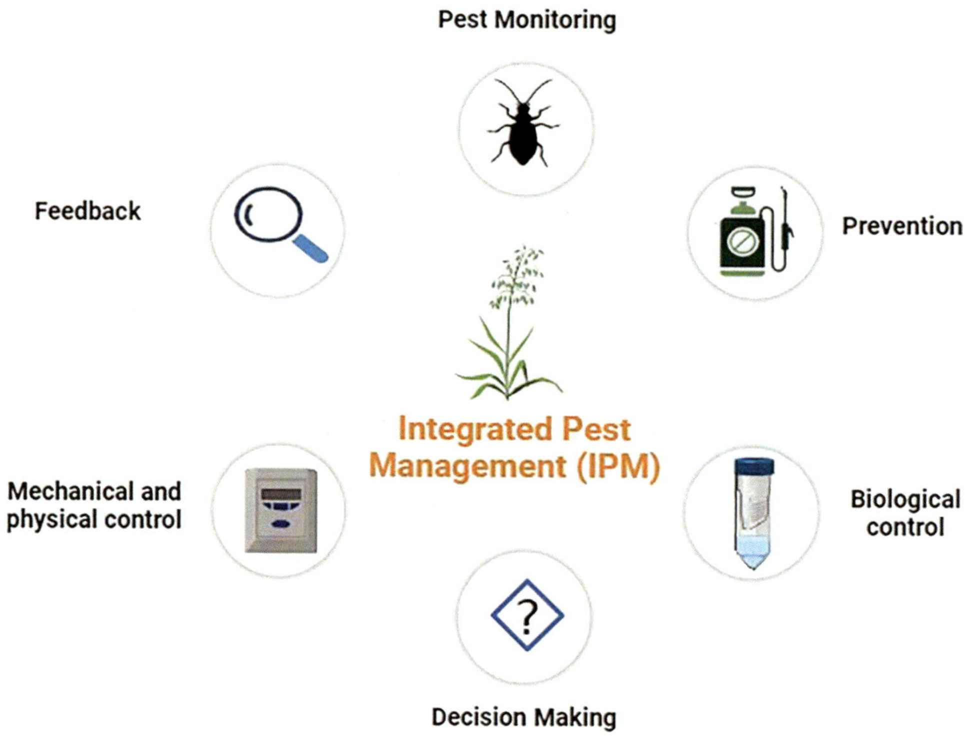

FIGURE 3.4 Integrated pest management (IPM) with biological control. The diagram shows how natural enemies like predators, parasitoids, and pathogens are used alongside other IPM strategies to manage pest populations sustainably and reduce reliance on chemical pesticides.

requirement for synthetic pesticides. It is a long-standing, environmentally friendly method of managing pests. It means devising strategies that are practical, affordable, and less harmful to the environment (Dara 2019). IPM keeps insect numbers below a threshold where they cannot negatively impact the economy (Hagstrum and Flinn 2018). IPM is the term for crop management that employs a range of techniques to keep pest levels below a predetermined threshold for economic benefit (Prokopy and Kogan 2009). It is a systematic process that incorporates a variety of pest control methods into one program. It is necessary to incorporate cultural, biological, genetic, physical, legal, and mechanical constraints in order to reduce reliance on pesticides (Oguh et al. 2019). A comprehensive plan for managing weeds that incorporates several chemical, cultural, physical, and preventive weed control techniques would guarantee a sustained availability of food. A producer's collection of cultural practices, such as crop density, the use of challenging cultivars, or the strategic use of mineral fertilizer, can readily alter agronomic weed control as part of integrated weed management strategies. Especially in situations where applying herbicides is not a possibility (Korres et al. 2019).

3.7.1 Limitation of IPM

IPM is reinterpreted for the modern era, when food production and consumption heavily rely on enhanced agricultural practices and communication channels. IPM is a component of agriculture, which is a sector increasingly centered on providing customers with high-quality, safe, and sustainable products. Agriculture is a component of global commerce that is influenced by a number of different elements (Dara 2019). Compared to their rural counterparts, small-scale urban farmers are significantly more frequently identified as missing basic horticultural knowledge, perhaps because they are more likely to have non-farming experiences (Surls et al. 2015) and can be negatively impacted by numerous additional factors, like linguistic constraints. Although managing weeds and pests is often cited as one of the biggest challenges faced by these farmers, agriculture education usually provides a rudimentary understanding of pest control (Opitz et al. 2016). A bigger problem than advisory services availability might be the lack of IPM research relevant to subsistence farming (Quarcoo et al. 2017). In addition to the reason that certain inputs could not be accessible in packaging quantities appropriate for their range, local farmers might not be able to use particular pest management measures because of cost alone (Grasswitz and Yao 2016). Pheromone transmitters are currently available in one-time package quantities that are deemed insufficient to satisfy the needs of farmers managing extensive plants. One further problem is that local farmers would need to use greater negotiation skills when interacting with farming equipment sellers (Quarcoo et al. 2017).

The widespread use of these extremely deadly pesticides, particularly in agricultural operations, has raised concerns about the actual negative consequences of pesticides on humans and the ecology. Pest classification has changed significantly over time because of shifts in farming methods, biodiversity, and ecology, both historically and regionally. Outbreaks are frequently caused by many pests spreading globally and becoming resistant to pesticides (Adhikari 2022). Each of the physical, chemical, biological, and cultural approaches has limits that influence how well they control a specific pest from a technological and theoretical perspective. Strict laws restrict our options even as we move from production to consumption. A range of physical control techniques are used, taking postharvest conditions into account. Chemical procedures have well-defined and limited mechanisms of operation. Modern infrastructure is frequently associated with affordable costs. Determining the appropriate thresholds for the standards' requirements on excellent hygienic and sanitary practices is another obstacle to the implementation of IPM (Thakur, Sharma, and Sharma 2021). The present problems with the use of conventional pesticides and the emphasis on low-cost, sustainable farming have led to an increase in the use of bacterial agents in IPM techniques (Kachhawa 2017). In order to proliferate and disseminate illness, entomopathogenic fungi (EPF) need particular climatic conditions. For commercial use, EPF are very costly to construct and have a limited lifespan (Maina et al. 2018).

3.8 FUTURE DIRECTIONS

Given the well-established issues about using chemical pesticides, appropriate solutions are desperately needed. Physical methods are used to alter the pest's physical environment in order to ensure that the insects do not provide a risk to the plant. There is currently a widespread demand to increase IPM treatments in order to reduce pest infestations at all phases of the pre- and postharvest food production processes (Maina et al. 2018; Thakur, Sharma, and Sharma 2021). The search for undiscovered natural enemies, such as microorganisms and novel strains of entomopathogens, requires constant research and the development of new biocontrol agents to target a wider variety of pests, particularly those resistant to current techniques. New habitats and ecosystems must also be explored (Li et al. 2024). Increasing the specificity and efficacy of biocontrol agents is essential to increasing the dependability and economy of biological control. To improve the virulence, adaptability, and specificity of biocontrol organisms, CRISPR and other gene-editing technologies can be applied (Komal et al. 2023). Most illnesses can multiply both spatially and temporally, which would help precision farming (Kachhawa 2017). Mycoinsecticides have the potential to be a crucial component of an integrated pest management (IPM) program for the exhaustive and usually risk-free control of insect pests in agricultural output. In order to do this, new interdisciplinary efforts are needed to address the pathogenic aggressiveness and range of action of the products, as well as their effectiveness in unfavorable climate conditions and compositions that would increase permanence, have a longer lifespan, and be easy to use. The activities of EPF have a significant impact on the sustainability of pest abundance and dispersion in ecological systems. Understanding this role prompted research into the possible application of infectious fungi in the management of insect pests (Maina et al. 2018).

Adaptive solutions are required because pest dynamics and the efficiency of biological control agents are changing due to climate change. Researching and creating biocontrol agents and techniques that are climate-resilient and flexible enough to adjust to shifting environmental conditions can be advantageous (Subedi, Poudel, and Aryal 2023). Clear legal frameworks and widespread public backing are becoming more and more necessary to facilitate the use of cutting-edge biological control technologies. The development of standardized risk assessment methodologies and public education campaigns is necessary to increase public confidence in and comprehension of biological control techniques. Biological control methods must be optimized by utilizing contemporary technology as well, such as artificial intelligence and remote sensing (Kebe et al. 2023). The development of better manufacturing, conceptualization, preservation, and surface treatments is necessary in light of increased efficacy, usefulness, and cost. Microbiological genomes can be included into traditional breeding for significant crop plants.

3.9 CONCLUSION

The domain of biological control has had notable progressions, driven by breakthroughs in biotechnology, ecology, and integrated pest management approaches. These advancements have added new weapons to the toolbox for managing insect populations in a sustainable manner, providing an alternative to traditional chemical pesticides. The development of genomic technologies, like RNA interference and CRISPR, has made it possible to increase the precision and efficacy of biocontrol agents, and the area is still being expanded by the discovery of novel endophytic and microbial solutions. Furthermore, biological control is starting to change as a result of the digital revolution. AI-driven technologies and precision agriculture allow for the more effective and focused application of biocontrol agents. But as these technologies develop, it becomes more and more important that strong legal frameworks and public participation are in place. Another specific difficulty posed by climate change is the need for adaptive measures to maintain biological control's efficacy in the face of changing environmental conditions. Biological control's future rests on the continuous search for new agents, the improvement of existing approaches, and the smooth incorporation of these strategies into all-encompassing sustainable agricultural practices. Global food security and environmental sustainability can be greatly enhanced by biological management by utilizing the most recent developments and tackling these new difficulties.

REFERENCES

Adhikari, Upama. 2022. "Insect pest management: Mechanical and physical techniques." *Reviews in Food and Agriculture (RFNA)* 3 (1):48–53.

Aguilar-Marcelino L. 2012. *Microorganismos con uso potencial contra el nemátodo de ovinos Haemonchus contortus.* Tesis presentada para obtener el grado de doctor en ciencias. Texcoco.

Aguilar-Marcelino, Lb, M. T. Quintero-Martínez, P. Mendoza De Gives, M. E. López-Arellano, E. Liébano-Hernández, G. Torres-Hernández, J. M. González-Camacho, and I. Cid Del Prado 2014. "Evaluation of predation of the mite Lasioseius penicilliger (Aracnida: Mesostigmata) on Haemonchus contortus and bacteria-feeding nematodes." *Journal of Helminthology* 88 (1):20–23.

Al-Ani, Laith Khalil Tawfeeq, and Shaymaa Fadhel Abbas Albaayit. 2018. "Antagonistic of some Trichoderma against Fusarium oxysporum sp. f. cubense tropical race 4 (FocTR4)." *The Eurasia Proceedings of Science Technology Engineering and Mathematics* (2):35–38.

Al-Ani, Laith Khalil Tawfeeq, and Edson Luiz Furtado. 2020. "The effect of incompatible plant pathogens on the host plant." In *Molecular aspects of plant beneficial microbes in agriculture*, 47–57. Elsevier.

Alphey, Luke S, Andrea Crisanti, Filippo Randazzo, and Omar S Akbari. 2020. "Standardizing the definition of gene drive." *Proceedings of the National Academy of Sciences* 117 (49):30864–30867.

Backus, Gregory A, and Jason A Delborne. 2019. "Threshold-dependent gene drives in the wild: Spread, controllability, and ecological uncertainty." *BioScience* 69 (11):900–907.

Barratt, BIP, VC Moran, Franz Bigler, and JC Van Lenteren. 2018. "The status of biological control and recommendations for improving uptake for the future." *BioControl* 63:155–167.

Bekal, Sadia, Leslie L Domier, Biruk Gonfa, Nancy K McCoppin, Kris N Lambert, and Kaustubh Bhalerao. 2014. "A novel flavivirus in the soybean cyst nematode." *Journal of General Virology* 95 (6):1272–1280.

Brown, Colin R, and Durham K Giles. 2018. "Measurement of pesticide drift from unmanned aerial vehicle application to a vineyard." *Transactions of the ASABE* 61 (5):1539–1546.

Champer, Jackson, Emily Yang, Esther Lee, Jingxian Liu, Andrew G Clark, and Philipp W Messer. 2020. "A CRISPR homing gene drive targeting a haplolethal gene removes resistance alleles and successfully spreads through a cage population." *Proceedings of the National Academy of Sciences* 117 (39):24377–24383.

Chen, Mengbin, Qikun Liu, Shu-Shan Gao, Abbegayle E Young, Steven E Jacobsen, and Yi Tang. 2019. "Genome mining and biosynthesis of a polyketide from a biofertilizer fungus that can facilitate reductive iron assimilation in plant." *Proceedings of the National Academy of Sciences* 116 (12):5499–5504.

Chen, Xiao Hua, Alexandra Koumoutsi, Romy Scholz, Andreas Eisenreich, Kathrin Schneider, Isabelle Heinemeyer, Burkhard Morgenstern, Björn Voss, Wolfgang R Hess, and Oleg Reva. 2007. "Comparative analysis of the complete genome sequence of the plant growth – promoting bacterium Bacillus amyloliquefaciens FZB42." *Nature Biotechnology* 25 (9):1007–1014.

Cheng, F, and Z Cheng. 2015. "Research progress on the use of plant allelopathy in agriculture and the physiological and ecological mechanisms of allelopathy." *Frontiers in Plant Science* 6:1020.

Cotton, James A, Sascha Steinbiss, Toshiro Yokoi, Isheng J Tsai, and Taisei Kikuchi. 2016. "An expressed, endogenous Nodavirus-like element captured by a retrotransposon in the genome of the plant parasitic nematode Bursaphelenchus xylophilus." *Scientific Reports* 6 (1):39749.

Crovadore, Julien, Shoutao Xu, Romain Chablais, Bastien Cochard, Delvia Lukito, Gautier Calmin, and François Lefort. 2017. "Metagenome-assembled genome sequence of Rhodopseudomonas palustris strain eli 1980, commercialized as a biostimulant." *Genome Announcements* 5 (18):e00221–17.

Dalakouras, Athanasios, Wolfgang Jarausch, Guenther Buchholz, Alexandra Bassler, Mario Braun, Thorsten Manthey, Gabi Krczal, and Michael Wassenegger. 2018. "Delivery of hairpin RNAs and small RNAs into woody and herbaceous plants by trunk injection and petiole absorption." *Frontiers in Plant Science* 9:1253.

Dalakouras, Athanasios, Venetia Koidou, and Kalliope Papadopoulou. 2024. "DsRNA-based pesticides: Considerations for efficiency and risk assessment." *Chemosphere*:141530.

Dara, Surendra K. 2019. "The new integrated pest management paradigm for the modern age." *Journal of Integrated Pest Management* 10 (1):12.

Dash, Jonathan P, Michael S Watt, Thomas SH Paul, Justin Morgenroth, and Robin Hartley. 2019. "Taking a closer look at invasive alien plant research: A review of the current state, opportunities, and future directions for UAVs." *Methods in Ecology and Evolution* 10 (12):2020–2033.

Dedrick, Rebekah M, Carlos A Guerrero-Bustamante, Rebecca A Garlena, Daniel A Russell, Katrina Ford, Kathryn Harris, Kimberly C Gilmour, James Soothill, Deborah Jacobs-Sera, and Robert T Schooley. 2019. "Engineered bacteriophages for treatment of a patient with a disseminated drug-resistant Mycobacterium abscessus." *Nature Medicine* 25 (5):730–733.

Del Amo, Víctor López, Brittany S Leger, Kurt J Cox, Shubhroz Gill, Alena L Bishop, Garrett D Scanlon, James A Walker, Valentino M Gantz, and Amit Choudhary. 2020. "Small-molecule control of super-Mendelian inheritance in gene drives." *Cell Reports* 31 (13).

Deroulers, P, and V Bretagnolle. 2019. "The consumption pattern of 28 species of carabid beetles (Carabidae) to a weed seed, Viola arvensis." *Bulletin of Entomological Research* 109 (2):229–235.

Dhole, Sumit, Alun L Lloyd, and Fred Gould. 2020. "Gene drive dynamics in natural populations: The importance of density dependence, space, and sex." *Annual Review of Ecology, Evolution, and Systematics* 51 (1):505–531.

Doty, Sharon L, Pierre M Joubert, Andrea Firrincieli, Andrew W Sher, Robert Tournay, Carina Kill, Shruti S Parikh, and Patricia Okubara. 2022. "Potential biocontrol activities of Populus endophytes against several plant pathogens using different inhibitory mechanisms." *Pathogens* 12 (1):13.

Dubrovina, Alexandra S, and Konstantin V Kiselev. 2019. "Exogenous RNAs for gene regulation and plant resistance." *International Journal of Molecular Sciences* 20 (9):2282.

Eastman, Alexander W, David E Heinrichs, and Ze-Chun Yuan. 2014. "Comparative and genetic analysis of the four sequenced Paenibacillus polymyxa genomes reveals a diverse metabolism and conservation of genes relevant to plant-growth promotion and competitiveness." *BMC Genomics* 15:1–22.

Faiçal, Bruno S, Heitor Freitas, Pedro H Gomes, Leandro Y Mano, Gustavo Pessin, André CPLF de Carvalho, Bhaskar Krishnamachari, and Jó Ueyama. 2017. "An adaptive approach for UAV-based pesticide spraying in dynamic environments." *Computers and Electronics in Agriculture* 138:210–223.

Fan, Ben, Cong Wang, Xiaofeng Song, Xiaolei Ding, Liming Wu, Huijun Wu, Xuewen Gao, and Rainer Borriss. 2018. "Bacillus velezensis FZB42 in 2018: The gram-positive model strain for plant growth promotion and biocontrol." *Frontiers in Microbiology* 9:2491.

Fausti, Scott W. 2015. "The causes and unintended consequences of a paradigm shift in corn production practices." *Environmental Science & Policy* 52:41–50.

Furuya, Seiichi, Mai Mochizuki, Yoshinao Aoki, Hironori Kobayashi, Tsutomu Takayanagi, Masafumi Shimizu, and Shunji Suzuki. 2011. "Isolation and characterization of Bacillus subtilis KS1 for the biocontrol of grapevine fungal diseases." *Biocontrol Science and Technology* 21 (6):705–720.

Garcia Ortiz, N, L Aguilar Marcelino, P Mendoza de Gives, ME Lopez Arellano, CR Bautista Garfias, and R Gonzalez Garduno. 2015. "In vitro predatory activity of Lasioseius penicilliger (Arachnida: Mesostigmata) against three nematode species: Teladorsagia circumcincta, Meloidogyne sp. and Caenorhabditis elegans." *Veterinaria México OA* 2 (1):01–09.

Gerbore, Jonathan, Aline Brutel, Arnaud Lemainque, Barbara Mairey, Claudine Médigue, David Vallenet, François Lefort, and Damien Grizard. 2016. "Complete genome sequence of Bacillus methylotrophicus strain B25, a potential plant growth-promoting rhizobacterium." *Genome Announcements* 4 (2):e00058–16.

Glandorf, Debora CM. 2019. "Re-evaluation of biosafety questions on genetically modified biocontrol bacteria." *European Journal of Plant Pathology* 154:43–51.

Grasswitz, TR, and S Yao. 2016. "Efficacy of pheromonal control of peachtree borer (Synanthedon exitiosa (Say)) in small-scale orchards." *Journal of Applied Entomology* 140 (9):669–676.

Guo, Yusong R, Corey F Hryc, Joanita Jakana, Hongbing Jiang, David Wang, Wah Chiu, Weiwei Zhong, and Yizhi J Tao. 2014. "Crystal structure of a nematode-infecting virus." *Proceedings of the National Academy of Sciences* 111 (35):12781–12786.

Hagstrum, David W, and Paul W Flinn. 2018. "Integrated pest management." In *Integrated management of insects in stored products*, 399–407. CRC Press.

Hammond, Andrew, Xenia Karlsson, Ioanna Morianou, Kyros Kyrou, Andrea Beaghton, Matthew Gribble, Nace Kranjc, Roberto Galizi, Austin Burt, and Andrea Crisanti. 2021. "Regulating the expression of gene drives is key to increasing their invasive potential and the mitigation of resistance." *PLoS Genetics* 17 (1):e1009321.

Heap, Ian. 2014. *Herbicide resistant weeds*. Springer.

Heimpel, George E, and Nicholas J Mills. 2017. *Biological control*. Cambridge University Press.

Herrera, Silvana Díaz, Cecilia Grossi, Myriam Zawoznik, and María Daniela Groppa. 2016. "Wheat seeds harbour bacterial endophytes with potential as plant growth promoters and biocontrol agents of Fusarium graminearum." *Microbiological Research* 186:37–43.

Hrabáková, Lenka, Igor Koloniuk, and Karel Petrzik. 2017. "Phomopsis longicolla RNA virus 1–Novel virus at the edge of myco-and plant viruses." *Virology* 506:14–18.

Hu, Zijin, Songsong Wu, Jiasen Cheng, Yanping Fu, Daohong Jiang, and Jiatao Xie. 2014. "Molecular characterization of two positive-strand RNA viruses co-infecting a hypovirulent strain of Sclerotinia sclerotiorum." *Virology* 464:450–459.

Iost Filho, Fernando H, Wieke B Heldens, Zhaodan Kong, and Elvira S de Lange. 2020. "Drones: Innovative technology for use in precision pest management." *Journal of Economic Entomology* 113 (1):1–25.

Isman, Murray B. 2006. "Botanical insecticides, deterrents, and repellents in modern agriculture and an increasingly regulated world." *Annual Review of Entomology* 51 (1):45–66.

Ivezić, Aleksandar, Paul F Rugman-Jones, Thibaut Malausa, Nicolas Ris, and Aleksandra Ignjatović-Ćupina. 2021. "Molecular identification of Trichogramma species parasitizing Ostrinia nubilalis in corn and pepper in south – east border of Europe." *International Journal of Pest Management* 67 (4):346–357.

Ivezić, Aleksandar, Paul F Rugman-Jones, and Branislav Trudić. 2021. "Rapid molecular identification of Trichogramma (Hymenoptera: Trichogrammatidae) parasitizing the eggs of the European corn borer, Ostrinia nubilalis (Lepidoptera: Crambidae) in Serbia." *Egyptian Journal of Biological Pest Control* 31:1–8.

Ivezić, Aleksandar, Branislav Trudić, and Aleksandra Ignjatović-Ćupina. 2022. "Predstavnici roda Trissolcus (Hymenoptera: Scelionidae) prirodni neprijatelji štetnih stenica (Hemiptera: Pentatomidae) i preliminarna istraživanja u Srbiji." *BILJNI LEKAR* 50 (1):15.

James, Stephanie, Frank H Collins, Philip A Welkhoff, Claudia Emerson, H Charles J Godfray, Michael Gottlieb, Brian Greenwood, Steve W Lindsay, Charles M Mbogo, and Fredros O Okumu. 2018. "Pathway to deployment of gene drive mosquitoes as a potential biocontrol tool for elimination of malaria in sub-Saharan Africa: Recommendations of a scientific working group." *The American Journal of Tropical Medicine and Hygiene* 98 (6 Suppl):1.

James, Stephanie L, John M Marshall, George K Christophides, Fredros O Okumu, and Tony Nolan. 2020. "Toward the definition of efficacy and safety criteria for advancing gene drive-modified mosquitoes to field testing." *Vector-Borne and Zoonotic Diseases* 20 (4):237–251.

Jensen, Birgit, Peter S Lübeck, and Hans JL Jørgensen. 2016. "Clonostachys rosea reduces spot blotch in barley by inhibiting prepenetration growth and sporulation of Bipolaris sorokiniana without inducing resistance." *Pest Management Science* 72 (12):2231–2239.

Jing, Tao, Dengbo Zhou, Miaoyi Zhang, Tianyan Yun, Dengfeng Qi, Yongzan Wei, Yufeng Chen, Xiaoping Zang, Wei Wang, and Jianghui Xie. 2020. "Newly isolated Streptomyces sp. JBS5–6 as a potential biocontrol agent to control banana Fusarium wilt: Genome sequencing and secondary metabolite cluster profiles." *Frontiers in Microbiology* 11:602591.

Jordan, Kieran, Marion Dalmasso, Juergen Zentek, Anneluise Mader, Geert Bruggeman, John Wallace, Dario De Medici, Alfonsina Fiore, Estella Prukner-Radovcic, and Maja Lukac. 2014. "Microbes versus microbes: Control of pathogens in the food chain." *Journal of the Science of Food and Agriculture* 94 (15):3079–3089.

Kachhawa, D. 2017. "Microorganisms as a biopesticides." *Journal of Entomology and Zoology Studies* 5 (3):468–473.

Kaldis, Athanasios, Margarita Berbati, Ourania Melita, Chrysavgi Reppa, Maria Holeva, Patricia Otten, and Andreas Voloudakis. 2018. "Exogenously applied dsRNA molecules deriving from the Zucchini yellow mosaic virus (ZYMV) genome move systemically and protect cucurbits against ZYMV." *Molecular Plant Pathology* 19 (4):883–895.

KaramiNejadRanjbar, Mohammad, Kolja N Eckermann, and Hassan MM Ahmed. 2018. "HM Sánchez C, S. Dippel, JM Marshall, EA Wimmer, Consequences of resistance evolution in a Cas9-based sex conversion-suppression gene drive for insect pest management." *Proceedings of the National Academy of Sciences* 115:6189–6194.

Kebe, Alhaji Alusine, Shumaila Hameed, Muhammad Sohail Farooq, Abu Sufyan, Muhammad Babar Malook, Sayed Awais, Muhammad Riaz, Muhammad Waseem, Usman Amjad, and Nasir Abbas. 2023. "Enhancing crop protection and yield through precision agriculture and integrated pest management: A comprehensive review." *Asian Journal of Research in Crop Science* 8 (4):443–453.

Kettles, Graeme J, Bernhard J Hofinger, Pingsha Hu, Carlos Bayon, Jason J Rudd, Dirk Balmer, Mikael Courbot, Kim E Hammond-Kosack, Gabriel Scalliet, and Kostya Kanyuka. 2019. "sRNA profiling combined with gene function analysis reveals a lack of evidence for cross-kingdom RNAi in the wheat – Zymoseptoria tritici pathosystem." *Frontiers in Plant Science* 10:892.

Köhl, Jürgen, Rogier Kolnaar, and Willem J Ravensberg. 2019. "Mode of action of microbial biological control agents against plant diseases: Relevance beyond efficacy." *Frontiers in Plant Science* 10:845.

Komal, J, HR Desai, Ipsita Samal, Andrea Mastinu, RD Patel, PV Dinesh Kumar, Prasanta Kumar Majhi, Deepak Kumar Mahanta, and Tanmaya Kumar Bhoi. 2023. "Unveiling the genetic symphony: Harnessing CRISPR-Cas genome editing for effective insect pest management." *Plants* 12 (23):3961.

Konakalla, Naga Charan, Athanasios Kaldis, Margarita Berbati, Hema Masarapu, and Andreas E Voloudakis. 2016. "Exogenous application of double-stranded RNA molecules from TMV p126 and CP genes confers resistance against TMV in tobacco." *Planta* 244:961–969.

Korres, Nicholas E, Nilda R Burgos, Ilias Travlos, Maurizio Vurro, Thomas K Gitsopoulos, Vijaya K Varanasi, Stephen O Duke, Per Kudsk, Chad Brabham, and Christopher E Rouse. 2019. "New directions for integrated weed management: Modern technologies, tools and knowledge discovery." *Advances in Agronomy* 155:243–319.

Koskimäki, Janne J, Anna Maria Pirttilä, Emmi-Leena Ihantola, Outi Halonen, and A Carolin Frank. 2015. "The intracellular scots pine shoot symbiont Methylobacterium extorquens DSM13060 aggregates around the host nucleus and encodes eukaryote-like proteins." *mBio* 6 (2):e00039–15.

Kunte, Nitish, Erin McGraw, Sydney Bell, David Held, and Luz-Adriana Avila. 2020. "Prospects, challenges and current status of RNAi through insect feeding." *Pest Management Science* 76 (1):26–41.

Lacey, LA. 2016. *Microbial control of insect and mite pests: From theory to practice*. Academic Press.

Lafi, Feras F, Juan S Ramirez-Prado, Intikhab Alam, Vladimir B Bajic, Heribert Hirt, and Maged M Saad. 2017. "Draft genome sequence of plant growth – promoting Micrococcus luteus strain K39 isolated from Cyperus conglomeratus in Saudi Arabia." *Genome Announcements* 5 (4):e01520–16.

Lake, Ellen C, Aaron S David, Thomas M Spencer, Victor L Wilhelm Jr, Travis W Barnett, Anwar A Abdel-Kader, Andrea Carmona Cortes, Adrian Acuna, Elizabeth D Mattison, and Carey R Minteer. 2021. "First drone releases of the biological control agent Neomusotima conspurcatalis on Old World climbing fern." *Biocontrol Science and Technology* 31 (1):97–106.

Landry, Markita P, and Neena Mitter. 2019. "How nanocarriers delivering cargos in plants can change the GMO landscape." *Nature Nanotechnology* 14 (6):512–514.

Larran, Silvina, Maria Rosa Simon, María Virginia Moreno, MP Santamarina Siurana, and Analía Perelló. 2016. "Endophytes from wheat as biocontrol agents against tan spot disease." *Biological Control* 92:17–23.

Leftwich, Philip T, Matthew P Edgington, Tim Harvey-Samuel, Leonela Z Carabajal Paladino, Victoria C Norman, and Luke Alphey. 2018. "Recent advances in threshold-dependent gene drives for mosquitoes." *Biochemical Society Transactions* 46 (5):1203–1212.

Lemus-Minor, Carlos G, M Carmen Cañizares, María D García-Pedrajas, and Encarnación Pérez-Artés. 2015. "Complete genome sequence of a novel dsRNA mycovirus isolated from the phytopathogenic fungus Fusarium oxysporum f. sp. dianthi." *Archives of Virology* 160:2375–2379.

Levy, Asaf, Isai Salas Gonzalez, Maximilian Mittelviefhaus, Scott Clingenpeel, Sur Herrera Paredes, Jiamin Miao, Kunru Wang, Giulia Devescovi, Kyra Stillman, and Freddy Monteiro. 2018. "Genomic features of bacterial adaptation to plants." *Nature Genetics* 50 (1):138–150.

Li, Xiangsong, Pengfei He, Pengbo He, Yongmei Li, Yixin Wu, Chan Mu, Shahzad Munir, and Yueqiu He. 2023. "Native endophytes from maize as potential biocontrol agents against bacterial top rot caused by cross-kingdom pathogen Klebsiella pneumoniae." *Biological Control* 178:105131.

Li, Xiaojiao, Shixian Zeng, Michael Wisniewski, Samir Droby, Longfeng Yu, Fuquan An, Yan Leng, Chaowen Wang, Xiaojun Li, and Min He. 2024. "Current and future trends in the biocontrol of postharvest diseases." *Critical Reviews in Food Science and Nutrition* 64 (17):5672–5684.

Lopes, Ralf, Louise Cerdeira, Grace S Tavares, Jeronimo C Ruiz, Jochen Blom, Elvira CA Horácio, Hilário C Mantovani, and Marisa Vieira de Queiroz. 2017. "Genome analysis reveals insights of the endophytic Bacillus toyonensis BAC3151 as a potentially novel agent for biocontrol of plant pathogens." *World Journal of Microbiology and Biotechnology* 33:1–15.

Luo, Yang, Meng Zhou, Qi Zhao, Fang Wang, Jiangli Gao, Hongmei Sheng, and Lizhe An. 2020. "Complete genome sequence of Sphingomonas sp. Cra20, a drought resistant and plant growth promoting rhizobacteria." *Genomics* 112 (5):3648–3657.

Maina, UM, IB Galadima, FM Gambo, and DJJOE Zakaria. 2018. "A review on the use of entomopathogenic fungi in the management of insect pests of field crops." *Journal of Entomology and Zoology Studies* 6 (1):27–32.

Malvi, Sanhita, Jayant Bhatt, Ashis K Das, Pratik Pali, Ashok Kumar, Aatmika Chouhan, Shubham Chichghare, and Sanjay Kharte. 2023. "Identification and characterization of endophytic bacteria isolated from citrus roots with biocontrol potential against Phytophthora nicotianae." *Indian Phytopathology* 76 (1):191–199.

Mark, Genevieve L, John P Morrissey, P Higgins, and Fergal O'gara. 2006. "Molecular-based strategies to exploit Pseudomonas biocontrol strains for environmental biotechnology applications." *FEMS Microbiology Ecology* 56 (2):167–177.

Marshall, John M, and Váleri N Vásquez. 2021. "Field trials of gene drive mosquitoes: Lessons from releases of genetically sterile males and Wolbachia-infected mosquitoes." *Genetically Modified and other Innovative Vector Control Technologies: Eco-bio-social Considerations for Safe Application*:21–41.

Meccariello, Angela, Shibo Hou, Serafima Davydova, James Daniel Fawcett, Alexandra Siddall, Philip T Leftwich, Flavia Krsticevic, Philippos Aris Papathanos, and Nikolai Windbichler. 2024. "Gene drive and genetic sex conversion in the global agricultural pest Ceratitis capitata." *Nature Communications* 15 (1):372.

Meena, Ram Swaroop, Vinod Vijayakumar, Gulab Singh Yadav, and Tarik Mitran. 2018. "Response and interaction of Bradyrhizobium japonicum and arbuscular mycorrhizal fungi in the soybean rhizosphere." *Plant Growth Regulation* 84:207–223.

Mesnage, Robin, Nicolas Defarge, Joël Spiroux de Vendômois, and Gilles-Eric Séralini. 2014. "Major pesticides are more toxic to human cells than their declared active principles." *BioMed Research International* 2014 (1):179691.

Mitter, Neena, Elizabeth A Worrall, Karl E Robinson, Peng Li, Ritesh G Jain, Christelle Taochy, Stephen J Fletcher, Bernard J Carroll, GQ Lu, and Zhi Ping Xu. 2017. "Clay nanosheets for topical delivery of RNAi for sustained protection against plant viruses." *Nature Plants* 3 (2):1–10.

Mohammed, AM, LKT Al-Ani, and B Salleh. 2013. "Potential management of Fusarium oxysporum f. sp. cubense, the banana wilt pathogen by using pseudomonas and beta-amino-butyric acid (BABA)." *International Symposium on Tropical Fungi, ISTF*, IPB International Convention Center, Bogor, Indonesia.

Mohammed, AM, LKT Al-Ani, B Salleh, and AMA Ghazali. 2014. "Determining plant growth promoting and biocontrol factor of bacterial culture media." *The 3rd Conference on Pests Management*, Crop Protection Research Centre, Sudan.

Muller, Adrian, Christian Schader, Nadia El-Hage Scialabba, Judith Brüggemann, Anne Isensee, Karl-Heinz Erb, Pete Smith, Peter Klocke, Florian Leiber, and Matthias Stolze. 2017. "Strategies for feeding the world more sustainably with organic agriculture." *Nature Communications* 8 (1):1–13.

Muñoz, Isabel Vicente, Sabrina Sarrocco, Luca Malfatti, Riccardo Baroncelli, and Giovanni Vannacci. 2019. "CRISPR-Cas for fungal genome editing: A new tool for the management of plant diseases." *Frontiers in Plant Science* 10:135.

Mushtaq, Sehrish, Muhammad Shafiq, Muhammad Rizwan Tariq, Adnan Sami, Muhammad Shah Nawaz-ul-Rehman, Muhammad Hamza Tariq Bhatti, Muhammad Saleem Haider, Saleha Sadiq, Muhammad Taqqi Abbas, and Mujahid Hussain. 2023. "Interaction between bacterial endophytes and host plants." *Frontiers in Plant Science* 13:1092105.

Muzell Trezzi, M, RA Vidal, AA Balbinot Junior, H von Hertwig Bittencourt, and AP da Silva Souza Filho. 2016. "Allelopathy: Driving mechanisms governing its activity in agriculture." *Journal of Plant Interactions* 11 (1):53–60.

Myers, John Peterson, Michael N Antoniou, Bruce Blumberg, Lynn Carroll, Theo Colborn, Lorne G Everett, Michael Hansen, Philip J Landrigan, Bruce P Lanphear, and Robin Mesnage. 2016. "Concerns over use of glyphosate-based herbicides and risks associated with exposures: A consensus statement." *Environmental Health* 15 (1):19.

Naidoo, Kubendran, and Shüné V Oliver. 2024. "Gene drives: An alternative approach to malaria control?" *Gene Therapy*:1–13.

Nian, Linyu, Yao Xie, Haozheng Zhang, Mengjun Wang, Biao Yuan, Shujie Cheng, and Chongjiang Cao. 2023. "Vishniacozyma victoriae: An endophytic antagonist yeast of kiwifruit with biocontrol effect to Botrytis cinerea." *Food Chemistry* 411:135442.

Ninkovic, Velemir, Dimitrije Markovic, and Iris Dahlin. 2016. "Decoding neighbour volatiles in preparation for future competition and implications for tritrophic interactions." *Perspectives in Plant Ecology, Evolution and Systematics* 23:11–17.

Noble, Charleston, Ben Adlam, George M Church, Kevin M Esvelt, and Martin A Nowak. 2018. "Current CRISPR gene drive systems are likely to be highly invasive in wild populations." *Elife* 7:e33423.

Oberhofer, Georg, Tobin Ivy, and Bruce A Hay. 2018. "Behavior of homing endonuclease gene drives targeting genes required for viability or female fertility with multiplexed guide RNAs." *Proceedings of the National Academy of Sciences* 115 (40):E9343–E9352.

Oguh, CE, CO Okpaka, CS Ubani, U Okekeaji, P St Joseph, and EU Amadi. 2019. "Natural pesticides (biopesticides) and uses in pest management-a critical review." *Asian Journal of Biotechnology and Genetic Engineering* 2 (3):1–18.

Okrikata, Emmanuel. 2021. "Present views, status and updates in biopesticide usage." In *Biopesticides: Botanicals and microorganisms for improving agriculture and human health*, Timothy O. Adejumo and Ralf T. Vögele (Editors), 249–280. Logos Verlag Berlin.

Opitz, Ina, Regine Berges, Annette Piorr, and Thomas Krikser. 2016. "Contributing to food security in urban areas: Differences between urban agriculture and peri-urban agriculture in the Global North." *Agriculture and Human Values* 33:341–358.

Parra, José Roberto Postali. 2014. "Biological control in Brazil: An overview." *Scientia Agricola* 71:420–429.

Pedrini, Nicolás, Almudena Ortiz-Urquiza, Carla Huarte-Bonnet, Shizhu Zhang, and Nemat O Keyhani. 2013. "Targeting of insect epicuticular lipids by the entomopathogenic fungus Beauveria bassiana: Hydrocarbon oxidation within the context of a host-pathogen interaction." *Frontiers in Microbiology* 4:24.

Pii, Youry, Tanja Mimmo, Nicola Tomasi, Roberto Terzano, Stefano Cesco, and Carmine Crecchio. 2015. "Microbial interactions in the rhizosphere: Beneficial influences of plant growth-promoting rhizobacteria on nutrient acquisition process. A review." *Biology and Fertility of Soils* 51:403–415.

Pirttilä, Anna Maria, Habibollah Mohammad Parast Tabas, Namrata Baruah, and Janne J Koskimäki. 2021. "Biofertilizers and biocontrol agents for agriculture: How to identify and develop new potent microbial strains and traits." *Microorganisms* 9 (4):817.

Pisa, Lennard W, Vanessa Amaral-Rogers, Luc P Belzunces, Jean-Marc Bonmatin, Craig A Downs, Dave Goulson, David P Kreutzweiser, Christian Krupke, Matthias Liess, and Melanie McField. 2015. "Effects of neonicotinoids and fipronil on non-target invertebrates." *Environmental Science and Pollution Research* 22:68–102.

Price, Tom AR, Nikolai Windbichler, Robert L Unckless, Andreas Sutter, Jan-Niklas Runge, Perran A Ross, Andrew Pomiankowski, Nicole L Nuckolls, Catherine Montchamp-Moreau, and Nicole Mideo. 2020. "Resistance to natural and synthetic gene drive systems." *Journal of Evolutionary Biology* 33 (10):1345–1360.

Prokopy, Ronald, and Marcos Kogan. 2009. "Integrated pest management." In *Encyclopedia of insects*, 523–528. Elsevier.

Pushpalatha, HG, J Naveen, N Geetha, G Hithamani, and H Shekar Shetty. 2023. "Plant growth promotion and biological control of Sclerospora graminicola in pearl millet by endophytic Streptomyces spp." *Indian Phytopathology* 76 (2):521–530.

Quarcoo, Franklin, Conrad Bonsi, David Nii O Tackie, Walter A Hill, Gertrude Wall, and George Hunter. 2017. "Economies of scale in integrated pest management in vegetable and fruit production." *Professional Agricultural Workers Journal (PAWJ)* 5 (1):53–68.

Raban, Robyn R, John M Marshall, and Omar S Akbari. 2020. "Progress towards engineering gene drives for population control." *Journal of Experimental Biology* 223 (Suppl_1):jeb208181.

Ruark, Casey L, Michael Gardner, Melissa G Mitchum, Eric L Davis, and Tim L Sit. 2018. "Novel RNA viruses within plant parasitic cyst nematodes." *PLoS One* 13 (3):e0193881.

Sayre, Richard M, and William P Wergin. 1989. "Morphology and fine structure of the trophozoites of Theratromyxa weberi (Protozoa: Vampyrellidae) predacious on soil nematodes." *Canadian Journal of Microbiology* 35 (5):589–602.

Schmidt, Hanno, Travis C Collier, Mark J Hanemaaijer, Parker D Houston, Yoosook Lee, and Gregory C Lanzaro. 2020. "Abundance of conserved CRISPR-Cas9 target sites within the highly polymorphic genomes of Anopheles and Aedes mosquitoes." *Nature Communications* 11 (1):1425.

Shelake, Rahul Mahadev, Dibyajyoti Pramanik, and Jae-Yean Kim. 2019. "Exploration of plant-microbe interactions for sustainable agriculture in CRISPR era." *Microorganisms* 7 (8):269.

Siddiqui, A, Paul SC Omkar, and G Mishra. 2015. "Predatory responses of selected lines of developmental variants of ladybird, Propylea dissecta (Coleoptera: Coccinellidae) in relation to increasing prey and predator densities." *Biocontrol Science and Technology* 25 (9):992–1010.

Simoni, Alekos, Andrew M Hammond, Andrea K Beaghton, Roberto Galizi, Chrysanthi Taxiarchi, Kyros Kyrou, Dario Meacci, Matthew Gribble, Giulia Morselli, and Austin Burt. 2020. "A male-biased sex-distorter gene drive for the human malaria vector Anopheles gambiae." *Nature Biotechnology* 38 (9):1054–1060.

Smith, Sandy M. 1996. "Biological control with Trichogramma: Advances, successes, and potential of their use." *Annual Review of Entomology* 41 (1):375–406.

Song, Xiu-Shi, Kai-Xin Gu, Xiao-Xin Duan, Xue-Mei Xiao, Yi-Ping Hou, Ya-Bing Duan, Jian-Xin Wang, Na Yu, and Ming-Guo Zhou. 2018. "Secondary amplification of siRNA machinery limits the application of spray-induced gene silencing." *Molecular Plant Pathology* 19 (12):2543–2560.

St. Leger, Raymond J, and Chengshu Wang. 2010. "Genetic engineering of fungal biocontrol agents to achieve greater efficacy against insect pests." *Applied Microbiology and Biotechnology* 85:901–907.

Stehle, Sebastian, and Ralf Schulz. 2015. "Agricultural insecticides threaten surface waters at the global scale." *Proceedings of the National Academy of Sciences* 112 (18):5750–5755.

Stenberg, Johan A. 2017. "A conceptual framework for integrated pest management." *Trends in Plant Science* 22 (9):759–769.

Stone, Wesley W, Robert J Gilliom, and Karen R Ryberg. 2014. *Pesticides in US streams and rivers: Occurrence and trends during 1992–2011*. ACS Publications.

Subedi, Bijay, Anju Poudel, and Samikshya Aryal. 2023. "The impact of climate change on insect pest biology and ecology: Implications for pest management strategies, crop production, and food security." *Journal of Agriculture and Food Research* 14:100733.

Surls, Rachel, Gail Feenstra, Sheila Golden, Ryan Galt, Shermain Hardesty, Claire Napawan, and Cheryl Wilen. 2015. "Gearing up to support urban farming in California: Preliminary results of a needs assessment." *Renewable Agriculture and Food Systems* 30 (1):33–42.

Talaviya, Tanha, Dhara Shah, Nivedita Patel, Hiteshri Yagnik, and Manan Shah. 2020. "Implementation of artificial intelligence in agriculture for optimisation of irrigation and application of pesticides and herbicides." *Artificial Intelligence in Agriculture* 4:58–73.

Teske, April L, Gang Chen, Christian Nansen, and Zhaodan Kong. 2019. "Optimised dispensing of predatory mites by multirotor UAVs in wind: A distribution pattern modelling approach for precision pest management." *Biosystems Engineering* 187:226–238.

Thakur, Kushal, Abhishek Sharma, and Keshav Sharma. 2021. "Management of agricultural insect pests with physical control methods." *The Pharma Innovation Journal* 10 (6):306–314.

Tiwari, S., R. Pudasaini, L. Kafle, S. Bhattarai, M. P. Ali, T. K. Babar, S. Sharma, G. Shrestha, and G. V. P. Reddy. 2019. "Trap cropping in South Asia: Concepts, limitations, and future strategy." *Annals of the Entomological Society of America* 112 (4):340–347.

Van Der Voort, Menno, Harold JG Meijer, Yvonne Schmidt, Jeramie Watrous, Ester Dekkers, Rodrigo Mendes, Pieter C Dorrestein, Harald Gross, and Jos M Raaijmakers. 2015. "Genome mining and metabolic profiling of the rhizosphere bacterium Pseudomonas sp. SH-C52 for antimicrobial compounds." *Frontiers in Microbiology* 6:693.

Van Driesche, RG. 2012. "The role of biological control in wildlands." *BioControl* 57 (2):131–137.

van Lenteren, Joop C, Karel Bolckmans, Jürgen Köhl, Willem J Ravensberg, and Alberto Urbaneja. 2018. "Biological control using invertebrates and microorganisms: Plenty of new opportunities." *BioControl* 63:39–59.

Van Nood, Els, Anne Vrieze, Max Nieuwdorp, Susana Fuentes, Erwin G Zoetendal, Willem M de Vos, Caroline E Visser, Ed J Kuijper, Joep FWM Bartelsman, and Jan GP Tijssen. 2013. "Duodenal infusion of donor feces for recurrent Clostridium difficile." *New England Journal of Medicine* 368 (5):407–415.

Velasco, Leonardo, Isabel Arjona-Girona, María T Ariza-Fernández, Enrico Cretazzo, and Carlos López-Herrera. 2018. "A novel hypovirus species from xylariaceae fungi infecting avocado." *Frontiers in Microbiology* 9:778.

Vijitrpanth, Atpakorn, Arom Jantasorn, and Tida Dethoup. 2023. "Potential and fungicidal compatibility of antagonist endophytic Trichoderma spp. from rice leaves in controlling dirty panicle disease in intensive rice farming." *BioControl* 68 (1):61–73.

Vimal, Shobhit Raj, Jaya Gupta, and Jay Shankar Singh. 2018. "Effect of salt tolerant Bacillus sp. and Pseudomonas sp. on wheat (Triticum aestivum L.) growth under soil salinity: A comparative study." *Microbiology Research* 9 (1).

Wang, Junrong, Shun Qin, Ruidong Fan, Qiang Peng, Xiaojing Hu, Liu Yang, Zengliang Liu, Ivan Baccelli, Quirico Migheli, and Gabriele Berg. 2023. "Plant growth promotion and biocontrol of leaf blight caused by Nigrospora sphaerica on passion fruit by endophytic Bacillus subtilis strain GUCC4." *Journal of Fungi* 9 (2):132.

Wang, Junru, Juyuan Xing, Jiangkun Lu, Yingjiao Sun, Juanjuan Zhao, Shaohua Miao, Qin Xiong, Yonggang Zhang, and Guishan Zhang. 2019. "Complete genome sequencing of Bacillus velezensis WRN014, and comparison with genome sequences of other Bacillus velezensis strains." *Journal of Microbiology and Biotechnology* 29 (5):794–808.

Wang, Ming, Arne Weiberg, Feng-Mao Lin, Bart PHJ Thomma, Hsien-Da Huang, and Hailing Jin. 2016. "Bidirectional cross-kingdom RNAi and fungal uptake of external RNAs confer plant protection." *Nature Plants* 2 (10):1–10.

Wycke, MA, R Perrocheau, and E Darrouzet. 2018. "Sarracenia carnivorous plants cannot serve as efficient biological control of the invasive hornet Vespa velutina nigrithorax in Europe." *Rethinking Ecology* 3:41–50.

Xia, Yandong, Junang Liu, Zhikai Wang, Yuan He, Qian Tan, Zhuang Du, Anqi Niu, Manman Liu, Zhong Li, and Mengke Sang. 2023. "Antagonistic activity and potential mechanisms of endophytic Bacillus subtilis YL13 in biocontrol of Camellia oleifera anthracnose." *Forests* 14 (5):886.

Xu, Xiang-Ru Shannon, Emily A Bulger, Valentino M Gantz, Carissa Klanseck, Stephanie R Heimler, Ankush Auradkar, Jared B Bennett, Lauren Ashley Miller, Sarah Leahy, and Sara Sanz Juste. 2020. "Active genetic neutralizing elements for halting or deleting gene drives." *Molecular Cell* 80 (2):246–262. e4.

Yakhin, Oleg I, Aleksandr A Lubyanov, Ildus A Yakhin, and Patrick H Brown. 2017. "Biostimulants in plant science: A global perspective." *Frontiers in Plant Science* 7:2049.

Zhai, Yile, Zongze Shao, Minmin Cai, Longyu Zheng, Guangyu Li, Ziniu Yu, and Jibin Zhang. 2019. "Cyclo (l-Pro – l-Leu) of Pseudomonas putida MCCC 1A00316 isolated from antarctic soil: Identification and characterization of activity against Meloidogyne incognita." *Molecules* 24 (4):768.

Zhalnina, Kateryna, Katherine B Louie, Zhao Hao, Nasim Mansoori, Ulisses Nunes Da Rocha, Shengjing Shi, Heejung Cho, Ulas Karaoz, Dominique Loqué, and Benjamin P Bowen. 2018. "Dynamic root exudate chemistry and microbial substrate preferences drive patterns in rhizosphere microbial community assembly." *Nature Microbiology* 3 (4):470–480.

Zhang, Huan, Gozde S Demirer, Honglu Zhang, Tianzheng Ye, Natalie S Goh, Abhishek J Aditham, Francis J Cunningham, Chunhai Fan, and Markita P Landry. 2019. "DNA nanostructures coordinate gene silencing in mature plants." *Proceedings of the National Academy of Sciences* 116 (15):7543–7548.

4 Innovations in Chemical Solutions for Pest Management

Shreya Singh[1*] *and Shivendra Singh*[2]
[1]Department of Soil Science and Agriculture chemistry, Udai Pratap College, Bhojubir, Varanasi, Uttar Pradesh, India
[2]Department of Agriculture, Udai Pratap College (An Autonomous Institution) Bhojubir, Varanasi, Uttar Pradesh, India
*Corresponding author

4.1 INTRODUCTION

Integrated pest management (IPM) is a framework for overseeing nuisances in farming creation that utilizes numerous strategies in light of financial, natural, biological, and human well-being influences. Insect pests address a huge impediment for creation of many yields. Conventional dependence on pesticides brings critical monetary expenses and natural liabilities of off-target float, compound buildups, and obstruction (Food and Agriculture Organization [FAO], 2002). Recent technological advancements in pest management and public awareness as far as sanitation and a healthy living environment have prompted the better integration of IPM in current agribusiness. The development of pest-resistant transgenic crops offers great possibilities of establishing more reasonable, financially viable IPM with the combination of a scope of non-synthetic strategies and significantly less dependence on pesticides. Also, the main impetuses that prompted the improvement of the way of thinking are examined and assessed, including pest resistance; adverse effects on climate, biodiversity, environments, and human well-being; and public responses (Ha, 2014). Up until this point, IPM has been demonstrated as a reasonable system intended to support horticulture. This chapter provides knowledge on the significant changes and improvements of IPM procedures, developing from a particular degree of pest control that focuses on the suppression of target pests by adopting a more eco-friendly or systematic approach of pest management where a range of non-synthetic options are used, including the reasonable utilization of pesticides (Barzman et al., 2015).

Insect pests are a significant constraint on the production of many crops overall, through decreases in direct yield and quality, through damage in storage, and through expenses related to control measures (Philips et al., 2014). Pest management utilizing conventional approaches that rely heavily on pesticides is often viable yet causes huge financial, natural, and social expenses and the added risk of insecticide resistance, which should be managed as well. Integrated pest management has, for some time, been proposed as a more manageable approach for some conditions; nevertheless, the response to an integrated pest management approach has been quite variable (Douglas, 2018). All the more frequently, IPM is linked with pest management using pesticides. Comprehensively, IPM can be characterized as the cautious consideration of all available pest management approaches and the subsequent integration of appropriate measures that prevent the spread of pest population, keep pesticides and similar substances to levels that are economically viable, and reduce or limit the dangers to human well-being and the environment (Sarfraz, Dosdall, & Keddie, 2005). IPM emphasizes crop production with the most minimal disruption to agro-ecosystems and strengthens reliable pest management approaches. The barriers to the improvement of IPM approaches include pest resistance to pesticides, adverse natural and environmental effects because of pesticide use and abuse, and lastly, public pressure (Dodia, Patel, & Patel, 2010). Initially, pest resistance was observed because of an overreliance on artificial substances, and subsequently, novel approaches would be expected for the more effective management of pests. Second, because of the extensive use of artificial pesticides, the world has seen significant effects regarding environmental damage, loss of biodiversity, and ultimately, human health risks. These have prompted strong responses from both government and international organizations ([IRAC], 2018). Thus, interventions at both community and international levels have been developed to reduce the use of harmful substances and advance new techniques in plant protection. Nonetheless, today IPM represents a more comprehensive approach that seeks to minimize pesticide use and incorporate a wider range of strategies, for example, the introduction of pest-resistant varieties, conservation and augmentation of beneficial pest population, use of selective and quick-acting pesticides, recognition of the compensatory growth of the plant, and different cultural control practices which have long been associated with the IPM concept. Recent technological advances in crop protection, such as computer-aided forecasting, biotechnology use in

DOI: 10.1201/9781003516538-4

developing resistant cultivars, semiochemicals, natural enemies, selective pesticides, trap cropping, and habitat management, are also being explored as new trends in modern crop protection (Sparks & Nauen, 2015).

Chemical control commonly alludes to the utilization of synthetic pesticides. Nonetheless, to be truly effective, substance control ought to incorporate manufactured synthetics as well as synthetics of microbial or natural origin. Although organic substances like azadirachtin, pyrethrins, and other microorganisms determined poisonous metabolites – for example, avermectin and spinosad are considered as biological – they are as yet compound atoms, like manufactured synthetics, and have a significant number of human and environmental risks as chemical pesticides (Gordon & Waterhouse, 2007). Chemical pesticides are categorized into different groups based on their mode of action, and rotating chemicals from different mode of action groups is recommended to decrease the risk of developing resistance. Government regulations restrict the timing and amount of certain chemical pesticides and help mitigate the associated risks (Mamta & Rajam, 2017).

4.1.1 Conventional Biopesticides

Biopesticides are grouped into two categories:

1. Botanicals – derived from plants or their products
2. Microbials – derived from microorganisms or their products

Plants and plant-derived substances have been widely utilized in farming for insect pest management well before knowledge of the composition of chemicals present in them became available. Extracts from dried flowers of *Chrysanthemum cinerariifolium* or related *Asteraceae* species contain active compounds (0.7–3%) known as pyrethrins, which affect the insect's nervous system. Biopesticides are similarly developed from the Meliaceae family, including *Azadirachta indica* and *Melia azadirachta*. The most active basic compound is azadirachtin, which is a complex tetranortriterpenoid. Numerous beneficial effects of neem compounds, such as insect repellence, feeding deterrence, disruption in growth and development, interference with reproduction, and ovipositional deterrence, have been observed against both sucking and chewing insect pests. Neemjal, Mahaneem, and Neem Shakti are some commercially available neem-based pesticides. The alkaloid derived from tobacco plants, particularly *Nicotiana tabacum* and *N. rustica* of Solanaceae, is known as nicotine, which acts as a nerve toxin and is effective against many sucking pests. Derived from the roots or rhizomes of *Lonchocarpus*, *Derris*, or again *Tephrosia*, rotenone is an isoflavonoid compound that has insecticidal and acaricidal properties. It is available in 1–5% dust formulations. It is a contact and stomach poison effective against many pests, including caterpillars, aphids, thrips, and other insects found on leafy foods.

A majority of microbial biopesticides are derived from microorganisms, such as bacteria, fungi, and viruses, or their bioactive compounds. *Bacillus thuringiensis* (Bt)–based insecticides, for example, Delfin and Dipel, are the most well-known biopesticides, with a nearly 90% dominance in the microbial pesticides market. Bt items are organically certified and utilized on various vegetables, fruits trees, and commercial crops. Virus-based biopesticides are used to control caterpillars in the field and in greenhouses. A few hours after pesticide application, the larvae die from polyhedral diseases, with their inner tissues liquified, containing numerous polyhedral bodies. Several formulations based on NPVs of *Helicoverpa* (Biokill-H and Samstar-Ha) and *Spodoptera* (Spodocide and Biokill-S) are commercially available. Fungi-based formulations of *Beauveria bassiana* (Poison and BB-Power), *Metarhizium anisopliae* (Metacide and Metawell), and *Lecanicillium lecanii* (Verticoz and Verticill) are commercially produced and widely used. They target insect pests belonging to Diptera, Hemiptera, Lepidoptera, and Coleoptera by penetrating and developing mycelium within the insect.

4.1.1.1 Constraints of Biopesticides

Large portions of biopesticides are used specifically for a target species or group of pests and other types of pests that might keep on causing damage. They quickly break down when exposed to moisture, to heat, or in broad daylight, and their effectivity is reduced. Microbials, such as fungi, require specific relative humidity to be effective. Proper timing of applications, storage conditions, and methods of pesticide application severely limit their efficacy. Being target-specific, the products have an unlimited market.

Some products, however, may not be commercially available or are somewhat expensive. The most widely used Bt biopesticides have led to increased pest resistance, raising concerns among organic farmers. Botanicals may have complex structures and may not be easily standardized. They also have slow effectivity, leading to situations where farmers perceive the pesticides as ineffective in pest management. Majority of botanicals have not been fully described for their human and environmental toxicity. Even the least toxic organic pesticides may require high dosage rates and frequencies. The use of pyrethrins is limited since it adversely affects beneficial insects, such as predatory mites and parasitoids. Insecticides, for example, rotenones, are toxic to fish, humans, and other well evolved creatures.

4.1.1.2 Evaluations on Ecological Effects of Biopesticides

Most experts guarantee that organic farming provides comparable yields to conventional farming, is more sustainable, and has significantly less negative environmental impact. However, basically, examining impact assessment results with various other evaluation methods demonstrates that the results are highly subjective, depending on the environmental risks being considered, and may have comparable impact scores. The selection of pesticides frequently emphasizes naturally derived products and organically certified pesticides on account of prevailing public opinion. Studies that focused on the environmental impacts of a few novel insecticides and also compared to organic insecticides in pesticide effects confirmed that the latter has reduced effectivity. Organically approved insecticides also had comparable or even greater adverse effects on many natural enemies under laboratory conditions and, furthermore, were detrimental to biological control organisms in field conditions, along with higher environmental impact residues. Hence, truly effective pest management can only be accomplished when synthetic environmentally safer insecticides are used in combination with other methods. The use of most synthetic pesticides is restricted by organic standards. However, under exceptional circumstances, the national organic program permits the use of synthetic pesticides under controlled applications. Synthetic substances are permitted, provided that:

1. They are proven to not be harmful to human or environmental health.
2. Conventional or other natural options are not accessible.
3. They conform to regular standards.

4.2 NOVEL INSECTICIDES

There is a significant role played by traditionally used neuroactive pesticides in managing pest population in light of the fact that they are broad-spectrum. However, the repeated use of neuroactive insecticides has led to issues in insect pest population management, including insecticide resistance, pest resurgence, secondary pest outbreaks, and residue-related issues and toxic effects on people (Ausborn et al., 2005). Replacing traditionally used neuroactive insecticides has become essential to manage insect pest population (Bretschneider et al., 2003). Emphasis is placed on searching for green chemistry insecticides with various mechanisms of action. Newly discovered substances are pest-specific, targeting particular biochemical sites present in specific pest groups, and have various properties present in different insecticides (Bynum, Archer, & Plapp, 1997). They have different modes of action, so there is little to no chance of developing resistance in insect pests, and they act as rotational alternatives, providing unique, specific chemicals for controlling pest infestations and offering more effective control in insect pest management. These newly discovered pesticides significantly have lower impact on non-target pests, natural enemies, vertebrates, and the environment due to their green chemical origin and low application doses (Campos & Omoto, 2002). There are three major strategies observed in resistance management. These are moderate use of insecticides at lower doses, increasing the efficacy of chemical toxicity on target pests through enhancing uptake of compound by adding attractants or inhibiting detoxifying enzymes of resistant pests and using different insecticide combinations or using alternative insecticides that have various modes of action in pest management (Caboni, Sammelson, & Casida, 2003).

TABLE 4.1
Overview of Conventional Biopesticides, Their Sources, Target Pests, and Limitations

Category	Biopesticide	Source	Target Pests	Limitations
Botanicals	Pyrethrins	Extract from *Chrysanthemum cinerariifolium*	Nervous system disruption in various insects	Breakdown in sunlight; affects beneficial insects, such as predatory mites; expensive; and limited availability
	Azadirachtin (neem)	*Azadirachta indica*, *Melia azadirach*	Repels pests, inhibits growth and reproduction	Slow effectivity, complex composition, high dosage required, not fully characterized for human toxicity
	Nicotine	*Nicotiana tabacum*, *Nicotiana rustica*	Sucking pests, nerve poison	Toxic to humans, fishes, and mammals; affects beneficial insects
	Rotenone	Roots of *Lonchocarpus*, *Derris*, or *Tephrosia*	Caterpillars, aphids, thrips, and more	Toxic to fishes, humans, and mammals; decomposes rapidly in sunlight
Microbials	*Bacillus thuringiensis* (Bt)	*Bacillus thuringiensis*	Caterpillars (larvae)	Rapid pest resistance development, requires specific environmental conditions for effectiveness
	Nuclear polyhedrosis virus (NPV)	Virus-based formulations for *Helicoverpa* and *Spodoptera*	Caterpillars, larvae	Requires specific environmental conditions; reduced efficacy in bright sunlight or heat
	Beauveria bassiana	Fungus *Beauveria bassiana*	Insects from Diptera, Hemiptera, and Lepidoptera	Requires specific humidity for effectiveness; narrow pest range
	Metarhizium anisopliae	Fungus *Metarhizium anisopliae*	Diptera, Hemiptera, Lepidoptera, and Coleoptera	Environmental conditions restrict use; low impact on broader pest spectrum
	Lecanicillium lecanii	Fungus *Lecanicillium lecanii*	Insects in Hemiptera	Environmental sensitivity (moisture); limited target range

Newly discovered novel insecticides have different modes of action that affect the development of pests by inhibiting chitin synthesis, juvenile hormone mimics, and ecdysone agonists. Some developed compounds have specific effects on pest groups through biochemical site activity inhibition or enhancement, such as respiration (Diafenthiuron), acetylcholine receptor activation (Neonicotinoids), or GABA receptor (avermectins) (Cordova et al., 2006).

Resistance. Pesticide resistance is the reduction in the effectiveness of a pesticide against a pest population that was previously susceptible. As per IRAC 2007, there are a number of methods by which pests develop pesticide resistance (IRAC, 2007).

Behavioral resistance. In this type of resistance, pests become aware of the presence of harmful substances on plants and try to avoid them (Caboni, Sammelson, & Casida, 2003).

Penetration resistance. Pests modify their exoskeleton to reduce insecticide penetration from the cuticle to their body (Ghanim & Ishaaya, 2016).

Target site resistance. In this type of resistance, pests genetically alter the target site where the insecticide acts, preventing the insecticide from interacting with the site of action and reducing its effectiveness (Hajimichael et al., 1996).

Metabolic resistance. Pests rapidly detoxify specific pesticide by increasing metabolic activity, like excretion, which transports insecticides from the digestive system and avoids chemical toxicity. Resistance has become a serious issue against commonly used insecticides, like organophosphates, carbonates, and pyrethroids, because of their repeated application and similar mode of action (Hardy, 2014). To prevent or delay resistance, there is a need to develop pesticides with different modes of action. All newly developed novel insecticides have different modes of action; they are less toxic to animals and the environment and are relatively safe for natural enemies. The aim of this review is to discuss the modes of action of various groups of novel or newer insecticides and their uses to prevent/delay the development of resistance in pests (Kinley et al., 2002).

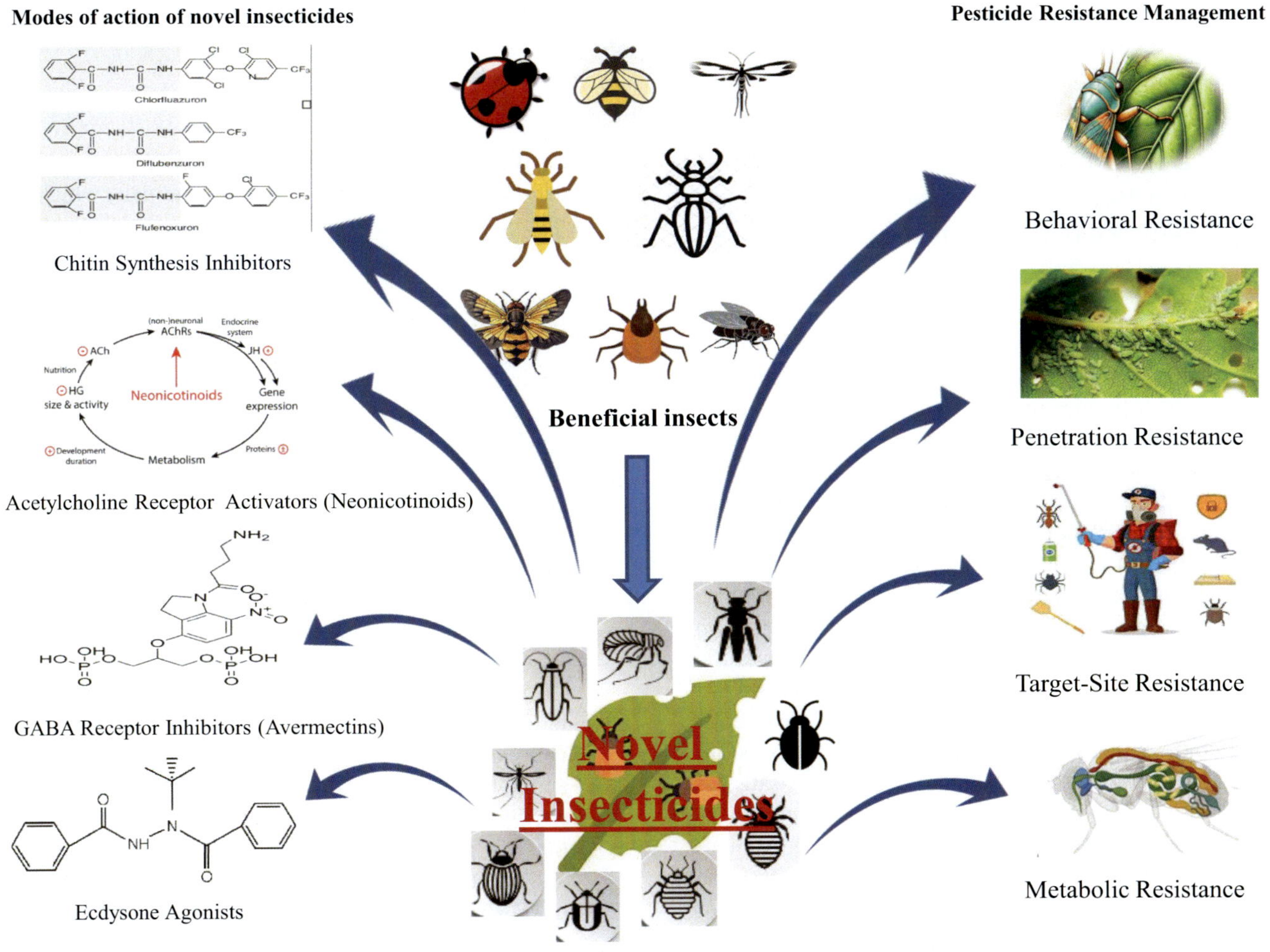

FIGURE 4.1 Various modes of action and resistance mechanisms of novel insecticides in pest management. The figure includes labeled sections that depict different insecticidal modes of action, such as neurotoxicity and metabolic disruption, alongside visual representations of resistance mechanisms, including target site insensitivity and enhanced detoxification. Arrows and flowcharts indicate the relationships between these actions and mechanisms, highlighting their significance in effective pest control strategies.

TABLE 4.2
Novel Insecticides Recommended for Key Pests in Various Crops

Insecticides	Crop	Target Pests	Mode of Action	Formulation
Spinosad 45 SC	Okra, legumes, chili	Fruit borer, pod borer, thrips	Disrupts nicotinic acetylcholine receptors (nAChR)	Suspension Concentrate
Cartap hydrochloride 50 SP	Paddy	Stem borer, leaf folder	Sodium channel blocker	Soluble powder
Emamectin benzoate 5 SG	Tomato, brinjal, legumes	Fruit borer, shoot and fruit borer, pod borer	Activates glutamate-gated chloride channels	Water-soluble granules
Buprofezin 25 SC	Paddy	Brown planthopper	Inhibits chitin synthesis	Suspension Concentrate
Lufenuron 5.4 EC	Cabbage, pigeon pea	Diamondback moth, pod borer	Inhibits chitin biosynthesis	Emulsifiable Concentrate
Novaluron 10 EC	Cabbage, tomato, chili	Diamondback moth, fruit borer	Inhibits chitin synthesis	Emulsifiable Concentrate
Hexythiazox 5.45 EC	Tea	Mites	Mitochondrial electron transport inhibitor	Emulsifiable Concentrate
Flonicamid 50 WG	Paddy	Brown planthopper, green leafhopper	Inhibits aphid feeding	Water-dispersible granules
Pymetrozine 50 WG	Paddy	Brown planthopper	Inhibits feeding of sucking insects	Water-dispersible granules
Chlorantraniliprole 18.5%	Paddy, cabbage, okra	Stem borer, leaf folder, diamondback moth, fruit borer	Activates ryanodine receptors	Suspension Concentrate
SPLAT	Cotton, brinjal, mango	Pink bollworm, shoot and fruit borer, fruit fly	Pheromone disruption	Specialized formulation
Sucrashield	Vegetables	Aphids, leafhoppers, mites, mealybugs, thrips	Coats pests to restrict feeding	Liquid concentrate

4.2.1 Different Groups of Novel Insecticides and Their Mode of Action Based on Insecticides with Synthetic Origin

Neonicotinoids. These are engineered organic insecticides developed through modification of regular substances that have both synthetic and organic properties. In modern crop assurance Neonicotinoids pesticides are the quickest developing class of insecticides. They are predominantly foundational in nature and quickly absorbed by plants and primarily used for managing piercing, sucking, and certain chewing insect pests (Hemingway et al., 2004). Because of their primary mode of action, they are rapidly absorbed by the spouting seeds of plants and provide long-term protection from target pest. In pest nervous system, nicotinic acetylcholine receptors (nAChRs) are the main target site of neonicotinoids insecticides. Neonicotinoid pesticides applied at low concentrations are remarkably target-specific and therefore safe for non-target organisms and the environment, making them a significant class of insecticide for managing insect pest resistance (Kinley et al., 2002).

Oxadiazines. To overcome issues like the development of resistance in pests, another insecticide of biological origin, Oxadiazine, was developed. Oxadiazine is a broad-spectrum insecticide that acts as a sodium channel blocker and has translaminar activity through the mesophyll of plants. Pests exhibit coloration because of the obstruction of insecticide particles on sodium channels, which hinders the movement of sodium particles through the application of spray solutions or the consumption of compound-treated plant parts. Upon the ingestion of synthetic atom, the suspension effectively incapacitates and kills pests. Oxadiazines do not adversely affect birds, worms, and amphibians (Kodanram, Rai, & Halder, 2010).

Diamides. Diamides are a new class of insecticides that act on the ryanodine receptor, a novel target in pest management. Muscular contractions kill the pests due to the release of stored calcium from the sarcoendoplasmic reticulum (Kundoo et al., 2017).

Ketoenols. To manage resistant pest and insect populations, ketoenols, a novel class of pesticides and acaricides, was created. Ketoenols inhibit acetyl-CoA carboxylase, a crucial enzyme involved in fatty acid synthesis in insects and pests. Ketoenols were significantly developed to manage whitefly, aphid, and other sucking pests.

Phenylpyrazoles. Phenylpyrazoles are insecticides that are remarkably powerful in managing insect and pest population by concentrating on the GABA-gated chloride channel. Insects and pests die because of the obstruction of the GABA-gated chloride channel, an ionotropic receptor (Morita et al., 2007).

Pyridine. Pyridine is a group of pesticides that plays a significant part in managing sucking pests, particularly aphid, whitefly, and plant hoppers, entering the plant structure via translaminar activity. Pyridine pesticides are highly target-specific. The obstruction of the stylet entrance due to pesticide effects results in the cessation of feeding in pests, causing starvation and death (Sarwar & Salman, 2015).

Flonicamid. Flonicamid is a novel insecticide containing the pyridinecarboxamide compound, which is highly effective against Hemiptera and Thysanoptera insect pests. Flonicamid causes an obstruction of the stylet entrance in plant tissue. Unable to feed because of the obstruction, pests die due to starvation (Sinha, Paul, & Ram, 2009).

METI. METI, a class of insecticides and acaricides, plays a significant part in managing pest and insect populations. METI insecticides and acaricides hinder the mitochondrial electron transport in insect pests. They are highly effective in managing pests and sucking insect population (Tomar, Sharma, & Malik, 2017).

Tetrazines. Tetrazine is a highly effective acaricide against phytophagous pests that does not adversely affect the environment. Tetrazine, a contact acaricide, has ovicidal properties as well as translaminar movement (Tomizawa & Casida, 2005).

Thiazolidinones. Thiazolidinone is used to control mites, thrips, and leafhoppers due to their broad-spectrum toxic effects on eggs, larvae, and nymphal stages. These are highly effective in managing insect population and can be used at any stage of development of plants (Kinley et al., 2002).

Nereistoxin. Nereistoxin is a broad-spectrum insecticide effective in managing sucking and chewing pests at all stages of development. It blocks the activity of nicotinic acetylcholine receptor channel in pests (Morita et al., 2007).

Formamidines. Having both acaricidal and insecticidal properties, Formamidine is used to control red mites, two-spotted spider mites, aphid, leaf miners, and scale insects. It acts on adrenoreceptors agonists and inhibits protein monoamine oxidase (Cordova et al., 2006).

4.3 COMBINATION OF FUNGICIDES WITH ORGANIC RESISTANCE INDUCERS

Biocontrol agents (BCAs) manage plant diseases by directly opposing impacts on plant microbes through parasitism, antibiosis, or competition for nutrients or space. Parasitism is a relationship between two organisms where one directly gains nutrients from the other. A particular kind of parasitism notable in the biocontrol field is mycoparasitism, in which contagious plant microorganisms are parasitized by biocontrol growths, purportedly frequently *Trichoderma* spp. The subsequent component, antibiosis, happens between two creatures when one produces antimicrobial metabolites that directly influence the development or digestion of the other organic entity (Barakat & Al-Masri, 2017). These antimicrobial products are produced at very low concentrations, are locally distributed, and have a short lifespan, thus posing low toxicological risks to humans. At last, competition happens when two organisms require the same limited nutrients or space. These protective mechanisms of direct BCAs are often complex and depend on various and multilayered modes of action, as expected to lower the chances of resistance development (Dubey et al., 2015). Another advantage of direct BCAs is the possibility of studying inhibitory impacts by means of basic bioassays between just the main antagonist and the microorganism, which is clearer than exploring the adverse impacts between the microbe, the cause of resistance, and the plant (Galindo et al., 2013). Nonetheless, total infection prevention is achieved when BCAs are combined with fungicides. In the following passages, such combinations will be discussed; however, no differentiation will be made between rivalry, hostility, and parasitism, since the mode of action isn't generally clear; all things considered, a qualification will be made in light of their time (either pre- or post-collection) or strategy for application or beginning (Abd-El-Khair et al., 2019).

Direct-acting microbial antagonists have reportedly been combined with fungicides to manage post-harvest diseases. An advantage of post-harvest use of antagonists and fungicides is direct treatment by dropping the harvested fruits in a single procedure (Barnett, 1963). Nonetheless, as referenced previously, the combination implies that the main antagonist is naturally protected against specific fungicides. In that capacity, the further development in managing harvest decay (brought about by *Alternaria alternata*) was obtained when harvested fruits were dipped in a combination of fungicide-safe *Trichoderma* spp. And different foundational and non-fundamental fungicides at low dosages of 50 or 100 ppm, when contrasted with the ten-times-higher portions normally applied for fungicides on fruits (Droby et al., 2009). Some *Trichoderma secludes* caused an inactive disease, which was totally inhibited with the combination. In another study on stored apples, a combination of the biocontrol yeast *Cryptococcus laurentii* and thiabendazole, at 10% of the standard portion, gave the most notable and the longest control of another significant post-harvest microbe, *B. cinerea*. The combination was considerably more effective against a thiabendazole-safe segregates of *B. cinerea*, additionally giving longer-lasting protection in contrast with treatment with the biocontrol yeast alone (Gilardi et al., 2008). In this way, BCA–fungicide combinations have potential desired effects against fungicide-sensitive and fungicide-safe populations, which are turning out to be increasingly more pervasive. Likewise, on harvested apples, however, utilizing more up-to-date fungicides, an answer of the biocontrol yeasts (*Rhodosporidium kratochvilovae* or *C. laurentii*) with a low portion of either boscalid or cyprodinil was more effective against blue form brought about by *Penicillium expansum* than the treatment without aid from another (Eisa & El-Feky, 2014). Curiously, lower fungicide buildups were seen with the combined treatment in any event, when contrasted with single treatment with a similar fungicide at a similar low portion. Best post-harvest treatments include the consolidated use of biocontrol yeasts and fungicides, which is logical because of the capacity of yeast to endure outrageous natural circumstances, making them attractive for food application (Gilardi et al., 2008). All things considered, yeast is sufficient in common conditions, including low oxygen levels, low temperatures, and UV radiation, yet should be added to in situations well defined for food sources, for example, low pH and high sugar concentrations. In some cases, the pre-harvest utilization of fungicides is more effective against post-harvest microorganisms, yet in addition, this approach could be improved through combination with BCAs (Hahn, 2014). Accordingly, the consolidated pre-harvest utilization of *Epicoccum nigrum* and different fungicides could lessen the fungicide portion threefold without influencing the administration of earthy-colored decay (brought about by *Monilinia* spp.) on collected peaches during four different field preliminaries. Illness decrease was best in years with lower sickness seriousness (Larena et al., 2005).

Contagious adversaries likewise further develop protection when combined with conventional fungicides against pre-harvest microorganisms. Specifically, the use of *Trichoderma* spp. against soilborne microbes is known. *Trichoderma* spp. are innately safe against certain fungicides, permitting their combined (Jambhulkar et al., 2018). Such a blend of *T. virens* and thiophanate-methyl was viewed as effective and highly compelling than either treatment alone against *Fusarium solani* and *Fusarium oxysporum* in field experiments of dry bean creation (Lima et al., 2006). The vegetative development of the plants and yield was likewise fundamentally expanded for the mix in contrast with single treatment. Essentially, the consolidated utilization of *Trichoderma* spp. with a low portion of fluazinam was viewed as more powerful in controlling avocado

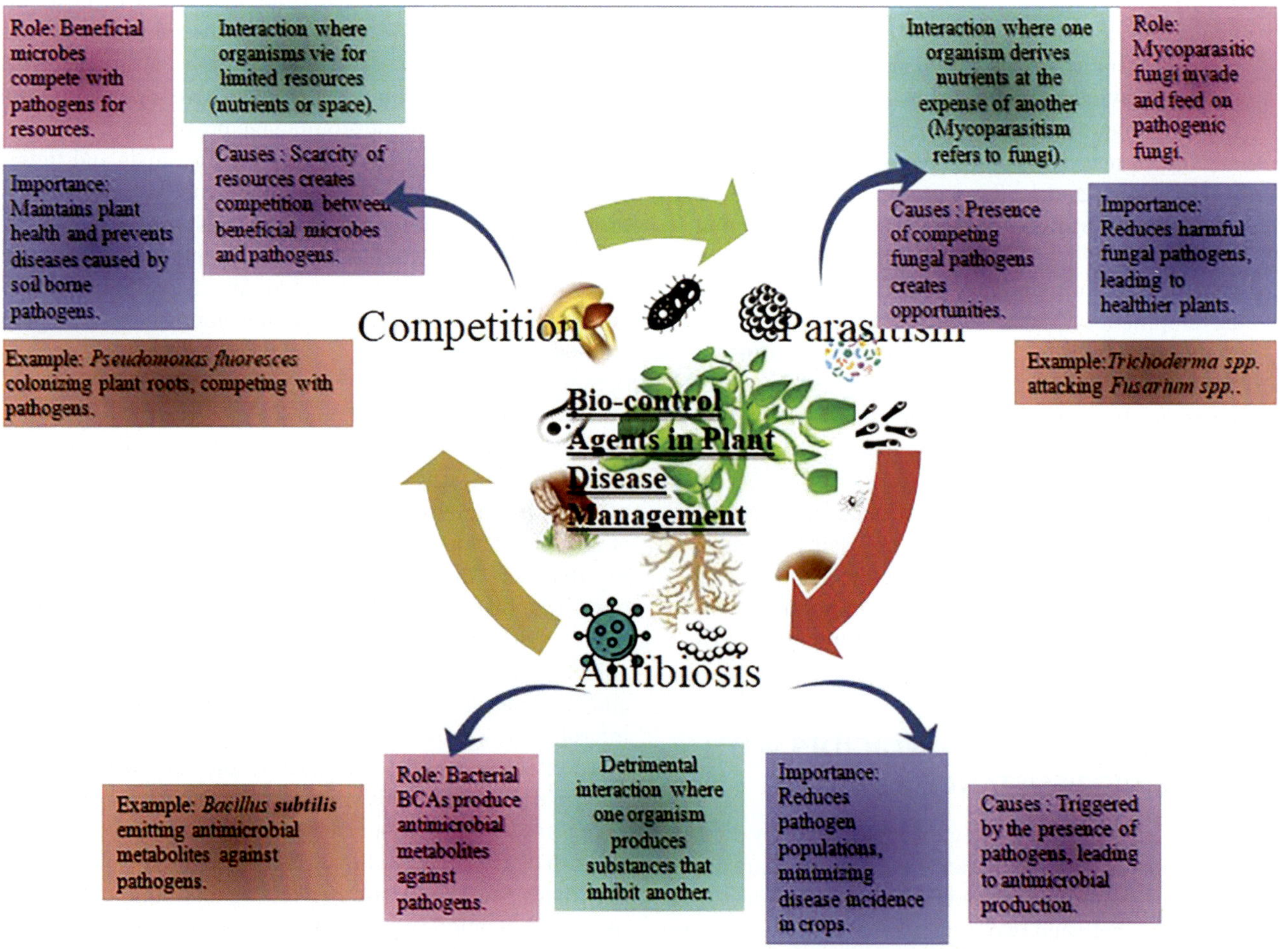

FIGURE 4.2 Mechanisms by which biocontrol agents contribute to plant disease management. The figure includes labeled sections depicting various biocontrol strategies, such as competition, predation, and parasitism. Arrows indicate the interactions between biocontrol agents and pathogens, showcasing processes like the inhibition of pathogen growth, the induction of plant defenses, and the enhancement of plant resilience. Visual elements may include representations of different biocontrol organisms, such as bacteria, fungi, and nematodes, demonstrating their roles in managing plant diseases.

white decay (brought about by *Rosellinia necatrix*) than either treatment alone. At long last, however, *Trichoderma* spp. were observed to not be as effective against *F. oxysporum* and *Acremonium strictum* in an in vitro study; combining them with a low concentration of the broad-spectrum fungicide tolclofos-methyl was better than the fungicide alone (Lima et al., 2011).

The rhizosphere of plants shapes a wellspring of microscopic organisms relevant for plant resistance as well as for direct biocontrol in pre-harvest applications. Like parasitic adversaries, these bacterial antagonists predominantly further develop infection protection against soilborne microorganisms. For instance, the inadequate infection protection offered by *Bacillus megaterium* against *F. oxysporum* on tomato could be improved when it is combined with a low dose of the fungicide carbendazim in plant packs (Moutou et al., 2004). The combination provides full protection in any event, outflanking application with the fungicide at a 10-overlay higher portion. Essentially, in a similar setup, the combination of rhizobacteria *P. fluorescens* and a 10-overlay reduced amount of benomyl was more effective than treatment with either alone and decreased the infection as much as a full portion of the fungicide alone (Nallathambi et al., 2009). Strangely, a few natural enemies can get on the leaves of plants, which allows splash application against foliar microbes. Thus, *Bacillus subtilis*, a rhizobacterium that has been generally employed in the development of antimicrobials, influences the cell mass of plant microorganisms. In numerous nursery trials, the foliar utilization of *B. subtilis* with azoxystrobin gave the best return and the best infectious prevention against fine buildup (brought about by *Podosphaera xanthii*) on zucchini, in contrast with either medicines alone (Nobutaka et al., 2006).

Fungicides have likewise been combined with various contagious and bacterial BCAs to improve disease management pre-harvest. A blend of *P. fluorescens*, *Mesorhizobium cicero*, and *T. harzianum* with the fungicide Vitavax® (dynamic fixings: carboxin and thiram; Haryana, India) gave the most noteworthy seed germination, grain yield, and the least shrink rate (brought about by *F. oxysporum*) in pot and field tests of chickpea (Spadaro & Droby, 2016). Besides, in field analyses of rice, the combination of *T. harzianum*, *P. fluorescens*, and carbendazim was more powerful against *Magnaporthe oryzae*, in contrast with their singular application (Omar et al., 2006).

Subsidiaries of living life-forms like plant separates are otherwise called direct BCAs, which can be combined with fungicides. Collaboration was seen between either CHOS or chitosan and different engineered fungicides on strawberry blossoms (Ruano-Rosa et al., 2018). A portion of the fungicide mixture at a 100 overlay was reduced, and chitosan or CHOS yielded a degree of protection against *B. cinerea*, like the fungicide at full strength. A comparative blend with a 10-overlay decrease of the fungicide dithianon was more compelling in controlling scab (brought about by *Venturia inaequalis*) than the fungicide alone at the suggested portion in field preliminaries of apple. In addition, plant concentrates of *Inula viscosa* combined with the fungicide iprodione at a decreased rate were as powerful against *B. cinerea* on bean plants as the full portion of fungicide (Rahman et al., 2014).

4.4 FUTURE DIRECTION FOR THE CHEMICAL CONTROL OF PEST MANAGEMENT

The eventual fate of synthetic insecticides is probably toward additional cost-effective and environmentally safe practices (Douglas, 2018). This development is driven by rising concerns about its environmental effects, the human health risks, and the development of resistance among pests to conventional pesticides. Here are a few critical directions for the future:

1. *Development of novel pesticides.* The emphasis will be on making pesticides with novel methods of activity that target pests all the more explicitly, limiting damage to non-target life-forms. This incorporates investigating new biochemical pathways in irritations and utilizing propels in genomics and sub-atomic science Bretschneider et al. (2003).
2. *Integrated pest management (IPM).* Synthetic mechanisms will increasingly be integrated with other pest management approaches, like organic regulation, community participation, and technological approaches. This integrated approach reduces the dependence on compound pesticides and advances a more feasible type of pest management Mamta and Rajam (2017).
3. *Precision agriculture.* Advances in innovation, like robots, sensors, and artificial intelligence, will enable a more responsible utilization of pesticides, decreasing the amounts utilized and limiting adverse reaction. Precision agriculture takes into consideration the designated pest control, reducing waste and environmental consequences.
4. *Resistance management.* To combat the emerging matter of pesticide resistance, future techniques must include rotating pesticides with different modes of action and integrating them with other pest control approaches. This approach helps prevent the development of resistance and enhance the effectivity of existing pesticides (Barzman et al., 2015).
5. *Biopesticides and regular products.* There is increased emphasis on developing and utilizing biopesticides derived from natural sources, like plants, microbes, and parasites. These options are often not as toxic, but rather are safer, to the environment than manufactured synthetic compounds (Philips et al., 2014).
6. *Regulatory and strategy changes.* States and their governments should enforce stricter guidelines on pesticide use, opting for the reception of safer, more cost-effective alternatives. Strategies that encourage a reduction in synthetic pesticide

use and promote IPM approaches play a significant part in molding the fate of pest management Sparks and Nauen (2015).

7. *Consumer and market demands.* As consumers become more aware of the environmental and health effects of pesticides, there will be a more notable demand for organic and ethically sourced food. This shift in consumer preference will drive the market toward integrated pest management approaches Tomizawa and Casida (2005).

4.5 CONCLUSION

The future of pest management is moving toward additional substantial approaches with an emphasis on lessening dependence on traditional synthetic pesticides. Advancements in novel insecticides, together with the integration of integrated pest management (IPM) and precision agriculture, are crucial in minimizing environmental impact and further improving effectiveness. Advances in biopesticides and natural products, along with management changes, are directing the industry towards safer and more effective pest control methods. Emphasizing resistance management and integrating diverse control strategies will be crucial in addressing the challenges presented by pest resistance and environmental concerns. Eventually, a balanced approach that integrates technology, policy, and environmental standards will illustrate improvements in pest management.

REFERENCES

Abd-El-Khair, H., Elshahawy, I. E., & Haggag, H. E. K. (2019). Field application of *Trichoderma* spp. combined with thiophanate-methyl for controlling *Fusarium solani* and *Fusarium oxysporum* in dry bean. *Bulletin of the National Research Centre*, *43*(19). https://doi.org/10.1186/s42269-019-0064-1

Ausborn, J., Wolf, H., Mader, W., & Kayser, H. (2005). The insecticide pymetrozine selectively affects chordotonal mechanoreceptors. *The Journal of Experimental Biology*, *208*, 4451–4466.

Barakat, R. M., & Al-Masri, M. I. (2017). Effect of clammy inula (*Inula viscosa*) plant extract in combination with a low dose of the fungicide iprodione on *Botrytis*. *American Journal of Plant Sciences*, *8*(5), 651–665. https://doi.org/10.4236/ajps.2017.85045

Barnett, H. L. (1963). The nature of mycoparasitism by fungi. *Annual Review of Microbiology*, *17*, 1–14. https://doi.org/10.1146/annurev.mi.17.100163.000245

Barzman, M., Bàrberi, P., Nicholas, A., Birch, E., Boonekamp, P., Dachbrodt-Saaydeh, S., . . . Lamichhane, J. R. (2015). Eight principles of integrated pest management. *Agronomy for Sustainable Development*, *35*(4), 1199–1215. https://doi.org/10.1007/s13593-015-0327-9

Bretschneider, T., Benet, B. J., Fischer, R., & Nauen, R. (2003). Spirodiclofen and spiromesifen – novel acaricidal and insecticidal tetronic acid derivatives with a new mode of action. *Chimia*, *57*(12), 697–701. https://doi.org/10.2533/000942903777679609

Bynum, J., Archer, E., & Plapp, J. (1997). Comparison of banks grass mite and two-spotted spider mite (acari: tetranychidae): Responses to insecticides alone and in synergistic combinations. *Journal of Economic Entomology*, *90*(6), 1125–1130.

Caboni, P., Sammelson, R. E., & Casida, J. E. (2003). Phenylpyrazole insecticide photochemistry, metabolism, and gabaergic action: Ethiprole compared with fipronil. *Journal of Agricultural and Food Chemistry*, *51*(24), 7055–7061.

Campos, F. J., & Omoto, C. (2002). Resistance to hexythiazox in Brevipalpus phoenicis (Acari: Tenuipalpidae) from Brazilian citrus. *Experimental and Applied Acarology*, *26*(3–4), 243–251.

Cordova, D., Benner, E. A., Sacher, M. D., Rauh, J. J., Sopa, J. S., Lahm, G. P., . . Tao, Y. (2006). Anthranilic diamides: A new class of insecticides with a novel mode of action, ryanodine receptor activation. *Pesticide Biochemistry and Physiology*, *84*(3), 196–214.

Dodia, D. A., Patel, I. S., & Patel, G. M. (2010). *Botanical pesticides for pest management*. Scientific Publishers (India).

Douglas, A. E. (2018). Strategies for enhanced crop resistance to insect pests. *Annual Review of Plant Biology*, *69*, 637–660.

Droby, S., Wisniewski, M., Macarisin, D., & Wilson, C. (2009). Twenty years of postharvest biocontrol research: Is it time for a new paradigm? *Postharvest Biology and Technology*, *52*(2), 137–145. https://doi.org/10.1016/j.postharvbio.2008.11.009

Dubey, S. C., Singh, V., Priyanka, K., Upadhyay, B. K., & Singh, B. (2015). Combined application of fungal and bacterial bioagents, together with fungicide and *Mesorhizobium* for integrated management of *Fusarium* wilt of chickpea. *BioControl*, *60*(4), 413–424. https://doi.org/10.1007/s10526-015-9650-3

Eisa, O. A., & El-Feky, N. M. (2014). Impact of the fungicide Rizolix T50% on the antagonistic activity of *Trichoderma harzianum* and *Trichoderma koningii*. *International Journal of Science and Research*, *3*(7), 1767–1773. www.ijsr.net/archive/v3i7/MDIwMTQxNzQ=.pdf

Food and Agriculture Organization (FAO). (2002). *www.fao.org*. Retrieved from www.fao.org

Galindo, E., Serrano-Carreón, L., Gutiérrez, C. R., Allende, R., Balderas, K., Patiño, M., Trejo, M., Wong, M. A., Rayo, E., & Isauro, D. (2013). The challenges of introducing a new biofungicide to the market: A case study. *Electronic Journal of Biotechnology*, *16*. https://doi.org/10.2225/vol16-issue3-fulltext-5

Ghanim, M., & Ishaaya, I. (2016). Insecticides with novel modes of action mechanism and resistance management. In *Tolerance to environmental contaminants* (pp. 385–407). CRC Press.

Gilardi, G., Manker, D. C., Garibaldi, A., & Gullino, M. L. (2008). Efficacy of the biocontrol agents *Bacillus subtilis* and *Ampelomyces quisqualis* applied in combination with fungicides against powdery mildew of zucchini. *Journal of Plant Diseases and Protection*, *115*(4), 208–213. https://doi.org/10.1007/BF03356274

Gordon, K. H., & Waterhouse, P. M. (2007). RNAi for insect-proof plants. *Nature Biotechnology*, *25*, 1231–1232.

Ha, T. M. (2014). A review on the development of integrated pest management and its integration in modern agriculture. *Asian Journal of Agriculture and Food Science*, *2*(4), 336–340.

Hahn, M. (2014). The rising threat of fungicide resistance in plant pathogenic fungi: *Botrytis* as a case study. *Journal of Chemical Biology*, *7*(4), 133–141. https://doi.org/10.1007/s12154-014-0111-9

Hajimichael, J., Horvath, G., Pap, L., & Szekely, I. (1996). Chemical evaluation of SZI-121, a tetrazine type miticide. *Journal of Environmental Science and Health, Part B*, *31*(3), 515–519.

Hardy, M. C. (2014). Resistance is not futile: It shapes insecticide discovery. *Insects*, *5*, 227–242.

Hemingway, J., Hawkes, N. J., Carroll, M. C., & Ranson, H. (2004). The molecular basis of insecticide resistance in mosquitoes. *Insect Biochemistry and Molecular Biology*, *34*(7), 653–665.

Insecticide Resistance Action Committee (IRAC). (2007). *Resistance management for sustainable agriculture and improved public health*. Retrieved from https://croplife.org/wp-content/uploads/2015/11/IRAC-Resistant-Management-for-Sustainable-Agriculture-and-Improved-Public-Health_2007.pdf

Insecticide Resistance Action Committee (IRAC). (2018). *IRAC mode of action classification scheme*. Retrieved from www.irac-online.org

Jambhulkar, P. P., Sharma, P., Manokaran, R., Lakshman, D. K., Rokadia, P., & Jambhulkar, N. (2018). Assessing synergism of combined applications of *Trichoderma harzianum* and *Pseudomonas fluorescens* to control blast and bacterial leaf blight of rice. *European Journal of Plant Pathology*, *152*(3), 747–757. https://doi.org/10.1007/s10658-018-1506-7

Kinley, N. M., Kijima, S., Cook, G., & Sherrod, D. (2002). *Avaunt (Indoxacarb): A new mode of action insecticide for control of several key orchard pests* (pp. 9–11). Pullman, WA: Washington State University.

Kodanram, M. H., Rai, A. B., & Halder, J. (2010). Novel insecticides for management of insect pests in vegetable crops: A review. *Vegetable Science*, *37*(2), 109–123.

Kundoo, A. A., Dar, S. A., Mushtaq, M., Bashir, Z., Dar, M. S., Gul, S., & Gulzar, S. (2017). Role of neonicotinoids in insect pest management: A review. *Journal of Entomology and Zoology Studies*, *6*(1), 333–339.

Larena, I., Torres, R., De Cal, A., Liñán, M., Melgarejo, P., Domenichini, P., Bellini, A., Mandrin, J. F., Lichou, J., & De Eribe, X. O. (2005). Biological control of postharvest brown rot (*Monilinia* spp.) of peaches by field applications of *Epicoccum nigrum*. *Biological Control*, *32*(3), 305–310. https://doi.org/10.1016/j.biocontrol.2004.10.011

Lima, G., Castoria, R., De Curtis, F., Raiola, A., Ritieni, A., & De Cicco, V. (2011). Integrated control of blue mold using new fungicides and biocontrol yeasts lowers levels of fungicide residues and patulin contamination in apples. *Postharvest Biology and Technology*, *60*(2), 164–172. https://doi.org/10.1016/j.postharvbio.2011.01.003

Lima, G., De Curtis, F., Piedimonte, D., Spina, A. M., & De Cicco, V. (2006). Integration of biocontrol yeast and thiabendazole protects stored apples from fungicide sensitive and resistant isolates of *Botrytis cinerea*. *Postharvest Biology and Technology*, *40*(3), 301–307. https://doi.org/10.1016/j.postharvbio.2006.01.015

Mamta, B., & Rajam, M. V. (2017). RNAi technology: A new platform for crop pest control. *Physiology and Molecular Biology of Plants*, *23*(3), 487–501.

Morita, M., Ueda, T., Yoneda, T., Koyanagi, T., & Haga, T. (2007). Flonicamid, a novel insecticide with a rapid inhibitory effect on aphid feeding. *Pest Management Science*, *63*(10), 969–973.

Moutou, K. A., Panagiotaki, P., & Mamuris, Z. (2004). A new soft-rot of Ber caused by *Phomopsis natsume* and its chemical control. *Indian Phytopathology*, *57*(3), 361–363.

Nallathambi, P., Umamaheswari, C., Thakore, B. B. L., & More, T. A. (2009). Post-harvest management of ber (*Ziziphus mauritiana* Lamk) fruit rot (*Alternaria alternata* Fr. Keissler) using *Trichoderma* species, fungicides and their combinations. *Crop Protection*, *28*(6), 525–532. https://doi.org/10.1016/j.cropro.2009.02.015

Nobutaka, S., Kenichi, T., Takanobu, Y., Masako, T. N., & Hiroyuki, S. (2006). Combined use of the biocontrol bacterium *Pseudomonas fluorescens* strain LRB3W1 with reduced fungicide application for the control of tomato fusarium wilt. *BioControl*, *11*(2), 75–80. https://doi.org/10.1007/s10526-005-7918-x

Omar, I., O'Neill, T. M., & Rossall, S. (2006). Biological control of fusarium crown and root rot of tomato with antagonistic bacteria and integrated control when combined with the fungicide carbendazim. *Plant Pathology*, *55*(1), 92–99. https://doi.org/10.1111/j.1365-3059.2005.01281.x

Philips, C. R., Kuhar, T. P., Hoffmann, M. P., Zalom, F. G., Hallberg, R., Herbert, D. A., Gonzales, C., & Elliott, S. (2014). Integrated pest management. *eLS*. John Wiley & Sons Ltd. Retrieved from https://doi.org/10.1002/9780470015902.a0003248.pub2.

Rahman, M. H., Shovan, L. R., Hjeljord, L. G., Aam, B. B., Eijsink, V. G. H., Sørlie, M., & Tronsmo, A. (2014). Inhibition of fungal plant pathogens by synergistic action of chito-oligosaccharides and commercially available fungicides. *PLoS ONE*, *9*(7), e102847. https://doi.org/10.1371/journal.pone.0102847

Ruano-Rosa, D., Arjona-Girona, I., & López-Herrera, C. J. (2018). Integrated control of avocado white root rot combining low concentrations of fluazinam and *Trichoderma* spp. *Crop Protection*, *112*, 363–370. https://doi.org/10.1016/j.cropro.2018.06.010

Sarfraz, M., Dosdall, L. M., & Keddie, B. A. (2005). Spinosad: A promising tool for integrated pest management. *Outlooks on Pest Management*, *16*, 78–84.

Sarwar, M., & Salman, M. (2015). Insecticide's resistance in insect pests or vectors and development of novel strategies to combat its evolution. *International Journal of Bioinformatics and Biomedical Engineering*, *1*(3), 344–351.

Sinha, S. N., Paul, A. V. N., & Ram, S. (2009). Bio-efficacy of insecticides against two major insect pests of brinjal. *Annals of Plant Protection Sciences*, *17*(1), 55–57.

Spadaro, D., & Droby, S. (2016). Development of biocontrol products for postharvest diseases of fruit: The importance of elucidating the mechanisms of action of yeast antagonists. *Trends in Food Science & Technology*, *47*, 39–49. https://doi.org/10.1016/j.tifs.2015.11.003

Sparks, T. C., & Nauen, R. (2015). IRAC: Mode of action classification and insecticide resistance management. *Pesticide Biochemistry and Physiology*, *121*, 122–128.

Tomar, S., Sharma, S., & Malik, K. (2017). Diafenthiuron (a novel insecticide) for management of whitefly, Bemisia tabaci. *Indian Journal of Scientific Research*, *16*(1), 120–122.

Tomizawa, M., & Casida, J. E. (2005). Neonicotinoid insecticide toxicology: Mechanisms of selective action. *Annual Review of Pharmacology and Toxicology*, *45*, 247–268.

5 Advances in Pest Control in Agriculture and Food Processing

Sanhita Chowdhury[1]*, *Dharanikota Lalithambica Devi*[2], *Sristi Choudhury*[1], *Nikhil Kumar*[1], *Akash Sharma*[1], *and Anureet Kaur Chandi*[1]

[1]Department of Entomology, College of Agriculture, Punjab Agricultural University, Ludhiana, Punjab, India

[2]Department of Entomology, GB Pant University of Agriculture and Technology, Pantnagar, Uttarakhand, India

*Corresponding author

5.1 INTRODUCTION

There is a spike in the demand for food due to the growing worldwide population. Food security and safety are therefore among the world's greatest challenges nowadays (Guru et al., 2022). But insects are our worst rivals for food and fiber. They transmit some of the most dangerous illnesses and are frequently the cause of global food insecurity. Farmers use pesticides for the efficient advantage in managing pest population. Chemical pesticides are frequently employed to protect crops, and many of the most widely used pesticides have experienced an increase in insecticide resistance as a result of their frequent application. Furthermore, the risk that many of these chemicals bring to the environment and human health is causing the public to become more concerned (Nicolopoulou-Stamati et al., 2016). All these problems make it very necessary to identify safer, more insect-specific solutions for controlling pest insects. In the past few years, several different kinds of advancement in managing pest population have been developed. Improvements in knowledge of the factors influencing pest and disease will be necessary to utilize new technologies to produce more resistant crops, control pests sustainably, and enhance food processing. This chapter discusses the different strategies involved in pest control in agriculture and the food processing sector (**Figure 5.1**).

5.1.1 Host Plant Resistance

Plant resistance may be defined as the sum of the genetically inherited qualities that determine the ultimate degree of damage (yield loss) done to the plant by the herbivore (Painter, 1951; Smith and Clement, 2012). When we talk about host plant resistance, we are concerned with the mechanism of plants, that is, how they undergo changes to avoid/reduce injury caused to them by invaders, like pathogens and/or insect pests. Host plant resistance is an ideal pest management tactic due to its constant, cumulative effects, simplicity, and compatibility with other tactics. As crop production is increasing and sustainable practices are becoming more important, arthropod-resistant varieties can play a greater role in pest management.

Plant resistance does not have the negative environmental effects associated with the use of insecticides, and in fact, the use of resistant plant varieties can reduce insecticide use (Wilde, 2002).

Although R. H. Painter (1951) originally described the "mechanisms" involved as antibiosis, antixenosis, and tolerance, which are better thought of as types or "functional categories" of effects (Smith, 2005), he developed this trichotomous framework for classifying resistance when much less was known about the causal bases of resistance. But there is some question about the continuing utility of these categories as there are also some mechanisms of plant resistance that cannot be explicitly assigned to any of the aforementioned categories, such as herbivore-induced emission of volatiles that attract parasitoids (Stout, 2013).

The traditional approach to host plant resistance in integrated pest management involves screening, categorization, breeding, and implementation. Newer approaches have improved analytical tools and genetic manipulation techniques, making it more efficient to alter plant phenotypes. This has led to unprecedented insights into resistance mechanisms and their effects on herbivores (Zheng and Dicke, 2008). For example, virus-induced gene silencing in wheat increased susceptibility to aphid *Diuraphis noxia* (van Eck et al., 2010) and had an impact on interactions with herbivores, while silencing genes responsible for volatile compounds in rice significantly affected herbivore interactions (rice BPH and rice leaf folder). Moreover, the affordability of current omics technologies is going to enhance our understanding of genotype–phenotype relationships, leading to significant advancements in plant

DOI: 10.1201/9781003516538-5

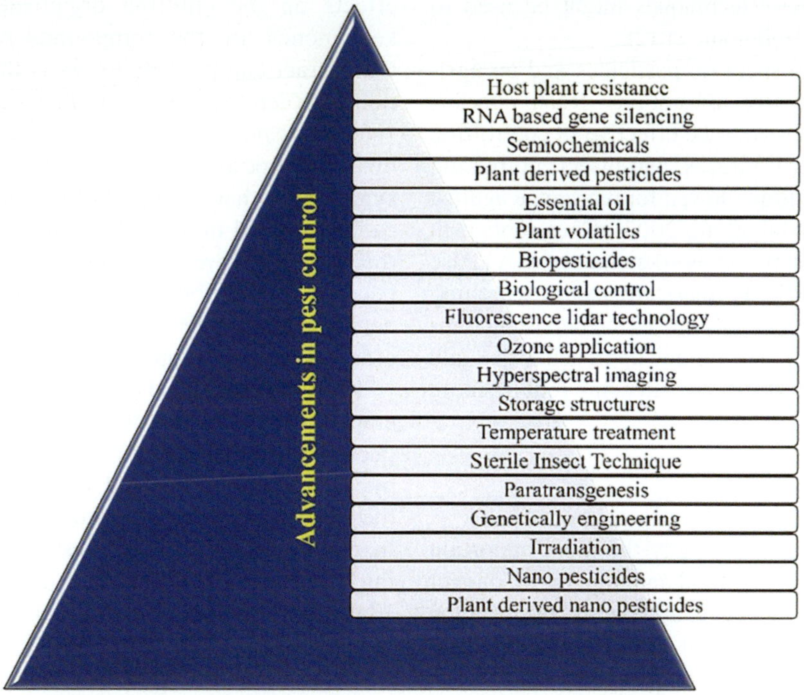

FIGURE 5.1 Different strategies of pest control beyond the conventional practice, which is the use of chemical pesticides. The advanced strategies include HPR, gene silencing, use of storage structures, nanopesticides, ozone gas, biopesticides, biological control, paratransgenesis, etc.

insect defense mechanisms. On the contrary, traditional approaches have limitations, such as breeding complexes and polygenic resistance. The scope of HPR lies in the refinement of analytical and molecular genetic tools which have improved understanding of plant resistance, leading to new targets for breeding, marker-assisted selection, and genetic engineering.

5.1.2 RNA-Based Gene Silencing

The process of improving crop plants through traditional plant breeding programs requires a lot of time and labor, so over the past 20 years, researchers have shifted their focus to biotechnological approaches. In this biological process, small RNA interferes with the translation of the target mRNA transcript, ultimately suppressing the expression of the gene (Table 5.1). Small non-coding RNAs are the result of the cleavage of dsRNA called microRNA (miRNA) and small interfering RNA (siRNA), which is carried out by a ribonuclease known as DICER, or Dicer-like enzyme (Pare and Hobman, 2007). The RNAi phenomenon is caused by short non-coding RNAs in conjunction with the RNA-induced silencing complex (RISC) (Redfern et al., 2013; Wilson and Doudna, 2013), Argonaute (AGO) (Riley et al., 2012), and other effector proteins. RNA-based gene silencing has been used as a potent tool for functional genomics and genetic engineering. These days, RNA interference (RNAi) makes it simple to manipulate gene expression for quality attributes in crops. It may be used by determining the target gene or genes, creating RNAi construct vectors, modifying the plant, and then screening and assessing its qualities. An essential source of knowledge for effective anti-resistance tactics is the evaluation of RNAi, from greenhouse experiments to field trials. By enhancing the defensive mechanisms

TABLE 5.1
List of Insects against Which RNAi Has Been Devised

Insect	Gene Targeted	References
Western corn rootworm, *Diabrotica virgifera virgifera*	V-ATPase A	Baum et al., 2007
American bollworm, *Helicoverpa armigera*	CYP6AE14	Mao et al., 2007
Egyptian cotton leafworm, *Spodoptera littoralis*	β-actin	Gvakharia et al., 2003
Egyptian cotton leafworm, *Spodoptera littoralis*	Per	Kotwica et al., 2009
Brown planthopper, *Nilaparvata lugens*	TPS	Chen et al., 2010
Triatomine bug, *Rhodnius prolixus*	NP2	Araujo et al., 2006
Beet armyworm, *Spodoptera exigua*	Chitin synthase gene	Chen et al., 2008
American locust, *Schistocerca americana*	Eye color gene vermilion	Dong and Friedrich, 2005
Light brown apple moth, *Epiphyas postvittana*	EposCXE1 and EposPBP1	Turner et al., 2006

in agricultural plants, these techniques might be used to manage biotic stressors (Hollomon, 2012).

Billions of dollars are spent on pesticides and on agricultural losses due to insect infestations. RNA interference (RNAi) has been shown to be effective in controlling nematodes, bacteria, fungi, viruses, and insects. Targeting the CYPAE14 gene in cotton allowed for resistance against *Helicoverpa armigera* (Mao et al., 2007). Corn rootworm resistance was achieved in maize by silencing the V-ATPase A gene (Baum et al., 2007). Farmers are always searching for alternate pest control methods due to the persistent danger of pesticide resistance. The unique approaches that add up to effective and long-lasting gene silencing require more investigation.

5.1.3 Semiochemicals

Chemical communication is probably the most important signal to find a mate, host, and food and to perceive danger as well as repel enemies in insects and plants. Information exchange between an organism that emits a signal (emitter/transmitter) and one that receives it (recipient), resulting in physiological/behavioral change, characterizes communication among living organisms. The signals emitted may be sound-producing, visual, or chemical. Chemical signals have certain merits over visual and vibrational cues; they can be transmitted at night when no visuals are available, do not require sight contact, and can be sensed from both short and long distances (Price, 1997; Bradbury and Vehrencamp, 2011).

Semiochemicals are the means of chemical communication of organisms and can act in intra- or inter-specific interactions. Intra-specific communication through chemical compounds is mediated by semiochemicals called pheromones (Nordlund and Lewis, 1976; Weber et al., 2018). Pheromones are divided into different classes, depending on how they act, as trail, oviposition, alarm, marking, aggregation, and sex. When semiochemicals are related to the relationship of individuals of different species, they are then called allelochemicals, and they are classified as allomones, kairomones, and synomones according to the organisms receiving and having an advantage from this interaction. The molecule alkyl methoxypyrazine is synthesized by the convergent ladybird beetle, *Hippodamia convergens* (Coleoptera: Coccinellidae), and exerts a negative impact on predators, acting as an allomone that inhibits predation (Wheeler and Cardé, 2014). Other examples of allomones include soybean plants with high levels of amino acids, which delay the development of the aphid *Aphis glycines* (Hemiptera: Aphididae) (Chiozza et al., 2010). A good example of kairomones would thus be the release of volatile chemicals by plants that attract herbivores, similar to what was witnessed in the chemical communication between cotton plants and the boll weevil, *Anthonomus grandis* (Coleoptera: Curculionidae) (Magalhães et al., 2016), which has positive effects on the recipient and negative effects on the emitting organism. Other examples of kairomones are the compounds released by host eggs that attract egg parasitoids, as is the case in the association between the eggs of *E. heros* and the egg parasitoid *Telenomus podisi* (Hymenoptera: Platygastridae), studied by Tognon et al. (2014) and Michereff et al. (2016). The synomones that benefit both transmitter and recipient can be noticed in tritrophic interactions. Pheromones are the most widespread semiochemicals and are use in IPM because of the power of pheromones in attracting partners of pheromone transmitters, and thus, they are very efficient tools for monitoring and even controlling insects.

Pheromones are the most common semiochemicals employed in IPM due to their ability to attract mates of pheromone transmitters; they are extremely helpful instruments for monitoring and even managing insects. Since the 1970s, pheromone-baited traps have shown to be an incredibly effective global instrument for insect monitoring in a variety of crop, fruit, and forest agroecosystems (Baker and Heath, 2005; Witzgall et al., 2010). Pheromone traps, which contain aggregation pheromones, have proven particularly helpful in monitoring moths (Lepidoptera) as well as Coleoptera, particularly in the case of bark beetles (Curculionidae: Scolytinae) (Wermelinger, 2004), and stink bugs (Pentatomidae) (Borges et al., 2011; Tilman and Cottrel, 2017). The mass trap strategy, which tries to catch a sizable fraction of the population in pheromone-baited traps, is another way to reduce population numbers using pheromone traps. The "attract and kill" tactic is an alternative to this strategy, which involves using pheromones to attract insects and using a toxic material to kill them (El-Sayed et al., 2009). The management of the cotton-boll weevil, *A. grandis*, is an intriguing illustration of this strategy. The invention of the boll weevil bait stick (BWBS) was aided by an effective mass trapping method that was employed in the 1980s (McKibben et al., 1990). Grandlure®, an insecticide at a very low dose (30 mg), and the aggregation pheromone are all contained in the polyvinyl chloride (PVC) stakes that make up this device.

Pheromones are classified depending on their effects on insect behavior:

- *Alarm pheromones* transmit an alarm signal due to the presence of a predator, parasitoid, or threat by the induction of two different behaviors: recruitment of more individuals to the site of disturbance, with a consequent attack on the aggressor, as is observed in Apidae, Formicidae, and Isoptera, or escape by rapid dispersion of insects, as occurs in Homoptera, Heteroptera, and Coleoptera.
- *Sex pheromones* can be defined as the odor produced by a male or a female that stimulates change in the opposite sex's behavior, leading to attraction and closeness of both for mating proposes (Cook et al., 2007; Weber et al., 2018).
- *Aggregation pheromones*, produced by either sex, act in attraction of both sexes and could favor

protection, reproduction, and feeding, or a combination of all these factors.
- *Trail pheromones* of social insects serve to communicate information to conspecifics about food and shelter, the size of a tunnel to be dug, and the recruitment of workers for this job.
- *Marking pheromones* are found in ants and bees and serve to mark the area where the nest is going to be, thus indicating the entry of the nest and differentiating the nests of different colonies. Such pheromones serve in fruit flies and parasitoids to mark the area as already colonized in order to avoid competition between individuals.
- *Oviposition pheromones* have been described in hematophagous mosquitoes, attracting pregnant females to oviposition in an environment already selected by individuals of the same species, increasing the number of larvae, and reducing the risks of predation.

5.1.4 Plant-Derived Pesticides

These are natural plant products with bioactivity toward pests. There are more than 2,500 plant species belonging to 235 families which have demonstrated their biological activity against several types of pests (Walia et al., 2014). These compounds have important ecological activities in nature, such as antifeedant, attractant, nematicide, fungicide, repellent, insecticide, insect growth regulator, and allelopathic agents, acting as a promising source for novel pest control agents or biopesticides. They comprise several classes of molecules, for example, terpenes, flavonoids, alkaloids, polyphenols, cyanogenic glucosides, quinones, amides, aldehydes, thiophenes, amino acids, saccharides, and polyketides (Souto et al., 2021). We may call these *the botanicals*, which denote pesticides originating from plant sources only, while the term *biopesticide* seems general. Plants yield the following products that are used in insect control: (1) toxicants, (2) attractants (geraniol, eugenol), (3) repellents (oils of citronella and cedar), (4) solvents for insecticides (cottonseed oil), and (5) dust carriers (soyabean and walnut shell flours). They are low-priced, biodegradable, and eco-friendly, and they act by several mechanisms of action in a more specific way, signifying that they are less of a hazard to humans and the environment. As a substitute to synthetic pesticides, which degrade crop health and the environment over the years and lead to pest resistance problems, these stand as a great solution. In order to improve the efficiency of crop production and reduce food crisis while preserving consumers' health, plant-derived pesticides are a safer alternative.

They act on several targets, such as (1) voltage-gated sodium channels (e.g., pyrethrins, used against a wide range of insects and mites, including spider mites, flies, fleas, and beetles) (Casida et al., 1998); (2) voltage-gated calcium channels (e.g., ryanodine, which has insecticidal activity against caterpillars, worms, potato beetles, lace bugs, aphids, and squash bugs) (Shivanandappa and Rajashekar, 2014); (3) acetylcholinesterase enzyme (AChE) (e.g., coumaran, used against houseflies and stored grain pests) (Rajashekar et al., 2014); (4) nicotinic acetylcholine receptors (e.g., nicotine, used against soft-bodied insects and mites) (Green et al., 2013); (5) GABA receptors and octopamine receptors (e.g., α-Thujone or picrotoxine, which, by evoking taste cell responses on most herbivorous insects, may also be considered as antifeedant or deterrent compounds, affecting mostly aphids, lepidopterans, and beetles) (González-Coloma et al., 2002); (6) octopamine receptors (e.g., eugenol, used against *Triatoma infestans*) (Reynoso et al., 2019); (7) respiratory inhibitors (e.g., rotenone, applied to protect lettuce and tomato crops as it has a broad spectrum of activity against mite pests, including leaf-feeding beetles, lice, caterpillars, mosquitoes, ticks, fire ants, and fleas); (8) insect hormone mimics or inhibitors (e.g., precocene I and II, which promote a precocious metamorphosis of the larvae and the production of sterile, moribund, and dwarfish adults in certain reported lepidopteran pests (Kamboj and Saluja, 2008); (9) vegetable crude oils (of rice bran, cottonseed, and palm kernel, which may act by disrupting the protective waxy covering of insects, affecting the water balance of insects through a rapid water loss from the cuticle, leading to death by desiccation); (10) repellents (e.g., lemongrass, eucalyptus, rosemary, vetiver, clove, and thyme); the anthraquinone tectoquinone, which has been described as a repellent against termites (Rudman et al., 1958); (11) attractants (e.g., limonene from white pine, which causes attraction of white pine cone beetle, *Conophthorus coniperda*) (Miller, 2007); (12) antifeedants (e.g., azadirachtin, documented to cause a deterrent action toward food in *Drosophila*) (Qiao et al., 2014); (13) phytotoxins/herbicides (e.g., citronella oil, used against different weed species) (Lajide et al., 1995); and (14) phototoxins. These are light-activated compounds which, instead of losing their efficiency due to sunlight degradation, actually increase or activate it, and they do so mainly by interacting with ground-state O_2 located in the tissue of its target, generating singlet oxygen and enabling insecticidal activity. Several classes, such as quinones, furanocoumarins, substituted acetylenes, and thiophenes, are included in this category. Marchant and Cooper (1987) have reported an oil constituent from *Artemisia monosperma* which, under sunlight-induced conditions, exerts an activity similar to DDT against the housefly and cotton leaf worm larvae.

5.1.5 Essential Oils

Essential oils are volatile, fragrant compounds that have an oily viscosity. Using conventional techniques like hydro-distillation, steam distillation, dry distillation, or environmentally friendly techniques like supercritical fluid extraction, microwave-assisted extraction, ultrasound-assisted extraction, and others, they are extracted from various plant parts (leaf, stem, flower, bark, and fruit) (Boukroufa et al., 2015). Since terpenoids and phenylpropanoids make up the majority of EOs and they have unique qualities (antioxidant, antibacterial, and insecticidal), these make them useful in industries like agriculture, food, and medicine (Plata-Rueda et al., 2018). Aromatic plants are rich in volatile essential oil components and have broad-spectrum applications in many processes and facets of life. On a commercial scale, essential oils have been obtained from nearly 400 plant species belonging to 67 families comprising Lamiaceae, Asteraceae, Apiaceae, Fabaceae, Lauraceae, Rutaceae, Myrtaceae, Cupressaceae, Pinaceae, Zingiberaceae, Burseraceae, and Rubiaceae (Kalia, 2005). Depending on their chemical composition, pure essential oils are classified as either oxygenated or hydrocarbons. As a substitute to synthetic chemicals, these essential oils are used to repel a variety of insects (both in field and in storage). The most effective essential oil components that have the ability to kill storage pests by fumigant, contact, or ingestion are covered in this review.

These EOs hinder development, reproduction, and physiology in insects. While many chemicals are effective, some have more competitive and better affinity than others, viz., eucalyptol, camphor, eugenol (Obeng-Ofori and Reichmuth, 1999), linalool, limonene (Liang et al., 2022), methone, and anethole (Yeom et al., 2012). Furthermore, substances including isoeugenol, methyleugenol, α-pinene, β-caryophyllene, limonene, α-bisabolol, linalool, terpinen-4-ol, and geraniol have been reported to significantly reduce the number of insects that feed on stored commodities.

5.1.6 Plant Volatiles

These are organic compounds originating from plants that are highly volatile in nature. They are secondary metabolites that are toxic to a range of insects and serve as an alternative to synthetic pesticides. Approximately 1,700 volatile secondary metabolites have been identified so far from more than 90 plant families (Knudsen and Gershenzon, 2020a). Plants, as one of the primary producers of chemicals in nature, provide a variety of bioactive organic compounds that have been found to have a defensive role in plants, particularly against insect pests (Figure 5.2) (Maffei et al., 2011). These "plant volatile organic compounds" have been traditionally used by farmers to combat stored grain pests worldwide. However, with the advent of synthetic insecticides following World War II, farmers have drifted away from these ancient practices. Taking into account the risks associated with these chemicals, using plant components that pose no or very few threats constitutes the most practical

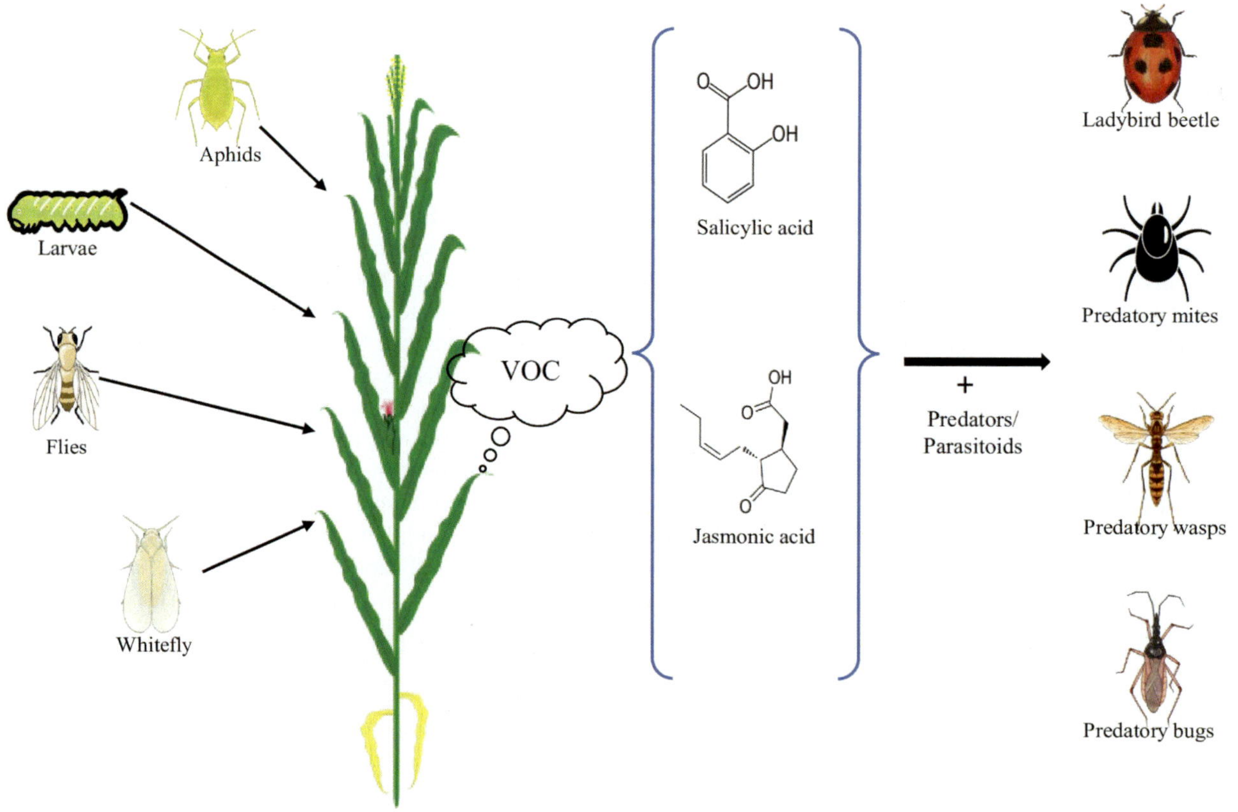

FIGURE 5.2 How plant volatiles like salicylic acid and jasmonic acid are produced upon the feeding of herbivores on plant. These volatiles are then responsible in attracting parasitoids and predators, like predatory wasps, predatory bugs, ladybird beetles, predatory mites, etc., which will then help minimize crop damage.

solution for effective use against various pests of tropical agriculture (Isman, 2008).Plant volatiles have a fumigant action because of the various functional groups that are attached to volatile compounds (Table 5.2), allowing them to stay longer in a closed environment (Copping and Menn, 2000). The insecticidal action of *Cinnamomum camphora* and *Pimpinella anisum* essential oils against *Trogoderma granarium* and *T. castaneum* has been reported (Nenaah and Ibrahim, 2011). Plant volatiles may also be developed into products for luring insects. For example, traps baited with volatile compounds, mainly phenylacetaldehyde obtained from Canada thistle, have been documented to attract the soybean looper, where more females are trapped than males (Stringer et al., 2008).

5.1.7 Biopesticides

The contribution of biopesticides to the insecticide sector accounts for about 15% of the total market of insecticides. Such a share is estimated to be around $3 billion in the world market, accounting for about 5% of total crop protection (Marrone, 2014). The growth rate of biopesticides is increased to 15%. At present, commercially, over 125 species of natural enemies are available as biopesticides. Sales of biopesticides are about 90% based on only one entomopathogenic bacteria, namely, *Bacillus thuringiensis*. The United States and the European Union and India are the major producers and users of biopesticides in the world. More than 200 Bt products are available in the US market, while in the EU market, around 60 Bt-based products and 60 analogous products are available (Kumar and Singh, 2015). The potential market under biopesticides in India is manifestly enormous as damage due to pests culminates to around USD 9,259 million annually (Rajamani and Negi, 2021). The present market share of biopesticides is 5% of the pesticide market. The CIBRC (Central Insecticides Board and Registration Committee) in India has registered around 970 microbial formulations. Besides, there are around 200 commercial formulations using entomopathogenic fungi. They include *Beauveria bassiana*, *B. brongniartii*, *Metarhizium anisopliae*, *Lecanicillium lecanii*, and *Hirsutella thompsonii*. A total of 45 products belonging to Bt are commercially available, and of the Bt group, *B. thuringiensis* subspecies *kurstaki* is commonly used against bollworms, loopers, and other lepidopterans, while *B. thuringiensis* subspecies *israelensis* and *sphaericus* are used against the mosquito. First time the use of the species of entomopathogenic nematodes *Steinernema carpocapsae* (DD-136 240 M. Rajamani and A. Negi strain) for the control of lepidopteran pests of rice, sugarcane, and apple in India. Several others described indigenous strains of entomopathogenic nematode, including *Heterorhabditis indica*, and four other species of *Steinernema*, namely, *S. carpocapsae*, *S. abbasi*, *S. thermophilum*, and *S. riobrave*, which were used for the management of field pests. A shelf life of 12 months is available in the commercial market in India (Kumar et al., 2019).

Based on classification by the Environmental Protection Agency on the ingredient used, biopesticides have been divided into the three main classes: (1) microbial, (2) biochemical, and (3) plant-incorporated protectants (Olson, 2015). Of the three categories, microbial pesticides feature the most massive class of pest-specific, broad-spectrum biopesticides. Microbial biopesticides include the use of microbes, such as bacteria, viruses, fungi, and protozoans, as active ingredients for the management of insect pests. They are relatively precise with their target species. Microbial biopesticides are self-perpetuating, host-specific, and environment-friendly. Among the most widely used microorganisms against insect pests is Bt. It is used to control a wide array of pests, including lepidopterans, coleopterans, and dipterans (Gill et al., 1992).

Plant-incorporated protectants (PIPs) are also referred to as genetically modified crops. Plants are made to produce pesticidal substances, and the genetic material used in producing such substances is introduced into the plants to protect them from pests. A typical example is the use of Bt protein to develop PIP through the genetic engineering process. The Bt toxin is host-specific, achieves quick mortality of the pests within 48 hours (Siegel, 2001), and poses no harmful effect on the ecosystem, as it does not harm vertebrates (Lacey et al., 2015).

Biochemical pesticides are also known as herbal pesticides (Pal et al., 2013). They are naturally occurring substances and secondary metabolites that control or inactivate pests. The most widely used biochemical pesticides are from neem and neem-based formulations, such as neem oil, neem seed kernel extract, neem extract concentrates from bark and leaves, which are available in India. Also, essential oils from canola, tea tree, lemongrass (*Cymbopogon citrates*), and pyrethrin from *Chrysanthemum cinerariaefolium* are used as biochemical pesticides.

5.1.8 Biological Control

The term *biological control* (or *biocontrol*) implies controlling/managing the population of one biological entity with another one. It has been applied in practice to almost all types of pests, including insect pests and pathogens of crops (Pertot et al., 2017), weeds, mosquitos (Ingabire et al., 2017), and rodents (Jäkel et al., 2019). Notably, the demand for biocontrol solutions has rapidly grown in recent years in parallel with the global endorsement of integrated pest management (IPM) as the future paradigm for crop protection (Stenberg, 2017). The term *biological control* has traditionally been used to describe actions to combat pests using other living agents. Cook and Baker (1983) defined it as "the reduction in the amount of inoculum or disease-producing activity of a pathogen accomplished by or through one or more organisms other than man." Their definition was restricted to control of pathogens. Later, Eilenberg et al. (2001) attempted to harmonize the terminology, where they emphasized that the term should exclusively be used for living agents. Biological control should only be recognized in

TABLE 5.2
List of Pronounced Plant Volatiles and Their Insecticidal Activity

Sr. No.	Plant Species	Family	Part of the Plant	Product	Active Principles	Bioassay Test	Target Insect	References
1.	*Acorus calamus*	Acoraceae	Rhizome	Extract	β- Asarone	Contact	*Bemesia tabaci*, *Sitophilus zeamais*	Wang et al., 2022; Yao et al., 2008
2.	*Aloysia citriodora*	Verbenaceae	Leaves	Essential oils	Citronellal and sabinene	Repellent, fumigant, contact	*Tribolium castaneum* and *T. confusum*	Benzi et al., 2014
3.	*Artemisia annua*	Asteraceae	Leaves	Essential oils	Artemisinin and 1,8-cineole	Repellent fumigant	*Cydia pomonella*, and *T. castaneum*	Durden et al., 2011; Deb and Kumar, 2020
4.	*Brugmansia suaveolens*	Solanaceae	Flowers	Fractions	Tropane alkaloids	Oviposition deterrent	*Zabrotes subfasciatus*	Gonçalves et al., 2019
5.	*Carum carvi*	Apiaceae	Leaves	Essential oils	Carvone Limonene, (E)- Anethole	Contact	*Rhizopertha dominica*, *S. oryzae*, and *S. zeamais*	Fang et al., 2010
6.	*Carum carvi*	Apiaceae	Seeds	Essential oils	Carvone	Repellent	*Meligethes aeneus*	Pavela, 2011
7.	*Carum copticum*	Apiaceae	Seeds	Essential oils	Thymol, γ-terpinene, o-cymene	Repellent	*Plutella xylostella*	Jamal et al., 2012
8.	*Chamaecyparis obtusa*	Cupressaceae	Leaves	Essential oils	Bornyl acetate	Contact, fumigant	*S. oryzae* and *Callosobruchus chinensis*	Park et al., 2003
9.	Citrus	Rutaceae	Fruit peel	Essential oils	Limonene, Eugenol	Contact, fumigant	*T. castaneum* and *S. oryzae*	Mishra et al., 2011
10.	*Coriander sativum*	Apiaceae	Seeds	Essential oils	Linalool	Contact	*S. oryzae*, *R. dominica*, and *C. pusillus*	López et al., 2008
11.	*Cupressus lusitanica*	Cupressaceae	Leaves	Essential oils	Umbellulone and α-pinene	Contact, fumigant	*T. castaneum*, *A. obtectus*, *S. cerealella*, and *S. zeamais*	Bett et al., 2016
12.	*Eucalyptus* spp.	Myrtaceae	Leaves	Essential oils	α-Terpinene 1, 8-Cineole α-pinene	Fumigant	*S. oryzae*	Lee et al., 2004
13.	*Illicium verum*	Illiciaceae	Fruits	Essential oils	Fumigant	Fumigant	*Reticulitermes speratus*	Park and Shin, 2005
14.	*Mentha citrata*	Lamiaceae	Aerial part	Essential oils	Carvone, Menthol, Linalool, Linalyl acetate	Contact, fumigant	*T. castaneum* and *C. maculatus*	Singh et al., 1989
15.	*Ocimum kilimandscharium*	Lamiaceae	Fresh aerial part	Essential oils	Camphor	Contact	*S. oryzae*	Bekele & Hassanali, 2001
16.	*Rosmarinus officinalis*	Lamiaceae	Leaves	Essential oils	Camphor	Fumigant	*S. oryzae*	Trivedi et al., 2017
17.	*Tagetes filifolia*	Asteraceae	Aerial parts	Essential oils	(E)-anethole and estragole	Fumigant	*T. castaneum*	Olmedo et al., 2015
19.	*Zingiber officinale*	Zingiberaceae	Rhizome	Essential oils	β-Zingiberene	Antifeedant, IGR	*T. castaneum*	Brito et al., 2021
20.	*Zataria multiflora*	Lamiaceae	Leaves and stems	Essential oils	7aα-nepetalactone	Fumigant	*Brevicoryne brassicae*	Motazedian et al., 2014

interactions between three separate players: (1) a pest, (2) a living biocontrol agent targeting the pest, and (3) a human stakeholder benefitting from the pest control service provided by the biocontrol agent (Stenberg et al., 2021).

Similarly, biological control works on three key principles: First, biological control involves a living agent directly or indirectly targeting a pest, thereby reducing damage from a human perspective. Second, damage must be reduced by pest control, rather than via general health improvement. Thus, identification of the mechanism(s) involved is important. In cases where a nature-based method to control pests does not meet the criteria for biological control, it is likely to fall within the other recognized area of bioprotection, called theoretically as the "bioprotection umbrella." The third principle is that all biological control methods can be conveniently classified in four main categories: natural biological control (if there is no deliberate human intervention), conservation biological control (involving human stimulation of resident agents of biological control), augmentative biological control (human addition of biocontrol agents, temporarily augmenting the population of biocontrol agents), and classical biological control (adding new biocontrol agents for proliferation and permanent establishment) (Stenberg et al., 2021).

5.1.9 Fluorescence Lidar Technology (FIT)

Radar techniques were introduced into the area of entomology throughout the previous several decades and are an effective means of tracking the flight path of pests. Harmonic radar was employed to investigate low-flying pests, whereas vertical-looking radar (VLR) is commonly utilized to monitor the movement of pests flying at high altitudes (Chapman et al., 2003). Radar methods are an effective means of tracking the flight path of pests. Harmonic radar was employed to track insects that flew at low altitude in order to study their behavior near pheromone traps. The fluorescence lidar technology has found extensive application in the identification of biological aerosols (Treaty, 2010), vegetation monitoring (Cecchi and Pantani, 1994), tracking bees to identify land mines (Shaw et al., 2005), and more recently, planthopper and moth monitoring (Mei et al., 2012); migration of the damselfly species *Calopteryx splendens* and *C. virgo* (Brydegaard et al., 2009); and movement of insects in the field (Guan et al., 2010). The natural fluorescence of insects is measured optically in an experimental setting as well as by remotely utilizing a fluorescence lidar device at a testing range of 50 m. Some insect specimens utilized in remote sensing data gathering were coated with fluorescent dyes, which are often employed in ecological research. This technique provides novel perspective on insect migration and movement patterns. For efficient pest monitoring during the day and at night, a portable harmonic radar system is a helpful instrument (Dwivedi et al., 2020).

5.1.10 Ozone Application

Ozone is a potent antioxidant that breaks down quickly into oxygen in the air, leaving behind little residue in the product and minimizing the need to extract gas from it. Ozone is the most promising green technology that might enhance food safety and quality since it is an eco-friendly process with non-residual features (Pandiselvam et al., 2020). Storage pests are controlled by ozone (Pandiselvam et al., 2019). In addition to controlling pests during storage, ozone is effectively utilized in the processing of grains to facilitate microbiological decontamination (Figure 5.3), mycotoxin degradation, starch modification, functional quality modifications, and germination ability (Sivaranjani et al., 2021). Once methyl bromide is phased out, ozone is used as a chemical fumigant substitute in storage grains, fruits, vegetables, and spices, achieving 100% death in the majority of pest species. Ozone has been documented to be effective against both external and internal feeders, including *Tribolium castaneum*, *Oryzaephilus surinamensis*, *Sitophilus granarius*, S. *zeamais*, and S. *oryzae* (Hansen et al., 2012). The concentration, fumigation duration, and moisture content of the grains were all affected by the fumigation of stored insects with ozone. Boopathy et al. (2021) found that while the adults in stored product insects were more susceptible to ozone fumigation, the eggs remained unharmed. Insects often respire sporadically after being exposed to ozone, which reduces oxidative tissue damage. This leads to membrane oxidation, DNA strand breaks, and changes in the insect's pulmonary function (Pandiselvam et al., 2019).

5.1.11 Hyperspectral Imaging

Hyperspectral imaging (HSI) is a powerful technique that combines spectra and images to obtain spectral and spatial information about the research object (Ram et al., 2024). HSI is a rapid, non-destructive detection technology. It can offer spatial imaging information and accurately capture the spectrum information of the whole sample surface, in contrast to NIR spectroscopy. Currently, agricultural product quality and safety detection make extensive use of HSI technology (Li et al., 2018). HSI has been widely used to find pesticide residues in fruits and vegetables (Sun et al., 2022; Gila et al., 2024; He et al., 2021). Besides this, HSI is employed for pest detection also. In addition to hyperspectral data, ground-collected data from unmanned aerial vehicles (UAVs) is utilized in the identification of pests, such as grape phylloxera (*Daktulosphaira vitifoliae* Fitch) in vineyards (Vanegas et al., 2018). An approach based on remote sensing (UAV) would improve the effectiveness of current monitoring methods (human inspectors and insect traps) in identifying pest infestations (such as grape phylloxera in vineyards). Advanced digital hyperspectral, multispectral, and RGB sensors are coupled with a UAV as

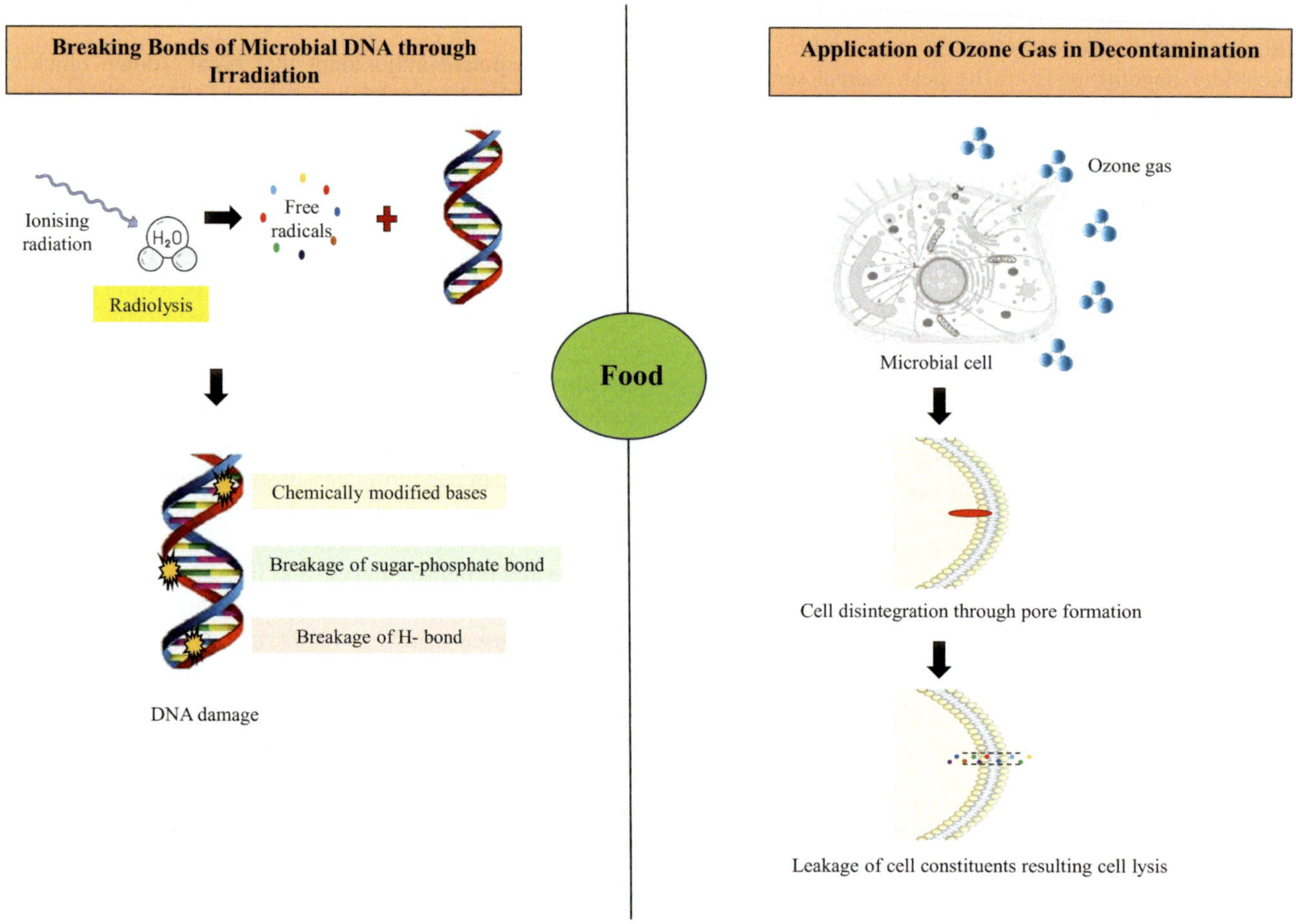

FIGURE 5.3 How irradiation forms free radicals through radiolysis that ultimately break the bonds of microbial DNA and cause DNA damage, while the second part describes the application of ozone gas. When microbial cell is exposed to ozone gas, it forms pores in the cell membrane, which cause leakage of cell constituents, leading to cell lysis.

part of this technology. A certain wavelength range of the electromagnetic spectrum, known as a spectral band, is represented by each image captured by a spectral sensor. For further processing and analysis, the images are merged to generate a three-dimensional hyperspectral data cube. Two spatial dimensions (x and y) plus a spectral dimension (λ) made up of various wavelengths make up the cube (Ali et al., 2025).

5.1.12 Storage Structures

For food grains to be delivered to customers on time, they must be preserved and stored safely. Currently, a number of structures – from little metal bins to large grain elevators/silos – ensure the safe storage of grains. Throughout the extended storage period, agricultural commodities are often vulnerable to biotic and abiotic contamination (Guru et al., 2022). Among the biotic agents, there is a significant loss in storage due to insects, mites, rodents, birds, and microbes. Approximately 10–20% of all storage losses are caused by a number of insect species that seriously damage commodities that are kept in storage (Esther et al., 2014). Insects that cause losses of stored goods of agricultural and animal origin include more than 600 species of beetles, 70 species of moths, and 355 species of mites (Rajendran and Sriranjini, 2008).

These days, low-pressure and hermetic storage structures are quite common. The concept of de-oxygenation inside the storage structure is used in airtight storage to lessen aerobic conditions because many traditional structures include internal oxygenation. Carbon dioxide (CO_2) concentration rises as a result of this decrease in oxygen (O_2) concentration, adversely affecting insects' metabolic processes and ultimately resulting in their death. Grain storage in hermetic seal storage facilities is a developing area of agricultural engineering study. The damaging potential of *T. castaneum* in various storage bags retaining wheat was investigated by Atta et al. (2020). At 30 days in an airtight bag, the results demonstrated 100% mortality of *T. castaneum*. In addition, a decrease in the rate of insect development (30%), damage to wheat grains (10.16%), and weight loss of grains (0.5–1.0%) were reported.

The concept of low-pressure storage is creating a grain storage structure with a lower internal pressure. It modifies the physical environment, which is often accomplished by providing vacuum and producing low-oxygen environments. The link between temperature and pressure in altered environmental circumstances was also noted by Mbata and Phillips (2001), who found that low pressure alone was less lethal to the various developmental stages of *T. castaneum*, *P. interpunctella*, and *R. dominica*. Moreover, mortality rose with longer exposure times to low pressure and rising temperatures throughout all life stages.

5.1.13 Temperature Treatment

Every insect needs a variety of ideal atmospheric conditions for normal growth and development, including temperature, relative humidity, and gas compositions. The metabolism of insects is adversely affected by deviations from their optimal level. The ideal temperature range for insect reproduction is 25 to 32°C, when the number of insects increases the most. In the range of 13 to 24°C and 33 to 35°C, development slows down. Insects cease feeding, grow more slowly, and die at temperatures below 13°C and over 35°C (Hamel et al., 2020). Greater success in controlling insects was achieved with either a rise or fall in temperature and helps in eradicating the infestation or preventing additional damage (Mason and Strait, 2019). The traditional method of applying thermal treatments for grain disinfestation involved regulated conditions and the use of hot/cool air–water combinations (Macana et al., 2018). *T. castaneum*, *T. confusum*, and *Oryzaephilus mercator* are generally the species that are most easily controlled by cold temperatures (Sinha and Watters, 1985), whereas *Trogloderma granarium*, *Sitophilus granarius*, *Ephestia kuehniella*, and *Plodia interpunctella* are more difficult to control. At 20°C and 40°C, *T. granarium* was unable to develop or complete its life cycle (Riaz et al., 2014). Also, radiofrequency (RF) heating aids in the management of pests in food crops (Costa and Marra, 2024). Using 40.68 MHz, 10 kW RF heating, Indumathi et al. (2023) assessed the disinfestation of *T. castaneum* (Herbst) in semolina, where 100% of *T. castaneum* in semolina was eliminated after 8 minutes of exposure, resulting in 100% mortality at 64.4°C. Also, heating at 54°C for 11 minutes was sufficient to kill the pest *R. dominica* (Hou et al., 2019).

5.1.14 Sterile Insect Technique (SIT)

For many years, broad-spectrum insecticides have been used to control insects and disease vectors. Sterile insect technique (SIT) is a modern approach of pest control without causing any adverse effect to the environment (Bourtzis and Vreysen, 2021). Sterile insect technique comes under the biological control method. While female insects' implantation is by the released method, radiation-sterilized males do not reproduce due to repeated releases of sterilized insects overcoming the population of pest. Wide-area pests control can be achieved when the population of sterile insects is systematically part of area-wide integrated pest management programs. SIT technique is non-polluting and resistance-free, invented by American entomologist Dr. Edward F. Knippling in the 1940s. SIT was successfully used for the first time for screwworm flies, then after that for tsetse flies, fruit flies, and moths, through genetic manipulation. This program is more advantageous by decreasing insecticide use and has also facilitated the use of biological control agents against secondary insect pests. SIT has been used in controlling various types of insect pest order, like Lepidoptera, Hemiptera, and livestock pests.

Lepidoptera are, in addition to fruit flies, major agricultural pests, and their populations can be very efficiently managed by AW-IPM programs that include a SIT component. The latter property has resulted in the development of the inherited sterility (IS) method, a SIT-derived technology, which is currently used to successfully suppress populations of several Lepidopteran pest species, whereas, Hemipterans are an unusual target insect group for SIT, in view that all development stages and potential release stages can cause damage to crops. Despite this, efforts have been made to assess whether populations of some Hemipteran pest species can be targeted by SIT. It has effectively controlled and, in some cases, completely eliminated tsetse populations that spread African trypanosomes, leading to sleeping sickness in humans and nagana in animals (Vreysen et al., 2021). It has been very successful in managing populations of plant and livestock pests worldwide. More recently, SIT has been considered as a tool to suppress mosquito species transmitting major human pathogens, such as the yellow fever mosquito.

In SIT, insects are sterilized by γ-radiation with a dose that induces sterility but which does not significantly impair the ability of sterile males to fly, mate, and transfer sperm to wild females. It is species-specific and does not require any exotic species, and it has some specific applications (Figure 5.4). Sterile insects are systematically released in a wide area for controlling pest population (Klassen et al., 2021). The capacity of the released sterile males to deliver functional sperm to wild females that fertilizes their eggs is essential to SIT's effectiveness. Biological SIT is also seen as a classical biological control.

5.1.15 Paratransgenesis

Paratransgenesis is a potential and innovative technique for reducing diseases transmitted by vectors. It utilizes the vector insect's genetically modified native microbiome (mutualistic, symbiotic, and commensal bacteria, fungi, and viruses) to suppress or kill the disease pathogen (Ratcliffe et al., 2022). This strategy involves engineering bacteria in vectors' guts to release anti-pathogen chemicals. *Asaia*, an *Alphaproteobacteria*, is prevalent in all developmental stages of the host and in various parts (of the host), making it a suitable option for paratransgenesis. Bacteria, fungi, and viruses can be used as excellent candidates for

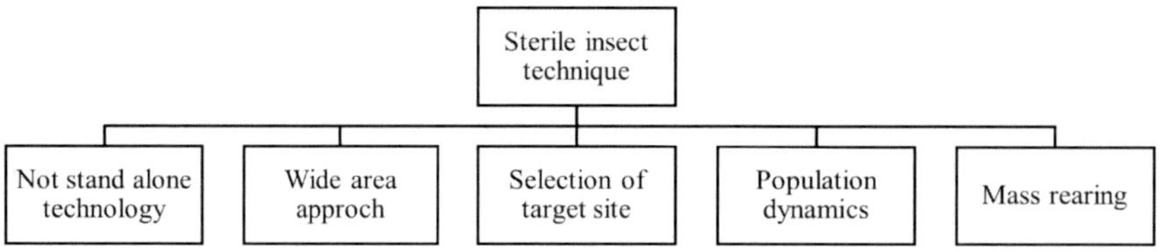

FIGURE 5.4 Different applications of sterile insect technique in the field of agriculture.

paratransgenesis. This strategy reduces the harmful effects of insecticides on non-target species, such as humans, the environment, soil, and water, while also preventing mosquito resistance (Jayakrishnan et al., 2018).

Native symbionts or commensals obtained from the vector are genetically modified in vitro to create antipathogen factors, which are subsequently reintroduced into the insect to disrupt the disease organism's life cycle (Ratcliffe et al., 2022). This approach has an advantage over genetically changed mosquitoes as the converted bacteria/fungi/viruses may colonize various insect vector strains and species. To avoid disease transmission, transgenic mosquitoes may need to transform each strain or species. Transformed bacteria are easier to manufacture in huge quantities than transgenic mosquitoes (Wilke and Marrelli, 2015). Beard, Durvasula, and colleagues created paratransgenesis in *Rhodnius prolixus* to inhibit the spread of the protozoan parasite *Trypanosoma cruzi*, which causes Chagas disease. This pioneering study established a paradigm for paratransgenesis research in other vector insects that spread diseases to animals, people, and agricultural plants (Ratcliffe et al., 2022). *Wolbachia pipientis*, a bacterium from the Rickettsiaceae family, is a well-studied example of an intracellular disease that spreads quickly across insect populations. Wolbachia infections in insects can induce incompatibility between eggs and sperm, resulting in infertility, feminization, parthenogenesis, and male death (Rajashekhar et al., 2016), and these outcomes can be used to control vectors. Transmitting the *Wolbachia* wMelPop-CLA strain into *Aedes aegypti* decreases the mosquito's lifespan and capacity to transmit chikungunya and dengue (Scolari et al., 2019). The CRISPR system can transform microbes for paratransgenesis and also mediates gene silencing, resulting in more consistent and robust knockdowns with fewer off-target events than RNA interference (Boettcher and Manus, 2015).

5.1.16 Genetic Engineering

Genetic engineering, also referred to as genetic modification, involves the deliberate, precise, and artificial alteration of an organism's genetic material. This technology enables scientists to modify the genetic makeup of cells by manipulating their genes directly, which can involve transferring genes across species boundaries to create organisms with enhanced or novel traits. In plant genetic engineering, this process involves incorporating a foreign gene into the genome of a host crop, with the expression of the gene being controlled by specific regulatory sequences. There are several variations of genetic engineering based on the relationship between the gene donor and the recipient. These diverse techniques enable the development of crops with desired characteristics, such as improved resistance to pests and diseases, enhanced nutritional content, and better adaptability to environmental stresses, while also raising significant ethical and ecological considerations (Gould, 1988).

Recent developments in cellular and molecular biology have paved the way for the creation of genetically engineered (transgenic) plants with novel genetic traits. These genetically modified (GM) crops present a valuable opportunity to significantly enhance integrated pest management (IPM) strategies (Babu et al., 2003). Transgenic plants have been developed in key crops, including maize, wheat, barley, rice, cotton, potato, tomato, and soybean. Various transformation techniques have been used to introduce genes that control important agronomic traits (Brar & Khush, 1993). Some prominent examples include Bt and protease inhibitor genes for resistance to insects; coat protein (CP) genes, antisense-CP, and satellite RNA for virus resistance; chitinase and ribosome-inactivating protein (RIP) genes for resistance to fungi; and bar and ALS genes for resistance to herbicides (Brown, 2000).

Advancements in genetic engineering have greatly expanded our understanding of gene silencing in transgenic plants, shedding light on how genomes are organized and how gene expression is regulated. The creation of transgenic plant varieties with enhanced resistance to pests, particularly those with potent insecticidal properties, has sparked concerns. These concerns include the potential for pests to develop resistance to these measures and the broader environmental impact, such as unintended effects on non-target species and ecosystems. This ongoing debate highlights the need for careful management and regulation of genetically modified organisms (GMOs) to balance agricultural benefits with ecological considerations (Stewart et al., 2001). The introduction of genetically modified organisms (GMOs) with traits like insect resistance has significantly decreased pest populations and reduced the need for pesticide applications. This has been particularly evident in crops engineered to produce Bt toxins, which are highly effective in controlling Lepidopteran pests. As a result, these GMOs have not only contributed to increased crop yields but also minimized the environmental impact associated with pesticide use. For example, the widespread adoption of Bt crops has lessened the reliance on chemical insecticides, thereby reducing the ecological footprint of agriculture and promoting a more sustainable farming approach. These benefits underscore the potential of GMOs to enhance agricultural productivity while also supporting environmental conservation efforts (Bale et al., 2008).

5.1.17 Irradiation

In order to manage *Lasioderma serricorne*, irradiation was initially used in cigars in 1929 as a commercial pest control method. However, the X-ray equipment that was utilized proved to be inadequate for continuous processing (Bakri et al., 2005). The mechanism of action of irradiation, whether from isotopes or artificial sources (e-beams or X-rays), is the same: The electrons, gamma rays, or X-rays disrupt electron orbits, resulting in the creation of ions. An electron shower is produced when the free electrons continue to be interfered with additional electrons. When radicals are created, they damage big organic molecules, like DNA, which halts the growth of irradiated organisms (Figure 5.3) (Hallman, 2011). The most common method for sterilizing mass-reared males for the sterile insect technique currently is gamma irradiation. The success of SIT is largely dependent on the generation of high-quality sterile males that are released into the intended wild populations. It is crucial that irradiation treatments achieve sufficient levels of sterility in order to guarantee that liberated males are efficient in causing reproductive failure in their female counterparts (Espo et al., 2015). In commercial application, gamma irradiation is the most used method. Large, bulky materials can be penetrated by gamma rays; therefore, gamma irradiation is a good option for treating entire pallet loads of products and administering treatment doses that stay below a maximum dose required by law or technology, all the while ensuring that every component of the product receives at least the minimal dose required to fulfil the phytosanitary purpose (Hallman et al., 2016). Phytosanitary irradiation differs from all other commercial treatments in one important technical consideration: The end point of the treatment is not acute mortality but prevention of further biological development and reproduction (Hallman, 2011). The treated commodities must be able to resist the phytosanitary treatment for it to be economically feasible. More fresh fruits and vegetables can withstand radiation at the 150–400 Gy dosages used for phytosanitary irradiation than any other commercial therapy (Heather and Hallman, 2008).

5.1.18 Nanopesticides

Pest management is a critical aspect of agriculture, and the use of pesticides is a common practice to control pests. However, traditional pesticides have several limitations, including environmental pollution, human health risks, and pesticide-resistant pest populations (Pimentel, 2009). Nanopesticides have emerged as a promising solution to these issues, offering a targeted, efficient, and environmentally friendly approach to pest management (Kah et al., 2013).

Nanopesticides are defined as pesticides that use nanoparticles as active ingredients or carriers (Owolade et al., 2008). They can be classified into two main types: nanoencapsulated pesticides and nanoparticle-based pesticides (Lowry et al., 2019). Nanoencapsulated pesticides involve the encapsulation of active ingredients in nanoparticles, which enhances their solubility, stability, and bioavailability (Shakiba et al., 2020). Nanoparticle-based pesticides, on the other hand, use nanoparticles as active ingredients, which can interact with pests at the nanoscale, leading to increased efficacy and reduced non-target effects (Bouwmeester et al., 2009). The use of nanopesticides has shown significant potential in controlling various pests, including insects, fungi, and weeds (Bouwmeester et al., 2009). For instance, nanoformulations of insecticides like pyrethroids and neem oil have demonstrated enhanced insecticidal activity against target pests (Yadav et al., 2021). Additionally, nanopesticides can be designed to target specific pests, reducing non-target effects and environmental impact. However, the development and application of nanopesticides also raise concerns regarding their environmental fate, toxicity, and potential impacts on non-target organisms (Kah et al., 2013). Further research is necessary to address these concerns and ensure the safe and effective use of nanopesticides in pest management.

5.1.19 Plant-Derived Nanopesticides

Nanopesticides are innovative formulations consisting of organic and inorganic components containing nanosized particles of a pesticide's active ingredient (AI) or other small engineered structures with pesticidal properties. Nanopesticides are particularly appealing because their tiny particles can enhance the delivery and effectiveness of active ingredients, leading to more precise targeting of pests and potentially reducing the amount of pesticide needed. This precision not only improves pest management outcomes but also minimizes the environmental footprint associated with traditional pesticides (Bergeson, 2010). Compared to conventional bulk pesticides, plant-derived nanopesticides have proven to be more effective in reducing pest populations and plant infestation levels due to their controlled and sustained release of toxicants.

Among these, polymer-based nanoformulations have been particularly successful in commercial applications, especially for encapsulating neem-derived products, such as seed kernel oil; extracts from seeds, leaves, or bark gum; and azadirachtin. These nanoformulations enhance the efficacy of the active ingredients and improve their environmental stability. However, significant concerns remain regarding the potential risks to environmental and human health, as well as the impact on non-target organisms.

5.2 CONCLUSION

Pests become the utmost threat for the whole agriculture and processing industry in this era of growing food demand. Use of chemical pesticides is the ultimate boon for us in managing damaging pest population. But these are causing a harmful impact on the environment and enters into the food chain. Challenges remain in the dissemination of newly discovered techniques and information to resource-poor farmers in developing countries, and the rapid expansion of social media by mobile phones and other forms of communication network will provide new opportunities. New technology creates more resistant crop, and sustainable pest control measures and improved food processing will require improved understanding of the elements driving pests and pathogens.

REFERENCES

Ali Habib., and Xiao, Y. (Eds.). (2025). Genome Editing for Pest Management (1st ed.). CRC Press. https://doi.org/10.1201/9781003478706

Araujo, R. N., Santos, A., Pinto, F. S., Gontijo, N. F., Lehane, M. J., and Pereira, M. H. 2006. RNA interference of the salivary gland nitrophorin 2 in the triatomine bug *Rhodnius prolixus* (Hemiptera: Reduviidae) by dsRNA ingestion or injection. *Insect Biochemistry and Molecular Biology* 36(9): 683–693.

Atta, B., Rizwan, M., Sabir, A. M., Gogi, M. D., and Ali, K. 2020. Damage potential of Tribolium castaneum (Herbst) (Coleoptera: Tenebrionidae) on wheat grains stored in hermetic and non-hermetic storage bags. *International Journal of Tropical Insect Science* 40: 27–37. https://doi.org/10.1007/s42690-019-00047-0

Babu, R. M., Sajeena, A., Seetharaman, K., and Reddy, M. S. 2003. Advances in genetically engineered (transgenic) plants in pest management – an over view. *Crop Protection* 22(9): 1071–1086.

Baker, T. C., and Heath, J. J. 2005. Pheromones – function and use in insect control. In L. I. Gilbert, K. Iatro, and S. S. Gill (Eds.), *Molecular insect science* (Vol. 6, pp. 407–460). London: Elsevier/Academic Press.

Bakri, A., Heather, N., Hendrichs, J., and Ferris, I. 2005. Fifty years of radiation biology in entomology: Lessons learned from IDIDAS. *Annals of the Entomological Society of America* 98(1): 1–12.

Bale, J. S., Van Lenteren, J. C., and Bigler, F. 2008. Biological control and sustainable food production. *Philosophical Transactions of the Royal Society B: Biological Sciences* 363(1492): 761–776.

Baum, J. A., Bogaert, T., Clinton, W., Heck, G. R., Feldmann, P., Ilagan, O., Johnson, S., Plaetinck, G., Munyikwa, T., Pleau, M., and Vaughn, T. 2007. Control of coleopteran insect pests through RNA interference. *Nature Biotechnology* 25(11): 1322–1326.

Bekele, J., and Hassanali, A. 2001. Blend effects in the toxicity of the essential oil constituents of *Ocimum kilimandscharicum* and *Ocimum kenyense* (Labiateae) on two post-harvest insect pests. *Phytochemistry* 57(3): 385–391.

Benzi, V., Stefanazzi, N., Murray, A. P., Werdin González, J. O., and Ferrero, A. 2014. Composition, repellent, and insecticidal activities of two South American plants against the stored grain pests *Tribolium castaneum* and *Tribolium confusum* (Coleoptera: Tenebrionidae). *International Scholarly Research Notices* 2014(1): 175827.

Bergeson, L. L. 2010. Nanosilver: US EPA's pesticide office considers how best to proceed. *Environmental Quality Management* 19(3). https://doi.org/10.1002/tqem.20255

Bett, P. K., Deng, A. L., Ogendo, J. O., Kariuki, S. T., Kamatenesi-Mugisha, M., Mihale, J. M., and Torto, B. 2016. Chemical composition of *Cupressus lusitanica* and *Eucalyptus saligna* leaf essential oils and bioactivity against major insect pests of stored food grains. *Industrial Crops and Products* 82: 51–62.

Boettcher, M., and McManus, M. T. 2015. Choosing the right tool for the job: RNAi, TALEN, or CRISPR. *Molecular Cell* 58(4): 575–585.

Boopathy, B., Rajan, A., & Radhakrishnan, M. 2022. Ozone: An alternative fumigant in controlling the stored product insects and pests: A status report. *Ozone: Science & Engineering* 44(1): 79–95. https://doi.org/10.1080/01919512.2021.1933899

Borges, M., Moraes, M. C. B., Peixoto, M. F., Pires, C. S. S., Sujii, E. R., and Laumann, R. A. 2011. Monitoring the Neotropical brown stink bug *Euschistus heros* (F.) (Hemiptera: Pentatomidae) with pheromone-baited traps in soybean fields. *Journal of Applied Entomology* 135(1–2): 68–80.

Boukroufa, M., Boutekedjiret, C., Petigny, L., Rakotomanomana, N., and Chemat, F. 2015. Bio-refinery of orange peels waste: A new concept based on integrated green and solvent free extraction processes using ultrasound and microwave techniques to obtain essential oil, polyphenols and pectin. *Ultrasonics Sonochemistry* 24: 72–79.

Bourtzis, K., and Vreysen, M. J. 2021. Sterile insect technique (SIT) and its applications. *Insects* 12(7): 638.

Bouwmeester, H., Dekkers, S., Noordam, M. Y., Hagens, W. I., Bulder, A. S., De Heer, C., and Sips, A. J. 2009. Review of health safety aspects of nanotechnologies in food production. *Regulatory Toxicology and Pharmacology* 53(1): 52–62.

Bradbury, J. W., and Vehrencamp, S. L. 2011. *Principles of animal communication* (2nd ed.). Sinauer Associates. Oxford University Press.

Brar, D. S., and Khush, G. S. 1993. Application of biotechnology in integrated pest management. *Journal of Insect Science* 6(1): 7–14.

Brito, V. D., Achimón, F., Pizzolitto, R. P., Ramírez Sánchez, A., Gómez Torres, E. A., Zygadlo, J. A., and Zunino, M. P. 2021. An alternative to reduce the use of the synthetic insecticide against the maize weevil *Sitophilus zeamais* through the synergistic action of Pimenta racemosa and Citrus sinensis essential oils with chlorpyrifos. *Journal of Pest Science* 94: 409–421.

Brown, J. K. 2000. Molecular markers for the identification and global tracking of whitefly vector – Begomovirus complexes. *Virus Research* 71(1–2): 233–260.

Brydegaard, M., Guan, Z., Wellenreuther, M., and Svanberg, S. 2009. Insect monitoring with fluorescence lidar techniques: Feasibility study. *Applied Optics* 48(30): 5668–5677. https://doi.org/10.1364/AO.48.005668

Casida, J., and Quistad, G. 1998. Golden age of insecticide research: Past, present, or future? *Annual Review of Entomology* 43: 1–16.

Cecchi, G., and Pantani, L. 1994. *International geoscience and remote sensing symposium* (vol. 1–4, p. 97).

Chapman, J. W., Reynolds, D. R., and Smith, A. D. 2003. Vertical-looking radar: A new tool for monitoring high-altitude insect migration. *Bioscience* 53: 503.

Chen, J., Zhang, D., Yao, Q., Zhang, J., Dong, X., Tian, H., Chen, J., and Zhang, W. 2010. Feeding-based RNA interference of a trehalose phosphate synthase gene in the brown planthopper, *Nilaparvata lugens. Insect Molecular Biology* 19(6): 777–786.

Chen, X., Tian, H., Zou, L., Tang, B., Hu, J., and Zhang, W. 2008. Disruption of *Spodoptera exigua* larval development by silencing chitin synthase gene A with RNA interference. *Bulletin of Entomological Research* 98(6): 613–619.

Chiozza, M. V., O'Neal, M. E., and MacIntosh, G. C. 2010. Constitutive and induced differential accumulation of amino acid in leaves of susceptible and resistant soybean plants in response to the Soybean Aphid (Hemiptera: Aphididae). *Environmental Entomology* 39: 856–864.

Cook, R. J., and Baker, K. F. 1983. *The nature and practice of biological control of plant pathogens*. St. Paul, MN: APS Press.

Cook, S. M., Khan, Z. R., Pickett, J. A. 2007. The use of push-pull strategies in integrated pest management. *Annual Review of Entomology* 52: 375–400.

Copping, L. G., and Menn, J. J. 2000. Biopesticides: A review of their action, applications and efficacy. *Pest Management Science* 56(8): 651–676. https://doi.org/10.1002/1526-4998(200008)56:8<651::AID-PS201>3.0.CO;2-U

Costa, J. M., and Marra, F. 2024. Advances in food processing through radio frequency technology: Applications in pest control, microbial and enzymatic inactivation. *Food Engineering Reviews* 1–19. https://doi.org/10.1007/s12393-024-09372-8

Deb, M., and Kumar, D. 2020. Bioactivity and efficacy of essential oils extracted from *Artemisia annua* against *Tribolium casteneum* (Herbst. 1797)(Coleoptera: Tenebrionidae): An eco-friendly approach. *Ecotoxicology and Environmental Safety* 189: 109988.

Dong, Y., and Friedrich, M. 2005. Nymphal RNAi: Systemic RNAi mediated gene knockdown in juvenile grasshopper. *BMC Biotechnology* 5: 1–7.

Durden, K., Sellars, S., Cowell, B., Brown, J. J., and Pszczolkowski, M. A. 2011. Artemisia annua extracts, artemisinin and 1, 8-cineole, prevent fruit infestation by a major, cosmopolitan pest of apples. *Pharmaceutical Biology* 49(6): 563–568.

Dwivedi, M., Shadab, M. H., and Santosh, V. R. 2020. Insect pest detection, migration and monitoring using radar and LiDAR systems. In *Innovative pest management approaches for the 21st century: Harnessing automated unmanned technologies* (pp. 61–76). Springer.

Eilenberg, J., Hajek, A., and Lomer, C. 2001. Suggestions for unifying the terminology in biological control. *Biocontrol* 46: 387–400.

El-Sayed, A. M., Suckling, D. M., Byers, J. A., Jang, E. B., and Wearing, C. H. 2009. Potential of "lure and kill" in long-term pest management and eradication of invasive species. *Journal of Economic Entomology* 102(3): 815–835.

Espo, E., Eyidozehi, K., and Ravan, S. 2015. Influence of gamma and ultraviolet irradiation on pest control. *Magnetic Research Report* 3: 319–326.

Esther, M., Sharon, M., Abirami, C. V. K., and Alagusundaram, K. 2014. Grain storage management in India. *Journal of Postharvest Technology* 2: 12–14. Available online at: http://jpht.in/MenuscriptFile/03b17047-fa10-48ed-b0cb-95fb9222798f.pdf

Fang, R., Jiang, C. H., Wang, X. Y., Zhang, H. M., Liu, Z. L., Zhou, L., Du, S. S., and Deng, Z. W. 2010. Insecticidal activity of essential oil of *Carum carvi* fruits from China and its main components against two grain storage insects. *Molecules* 15(12): 9391–9402.

Gila, D. M. M., Martínez, D. B., Martínez, S. S., Marchal, P. C., and García, J. G. 2024. Non-invasive detection of pesticide residues in freshly harvested olives using hyperspectral imaging technology. *Smart Agricultural Technology* 9: 100644. https://doi.org/10.1016/j.atech.2024.100644

Gill, S. S., Cowles, E. A., and Pietrantonio, P. V. 1992. The mode of action of Bacillus thuringiensis endotoxins. *Annual Review of Entomology* 37(1): 615–634.

Gonçalves, G. L. P., Crevelin, E. J., Lira, S. P. D., and Vendramim, J. D. 2019. Effects of Brugmansia suaveolens fractions on *Zabrotes subfasciatus* (Coleoptera: Chrysomelidae: Bruchinae). *Journal of Biopesticides* 12(1): 19–29.

González-Coloma, A., Valencia, F., Martín, N., Hoffmann, J. J., Hutter, L., Marco, J. A., and Reina, M. 2002. Silphinene sesquiterpenes as model insect antifeedants. *Journal of Chemical Ecology* 28:117–129.

Gould, F. 1988. Genetic engineering, integrated pest management and the evolution of pests. *Trends in Biotechnology* 6(4): S15–S18.

Green, B., Welch, K., Panter, K., and Lee, S. 2013. Plant toxins that affect nicotinic acetylcholine receptors: A review. *Chemical Research in Toxicology Journal* 26: 1129–1138.

Guan, Z., Brydegaard, M., Lundin, P., Wellenreuther, M., Runemark, A., Svensson, E. I., and Svanberg, S. 2010. Insect monitoring with fluorescence lidar techniques: Field experiments. *Applied Optics* 49(27): 5133–5142. https://doi.org/10.1364/AO.49.005133

Guru, P. N., Mridula, D., Dukare, A. S., Ghodki, B. M., Paschapur, A. U., Samal, I., Nikhil, R. M, Padala, V. K., Rajashekhar, M., and Subbanna, A. R. N. S. 2022. A comprehensive review on advances in storage pest management: Current scenario and future prospects. *Frontiers in Sustainable Food Systems* 6: 993341. https://doi.org/10.3389/fsufs.2022.993341

Gvakharia, B. O., Bebas, P., Cymborowski, B., and Giebultowicz, J. M. 2003. Disruption of sperm release from insect testes by cytochalasin and β-actin mRNA mediated interference. *Cellular and Molecular Life Sciences CMLS* 60: 1744–1751.

Hallman, G. J. 2011. Phytosanitary applications of irradiation. *Comprehensive Reviews in Food Science and Food Safety* 10(2): 143–151.

Hallman, G. J., Hénon, Y. M., Parker, A. G., and Blackburn, C. M. 2016. Phytosanitary irradiation: An overview. *Florida Entomologist* 1–13.

Hamel, D., Rozman, V., and Liška, A. 2020. Storage of cereals in warehouses with or without pesticides. *Insects* 11(12): 846. https://doi.org/10.3390/insects11120846

Hansen, L. S., Hansen, P., and Jensen, K. V. 2012. Lethal doses of ozone for control of all stages of internal and external feeders in stored products. *Pest Management Science* 68: 1311–1316. https://doi.org/10.1002/ps.3304

He, W., He, H., Wang, F., Wang, S., and Lyu, R. 2021. Non-destructive detection and recognition of pesticide residues on garlic chive (Allium tuberosum) leaves based on short wave infrared hyperspectral imaging and one-dimensional convolutional neural network. *Journal of Food Measurement and Characterization* 15(5): 4497–4507.

Heather, N. W., and Hallman, G. J. 2008. *Pest management and phytosanitary trade barriers*. CABI.

Hollomon, D. W. 2012. Do we have the tools to manage resistance in the future? *Pest Management Science* 68: 149–154.

Hou, L., Liu, Q., and Wang, S. 2019. Efficiency of industrial-scale radio frequency treatments to control Rhyzopertha dominica (Fabricius) in rough, brown, and milled rice. *Biosystems Engineering* 186:246–258. https://doi.org/10.1016/j.biosystemseng.2019.08.009

Indumathi, C., Manoj, D., Loganathan, M., and Shanmugasundaram, S. 2023. Radio frequency disinfestation of Tribolium castaneum (Herbst) in Semolina: An emerging thermal technique. *Journal of Food Process Engineering* 46:e14443. https://doi.org/10.1111/jfpe.14443

Ingabire, C. M., Hakizimana, E., Rulisa, A., Kateera, F., Van Den Borne, B., Muvunyi, C. M., Leon, M, van Vugt, M., Koenraadt, C. J. M., Takken, W., and Alaii, J. 2017. Community-based biological control of malaria mosquitoes using Bacillus thuringiensis var. israelensis (Bti) in Rwanda: Community awareness, acceptance and participation. *Malaria Journal* 16(1): 399.

Isman, M. B. 2008. Botanical insecticides: For richer, for poorer. *Pest Management Science* 64(1): 8–11. https://doi.org/10.1002/ps.1470

Jäkel, T., Promkerd, P., Sitthirath, R., Guedant, P., and Khoprasert, Y. 2019. Biocontrol of rats in an urban environment in Southeast Asia using Sarcocystis singaporensis. *Pest Management Science* 75: 2148–2157.

Jamal, M., Moharramipour, S., Zandi, M., and Negahban, M. 2012. Ovicidal activity of nano-encapsulated essential oil of Carum copticum on diamondback moth Plutella xylostella. *Proceedings of the 1st National Congress of Monitoring and Forecasting in Plant Protection,* February 14–15, 2012, Borujerd, Iran, pp. 128–129.

Jayakrishnan, L., Sudhikumar, A. V., and Aneesh, E. M. 2018. Role of gut inhabitants on vectorial capacity of mosquitoes. *Journal of Vector Borne Diseases* 55(2): 69–78.

Kah, M., Beulke, S., and Tiede, K. 2013. Nanopesticides: State of knowledge, environmental fate, and exposure modeling. *Critical Reviews in Environmental Science and Technology* 43(16): 1823–1867.

Kalia, A. N. 2005. *Text book of industrial pharmacognosy*. New Delhi: CBS Publishers and Distributers.

Kamboj, A., and Saluja, A. 2008. Ageratum conyzoides L.: A review on its phytochemical and pharmacological profile. *International Journal of Green Pharmacy* 2: 59–68.

Klassen, W., Curtis, C. F., and Hendrichs, J. 2021. History of the sterile insect technique. In *Sterile insect technique* (pp. 1–44). CRC Press.

Kotwica, J., Bebas, P., Gvakharia, B. O., and Giebultowicz, J. M. 2009. RNA interference of the period gene affects the rhythm of sperm release in moths. *Journal of Biological Rhythms* 24(1): 25–34.

Kumar, S., and Singh, A. (2015). Biopesticides: present status and the future prospects. *Journal of Fertilizers & Pesticides*, 6, Article No. e129. https://doi.org/10.4172/2471-2728.1000e129

Kumar, K. K., Sridhar, J., Murali-Baskaran, R. K., Senthil-Nathan, S., Kaushal, P., Dara, S. K., and Arthurs, S. 2019. Microbial biopesticides for insect pest management in India: Current status and future prospects. *Journal of Invertebrate Pathology* 165: 74–81.

Lacey, L. A., Grzywacz, D., Shapiro-Ilan, D. I., Frutos, R., Brownbridge, M., and Goettel, M. S. 2015. Insect pathogens as biological control agents: Back to the future. *Journal of Invertebrate Pathology* 132: 1–41.

Lajide, L., Escoubas, P., and Mizutani, J. 1995. Termite antifeedant activity in Xylopia aethiopica. *Phytochemistry* 40: 1105–1112.

Lee, B. H., Annis, P. C., and Choi, W. S. 2004. Fumigant toxicity of essential oils from the Myrtaceae family and 1, 8-cineole against 3 major stored-grain insects. *Journal of Stored Products Research* 40(5): 553–564.

Li, Y., Ma, B., Wang, H., Zhang, Y., Li, Y., and Yu, G. 2018. Detecting different pesticide residues on Hami melon surface using hyperspectral imaging combined with 1D-CNN and information fusion. *Frontiers in Plant Science* 14: 1105601. https://doi.org/10.3389/fpls.2023.1105601

Liang, J. Y., An, Y., Hou, Z. B., Wang, X. D., Zhou, F., Zhang, J., and Wang, J. L. 2022. Acute toxicity of Zanthoxylum bungeanum against two stored product insects and synergistic interactions between two major compounds limonene and linalool. *Journal of Environmental Science and Health B* 57(9): 739–744. https://doi.org/10.1080/03601234.2022.2107370

López, M. D., Jordán, M. J., and Pascual-Villalobos, M. J. 2008. Toxic compounds in essential oils of coriander, caraway and basil active against stored rice pests. *Journal of Stored Products Research* 44(3): 273–278.

Lowry, G. V., Avellan, A., and Gilbertson, L. M. 2019. Opportunities and challenges for nanotechnology in the agri-tech revolution. *Nature Nanotechnology* 14(6): 517–522.

Macana, R. J., Moirangthem, T. T., and Baik, O. D. 2018. 50-ohm RF technology based applicator design and fabrication for disinfestation of insect pests in stored grains. In *ASABE 2018 annual international meeting Saskatchewan.* American Society of Agricultural and Biological Engineers. https://doi.org/10.13031/aim.201801011

Maffei, M. E., Gertsch, J., and Appendino, G. 2011. Plant volatiles: Production, function and pharmacology. *Natural Product Reports* 28(8): 1359–1380.

Magalhães, D. M., Borges, M., Laumann, R. A., Woodcock, C. M., Pickett, J. A., Birkett, M. A., and Blassioli-Moraes, M. C. 2016. Influence of two acyclic homoterpenes (tetranorterpenes) on the foraging behavior of Anthonomus grandis Boh. *Journal of Chemical Ecology* 42: 305–313.

Mao, Y. B., Cai, W. J., Wang, J. W., Hong, G. J., Tao, X. Y., Wang, L. J., Huang, Y. P., and Chen, X. Y. 2007. Silencing a cotton bollworm P450 monooxygenase gene by plant-mediated RNAi impairs larval tolerance of gossypol. *Nature Biotechnology* 25: 1307–1313.

Marchant, Y. Y., and Cooper, G. K. 1987. Structure and function relationships in polyacetylene photoactivity. In *Light-activated pesticides* (Vol. 339, pp. 241–254). Washington, DC: American Chemical Society.

Marrone, P. G. 2014. The market and potential for biopesticides. In A. D. Gross, J. R. Coats, S. O. Duke, and J. N. Seiber (Eds.), *Biopesticides: State of the art and future opportunities* (pp. 245–258). Washington, DC: American Chemical Society.

Mason, L. J., and Strait, C. A. 2019. Stored product integrated pest management with extreme temperatures. In *Temperature sensitivity in insects and application in integrated pest management* (pp. 141–177). CRC Press.

Mbata, G. N., and Phillips, T. W. 2001. Effects of temperature and exposure time on mortality of stored-product insects exposed to low pressure. *Journal of Economic Entomology* 94: 1302–1307. https://doi.org/10.1603/0022-0493-94.5.1302

McKibben, G. H., Smith, J. W., McGovern, W. L. 1990. Design of an attract-and-kill device for the boll weevil (Coleoptera: Curculionidae). *Journal of Entomological Science* 25: 581–586.

Mei, L., Guan, Z. G., Zhou, H. J., Lv, J., Zhu, Z. R., Cheng, J. A., Chen, F. J., Lofstedt, C., and Somesfalean, G. 2012. Agricultural pest monitoring using fluorescence lidar techniques: Feasibility study. *Applied Physics B* 106: 733–740.

Michereff, M. F. F., Borges, M., Aquino, M. F. S., Laumann, R. A., Gomes, A. M., and Blassioli-Moraes, M. C. 2016. The influence of volatile semiochemicals from stink bug eggs and oviposition-damaged plants on the foraging behaviour of the egg parasitoid Telenomus podisi. *Bulletin of Entomological Research* 106(5): 663–671.

Miller, D. R. 2007. Limonene: Attractant kairomone for white pine cone beetles (Coleoptera: Scolytidae) in an eastern white pine seed orchard in western North Carolina. *Journal of Economic Entomology* 100: 815–822.

Mishra, B. B., Tripathi, S. P., and Tripathi, C. P. M. 2011. Contact toxicity of Essential oil of Citrus reticulata fruits peels against stored grain pests *Sitophilus oryzae* (Linnaeus) and *Tribolium castaneum* (Herbst). *World Journal of Zoology* 6(3): 307–311.

Motazedian, N., Aleosfoor, M., Davoodi, A., and Bandani, A. R. 2014. Insecticidal activity of five medicinal plant essential oils against the cabbage aphid, *Brevicoryne brassicae*. *Journal of Crop Protection* 3(2): 137–146.

Nenaah, G. E., and Ibrahim, S. I. 2011. Chemical composition and the insecticidal activity of certain plants applied as powders and essential oils against two stored-products coleopteran beetles. *Journal of Pest Science* 84: 393–402.

Nicolopoulou-Stamati, P., Maipas, S., Kotampasi, C., Stamatis, P., and Hens, L. 2016. Chemical pesticides and human health: The urgent need for a new concept in agriculture. *Frontiers in Public Health* 4: 148. https://doi.org/10.3389/fpubh.2016. 00148

Nordlund, D. A., and Lewis, W. J. 1976. Terminology of chemical releasing stimuli in intraspecific and interspecific interactions. *Journal of Chemical Ecology* 2: 211–220.

Obeng-Ofori, D., and Reichmuth, C. H. 1999. Plant oils as potentiation agents of monoterpenes for protection of stored grains against damage by stored product beetle pests. *International Journal of Pest Management* 45(2):155–159.

Olmedo, R. H., Lucini, E., Herrera, J. M., Zunino, M. P., Pizzolitto, R. P., Dambolena, J. S., and Zygadlo, J. A. 2015. Fumigant toxicity of Tagetes filifolia Lag essential oil and its main compounds against *Tribolium castaneum* Herbs and its relation to lipid peroxidation. *Agriscientia* 32(2): 113–121.

Olson, S. 2015. An analysis of the biopesticide market now and where it is going. *Outlooks Pest Management* 26(5): 203–206.

Owolade, O. F., Ogunleti, D. O., and Adenekan, M. O. 2008. Titanium dioxide affects biochemical parameters in rats. *International Journal of Environmental Research and Public Health* 5(4): 278–284.

Painter, R. H. 1951. *Insect resistance in crop plants*. Lawrence: The University Press of Kansas.

Pal, G. K., Kumar, B., and Shahi, S. K. 2013. Antifungal activity of some common weed extracts against phytopathogenic fungi Alternaria spp. *International Journal of Universal Pharmacy and Life Sciences* 3(2): 6–14.

Pandiselvam, R., Kaavya, R., Jayanath, Y., Veenuttranon, K., Lueprasitsakul, P., Divya, V., and Ramesh, S. V. 2020. Ozone as a novel emerging technology for the dissipation of pesticide residues in foods – a review. *Trends in Food Science & Technology* 97: 38–54. https://doi.org/10.1016/j.tifs.2019.12.017

Pandiselvam, R., Thirupathi, V., Mohan, S., Vennila, P., Uma, D., Shahir, S., and Anandakumar, S. 2019. Gaseous ozone: A potent pest management strategy to control Callosobruchusmaculatus (Coleoptera: Bruchidae) infesting green gram. *Journal of Applied Entomology* 143(4): 451–459. https://doi.org/10.1111/jen.12618

Pare, J. M., and Hobman, T. C. 2007. Dicer: Structure, function and role in RNA-dependent gene-silencing pathways In J. Polaina and A. P. MacCabe (Eds.), *Industrial enzymes* (pp. 421–438). Berlin: Springer.

Park, I. K., Lee, S. G., Choi, D. H., Park, J. D., and Ahn, Y. J. 2003. Insecticidal activities of constituents identified in the essential oil from leaves of *Chamaecyparis obtusa* against *Callosobruchus chinensis* (L.) and *Sitophilus oryzae* (L.). *Journal of Stored Products Research* 39(4): 375–384.

Park, I. K., and Shin, S. C. 2005. Fumigant activity of plant essential oils and components from garlic (*Allium sativum*) and clove bud (*Eugenia caryophyllata*) oils against the Japanese termite (*Reticulitermes speratus* Kolbe). *Journal of Agricultural and Food Chemistry* 53: 4388–4392.

Pavela, R. 2011. Insecticidal and repellent activity of selected essential oils against of the pollen beetle, *Meligethes aeneus* (Fabricius) adults. *Industrial Crops and Products* 34: 888–892.

Pertot, I., Giovannini, O., Benanchi, M., Caf, T., Rossi, V., and Mugnai, L. 2017. Combining biocontrol agents with different mechanisms of action in a strategy to control Botrytis cinerea on grapevine. *Crop Protection* 97: 85–93.

Pimentel, D. 2009. Environmental and economic costs of the application of pesticides primarily in the United States. *Environment, Development and Sustainability* 11(1): 55–74.

Plata-Rueda, A., Campos, J. M., da Silva Rolim, G., Martínez, L. C., Dos Santos, M. H., Fernandes, F. L., José Eduardo Serrão, J. E., and Zanuncio, J. C. 2018. Terpenoid constituents of cinnamon and clove essential oils cause toxic effects and behavior repellency response on granary weevil, Sitophilus granarius. *Ecotoxicology and Environmental Safety* 156: 263–270. https://doi.org/10.1016/j.ecoenv.2018.03.033

Price, P. W. 1997. *Insect ecology*. New York: Wiley, pp 73–138.

Qiao, J., Zou, X., Lai, D., Yan, Y., Wang, Q., Li, W., Deng, S., Xu, H.-H., and Gu, H. 2014. Azadirachtin blocks the calcium channel and modulates the cholinergic miniature synaptic current in the Central Nervous System of Drosophila. *Pest Management Science* 70: 1041–1047.

Rajamani, M., and Negi, A. 2021. Biopesticides for pest management. In V. Venkatramanan, S. Shah and R. Prasad (Eds.), *Sustainable bioeconomy: Pathways to sustainable development goals* (pp. 239–266). Springer, Singapore.

Rajashekar, Y., Raghavendra, A., and Bakthavatsalam, N. 2014. Acetylcholinesterase inhibition by biofumigant (coumaran) from leaves of Lantana camara in stored grain and household insect pests. *BioMed Research* 2014: 187019.

Rajashekhar, M., Dharavath, V., Savani, A. K., Raidu, L. C., and Kalia, V. 2016. Endosymbionts and paratransgenesis approach for pest management. *Journal of Progressive Research in Social Sciences* 11(V): 2822–2826.

Rajendran, S., and Sriranjini, V. 2008. Plant products as fumigants for stored-product insect control. *Journal of Stored Products Research* 44: 126–135. https://doi.org/10.1016/j.jspr.2007.08.003

Ram, B. G., Oduor, P., Igathinathane, C., Howatt, K., and Sun, X. 2024. A systematic review of hyperspectral imaging in precision agriculture: Analysis of its current state and future prospects. *Computers and Electronics in Agriculture* 222: 109037. https://doi.org/10.1016/j.compag.2024.109037

Ratcliffe, N. A., Furtado Pacheco, J. P., Dyson, P., Castro, H. C., Gonzalez, M. S., Azambuja, P., and Mello, C. B. 2022. Overview of paratransgenesis as a strategy to control pathogen transmission by insect vectors. *Parasites & Vectors* 15(1): 112.

Redfern, A. D., Colley, S. M., Beveridge, D. J., Ikeda, N., Epis, M. R., Li, X., Foulds, C. E., Stuart, L. M., Barker, A., Russell, V. J., Ramsay, K., Kobelke, S. J., Li, X., Hatchell, E. C., Payne, C., Giles, K. M., Messineo, A., Gatignol, A., Lanz, R. B., O'Malley, B. W., and Leedman, P. J. 2013. RNA induced silencing complex (RISC) proteins PACT, TRBP, and Dicer are SRA binding nuclear receptor coregulators.

Proceedings of the National Academy of Sciences of the United States of America 110: 6536–6541. https://doi.org/10.1073/pnas.1301620110.

Reynoso, M. M. N., Lucia, A., Zerba, E. N., and Alzogaray, R. A. 2019. The octopamine receptor is a possible target for eugenol-induced hyperactivity in the blood-sucking bug Triatoma infestans (Hemiptera: Reduviidae). *Journal of Medical Entomology* 57: 627–630. https://doi.org/10.1093/jme/tjz183

Riaz, T., Shakoori, F. R., and Ali, S. S. 2014. Effect of temperature on the development, survival, fecundity and longevity of stored grain pest, Trogoderma granarium. *Pakistan Journal of Zoology* 46(6): 1485–1489.

Riley, K. J., Yario, T. A., and Steitz, J. A. 2012. Association of Argonaute proteins and microRNAs can occur after cell lysis. *RNA* 18(9): 1581–1585.

Rudman, P., Da Costa, E. W. B., Gay, F. J., and Wetherly, A. H. 1958. Relationship of tectoquinone to durability in tectonagrandis. *Nature* 181: 721–722.

Scolari, F., Casiraghi, M., and Bonizzoni, M. 2019. Aedes spp. and their microbiota: A review. *Frontiers in Microbiology* 10: 2036.

Shakiba, S., Astete, C. E., Paudel, S., Sabliov, C. M., Rodrigues, D. F., and Louie, S. M. 2020. Emerging investigator series: Polymeric nanocarriers for agricultural applications: Synthesis, characterization, and environmental and biological interactions. *Environmental Science: Nano* 7(1): 37–67.

Shaw, J. A., Seldomridge, N. L., Dunkle, D. L., Nugent, P. W., Spangler, L. H., Bromenshenk, J. J., Henderson, C. B., Churnside, J. H., and Wilson, J. J. 2005. Polarization lidar measurements of honey bees in flight for locating land mines. *Optics Express* 13(15): 5853–5863.

Shivanandappa, T., and Rajashekar, Y. 2014. Mode of action of plant-derived natural insecticides. In D. Singh (Ed.), *Advances in plant biopesticides* (1st ed., pp. 323–345). New Delhi, India: Springer.

Siegel, J. P. 2001. The mammalian safety of Bacillus thuringiensis-based insecticides. *Journal of Invertebrate Pathology* 77(1): 13–21.

Singh, D., Siddiqui, M. S., and Sharma, S. 1989. Reproduction retardant and fumigant properties in essential oils against rice weevil (Coleoptera: Curculionidae) in stored wheat. *Journal of Economic Entomology* 82(3): 727–732.

Sinha, R. N., and Watters, F. L. 1985. *Insect pests of flour mills, grain elevators and feed mills and their control*. Ottawa, Canada: Canadian Govt. Publ. Centre, 290 pp.

Sivaranjani, S., Pandiselvam, R., Priyadarshini, S., Arjunan, K., and Chandrasekaran, M. 2021. Recent advances in applications of ozone in the cereal processing chain. *Cereal Chemistry* 98(3): 430–444. https://doi.org/10.1002/cche.10358

Smith, C. M. 2005. *Plant resistance to arthropods: Molecular and conventional approaches*. Springer.

Smith, C. M., Clement, S. L. 2012. Molecular bases of plant resistance to arthropods. *Annual Review of Entomology* 57: 309–328.

Souto, A. L., Sylvestre, M., Tölke, E. D., Tavares, J. F., Barbosa-Filho, J. M., and Cebrián-Torrejón, G. 2021. Plant-derived pesticides as an alternative to pest management and sustainable agricultural production: Prospects, applications and challenges. *Molecules* 26(16): 4835.

Stenberg, J. A. 2017. A conceptual framework for integrated pest management. *Trends in Plant Science* 22: 759–769.

Stenberg, J. A., Sundh, I., Becher, P. G., Björkman, C., Dubey, M., Egan, P. A., Friberg, H., Gil, J. F., Jensen, D. F., Jonsson, M., Karlsson, M., Khalil, S., Ninkovic, V., Rehermann, G., Vetukuri, R. R., and Viketoft, M. 2021. When is it biological control? A framework of definitions, mechanisms, and classifications. *Journal of Pest Science* 94(3): 665–676.

Stewart, S. D., Adamczyk Jr, J. J., Knighten, K. S., and Davis, F. M. 2001. Impact of Bt cottons expressing one or two insecticidal proteins of Bacillus thuringiensis Berliner on growth and survival of noctuid (Lepidoptera) larvae. *Journal of Economic Entomology* 94(3): 752–760.

Stout, M. J. 2013. Plant-mediated interactions between herbivores and natural enemies: Mechanisms, applications, and challenges. *Journal of Chemical Ecology* 39(6): 628–640. https://doi.org/10.1007/s10886-013-0305-0

Stringer, L. D., El-Sayed, A. M., Cole, L. M., Manning, L. A. M., and Suckling, D. M. 2008. Floral attractants for the female soybean looper, Thysanoplusia orichalcea (Lepidoptera: Noctuidae). *Pest Manag. Sci.* 64(12): 1218–1221. https://doi.org/10.1002/ps.1621

Sun, J., Hu, Y., Zou, Y., Geng, J., Wu, Y., Fan, R., and Kang, Z. 2022. Identification of pesticide residues on black tea by fluorescence hyperspectral technology combined with machine learning. *Food Science and Technology* 42: e55822.

Tilmann, P. G., and Cottrel, T. 2017. Use of pheromones for monitoring phytophagous stink bugs (Hemiptera: Pentatomidae) In A. Colk and M. Borges (Eds.), *Stink bugs: Biorational control based on communication* (pp. 210–225). Boca Raton: CRC Press.

Tognon, R., Sant'ana, J., and Jahnke, S. M. 2014 Influence of original host on chemotaxic behaviour and parasitism in Telenomus podisi Ashmead (Hymenoptera: Platygastridae). *Bulletin of Entomological Research* 104: 781–787.

Treaty, N. A. 2010. *Laser based stand-off detection of biological agents*. TR-SET-098. Brussels: NATO Research and Technology Organisation, p. 82. ISBN: 978-92-837-0086-9

Trivedi, A., Nayak, N., and Kumar, J. 2017. Fumigant toxicity study of different essential oils against stored grain pest Callosobruchus chinensis. *Journal of Pharmacognosy and Phytochemistry* 6(4): 1708–1711.

Turner, C. T., Davy, M. W., MacDiarmid, R. M., Plummer, K. M., Birch, N. P., and Newcomb, R. D. 2006. RNA interference in the light brown apple moth, *Epiphyas postvittana* (Walker) induced by double-stranded RNA feeding. *Insect Molecular Biology* 15(3): 383–391.

van Eck, L., Schultz, T., Leach, J. E., Scofield, S. R., Peairs, F. B., Botha, A., and Lapitan, N. L. V. 2010. Virus-induced gene silencing of WRKY53 and an inducible phenylalanine ammonia-lyase in wheat reduces aphid resistance. *Plant Biotechnology Journal* 8: 1023–1032. https://doi.org/10.1111/j.1467-7652.2010.00539.x

Vanegas, F., Bratanov, D., Powell, K., Weiss, J., and Gonzalez, F. 2018. A novel methodology for improving plant pest surveillance in vineyards and crops using UAV-based hyperspectral and spatial data. *Sensors* 18(1): 260. https://doi.org/10.3390/s18010260

Vreysen, M. J. B., Seck, M. T., Sall, B., Mbaye, A. G., Bassene, M., Fall, A. G., Lo, M., and Bouyer, J. 2021. Area-wide integrated management of a Glossina palpalis gambiensis population from the Niayes area of Senegal: A review of operational research in support of a phased conditional approach. In *Area-wide integrated pest management: Development and field application* (pp. 275–303). Boca Raton, FL: CRC Press.

Walia, S., Saha, S., and Rana, V. 2014. Phytochemical pesticides. In D. Singh (Ed.), *Advances in plant biopesticides* (pp. 295–322). New Delhi, India: Springer.

Wang, R., Fang, Y., Che, W., Zhang, Q., Wang, J., and Luo, C. 2022. The Toxicity, Sublethal Effects, and Biochemical

Mechanism of β-Asarone, a Potential Plant-Derived Insecticide, against *Bemisia tabaci*. *International Journal of Molecular Sciences* 23(18): 10462.

Weber, D. C., Khrimian, A., Blassioli-Moraes, M. C., and Millar, J. G. 2018. Semiochemistry of Pentatomoidea. In *Invasive stink bugs and related species (Pentatomoidea)* (pp. 677–726). CRC Press.

Wermelinger, B. 2004. Ecology and management of the spruce bark beetle Ips typographus—a review of recent research. *Forest Ecology and Management* 202(1–3): 67–82. https://doi.org/10.1016/j.foreco.2004.07.018

Wheeler, C. A., and Cardé, R. T. 2014. Following in their footprints: Cuticular hydrocarbons as overwinter ing aggregation site markers in Hippodamia convergens. *Journal of Chemical Ecology* 40: 418–428.

Wilde, G. 2002. Arthopod host plant resistant crops. In D. Pimentel (Ed.), *Encyclopedia of pest management* (pp. 33–35). Boca Raton, FL: CRC Press.

Wilke, A. B. B., and Marrelli, M. T. 2015. Paratransgenesis: A promising new strategy for mosquito vector control. *Parasites & Vectors* 8: 1–9.

Wilson, R. C., and Doudna, J. A. 2013. Molecular mechanisms of RNA interference. *Annual Review of Biophysics* 42: 217–239.

Witzgall, P., Kirsch, P., and Cork, A. 2010. Sex pheromones and their impact on pest management. *Journal of Chemical Ecology* 36: 80–100.

Yadav, J., Jasrotia, P., Kashyap, P. L., Bhardwaj, A. K., Kumar, S., Singh, M., and Singh, G. P. 2021. Nanopesticides: Current status and scope for their application in agriculture. *Plant Protection Science* 58(1): 1–17.

Yao, Y., Cai, W., Yang, C., Xue, D., and Huang, Y. 2008. Isolation and characterization of insecticidal activity of (Z)-asarone from *Acorus calamus* L. *Insect Science* 15(3): 229–236.

Yeom, H. J., Kang, J. S., Kim, G. H., and Park, I. K. 2012. Insecticidal and acetylcholine esterase inhibition activity of Apiaceae plant essential oils and their constituents against adults of German cockroach (Blattella germanica). *Journal of Agricultural and Food Chemistry* 60(29): 7194–7203. https://doi.org/10.1021/jf302009w

Zheng, S.-J., and Dicke, M. 2008. Ecological genomics of plant insect interactions: From gene to community. *Plant Physiology* 146: 812–817.

6 Integrated Pest Management in Industrial Forest Plantation

Focus on Acacia crassicarpa *and* Eucalyptus pellita

Toni Arya Dharma[1]* *and Felix Bast*[2]
[1]Department of Botany, Central University of Punjab, Bathinda, Punjab 151401, India (https://orcid.org/0009-0009-0318-5729)
[2]Department of Botany, Central University of Punjab, Bathinda, Punjab 151401, India https://orcid.org/0000-0002-5090-0109)
*Corresponding author

6.1 INTRODUCTION

Industrial forest plantations are expected to reach around 379 million hectares by 2070 as a result of rising market demand and ongoing plantation expansion (Nepal et al. 2019). Pest attacks are one of the challenges or impediments encountered in forest plantations. Since the seedling stage, *Eucalyptus pellita* and *Acacia crassicarpa* have been subject to pest assaults. This attack has been documented for a long time by a variety of insect pests and insect groups, including Lepidoptera, Hemiptera, Thysanoptera, and others (Kkadan et al. 2020). In reality, numerous new insect pest attacks, such as those by *Rhytiphora bankii* (Fabricius) (Coleoptera: Cerambycidae) and *Myllocerus scapularis* Roelofs (Coleoptera: Curculionidae), have been documented in forest plantations in various parts of Indonesia. This pest infestation, which varies in severity, takes place in *Acacia crassicarpa* plantations in North Sumatra and Riau, Indonesia. Attacks happen in nurseries on plants as young as two months, and the obvious signs of an assault are damage to the leaves from herbivore insect attacks (Sirait et al. 2020).

The issue of herbivorous insects can be solved in a number of ways, including chemical and non-chemical approaches. Despite the undeniable need for prompt action to eliminate insect pests, the forest plantation business typically employs chemical control. However, numerous publications also document the effective application of integrated pest management (IPM), entomopathogens, bioinsecticides, resistant cultivars, natural enemies as parasitoids and predators, and agricultural modification techniques. Because of the urgency of IPM, the environment can be engineered to support a variety of beneficial insects that can regulate the presence of herbivore insects. According to Dharma et al. (2018), organic farming practices can boost the number of parasitoids and predators in farmed crops.

In an effort to control insect pests, it is necessary to know the economic threshold number of each type of pest insect at the point where pest attacks cannot be tolerated any longer, which can have fatal consequences if identification and control measures are too late. Control using chemicals is indeed the easiest to use, but there are many side effects produced, such as changes in the ecosystem, insect pest resistance, resuscitation, and killing beneficial insects. So preventive measures are also the most important part of efforts to manage the presence of insect pests in industrial forest plantations. The plants *A. crassicarpa* and *E. pellita*, which are typically grown in industrial forest plantations, will be reviewed in this article along with the kinds of insect pests that are known to attack them. Additionally, a discussion of pest control methods that have been shown to be effective in controlling these two plants will be included, along with a variety of control methods that can be combined with one another to prevent and suppress insect pest attacks while giving priority to environmental and economic factors.

6.2 FOREST PLANTATION CROPS

The *Acacia crassicarpa*, an endemic tree found throughout Australia and Papua Island (Indonesia and Papua New Guinea), is one of the trees used in industrial forest plantations because it is known to have the potential to develop in a variety of ecosystem conditions and is typically cultivated in wet land areas that are rich in organic material but minimal in minerals. It is good for furniture, pulp, and paper because it grows quickly and has strong wood (Hanchor et al. 2016). The strength of wood has an air-dry density of 710 kg per cubic meter and a basic density of 620 kg per cubic meter. *A. crassicarpa* is the best species of wood tree to be used as pulp and paper. Other advantages are the high nitrogen fixation ability and the rhizobium nodulation ability. The seeds of natural populations of *A. crassicarpa* originating from Papua New Guinea and Australia have been distributed in various countries in Asia, Africa, America, and various island countries, such as the Leeward Islands,

DOI: 10.1201/9781003516538-6

Hawaii, Micronesia, and Madagascar, although there are also many reports that not all areas are able to cultivate this tree (Pasieczni and McDonald 2016).

Other than *Acacia crassicarpa*, *Eucalyptus pellita* is a forest plant that is grown on industrial forest plantations. The species most frequently utilized is *Eucalyptus pellita*, which is a raw material used worldwide in the pulp and paper industry. For instance, this species is used in mineral land areas in Indonesia. Because of its rapid growth, *E. pellita* is typically harvested after 7 years; at the moment, it is even harvested at 5 years of age (Muhdi et al. 2019). *E. pellita* is naturally distributed in Papua New Guinea, Australia, and Papua (Indonesia). Generally, *E. pellita* grows at an altitude of 0–800 m, with rainfall of 1,000 to 3,000 mm/year and a dry season of up to 5 months. For productivity, it is known that *E. pellita* will be more productive in areas with high rainfall. It is also known that the productivity of *E. pellita* is between 9.4 and 34.9 m^3/ha/year. Not only in high-rainfall areas but also in low-rainfall (1,000–1,200 mm/year) and lower-temperatures (22°C) regions *E. pellita* is able to grow, provided that many other factors will also influence its productivity, such as soil type, silvicultural practices, and season duration. Each of these factors will exert great influence; for example, a record increase of 100 mm of rainfall in Brazil can influence the biomass of the *E. grandis* × *E. urophylla* clone above ground to 2.3 Mg/ha/year (Hutapea et al. 2023)

6.3 INSECT PESTS OF FOREST PLANTATION

The *Acacia mangium* species was the most common species selected for mineral and wet land types at the start of the industrial plantation forest (HTI) sector. Attacks by pests and diseases are one of the reasons that *E. pellita* and *A. crassicarpa* are beginning to supplant this kind. However, it is currently undeniable that *E. pellita* and *A. crassicarpa* plants are also regularly attacked by insect pests. There have been numerous instances of fresh insect pest infestations on plants used for industrial forests. Sirait et al. (2020) reported pest attacks on *A. crassicarpa* in North Sumatra, Indonesia, and also in Riau, Indonesia. It has been reported that *Myllocerus scapularis* and *Rhytiphora bankii* attack many young plants in nursery and under-6-month-old plants. Previously, pests such as *M. scapularis* and *R. bankii*, known by the common names ash weevil and longhorn beetle, were only reported on soybean and cotton plants in South Asian countries, such as Sri Lanka and Bangladesh (Ali 2008; Thakur et al. 2021). Next, in Table 6.1, several types of insect pests that often attack industrial forest plantations will be presented, as well as several new reports that have been recorded as attacking and causing harm to industrial forest plantations in the species *A. crassicarpa* and *E. pellita*.

6.4 MANAGEMENT OF INSECT PESTS OF INDUSTRIAL FOREST PLANTATIONS

Insect pests on forest plants are managed using both chemical and non-chemical methods (Soylu et al., 2025). In industrial forest plantations, fungicide has been soaked on the seeds that will be planted in the field, and even treatment has been done on the seeds that will be utilized as seedlings. Control tactics include not only responding to an insect

TABLE 6.1
Synthetic Insecticides Used Against Major Insect Pests of *Acacia crassicarpa* and *Eucalyptus pellita*

Crop	Target Insect Pest	Synthetic Insecticide	References
A. crassicarpa	*Tetranychus* sp.	Amitraz	Lumbantobing et al. 2023
		Pyridaben	
		Propargite	
		Sulfur	
	S. exigua; *Scirtothrips* sp.	Spinetoram	Kkadan et al. 2020
	S. exigua; *Scirtothrips* sp.	Chlorpyrifos	
	X. crassiusculus	Dimehipo	Saputra and Aluyah 2019
		Cyantraniliprole + lufenuron	
	Strepsicrates sp.	Imidacloprid	Agustin et al. 2023
		Permetrin + abamectin	
		Dinotefuran	
		Tefluthrin	
		Fonofos	
E. pellita	Various termites	Fipronil	Wilckin et al. 2002
		Disulfoton + triadimenol	
		Acephate	
	Strepsicrates semicanella	Spinetoram	Kkadan et al. 2020
	Bemisia tabaci; *Strepsicrates macropetana*	Lamda sihalotrin	Jufenlin et al. 2023

pest attack but also proactively preventing such attacks on a daily and monthly basis. It is well-known that industrial forest plantations only grow uniform plants, making them extremely vulnerable to insect and pest infestations.

6.4.1 Application of Synthetic Insecticides

Industrial plantation woods are typically where synthetic insecticides are applied. Applications are made at different stages of plant development, beginning with seedlings and plants that are 12 months old. There is a wide range of pest attacks on *A. crassicarpa* and *E. pellita* throughout the nursery stage. Due to international laws governing pesticide applications, low-hazard insecticides are used in industrial plantation forests. Kkadan et al. (2020) conducted research on acacia and eucalyptus and found that the application of spinetoram was very effective in controlling *S. exigua* and thrips, with 100% and 95% effectiveness on each object tested using a spinetoram-based solution (0.12%). The application of spinetoram to *Eucalyptus* also shows high effectiveness against attacks by *Strepsicrates semicanella*. Insecticide applications are also often carried out in various combinations to measure the level of effectiveness on insect pests, as was done by Saputra and Aluyah (2019). They tried to combine thuricide insecticide, which has the active ingredient *Bacillus thuringiensis*, with cassava tape suspension. In the application of insecticides to several insect pests, it was carried out not only using the spraying technique but also using the fogging technique. It was also found that, in some cases, the fogging technique was more effective in reducing pest attacks. Jufenlin et al. (2023) stated that the application of the insecticide Matador 250 EC with the active ingredient Lamda sihalotrin using a spraying technique was able to reduce *B. tabaci* to 44%, while the fogging technique was 56%, with the fogging application on *Strepsicrates macropetana* to 41%, and to 26% on the spray technique. For *A. crassicarpa* and *E. pellita* in particular, Table 6.1 lists several kinds of pesticides commonly employed in industrial forest plantings.

6.4.2 Cultural Control

In industrial forest plantations, proper silvicultural measures have been adopted to prevent insect infestations. In order to prevent insect pest infestations, appropriate silviculture practices have been selected as a comprehensive component of preventative action. The following are some real-world examples that can be taken into consideration.

6.4.2.1 Land Preparation

Land preparation is a common technical practice in forest industry plantings. The general goal of tilling the soil is to loosen the soil mass so that there is enough space for plant roots to grow and thrive. The way the soil is prepared is one factor that can influence the emergence of diseases and insect infestations. In addition to influencing the occurrence of pests and diseases, this process can also suppress weeds. Nuri (2005) asserts that mechanical site preparation is the most effective method of preventing white root rot disease. This is due to the fact that stumps and remaining roots can be removed from the soil by machines.

6.4.2.2 Plant Population Manipulation

The regulation of planting distances for *A. crassicarpa* and *E. pellita* is based on more than just quantitative wood production; research findings indicate that planting distances significantly impact height and diameter growth, but not survival percentage or competition index, in plants that are 12 months old (Purwanti et al. 2022). Plant spacing is also considered a preventive measure to control insect pests. *A. crassicarpa*, which is cultivated in peat areas, has a planting distance of 3 × 2 m, while *E. pellita*, which is generally cultivated in mineral soil, has a planting distance of 3 × 2 m and also has other options, such as 3 × 2.5 m, 3.5 × 2 m, and 3.3 × 2.2 m. Asmaliyah and Rostiwati (2015) stated that a planting distance of 3 × 4 m is an effective planting distance for suppressing the development of the *Cycnotrachelus* sp. beetle pest and the pathogen *Cephaleuros* sp. on land island plants in the third year after planting. This condition is indicated by a decrease in the percentage of attacks by 0.57% and a reduction in the intensity of attacks by beetle pests by 2.48% and pathogenic diseases by 8.44%. In Figure 6.1, we can see an illustration of the planting distance implemented in the field.

6.4.2.3 Supplemental Application of Soil Nutrients

According to Herliyana et al. (2012), healthy soil plays a major impact in plant output as well as a plant's resistance to various insect and disease attacks. To improve soil fertility in industrial plantation forests, organic ingredients, such as ameliorant materials made from factory processing residues, are combined with other fertilizers and nutritional supplements. These substances enhance the soil's biological, chemical, and physical characteristics. Almiron material is usually utilized before *A. crassicarpa* is planted on peatlands in industrial planted forests. Applying the correct kind and quantity of fertilizer, however, requires caution because some fertilizers might make insect attacks worse.

6.4.2.4 Weed Control

Weeds will compete with the primary plants for sunshine, nutrients, water, and room to grow; hence, their existence will be an issue in industrial planted forests in addition to pest infestations (Wibowo 2006). It is commonly known that a number of weeds that resemble the primary crop typically serve as nesting sites for several insect pests (Capinera 2015). In plantations around *A. crassicarpa*, there are generally many other types of acacia that are not cultivated, such as *A. mangium*, which has been studied and can be a place for insect pests to colonize. So by understanding proper weed control, we can suppress the presence of insect pests, and unwanted weeds such as *A. mangium* can also become a diversionary plant from the main crop.

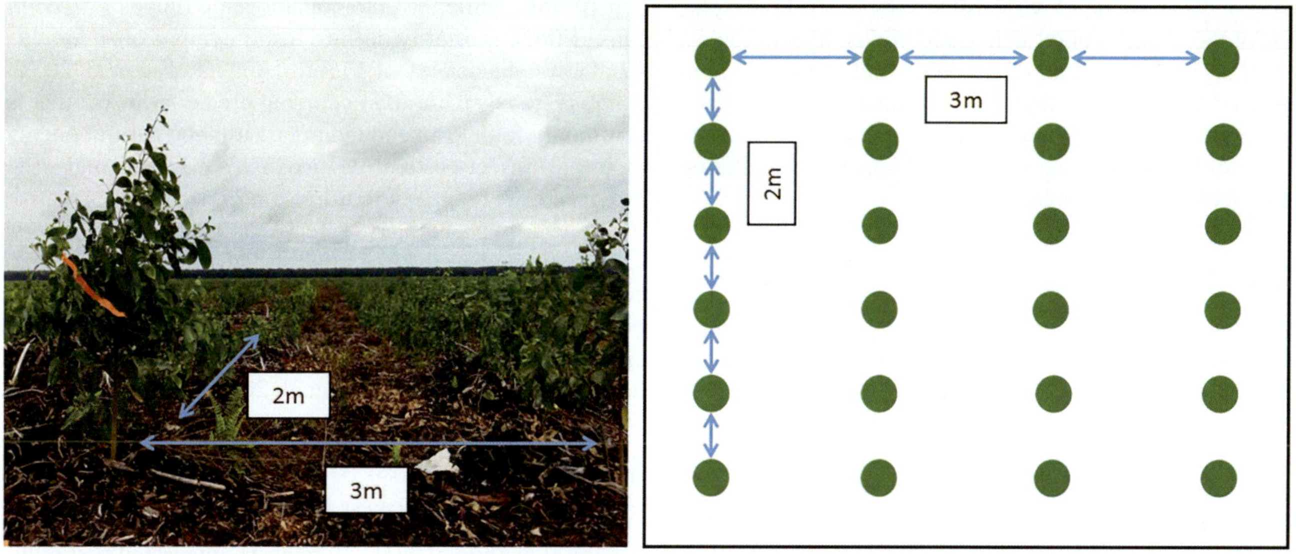

FIGURE 6.1 Planting distance commonly used for *Acacia crassicarpa* plantations, where trees are spaced in a systematic grid with a distance of 3 m between rows and 2 m between trees within a row. This arrangement results in a planting density of 1,666 trees per hectare, optimizing land use while allowing adequate space for tree growth and maintenance activities.

6.4.2.5 Trap Cropping

Insect pests can be managed by using cultural control techniques like trap plants, which are plants that are planted next to or before the main plant. Trap crops can increase the efficacy of pest control by concentrating pests in a single location. It is also possible to deploy biological control agents into trap crops, which act as a refuge for beneficial organisms that gradually trickle down to the main crop (Reddy 2017). Trap crops can perform especially well in reduced production settings when crop diversification and the decrease of synthetic inputs are prioritized over crop output alone.

6.4.2.6 Use of Bioinsecticide

Bioinsecticides come from a variety of natural sources, including pheromones, plant bioactive compounds, and microorganisms, such as bacteria, fungi, viruses, or protozoa (Ali et al., 2019). It is known that plants have a variety of chemical defense systems against insects. Phytochemicals found in plants act as growth regulators, poisons, insect repellents, and feeding deterrents. Small plants' whole bodies, a variety of plants, or parts of larger plants can be used as raw materials to make botanical or bioinsecticides (Şengül Demirak and Canporat 2022). According to Rusandi et al. (2016), mahogany seeds have the ability to regulate *Spodoptera litura* on *A. crassicarpa*; the optimal concentration for regulating *Spodoptera litura* F. is 30 g/L of water, which has the quickest initial death time of 2.75 hours. There are numerous different plants that can be employed as biopesticides, and the fastest time of death is 2 hours and 45 minutes.

6.4.2.7 Use of Insect-Resistant Varieties

One efficient strategy to lessen damage to industrial forest plantations is to grow plant clones that are completely or partially resistant to pest attacks so that numerous forestry firms are actively creating clones that are disease- and pest-resistant (Hutasuhut et al. 2015). A breeding population is created by selecting specific people with traits that align with the goal of breeding in order to produce resistant clones. It is possible to select for individuals with certain advantages from the breeding population. These individuals can then be crossed or hybridized to create new individuals with a combination of traits from both parents (Sunarti 2018).

6.4.3 Mechanical Control

The direct treatment of plants with specific instruments or by hand is known as mechanical control, which is physical pest control. The use of tools to fend against animals and illnesses is a practical use of this strategy. Chemicals like pesticides and medications are not necessary for mechanical methods. They instead make use of instruments like machetes, sticky yellow traps, lamp traps, sickles, handpicks, and plant scissors. Dudley et al. (2012) state that one mechanical engineering method is to treat the seeds, which entails filtering them of any impurities and enclosing them in laminated plastic bags that contain carbon dioxide for two weeks prior to storage. Methyl bromide is also sprayed on the seeds to avoid contamination. Hot water and acid scarification procedures are used to disrupt dormancy and provide further protection against insect pest attacks. Insect larvae isolation in large seed treatments, however, may make these methods less than perfect. You can store seeds in cans, jute bags, or baskets. The use of pheromone traps to regulate *Zeuzera* sp. imago in *E. pellita* is one type of mechanical control that has been implemented in the industry. *Zeuzera* sp. is responsible for broken lower portions of the plant, swollen stems, and broken upper crowns, according to Suheri et al. (2020). The study he carried out produced encouraging findings. The installation of delta traps and *Zeuzera* pheromone pyrine was used to manage pests. Still, the outcomes are insufficient, necessitating comprehensive pest management.

6.4.4 Biological Control

The use of biological control is rapidly growing in popularity as more people recognize how important it is to maintain environmental balance and sustainability. In actuality, natural organisms control about 98% of pest populations regardless of the environment. Chemical pesticides and insecticides are employed much less frequently these days due to their persistent nature, continuous environmental buildup, toxicity to biotic components of life, and possible health concerns to humans (Jaiswal et al. 2022; Yousaf et al., 2024). Numerous entomopathogenic organisms, including bacteria, nematodes, viruses, and fungi, have been employed to manage insect pests (Figure 6.2).

Biological control is the process of managing insect pests by using beneficial insects, such as parasitoids and predators, that are either naturally present in the ecosystem or intentionally distributed. Unwanted insects can also be controlled by pathogens, like viruses and fungi. *Helopeltis* sp., one of the main pests on *A. crassicarpa*, can be controlled by the entomopathogenic fungi *Beauveria bassiana* and *Lecanicillium lecanii*, both of which have a considerable mortality capability but are in distinct stages of insect pest life. Laboratory research revealed that the entomopathogenic fungus *B. bassiana* was successful in causing 100% mortality of *Helopeltis* sp. imago at conidia density of 10^8 and 10^9/mL, while the entomopathogenic fungus *L. lecanii* caused 100% mortality of third instar nymphs of *Helopeltis* sp. at a conidia density of 10^9/mL (Anggarawati et al. 2017). *Sycanus* sp. is a naturally occurring predatory insect that is commonly found in peatland areas. It is the natural enemy of several insect pests of *A. crassicarpa*. Interestingly, the presence of geronggang (*Cratoxylon arborescens*), a native tree species in peat areas that serves as a habitat for *Sycanus* sp., can maximize the potential of *Sycanus* sp. (Sucahyono et al. 2022). *S. litura*, which is a dangerous polyphagous pest in *A. crassicarpa* and *E. pellita* nurseries, has been capable of being controlled by utilizing the nuclear polyhedrosis virus (NPV) (Gafur et al. 2012; Yasin et al. 2020).

6.4.5 Integrated Pest Management in Industrial Forest Plantation

A natural method of controlling pests, integrated pest management (IPM) combines different forms of control, including biological, mechanical, and cultural (Figure 6.3), which seeks to be sustainable and ecologically friendly (He et al., 2025). In industrial forest plantations, IPM has been widely implemented on *A. crassicarpa* and *E. pellita* plantations. In order to effectively manage ambrosia pests on *A. crassicarpa* plants, Saputra and Aluyah (2019) utilized a combination of cassava extract and pesticides. It was investigated by Sucahyono et al. (2022) how *C. arborescens* plants affect the release of *Sycanus* sp., a natural enemy of numerous insect pests in industrial forest plantations, particularly in peatland environments. IPM strives for both temporal and spatial sustainability, according to Ali et al. (2023). The goal is to focus on natural preventative and control methods and to always act cautiously when implementing any control techniques that may have negative environmental

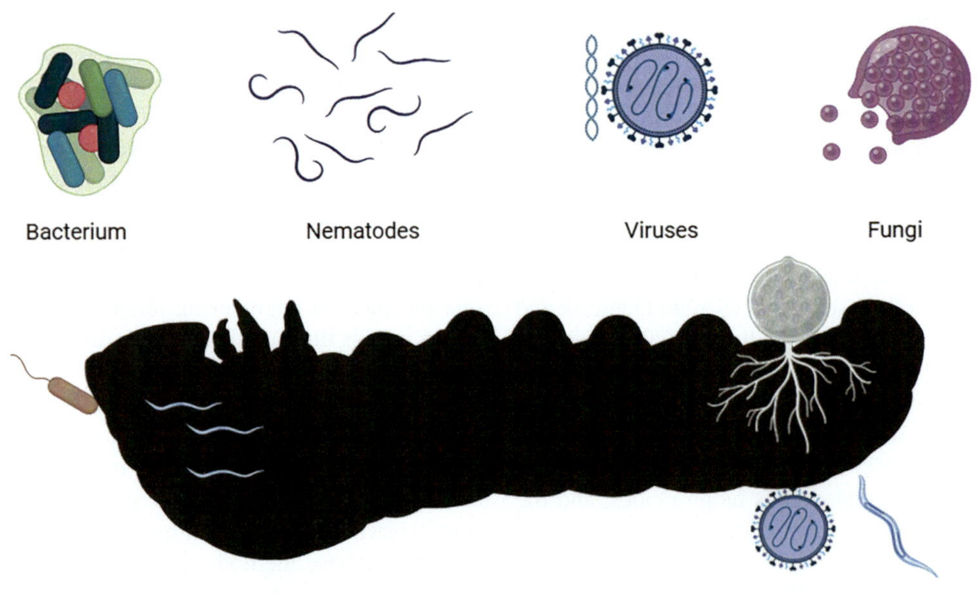

FIGURE 6.2 Microorganisms utilized as entomopathogens, showcasing various types, such as bacteria, fungi, viruses, and nematodes, that infect and control insect pests. The image highlights their mechanisms of action and examples of target pests.

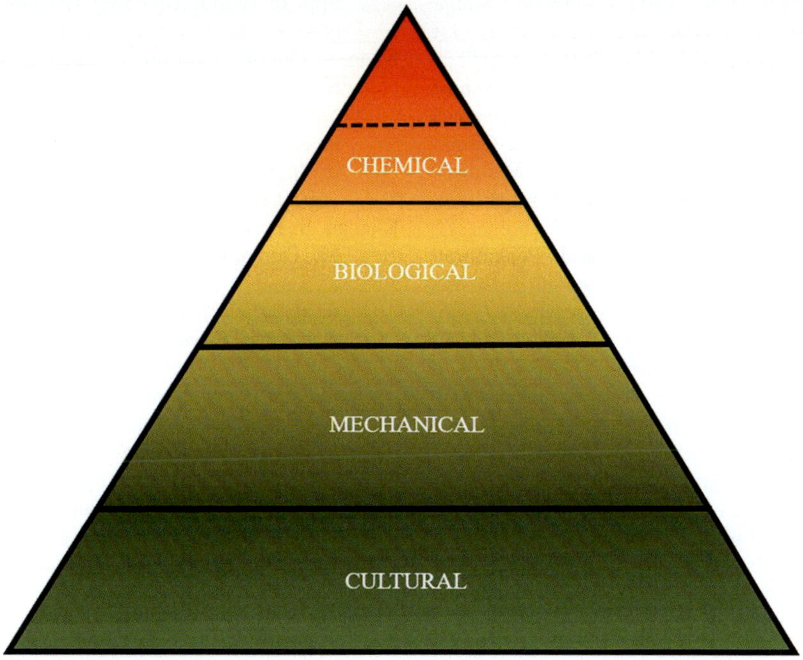

FIGURE 6.3 Pyramid of pest management tactics, illustrating a hierarchical approach to pest control strategies. The pyramid emphasizes preventive measures at the base, including cultural and biological controls, with progressively reactive methods, such as mechanical and chemical controls, at the top, representing last-resort actions for effective pest management.

effects. When the pest has surpassed the economic threshold, this has to be given priority as well. As a matter of fact, IPM has emerged as a benchmark for the management of industrial plantation forests, bolstered by numerous rules that prioritize the preservation of the ecosystem. The nongovernmental organization Forest Steward Council (FSC) forbids the use of specific chemicals in insect pest management programs (Lemes et al. 2017).

6.5 CONCLUSION

There are many insect pests that cause damage and losses to industrial plantation forests, starting from seedlings and also after being transplanted into the field. These pests attack at various stages of insect life. However, there are many solutions that can be utilized to control these pests, and integrated pest management (IPM) seeks to integrate various types of control while minimizing or excluding the application of chemical insecticides. Although chemical insecticides are highly effective, their overuse poses significant environmental risks, such as pest resistance, alterations in the behavior of other insect pests, and disruption of the natural ecological balance. Sustainability is a key aspect that must be prioritized in pest control strategies. Efforts to combine multiple control methods, such as developing pest-resistant clones, utilizing entomopathogens, and implementing biological control, should be continuously developed and refined. Furthermore, preventive actions and regular pest monitoring are not only more sustainable but also economically advantageous in the long term. These proactive measures reduce the likelihood of severe infestations and mitigate the need for intensive chemical control, thereby preserving the balance of natural enemies and fostering a healthier ecosystem. By prioritizing IPM approaches, the long-term sustainability of industrial plantation forests can be better ensured while minimizing negative environmental impacts

ACKNOWLEDGMENTS

The author, T. A. D., thanks the ICCR scholarship, which has supported his full scholarship while studying in India. The author also thanks the Department of Botany, Central University of Punjab, Bathinda, India. Also, thanks to BioRender for helping make the figure.

REFERENCES

Agustin, Syahrani, Prijono, Agus, & Kusumaningsih, dan Karti Rahayu (2023). Uji Efektivitas Beberapa Jenis Insektisida Terhadap Pengendalian Hama Ulat Penggulung Daun (*Strepsicrates* sp.) Pada Bibit Eucalyptus Hybrid. *Jurnal Mahasiswa Instiper (Agroforetech)*, 1(1), 810–815.

Ali, M. I. (2008). A survey of the insect pests of soybean in Northern Bangladesh, their damage and occurrence. *Tropical Pest Management*, 34(3), 328–330. https://doi.org/10.1080/09670878809371266

Ali, H., Muhammad, A., Sanda, N.B., Huang, Y., & Hou, Y. (2019). Pyrosequencing uncovers a shift in bacterial communities across life stages of Octodonta nipae (Coleoptera: Chrysomelidae). *Frontiers in microbiology*, 10, 466.

Ali, H., Hou, Y., & Tahir, M.B. (Eds.). (2023). *Climate Change and Insect Biodiversity: Challenges and Implications* (1st ed.). CRC Press. https://doi.org/10.1201/9781003382089

Anggarawati, S. H., Santoso, T., & Anwar, R. (2017). The use of entomopathogenic fungi *Beauveria bassiana* (Balsamo) Vuillemin and *Lecanicillium lecanii* (Zimm) Zare & Gams) for controlling *Helopeltis antonii* sign (Hemiptera: Miridae). *Journal of Tropical Silviculture*, 8(3), 197–202. https://doi.org/10.29244/j-siltrop.8.3.197-202

Asmaliyah, A., & Rostiwati, T. (2015). Pengaruh Pengaturan Jarak Tanam Terhadap Perkembangan Serangan Hama Dan Penyakit Pulai Darat (*Alstonia angustiloba*). *Jurnal Penelitian Hutan Tanaman*, 11(3), April, 41–50.

Capinera, J. L. (2015). Relationships between insect pests and weeds: An evolutionary perspective. *Weed Science*, 53(6), 892–901. https://doi.org/10.1614/WS-04-049R.1

Dara, S. K. (2017). *Entomopathogenic microorganisms: Modes of action and role in IPM*. Agriculture and Natural Blogs, University of California, 7p.

Dharma, T. A., Sitepu, S. F., Lubis, L., & Br. Girsang, S. S. (2018). Canopy insects abundance in shallot (Allium ascolanicum L.) cultivation with semi organic and conventional farm. *Jurnal Pertanian Tropik*, 5(2), 268–275. https://doi.org/10.32734/jpt.v5i2.3012

Dudley, T. L., & Bean, D. W. (2012). Tamarisk biocontrol, endangered species risk and resolution of conflict through riparian restoration. *BioControl*, 57(2), 331–347. https://doi.org/10.1007/s10526-011-9436-9

Gafur, A., Nasution, A., Tarigan, M., Tjahjono, B. (2012). Development of biological control agents to protect plantation forests in Sumatra, Indonesia. *Proceeding International Conference on the Impact of Climate Change to Forest Pests and Diseases in the Tropics*. UGM, Indonesia, Pages 200–206.

Hanchor, U., Maelim, S., Suanpaga, W., Park, J.-M., & Kang, K.-S. (2016). Growth performance and heritability estimation of *acacia crassicarpa* in a progeny trial in eastern thailand. *Silvae Genetica*, 65(2), 58–64. https://doi.org/10.1515/sg-2016-0017

He, Z., Ali, H., Wu, J., Liu, Z., Wei, X., & Zhuo, Z. (2025). Impact of climate change on the distribution of Isaria cicadae Miquel in China: predictions based on the MaxEnt model. *Frontiers in Microbiology*, 16, 1509882.

Herliyana, E. N., Achmad, Putra, A. (2012). Liquid organic fertilizer influence on Jabon (*Anthocephalus cadamba* miq.) seedling growth and its resistance to disease. *Jurnal Silvikultur Tropika*, 03(03), Hal. 168–173.

Hutapea, F. J., Weston, C. J., Mendham, D., & Volkova, L. (2023). Sustainable management of Eucalyptus pellita plantations: A review. *Forest Ecology and Management*, 537, 120941. https://doi.org/10.1016/j.foreco.2023.120941

Hutasuhut, J. H., Anna, M., & Siregar, E. B. M. (2015). Uji Infeksi *Cylindrocladium* SP. Pada Klon Hibrid *Eucalyptus Grandis* X *Eucalyptus Urophylla* (Infection test *Cylindrocladium* SP. on Hybrid Clones of *Eucalyptus Grandis* X *Eucalyptus Urophylla*). *Peronema Forestry Science Journal*, 4(3), 143–148.

Jaiswal, D. K., Gawande, S. J., Soumia, P. S., Krishna, R., Vaishnav, A., & Ade, A. B. (2022). Biocontrol strategies: An eco-smart tool for integrated pest and diseases management. *BMC Microbiology*, 22(1), 324. https://doi.org/10.1186/s12866-022-02744-2

Jufenlin, M., Prijono, A., & Sumardi, I. (2023). Uji Efektivitas Metode Semprot dan Pengkabutan terhadap Intensitas Serangan Hama Whiteflies (*Bemisia tabaci*) dan Leafroller (*Strepsicrates macropetana*) pada Tanaman Induk Eucalyptus Spp. di Indoor Breeding Orchard (Ibo) Kerinci Research Nursery. *Agroforetech*, 1(01).

Khan, S., Duran, A., Ikram, M., Hanjelina Br Sinulingga, N. G., De Souza Tavares, W., Sirait, B. A., Kkadan, S. K., & Tarigan, M. (2023). *Trichogramma yousufi* sp. Nov. employed for the management of *Spodoptera exigua* and *Spodoptera litura* in Indonesia. *Florida Entomologist*, 103(3). https://doi.org/10.1653/024.103.0307

Kkadan, S. K., Sirait, B. A., Asfa, R., Tavares, W. D. S. Tarighan, M., Duran, A., Wong, C. Y., Sharma, M,. (2020). Evaluation of a spinetoram based insecticide against lepidopteran and thrips infesting acacia and eucalyptus in Sumatra, Indonesia. *Journal of Entomology and Zoology Studies*, 8(2), 1345–1351.

Lemes, P. G., Zanuncio, J. C., Serrão, J. E., & Lawson, S. A. (2017). Forest Stewardship Council (Fsc) pesticide policy and integrated pest management in certified tropical plantations. *Environmental Science and Pollution Research*, 24(2), 1283–1295. https://doi.org/10.1007/s11356-016-7729-3

Lumbantobing, I. R. C., Wijayani, S., & Andayani, S. T. (2023). Uji Efektivitas Beberapa Akarisida untuk Pengendalian Hama Tungau Merah (Tetranychus Sp.) pada Tanaman Induk Acacia Crassicarpa. *Agrotechnology, Agribusiness, Forestry, and Technology: Jurnal Mahasiswa Instiper (Agroforetech)*, 1(1), 796–800.

Muhdi, Sahar, A., Hanafiah, D. S., Zaitunah, A., & Nababan, F. W. B. (2019). Analysis of biomass and carbon potential on eucalyptus stand in industrial plantation forest, North Sumatra, Indonesia. *IOP Conference Series: Earth and Environmental Science*, 374(1), 012054. https://doi.org/10.1088/1755-1315/374/1/012054

Nepal, P., Korhonen, J., Prestemon, J. P., & Cubbage, F. W. (2019). Projecting global planted forest area developments and the associated impacts on global forest product markets. *Journal of Environmental Management*, 240, 421–430. https://doi.org/10.1016/j.jenvman.2019.03.126

Nuri, S. F. (2005). Dampak Penyiapan Lahan *Acacia crassicarpa* Terhadap Serangan Penyakit Busuk Akar Putih. Theses. IPB University, Bogor.

Pasieczni, N., McDonald, D. (2016). Acacia crassicarpa (northern wattle). *CABI Compendium*. https://doi.org/10.1079/cabicompendium.2192

Purwanti, E., Nurjanto, H. H., & Widiyanto. (2022). Pertumbuhan Acacia Crassicarpa Umur 12 Bulan Pada Berbagai Jarak Tanam Di KHDTK Wanagama. Undergraduate Theses. Universitas Gadjah Mada, Yogyakarta.

Reddy, P. P. (2017). Trap cropping. In P. P. Reddy, *Agro-ecological approaches to pest management for sustainable agriculture* (pp. 133–147). Springer Singapore. https://doi.org/10.1007/978-981-10-4325-3_9

Rusandi, R., Mardhiansyah, M., & Arlita, T. (2016). The use of Mahagony seed extract as a botanical pesticide for controllinf *Spodoptera litura* F. of the *Acacia crassicarpa* A. Cunn. ex Benth. nursery. *Jurnal Online Mahasiswa Fakultas Pertanian Universitas Riau*, 3(1), February, 1–7.

Saputra, C., & Aluyah, C. (2019). Pengaruh dosis suspensi tape singkong dan jenis insektisida dalam mengendalikan kumbang penggerek batang ambrosia pada tanaman krasikarpa (Acacia crassicarpa) di pt. Bumi mekar hijau. *Sylva: Jurnal Ilmu-Ilmu Kehutanan*, 8(1), 13. https://doi.org/10.32502/sylva.v8i1.1855

Şengul Demirak, M. Ş., & Canpolat, E. (2022). Plant-based bioinsecticides for mosquito control: Impact on insecticide resistance and disease transmission. *Insects*, 13(2), 162. https://doi.org/10.3390/insects13020162

Sirait, B. A., Sinulingga, N. G. H. B., Samosir, M. N., Tavares, W. D. S., Kkadan, S. K., Tarigan, M., & Duran, A. (2020). First report of *Myllocerus scapularis* roelofs (Coleoptera: Curculionidae) and *Rhytiphora bankii* (Fabricius) (Coleoptera: cerambycidae) on commercial plantings of acacia crassicarpa (Fabaceae) in indonesia. *The Coleopterists Bulletin,* 74(2), 404. https://doi.org/10.1649/0010-065X-74.2.404

Soylu, S., Bozkurt, A., Ali, H. et al. (2025). Evaluation of resistance of bean genotypes to common bean bacterial blight disease agents and characterisation of plant defence mechanisms involved in resistance. *Eur J Plant Pathol* 173, 387–411. https://doi.org/10.1007/s10658-025-03066-5

Sucahyono, M. P., Hidayat, R., Gafur, A., Indrayadi, H., Wahno, I., Herdyantara, A. B., & Tjahjono, B. (2022). The potential of geronggang (*Cratoxylon arborescens* (vahl.) Blume) in conserving *Sycanus* sp. As natural enemy of insect pests in peatland areas. *IOP Conference Series: Earth and Environmental Science*, 1025(1), 012041. https://doi.org/10.1088/1755-1315/1025/1/012041

Suheri, M., Haneda, N. F., Jung, Y. H., Sukeno, S., & Moon, H. K. (2020). Effectiveness of pheromone traps for monitoring *Zeuzera* sp. (Lepidoptera: Cossidae) population on *Eucalyptus pellita* plantation. *IOP Conference Series: Earth and Environmental Science*, 468(1), 012016. https://doi.org/10.1088/1755-1315/468/1/012016

Sunarti, S. (2018). The role of biodiversity in forest plant breeding: A case study on the development of new Acacia hybrid varieties (*Acacia mangium* x *Acacia auriculiformis*). *Prosiding Seminar Nasional Masyarakat Biodiversitas Indonesia*, 4(1), 28–34. https://doi.org/10.13057/psnmbi/m040104

Thakur, K., Sharma, A., & Sharma, K. (2021). Management of agricultural insect pests with physical control methods. *The Pharma Innovation Journal*, 10(6S), 306–314. www.thepharmajournal.com/specialissue?year=2021&vol=10&issue=6s&ArticleId=6650

Wibowo, A. (2006). *Gulma di Hutan Tanaman dan Upaya Pengendaliannya.* Pusat Penelitian dan Pengembangan Hutan Tanaman. Badan Penelitian dan Pengembangan Kehutanan, Bogor.

Wilckin, C. F., Raetano, C. G., & Forti, L. C. (2002). Termite pests in eucalyptus forest of Brazil. *Sociobiology*, 40(01). http://hdl.handle.net/11449/66807

Yasin, M., Qazi, M. S., Wakil, W., & Qayyum, M. A. (2020). Evaluation of nuclear polyhedrosis virus (Npv) and E*mamectin benzoate* against *Spodoptera litura* (F.) (Lepidoptera: Noctuidae). *Egyptian Journal of Biological Pest Control*, 30(1), 88. https://doi.org/10.1186/s41938-020-00271-8

Yousuf, H.M., Yasin, M., Ali, H., Naveed, K., Riaz, A., AlGarawi, A.M., Hatamleh, A.A., & Shan, Y. (2024). Evaluation of different Kabuli chickpea genotypes against Helicoverpa armigera (Hübner)(Lepidoptera: Noctuidae) in relation to biotic and abiotic factors. *PeerJ*., 12, e16944. https://doi.org/10.7717/peerj.16944

7 Advancements in the Management of Household Pests

*Aquiny Befairlyne T. Mawthoh and Devina Seram**
Department of Entomology, School of Agriculture, Lovely Professional University, Phagwara144411, Punjab, India
*Corresponding author

7.1 INTRODUCTION

Homes serve as small ecosystems, offering conditions that attract a wide range of arthropods. These pests thrive in various household environments, finding food, shelter, and breeding grounds in even the cleanest of homes. Their presence can lead to structural damage, contaminate food, and trigger allergies or other health issues, making effective pest management essential in maintaining a healthy living space (Leong et al., 2020). The classification of household arthropods as pests is primarily influenced by the resident's perceptions, knowledge, and attitudes. Arthropods that threaten the health of humans and pets by acting as disease vectors, stinging or biting, damaging the home's structure or materials, contaminating food, or causing annoyance due to their high numbers, are often regarded as pests (Byrne et al., 1984). Common household pests include ants, cockroaches, mosquitoes, bedbugs, lice, fleas, termites, and ticks, all of which can cause significant problems. Termites, for instance, can inflict severe structural damage to a home and also act as vectors for various diseases. Cockroaches, on the other hand, are notorious for spreading illnesses and triggering allergies or asthma attacks. Mosquitoes are dangerous as they can transmit diseases like malaria, the West Nile virus and Zika virus. Ants, while small, can contaminate food supplies and deliver painful bites. Fleas and ticks not only cause itchy, irritating bites but also can spread serious diseases, such as Lyme disease and southern tick–associated rash illness. Bedbugs, though not known for transmitting diseases, can quickly infest a home, leading to significant discomfort and anxiety for the occupants (Schoelitsz et al., 2019). Given the wide range of issues these pests can cause, proper management is essential to protect our home and health. Pest control should go beyond merely eliminating the pests currently in the home; it should also address the root causes of the infestation. By actively eliminating the conditions that allow pests to enter, settle, and thrive, it is possible to prevent their development and the problems they cause (Hendrichs et al., 2007). When pests are present, immediate control measures might be needed to swiftly lower their numbers to acceptable levels. When modern pesticides were first introduced, they were used extensively, resulting in the rapid elimination of pests susceptible to these chemicals. However, the pests that were resistant survived and reproduced, leading to a rise in pesticide-resistant populations (Georghiou, 1972). This highlighted that relying solely on pesticides was not a comprehensive solution. Consequently, scientists began to develop a new method for pest control known as integrated pest management (IPM). This approach integrates various control methods, including mechanical, cultural, biological, and chemical, applied as needed. An IPM strategy allows for a controlled level of pests in the environment and is more effective in reducing pest populations through diverse methods (Kogan, 1998). Unlike simply eliminating the current pest population, IPM aims to disrupt and prevent the growth of future pest populations (Zalucki et al., 2009). IPM offers a structured approach involving multiple steps for sustainable pest control. It requires an understanding of pest biology and management strategies, and it relies on established threshold levels to determine when to implement these measures. IPM in household pest management can be approached through various methods, including traditional physical techniques, chemical methods, biological strategies, and modern technological innovations. The physical or mechanical method is the initial line of defense against household pests and involves maintaining a clean environment. This includes promptly cleaning up crumbs and spills, storing food in airtight containers, and reducing entry points by repairing screens, sealing cracks, and addressing gaps around pipes and vents (Ehler, 2006). Chemical methods, on the other hand, involve using pesticides and rodenticides to control household pests. Although these chemicals can effectively reduce pest populations, they raise concerns about environmental damage and health risks. Pesticides can leave residues in soil, water, and air and may adversely affect non-target organisms, such as pollinators, predators, and wildlife (Schoelitsz et al., 2019). Biological pest control offers a sustainable and environmentally friendly alternative. It involves using living organisms to suppress pest populations, such as introducing natural predators or parasites. This method includes the use of biopesticides derived from natural materials, like plants, bacteria, and minerals, which target specific pests and reduce reliance on chemical pesticides (Bale et al., 2008). As technology rapidly advances, pest control is becoming

DOI: 10.1201/9781003516538-7

increasingly precise and eco-friendly. New developments in biotechnology (Mawthoh et al., 2023a) and artificial intelligence (AI) are revolutionizing pest management by offering methods that are more targeted, effective, and environmentally conscious. These innovative techniques promise to redefine traditional pest control, offering exciting possibilities for revolutionizing how we manage and mitigate pest populations.

Even in the most carefully maintained houses, pest problems can still occur from time to time. The first and most crucial step in addressing any pest issue is accurate identification. Understanding the specific pest and the potential damage it can cause allows us to choose the most effective control methods, whether that involves natural remedies, traps, chemical treatments, or innovative techniques. Additionally, knowing the pest's behavior, habitat, and life cycle is essential in preventing future infestations and minimizing the need for repeated treatments. Regular inspections and prompt action are key to managing pests effectively, ensuring your home remains a safe and comfortable place (Srivani, 2019). Combining preventive measures, consistent monitoring, and timely responses will help keep your home pest-free and safeguard your health and property. This chapter will explore a range of techniques available, from conventional to advanced, in managing household pests.

7.2 UNDERSTANDING HOUSEHOLD PESTS

7.2.1 Common Household Pests, Their Causes, and Traditional Control Methods

Common household pests include a wide variety of unwelcome creatures that can infest residences and create distress for people. Insects are the most common household pests, including ants, cockroaches, and termites. Ants, such as odorous house ants and pavement ants, can contaminate food and transmit disease, whereas cockroaches, such as German and American cockroaches, can cause allergies and asthma symptoms. Termites, such as subterranean and dry wood termites, can cause extensive structural damage to homes by eating on wood and other cellulose-based materials. Other insect pests include bedbugs, which feed on human blood and can cause discomfort and worry, as well as fleas and ticks, which can infest dogs and spread diseases, such as typhus and Lyme disease. Rodents, such as mice and rats, are typical home pests that can transmit diseases, including salmonella and leptospirosis, *via* urine and faeces (Rabiee et al., 2018). Furthermore, rodents can inflict extensive property damage by chewing on wires, insulation, and other materials. Spiders, silverfish, and carpet beetles are some of the other household pests that can cause discomfort and property damage. Discovering an infestation by one of these pests at homes can be distressing, as they not only damage property but also pose serious health risks to residents. While many pests are relatively benign, some can transmit diseases and cause significant damage. Therefore, understanding these common household pests causing high infestations, their potential impacts, and related prevention and effective control strategies is essential in maintaining a safe and healthy living environment. Table 7.1 provides a comprehensive list of common household pests, detailing the damages they cause and some recommended methods in managing them.

7.2.2 Factors Influencing Household Pest Attraction to Houses

Pests are drawn to a certain location by a variety of conditions that also provide them access to the materials they

TABLE 7.1
Common Household Pests Along with Their Management Strategies

Pests (Order) (Sc. Name)	Damages/Disease Caused	Control Strategies
Cockroaches (Dictyoptera/Blattodea) (*Periplanata americana*, *Blattella germanica*, *Blatta orientalis*, *Supella longipalpa*, etc.)	They can carry pathogens on their bodies. Cockroach infestations have been linked to asthma, and living in homes with infestations can also contribute to psychological stress.	• Store all food in sealed, airtight containers. Place pet food bowls out only for the duration of the pets' meals, and put away any leftover food. Wash dirty dishes immediately after use, and transfer open bags of pet food into airtight containers. • Eliminate water sources, such as leaking faucets and pipes, plant saucers, beverages left out in open containers, and place pet water dishes on risers to keep them off the floor. • Reduce clutter in the home to minimize hiding spots for cockroaches. Seal cracks in walls, baseboards, and windowsills with caulk to limit their shelter options. • Use sticky traps throughout the home to identify areas with high cockroach activity, and place bait in these areas for effective control. Sticky traps can also help monitor for reinfestation after eradication. • Due to cockroaches' egg-laying habits, additional insecticide applications may be needed to ensure complete eradication.

TABLE 7.1 *(Continued)*
Common Household Pests Along with Their Management Strategies

Pests (Order) (Sc. Name)	Damages/Disease Caused	Control Strategies
Termites (Isoptera) (*Coptotermes* sp., *Ancistrotermes* sp., *Amitermes* spp., *Macrotermes* sp., etc.)	These pests can cause serious structural damage that poses significant risks to the people living in the building.	• Keep wood debris away from the foundation to help avoid infestations. • It is best to have a professional exterminator deal with this pest, as their hidden lifestyle makes them difficult to access and manage on our own.
Ants (Hymeoptera) (*Monomorium pharaonis*, *Monomorium minimum*, *Tapinoma sessile*, etc.)	Ants come in various sizes and colors, including species like the acrobat ant, black carpenter ant, little black ant, little fire ant, odorous house ant, pharaoh ant, red harvester ant, and little thief ant. These pests can be found both inside and around the house. While they typically do not damage sound wood, their presence suggests wet wood in the structure. They can contaminate household food.	• To effectively control and potentially eradicate an ant colony, apply an ant-specific pesticide to all entry points where ants enter the home or structure and where they search for food. Use caulk and screening to block ants from entering the home. • Place ant baits in areas where ants are active, as they can help in controlling infestation when used correctly. A mix of boric acid powder (3 tsp) and syrup (10 oz) in a small container can be effective. • Remove debris and wood piles around the structure to eliminate potential nesting and hiding spots.
Houseflies (Diptera) (*Musca domestica*)	Houseflies can spread diseases by transferring germs from surfaces they contact to food and feeding areas.	• Keep garbage cans tightly sealed, and discard spoiling foods promptly. Ensure window and door screens are intact to prevent flies from entering. Use sticky fly strips near windows indoors, and use a flyswatter to catch flies. • Check mousetraps daily, and dispose of dead mice in an outdoor trash can with a secure lid.
Pantry pests (Coleoptera, Lepidoptera) (*Tribolium confusum*, *T. castaneum*, *Lasioderma serricorne*, *Stegobium paniceum*, *Plodia interpunctella*, etc.)	These insects consume a variety of household items, including cereal, flour, cornstarch, dry pet food, pepper, spices, nuts, dried fruit, and grits. They also infest non-food items, like dried flower arrangements and plant-based stuffing in furniture and toys. Common pests include flour moths, flour beetles, weevils, mealworms, and bran bugs.	• Discard any food spills contaminated by pests, such as flour, cornstarch, nuts, and grits. • Store treated foods in tightly sealed, insect-resistant containers. Thoroughly clean the affected areas of food debris before restocking. Freeze dry foods for one week to kill any potential pests. • Alternatively, heat dry foods in a shallow container at 130° to 150°F for 30 minutes to eliminate pests.
Clothes moths (Lepidoptera) (*Tineola bisselliella*)	During their larval stage, these moths damage wool, mohair, fur, cotton, and other materials made from plants and animals. As adults, they might fly out of a closet when it is opened.	• Maintain good housekeeping practices, ensure thorough sanitation, and regularly clean garments to help minimize infestations.
Common fruit fly (Diptera) (*Drosophila melanogaster*)	They can spread harmful bacteria by landing on food.	• Keep all garbage cans tightly covered. Dispose of food waste properly. • Throw away any rotting fruits and vegetables. Store all fresh produce in the refrigerator when possible. • Keep outdoor trash cans away from doors and windows to prevent flies from entering the home. Ensure window and door screens are intact to keep flies out. Hang sticky fly strips indoors to capture flies.
Ticks (Acari) (*Ixodes scapularis*, *Rhipicephalus sanguineus*, *Amblyomma americanum*, etc.)	Ticks can spread illnesses, such as Lyme disease, and their bites can lead to skin irritation, allergic reactions, and secondary infections.	• Treat and replace pet bedding with clean, non-insecticide-treated bedding. Dogs and cats often require treatment with topical or systemic insecticides. Consult a veterinarian to select a product specifically labeled for tick control. • Apply insecticides to the yard, surrounding planter areas, and animal resting spots if necessary. Use aerosols, foggers, and sprays containing insect growth regulators (IGRs) on baseboards and in cracks and crevices to control brown dog ticks.

(*Continued*)

TABLE 7.1 ***(Continued)***
Common Household Pests Along with Their Management Strategies

Pests (Order) (Sc. Name)	Damages/Disease Caused	Control Strategies
Fleas (Siphonaptera) (*Xenopsylla cheopis*, *Ctenocephalides felis*, *Ctenocephalides canis*, etc.)	Fleas are typically brought into the home by dogs, cats, rats, mice, and other animals. Fleas can cause various diseases, such as plague, and allergic reactions.	• Consult a veterinarian for the safest pesticide options, as some pet breeds may be sensitive to certain systemic pesticides. • Thoroughly vacuum carpets before treatment, and dispose of the vacuum bags in a sealed garbage container. • Pyrethrins and DeltaDust can be used for treating lawns and dog kennels. Use products with pyrethrins and a synergist as directed on the label. • For severe flea infestations indoors or outdoors, it is advisable to hire a pest management professional.
Drain flies (Diptera) (*Psychoda alternata*, *Psychoda* spp., *Clogmia albipunctata*, etc.)	Drain flies, also known as moth or sewer flies, can cause respiratory problems, especially in those with asthma, and may trigger allergic reactions from inhaled particles. They can also spread bacteria, contaminating surfaces and food.	• Clear out gelatinous buildup from drains and sinks, and flush them with boiling water to remove breeding conditions. • Use clog-removing drain cleaners, including chemical and microbial products, to kill larvae and eliminate breeding sites. Take out indoor garbage daily to reduce attractants.
Mosquitoes (Diptera) (*Aedes aegypti*, *Aedes albopictus*, *Anopheles gambiae*, *Culex quinquefasciatus*, etc.)	Mosquitoes are vectors for various diseases globally, such as malaria, dengue, yellow fever, Chikungunya, and West Nile virus. They can transmit West Nile virus and are also carriers of dog heart worm, *Dirofilaria immitis*, an internal parasite that can be fatal for dogs, and sometimes cats, if left untreated.	• Minimize mosquito populations by eliminating standing water where they lay eggs. • Properly dispose of old tires, empty children's pools when not in use, and remove flowerpot saucers or other items that can collect rainwater. • Keep vegetation trimmed to reduce resting spots for adult mosquitoes. • Wear long-sleeved shirts and pants to reduce mosquito bites. • Ensure window screens are intact to prevent mosquitoes from entering the home.
Beetles (Coleoptera) (*Anthrenus verbasci*, *Dermestes ater*, etc.)	These pests can damage household items, including wooden structures and furniture, as well as natural fibers, like wool, silk, and cotton, and some contaminate stored food products.	• Regularly clean potential food sources, including laundering wool blankets and vacuuming rugs, to help control beetles. • Store items in sealed containers to prevent infestations. • Freeze taxidermied items to kill any larvae.
Silverfish (Thysanura/Zygentoma) (*Lepisma saccharina*)	These pests are mostly bothersome; they can also cause harm to wallpaper, fabrics, books, and stored items.	• Avoid storing books, paper, or cardboard in damp locations. Use a fan or dehumidifier to ventilate and dry infested areas, enhancing air circulation. • Remove residues of high-protein foods. • Seal cracks and crevices. • Sticky traps can help locate where these insects are residing. • Effective control involves drying out the areas where these insects are found.
Bedbugs (Hemiptera) (*Cimex lectularius*, *Cimex hemipterus*)	Bites can cause hives, rashes, and welts. Scratching the bites may result in secondary skin infections. Infestations can also lead to psychological stress for those affected.	• Prevention is the most effective control method. Avoid secondhand furniture when possible. • Wash newly purchased clothing immediately. • Inspect hotel rooms for signs of bedbugs before staying. • Insecticides are generally ineffective; heat treatment is the preferred method.
Centipedes and millipedes (Scutigeromorpha and Julida) (*Scutigera coleoptrata*, *Cylindroiulus* spp.)	Centipedes can inflict painful bites that lead to redness, swelling, and irritation. They feed on small insects inside the home, which can attract other pests and indirectly cause property damage. Millipedes, on the other hand, may harm indoor plants by consuming their roots and decaying plant material.	• For indoor management, treat thresholds, baseboards, and cracks with ready-to-use sprays or dusts. • For outdoor control, apply pesticides to entry points and areas where pests are active.

TABLE 7.1 *(Continued)*
Common Household Pests Along with Their Management Strategies

Pests (Order) (Sc. Name)	Damages/Disease Caused	Control Strategies
Crickets (Orthoptera) (*Acheta domesticus*, *Gryllus bimaculatus*)	Crickets can damage plants and create holes in paper, rubber, and fabrics made of cotton, linen, wool, or fur. Male crickets are notorious for their disruptive chirping, particularly at night. Additionally, they may contaminate food and surfaces with their droppings and shed skin.	• Treat doors, thresholds, baseboards, and cracks where crickets may hide. • Apply exterior perimeter insecticides, and treat the lawn to prevent crickets from entering the home. • Caulk and seal all small exterior openings and entry points. • Turn off outdoor lights at night to reduce cricket attraction and nuisance.
Head louse (Siphunculata/Anoplura) (*Pediculus humanus capitis*)	Lice infestations can lead to discomfort and itching, and scratching can result in secondary skin infections.	• Use over-the-counter medicated shampoos to kill lice. • Launder bedding, hats, and scarves to eliminate lice from these items.
Wasps, hornets (Hymenoptera) (*Vespula vulgaris*, *Vespa* spp.)	Paper wasps, hornets can be more aggressive than bees, especially if their nests are disturbed, and can deliver painful stings that may cause intense pain, swelling, and in some cases, allergic reactions.	• Apply insecticide directly to nests and resting area, ideally at night, and remove nests once no activity is observed. • For underground nests like yellow jackets, apply insecticide directly into the entrance and cover with soil immediately. • Use ready-to-use sprays and aerosols designed for wasps and hornets for both indoor and outdoor applications.
Springtails (Collembola/Oligogentoma) (*Orchesella* spp.)	These pest are generally not harmful to humans; however, they can become a nuisance in homes, particularly in damp areas, though they do not typically cause direct damage to household items or structures.	• Often found in damp areas, like kitchens, basements, bathrooms, around drains, and in the soil of potted plants. • Keep these areas dry when possible, and address any moisture issues.
Rodents (Rodentia) (*Rattus rattus*, *Mus musculus*, *Rattus norvegicus*, etc.)	Rodents can contaminate food and surfaces with their droppings, urine, and fur, posing health risks and causing food spoilage. They may also damage personal belongings, chew through building materials, insulation, and electrical wiring.	• Dispose of leftover food from dog or cat bowls. • Use traps like sticky traps cautiously, avoiding areas where children, pets, or native animals could be harmed.

Sources: Cranshaw (2011); Frederick (2015); Molly (2016); Wikipedia.

need to exist and flourish. For domestic bugs, the main draws are food, water, and refuge. Pet food and birdseed can also be used as a food source, and spills, crumbs, and open food containers can all attract insects and rodents. Leaky pipes, clogged drains, and standing water can all supply pests with the water they require to survive. Piles of paper, cardboard, and fabric can act as nesting grounds for bugs, while crowded storage rooms, stacked boxes, and clutter can provide them with cover and hiding places. Furthermore, particular parts of the house may attract pests more than others due to factors like warmth, dampness, and humidity. For instance, crawlspaces, basements, and attics can offer the perfect environment for pests like cockroaches, rats, and silverfish. Pests can also be brought in by unclean surfaces, filthy drains, and inadequate ventilation. In addition, openings surrounding doors, windows, and pipes can serve as entry places for pests to get in the house. Effective pest management and prevention depend on identifying and removing these attractants and habitats, which can lessen the chance of infestations and make the house less alluring to pests.

7.2.3 Household Pests Biology and Behavior

Household pests have distinct biological and behavioral traits that allow them to survive in residential environments. Understanding these characteristics is crucial for their effective management. One important component is their reproductive ability or fecundity. For example, a single female cockroach can lay up to 50 eggs at once, resulting in fast population increase (Li et al., 2021). Similarly, mice can reproduce all year and have up to 12 litters (Cranshaw, 2011). Pests may rapidly colonize and adapt to new areas due to their high reproduction rate. Pests also have distinct behavioral features that help them survive. Ants are highly social organisms that communicate using chemical signals to coordinate foraging and nesting activities (Buczkowski et al., 2014). Bedbugs, on the other hand, are nocturnal and possess a unique ability to smell carbon dioxide, which allows them to identify human hosts (Yturralde and Hofstetter, 2012). Household pests are adaptable and resilient. Rodents can fit through small gaps and thrive in a range of habitats (Molly, 2016), and cockroaches can endure high temperatures and radiation, making them highly resistant to eradication efforts (Ligonniere et al., 2024).

7.3 CHEMICAL CONTROL OF HOUSEHOLD PESTS

Pesticides, including insecticides, fungicides, rodenticides, and herbicides, are commonly used in households to manage pests, such as mosquitoes, ants, flies, cockroaches, mice, and termites. These chemical controls are generally reliable and can significantly lower pest populations (DuBois, 1958). The primary objective of chemical pest control is to swiftly eliminate pests, with modern insecticides providing rapid action, ease of use, and minimal toxicity to humans and pets. Surveys from developed nations highlight the prevalence of household pesticide use. In the United States, approximately 82% of households use these products, while in the United Kingdom, the figure ranges from 80% to 85%. In India, mosquito control products are the most widely used household pesticides, with mosquito coils being particularly popular (Daniels et al., 2001). Common chemical agents used for chemical control include organophosphates, pyrethrins, synthetic pyrethroids, neonicotinoids, phenyl pyrazoles, oxadiazines, pyrroles, and insect growth regulators, like methoprene and pyriproxyfen. These insecticides come in various forms, including sprays, dusts, gels, liquids, crystals, baits, and foggers, typically with active ingredients in concentrations of 0.5–1%. They are readily available at DIY stores and supermarkets. However, consumers often select pesticides based on perceived safety, past experience, or marketing rather than detailed information sources (Grossman, 1995; Chitra et al., 2013). While chemical pesticides are effective, their overuse, the improper cleaning of treated items, and the misuse of application techniques can lead to serious health issues, including fatalities. Studies from developed countries have even suggested a potential link between exposure to household pesticides and various cancers. Moreover, the use of banned or restricted products is sometimes reported, particularly when dealing with resistant pest populations. This misuse can result in increased pesticide application and disregard for safety instructions (Metcalf, 1989). To manage pests effectively, it is crucial to carefully select and use pesticides, regularly review product labels, and consider integrated pest management (IPM) strategies to reduce reliance on chemical controls alone. IPM emphasizes a combination of methods, including biological controls, habitat management, and minimal use of chemical pesticides, to achieve sustainable pest management.

7.4 NON-CHEMICAL CONTROL OF HOUSEHOLD PESTS

Due to the negative impacts of insecticides on health and the environment, as well as the rise of insecticide resistance, non-chemical control methods are essential in integrated pest management (IPM). Physical methods also play a role in non-chemical pest management. Physical pest control methods involve using equipment and techniques to either remove pests or prevent their entry. Key methods include pest-proofing, which creates barriers and eliminates nests to stop infestations; traps and bait stations, which capture pests like rodents and insects; and temperature control, where extreme temperatures kill pests or slow their growth. Regular cleaning and maintenance, such as vacuuming carpets to remove pests and their eggs and using sticky traps to monitor and capture insects, can significantly reduce pest populations (Weintraub, 2009). For example, trapping cockroaches with pheromone-based sticky traps and vacuums has proven as effective as traditional insecticide methods in reducing populations. Similarly, the resurgence of bedbugs has increased interest in temperature treatments, as heating effectively kills all bedbug stages. Washing laundry at 60°C and subsequently dry-cleaning can eliminate bedbugs, while whole-room treatments can eradicate infestations if done correctly. Low-temperature treatments, such as exposing bedbugs to 17°C for 1–2 hours, can also be effective (Fields et al., 2018). Another approach is biological control, which uses natural predators instead of chemicals to manage pest populations. This method involves introducing natural predators or parasites that target specific pests. For example, predatory beetles and mites can be employed to control populations of mites and other small insects. Biopesticides, derived from natural organisms or substances, present an alternative method for pest control. Essential oils, such as neem, eucalyptus, and lavender, possess insecticidal properties that can effectively repel or eliminate pests. Due to their low toxicity to mammals, essential oils and diatomaceous earth dust are promising options for household pest control. Both materials have proven effective in reducing insect populations and are easily accessible to the general public. For instance, various botanical plants have been used to control mosquito population (Sudakin, 2003). However, the use of biological control agents, which is more common in agriculture (fields), tends to be less popular in domestic or household environment. This is partly because some individuals may consider these biological agents to be nuisances themselves. The application of entomopathogenic nematodes, fungi, and bacteria for indoor pest control has not been extensively explored, but if embraced by the public, these methods could become valuable tools within integrated pest management (IPM) programs. Experimental studies have shown promising results, such as bait stations with entomopathogenic nematodes demonstrating effectiveness comparable to chemical bait stations in controlling cockroach populations. Entomopathogenic fungi have also shown potential in controlling pests like bedbugs, cockroaches, and termites. These fungi can be applied by suspending their spores in oil, which is then sprayed or fumigated in infested areas. The decision to use non-chemical methods largely depends on individuals' attitudes toward both non-chemical and chemical approaches, as well as their tolerance for specific pests. Integrating these non-chemical methods into pest management strategies can offer effective control while minimizing environmental impact and health risks (Hajek et al., 2018). By combining biological control, biopesticides, and physical methods, pest management efforts can be enhanced, leading to more sustainable and safer household environments.

7.5 NOVEL TECHNIQUES IN THE ADVANCEMENTS FOR HOUSEHOLD PESTS MANAGEMENT

7.5.1 Electronic and Ultrasonic Pest Repellents

Devices that generate sounds or electromagnetic fields to keep pests out of a specific area are known as electronic and ultrasonic pest repellents (Tiwari and Ansari, 2016). These gadgets are frequently promoted as an eco-friendly and non-toxic substitute for conventional pest control techniques. For instance, ultrasonic devices produce noises between 16 and 45 kHz, which is a frequency that pests find unattractive (Panthawong et al., 2021). Although their efficacy in keeping out rats, insects, and other pests is still up for discussion, these sounds are meant to do just that. While some studies have found no discernible effect, others have demonstrated that ultrasonic devices can repel specific pests, like rodents (Yturralde and Hofstetter, 2012). On the other side, some pests are repelled by electronic repellents, which work with electromagnetic fields. According to William (1978), these gadgets frequently imitate the electromagnetic fields generated by predators, which can keep pests away. For instance, cockroach infestations were lowered by 50% when using an electronic repellent device (Khan-Ahmadi et al., 2023). It is important to remember that several parameters, such the type of pest and the positioning and quality of the device, might affect how effective these devices are. Furthermore, some experts contend that because pests might grow acclimated to the sounds or electromagnetic fields, these devices could not be successful in the long run (Panthawong et al., 2021). The following are some of the examples:

1. Mouse repellents with ultrasonic technology, such as the Victor Ultrasonic Mouse Repellent
2. Electronic insect repellents (e.g., Bell+Howell Ultrasonic Pest Repeller)
3. Repellants based on electromagnetic fields, like the Pest A Clear gadget

7.5.2 Acoustics for Hidden Pests Detection

Acoustic technology has long been utilized in the study of insect communication and in monitoring insect populations, geographic distributions, and species diversity. It has also been used to detect hidden insects in environments such as soil, wood, container crops, and stored products. Various acoustic devices, differing in size and power, have effectively trapped insect pests that respond to sound cues, such as mosquitoes, midges, mole crickets, field crickets, moths, cockroaches, bedbugs, and Tephritidae fruit flies. The effectiveness of these traps depends on the insect's behavior, physiological condition, and age and can be influenced by environmental factors, like temperature and light (Mankin, 2012). Despite the potential, widespread adoption of acoustic traps has been limited due to the high costs of the equipment and the fact that only a small portion of insect populations, typically mate-seeking adults of a specific age range, are attracted to sound sources. However, trap effectiveness can often be enhanced by incorporating swarm markers, chemical attractants, or black lights and by carefully matching the timing and frequency of sound patterns to natural insect communication signals. Additionally, there is ongoing exploration into using ultrasonic bat calls to disrupt the behavior of nocturnal insects, though these signals tend to be ineffective on insects not typically preyed upon by bats (Walker, 1996). Future advancements in acoustic pest management could include the development of signals that interfere with vibrational communication, especially in Hemipteran insects, and the creation of control strategies that integrate pheromones with precisely timed sonic or vibrational signals.

7.5.3 Sterile Insect Technique (SIT)

For the past 60 to 70 years, the sterile insect technique (SIT) has been a crucial tool in large-scale pest management efforts aimed at controlling, containing, eradicating locally, or preventing the reinvasion of insect pest populations and disease vectors worldwide (Benedict, 2021). Developed by Edward F. Knipling, SIT was first used in the 1950s to manage the New World screwworm, *Cochliomyia hominivorax*. Knipling found that making insects sterile could be more effective inreducing populations than direct killing methods, like insecticides. Knipling also suggested that achieving sterility through alternative methods, such as using insect growth regulators, like pyriproxyfen, could be more cost-effective and simpler than releasing sterile males. SIT has proven to be one of the most successful genetic control strategies, notably eradicating the New World screwworm from North and Central America – a major achievement in pest control (Dyck et al., 2021). In SIT programs, insects are mass-produced, sterilized by radiation, and released to mate with wild insects, leading to no offspring and a gradual reduction in the pest population. The four primary strategic approaches for employing the sterile insect technique

(SIT) are suppression, containment, prevention, and eradication (Bourtzis et al., 2021). This approach is specific to the target species and has been widely adopted in area-wide pest management strategies around the globe. A recent study by Balatsos et al. (2024) in Greece applied the sterile insect technique (SIT) to combat the invasive dengue vector mosquito, *Aedes albopictus*. The findings revealed a substantial decline in egg density, with a 78% reduction maintained for 4 weeks after the release period, along with a consistently lower ovitrap index in the SIT-treated area compared to the control. This trial successfully demonstrated the practicality of transporting and releasing sterile mosquitoes over long distances, offering crucial insights for future SIT initiatives.

7.5.4 Genetically Modified (GM) Insects

Experimental releases of genetically modified (GM) insects are being tested in several countries, including Brazil, the Cayman Islands, France, Guatemala, India, Malaysia, Mexico, Panama, the Philippines, Singapore, Thailand, the United States, and Vietnam (Reeves et al., 2012). Field trials of GM mosquitoes, specifically *Aedes aegypti*, began in the Cayman Islands in 2009, with subsequent trials in Malaysia and Brazil. In recent years, there has been a notable increase in the experimental deployment of GM insects (Mumford, 2012). For instance, in November 2009, GM mosquitoes were released into populated areas of the Cayman Islands. Genetically modified insects are created by inserting new genes into their DNA, a process that has advanced significantly over the past decade. This genetic manipulation has enabled the engineering of various insects, including agricultural pests, like the Mediterranean fruit fly, and disease vectors, such as mosquitoes (Schwindenhammer, 2020). For example, transgenic mosquitoes have been developed with the aim of either replacing or suppressing wild populations to reduce the spread of diseases and improve public health. Figure 7.1 shows how GM male mosquitoes carrying sterile genes mate with wild females, producing offspring that die before maturity. This innovative process reduces mosquito populations over time, helping control the spread of mosquito-borne diseases. In 2022, the Environmental Protection Agency (EPA) approved the release of up to 2.4 million GM mosquitoes in certain areas of Florida and California, USA. This approval was granted to Oxitec, a UK-based biotech company specializing in genetically modified mosquitoes. The goal of these releases is to manage mosquito populations and reduce the transmission of diseases like dengue fever, Zika virus, and malaria (Schairer et al., 2021). This technology holds promise for a more targeted pest control and disease management, leveraging genetic modifications to achieve public health benefits while minimizing reliance on chemical pesticides.

7.5.5 RNA Interference (RNAi)

Young (2017) highlights the importance of developing new tools and strategies for household integrated pest management (IPM). There is a growing interest in creating insecticides based on nucleic acids, with RNA interference

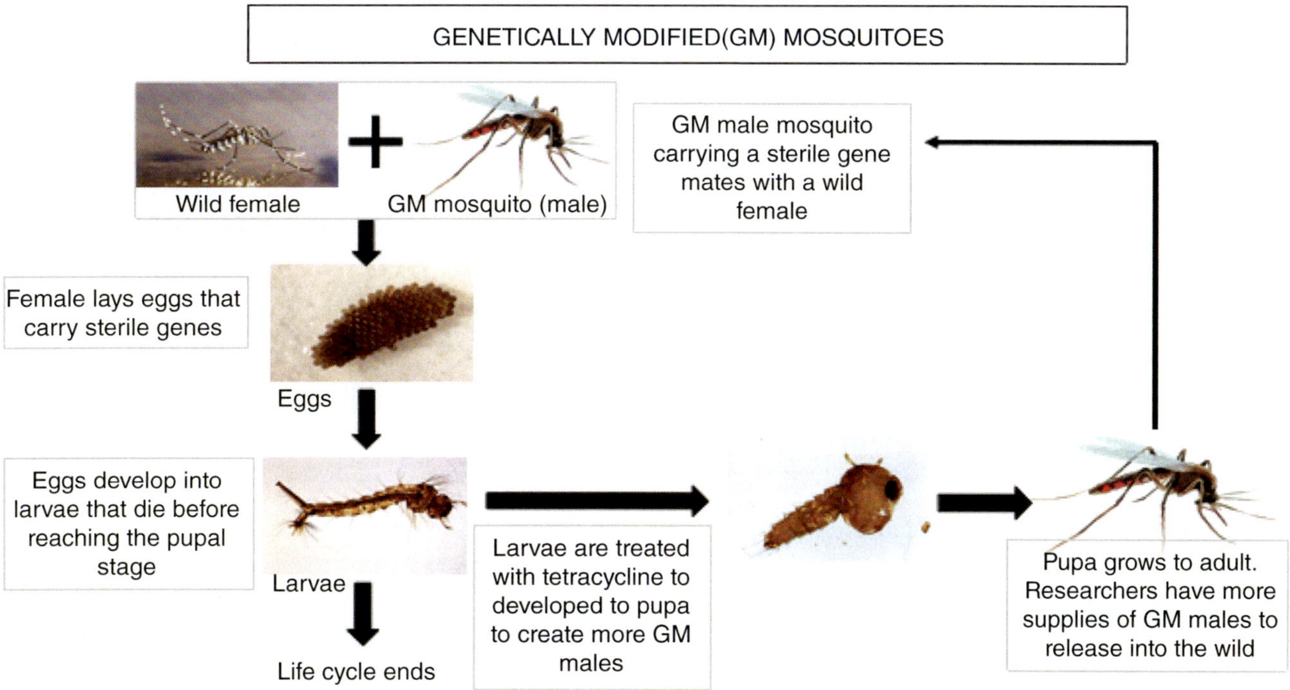

FIGURE 7.1 How genetically modified male mosquitoes carrying sterile genes mate with wild females, producing offsprings that die before maturity. On the other hand, larvae are also treated with tetracycline to create genetically modified males.

(RNAi) being a technique that is gaining traction for pest control. RNAi works by targeting specific genes critical for the growth, development, or reproduction of pest insects, which allows for selective pest elimination without harming non-target species (Wynant et al., 2014). RNAi is a natural cellular defense mechanism that operates through double-stranded RNA (dsRNA). In most eukaryotic cells, the presence of long dsRNA is seen as a potential threat, either from a viral infection or from the activity of transposons, which are segments of DNA that can move within the genome. To protect itself, the cell activates a series of processes to suppress these threats. The process begins with the enzyme Dicer-2, which cuts double-stranded RNA (dsRNA) into small interfering RNAs (siRNAs). These siRNAs are then transferred to the RNA-induced silencing complex (RISC), where they are separated into two strands. One strand is discarded, while the other, known as the guide strand, directs the RISC to bind to matching messenger RNAs (mRNAs). The RISC, with the help of an enzyme called Argonaute-2 (Ago2), then cuts the mRNA, effectively silencing the gene by blocking its ability to produce proteins (Figure 7.2a). Because of its sequence-specific action, many research groups and biotech companies are exploring the potential of dsRNA as a new, environmentally friendly, and species-specific insecticide. Some insects, particularly beetles (Coleoptera), are highly sensitive to dsRNA. However, not all insects respond as effectively to ingested dsRNA. Insects from the orders Lepidoptera (moths and butterflies), Diptera (flies and mosquitoes), and Hemiptera (aphids, hoppers, stinkbugs) show more variability in their responses to dsRNA compared to beetles (Christiaens et al., 2020). Feeding insects with double-stranded RNA (dsRNA) triggers an antiviral defense that cuts the dsRNA into small interfering RNAs (siRNAs). These siRNAs then bind to the corresponding messenger RNA (mRNA), leading to its cleavage and reduced gene expression. For example, injecting dsRNA into bedbugs significantly inhibited their reproduction. However, direct injection isnot practical, so the ingestion of dsRNA offers a more feasible approach. The effectiveness of RNA interference (RNAi) depends on factors like the target gene, the insect's life stage, and the length and concentration of dsRNA. Since dsRNA can degrade in the insect's midgut, methods like encapsulating dsRNA in liposomes have been developed to protect it, enhancing its effectiveness (Schoelitsz et al., 2019). RNAi is promising due to its specificity, but its application will likely be limited to certain insect species, as it requires precise identification of the target pest. In a recent study by Ligonniere et al. (2024) on American cockroaches (*Periplaneta americana*), RNA interference was used to inhibit the production of two proteins, hexamerin and allergen, which are essential for nutrient storage and reproduction. These proteins are involved in vitellogenesis, a vital process for female reproduction. The study found that blocking these proteins halted the development of eggs and ovaries in female cockroaches.

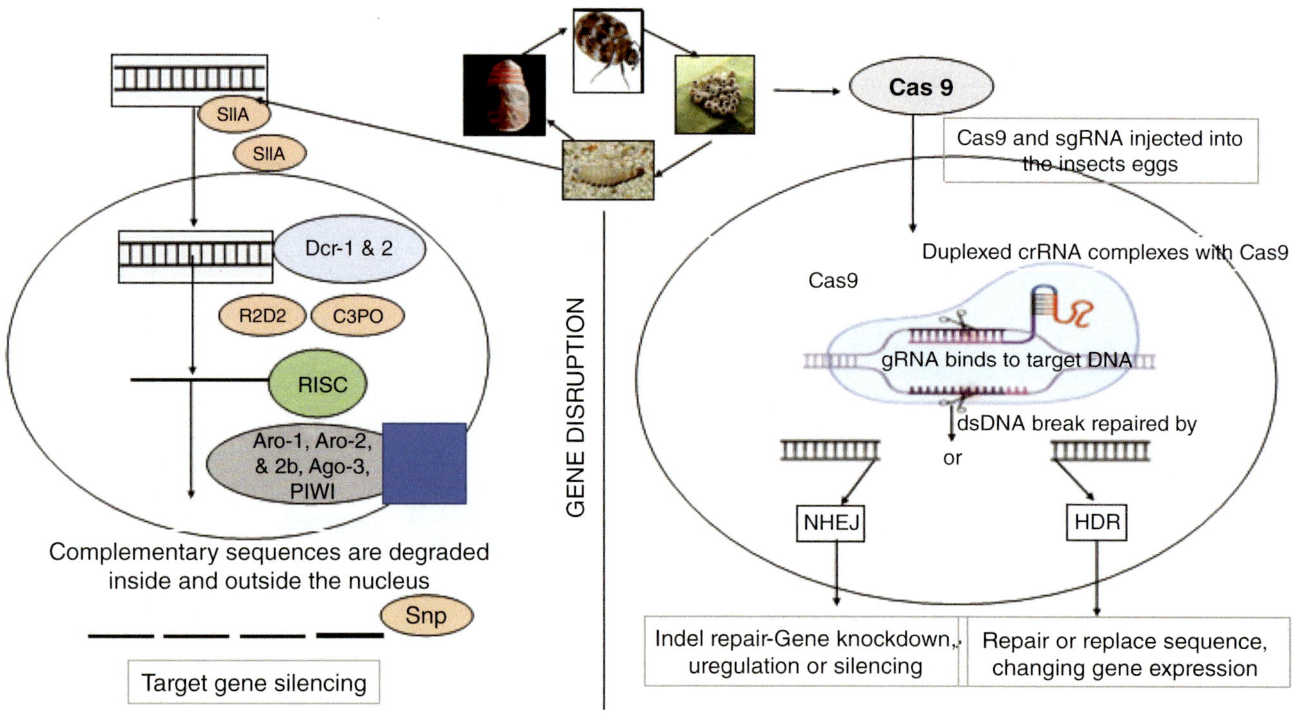

FIGURE 7.2 (a) How RNAi works by silencing the gene and blocking its ability to produce proteins. (b) Process of CRISPR-Cas in disrupting the insect gene for its successful management.

7.5.6 CRISPR-Cas Genome Editing (Clustered Regularly Interspaced Short Palindromic Repeats)

Recent years have seen a notable rise in research into insect genomics and gene functions. The i5k Project, also known as the "5000 Insect Genome Project," has been crucial in this field, aiming to sequence and analyze the genomes of 5,000 insect species to enhance our understanding of insect biology, evolution, and ecology (i5K Consortium, 2013). While traditional genetic control methods like the sterile insect technique (SIT) have effectively managed pests globally, they face challenges, such as decreased efficiency from bisexual releases and the need for frequent, large-scale releases. To overcome these issues, advanced technologies like CRISPR/Cas9 have been developed (Komal et al., 2023). CRISPR/Cas9 allows for precise genetic modifications through a structured process. Initially, scientists design a guide RNA (gRNA) that targets a specific DNA sequence. This gRNA, made up of CRISPR RNA (crRNA) and trans-activating CRISPR RNA (tracrRNA), guides the Cas9 enzyme to the target sequence. The Cas9 enzyme then makes a precise double-strand break (DSB) in the DNA (Figure 7.2b). The cell's repair mechanisms address this break through either non-homologous end joining (NHEJ), which can cause mutations, or homology-directed repair (HDR), which allows for specific genetic changes or corrections. This technology enables accurate genome editing, facilitating gene function studies, the development of new therapies, and the creation of genetically modified organisms (Barrangou et al., 2016). Table 7.2 presents a list of common household insects that have been studied for pest management using genome editing with the CRISPR/Cas system. It details the target genes, the genetic traits being modified, the findings from these studies, and the references for further information.

7.5.7 Nanotechnology

Nanoparticles, also known as *ultrafine particles*, are extremely small, ranging from 1 to 100 nm in size. Their tiny dimensions give them unique properties, including high mobility and a large surface area (Mawthoh et al., 2023b). Insecticidal or carrier nanoparticles can become

TABLE 7.2
Application of Genome Editing Using CRISPR/Cas in Controlling Some Common Household Pests

Insects	Target Gene	Genetic Trait	Findings	References
Anopheles gambiae Giles (mosquito)	Double sex (*Agdsx*)	Sex determination and differentiation	Disrupting the boundary between intron 4 and exon 5 using CRISPR/Cas9 led to a failure in producing the functional AgdsxF transcript. Females with two copies of the disrupted allele displayed an intersex phenotype, showing both male and female traits, and were entirely sterile. In contrast, males were unaffected in terms of development and fertility, highlighting that the AgdsxF transcript is crucial specifically for female differentiation.	Kyrou et al. (2018)
Drosophila melanogaster Meigen (common fruit fly)	Rosy, *DSH3PX1*	Pigmentation and mating	The research highlights that the CRISPR/Cas9 system is highly effective for genome editing in *Drosophila*. Compared to other methods, like ZFNs and TALENs, CRISPR/Cas9 is easier to use and quicker to implement, making it a valuable tool for genetic engineering.	Gratz et al. (2013)
Drosophila melanogaster Meigen (common fruit fly)	*LUBEL*	Growth and development	Flies with LUBEL mutations showed lower survival rates and impaired climbing ability when exposed to heat.	Asaoka et al. (2016)
Drosophila melanogaster Meigen (common fruit fly)	*Clamp*	Sex-specific	In CLAMP-deficient females, roX gene expression is increased, while in CLAMP-deficient males, it is decreased.	Urban et al. (2017)
Drosophila suzukii Meigen (Common fruit fly)	White, *Sxl*	Sex determination	Mutations in the *sex lethal* (Sxl) gene caused somatic mosaic females to develop abnormal genitalia and reproductive tissues.	Li and Scott (2016)

TABLE 7.2 *(Continued)*
Application of Genome Editing Using CRISPR/Cas in Controlling Some Common Household Pests

Insects	Target Gene	Genetic Trait	Findings	References
Tribolium castaneum (Herbst) (Flour beetle)	*E-cadherin*	E-cadherin gene, crucial for cell adhesion and vital for processes such as dorsal closure during embryonic development	Between 55% and 80% of individuals injected with CRISPR/Cas9 exhibited mutations, such as insertions or deletions (indels), due to non-homologous end joining (NHEJ). This included mosaic bi-allelic knockouts. Additionally, 71% to 100% of these individuals had mutations in their germ lines and successfully passed them on to their offspring.	Gilles et al. (2015)
Tetranychus urticae Koch (Mite)	*Phytoene desaturase*	Pigmentation	The study successfully applied CRISPR-Cas9 to create gene knockouts in the spider mite, *Tetranychus urticae*, resulting in two independent lines with mutations in the phytoene desaturase gene. These lines exhibited a recessive albino phenotype, demonstrating that CRISPR-Cas9 can be used for targeted genetic modifications in chelicerates.	Dermauw et al. (2020)

a crucial tool in pest management. Plant-derived nanoparticles have proven effective in managing urban pests, like mosquitoes and houseflies. Integrating nanotechnology with RNA interference (RNAi) could enhance pest control by combining their benefits. Specifically, negatively charged nanoparticles can deliver positively charged RNA molecules to target sites more efficiently, improving solubility, stability, and targeting precision (Athanassiou et al., 2018). While nanoparticle-based insecticides and carrier systems have been researched in various pest management contexts, their practical application remains uncertain. Further research is required to evaluate their effectiveness and scalability before these technologies can be approved for widespread use. Metal nanoparticles are gaining attention for their potential as insecticides in mosquito control. Research has shown these particles to be highly effective. In a study by Murugan et al. (2015), gold nanoparticles (AuNPs) were synthesized using leaf extract from lemongrass (*Cymbopogon citratus*) and tested on mosquito larvae and pupae. The study revealed that AuNPs were notably more effective than the plant extract alone, with LC50 values between 18.80 and 41.52 ppm (Blore et al., 2022).

7.5.8 Sex Pheromone Trap and Insect Growth Regulators

For over 50 years, the concept of using species-specific chemicals that alter insect behavior has been a key goal in managing harmful insects across agriculture, horticulture, forestry, and stored products, as well as in controlling disease vectors. Numerous pheromones and other semiochemicals have been identified and are now used to track insect populations and safeguard plants and animals from insect damage (Witzgall et al., 2010). Pheromones are chemical signals used in communication between individuals of the same species (intraspecific). Insects, particularly adult Lepidoptera, are well-known for using sex pheromones, which are released by females to attract males (Ando et al., 2004). These pheromones are produced commercially by synthesizing the necessary chemicals and are loaded into dispensers. These dispensers are then used in traps for various applications in agriculture, horticulture, forestry, and storage. Pheromone traps are widely used for several purposes: monitoring insect populations, mass trapping, and disrupting mating. By tracking the number of insects caught in these traps, it is possible to assess species abundance and potential damage. This data helps in field scouting, predicting egg laying, and timing control measures based on established action thresholds (Alam et al., 2023). Insect growth regulators (IGRs) are a relatively recent class of pest control agents designed to be safer alternatives to traditional insecticides and to address resistance issues (Graf, 1993). These synthetic hormones act as insecticides that manage pest populations by disrupting the molting process, preventing insects from reaching maturity. IGRs are less harmful to humans and the environment compared to conventional insecticides, and they are often used in combination with other chemicals, like pyrethrins and pyrethroids, to target adult insects (Tunaz and Uygun, 2004). IGRs only impact the developmental stages of insects and arthropods, not the adults. They are divided into two main categories: juvenile hormone analogs and insect development inhibitors. For instance, pyriproxyfen, an IGR found in NyGuard®, offers prolonged residual effects compared to typical adulticides, controlling current cockroach

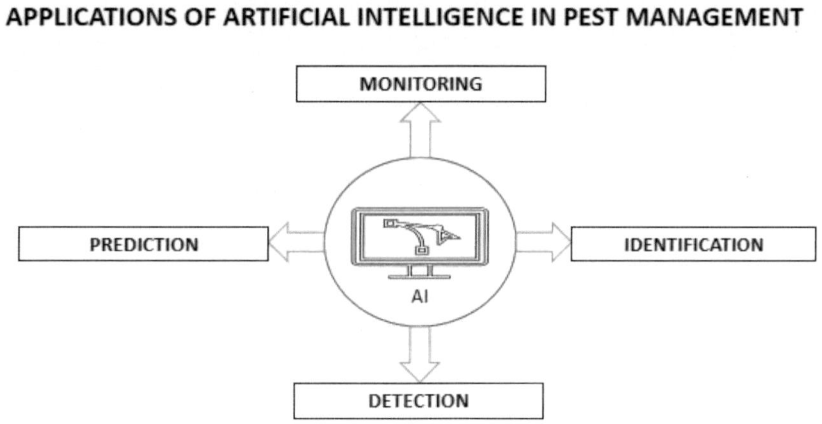

FIGURE 7.3 Various applications of AI in insect pest management.

populations and preventing new ones from establishing. IGRs like methoprene and hydroprene can be effective through ingestion or contact, while diflubenzuron, a chitin synthesis inhibitor, works only when ingested. Methoprene and hydroprene are approved for use in protecting stored commodities in various countries, including the USA, Australia, and the UK, and are effective against many stored product pests either alone or in combination with other control methods (Bell, 2014). Given their benefits and effectiveness against target species, IGRs are likely to become a key component of future insect control strategies.

7.5.9 Artificial Intelligence (AI)

Integrated pest management (IPM) seeks to reduce the environmental impact of pesticide use and minimize health risks to humans and animals. It focuses on two main strategies: preventing and monitoring diseases and pests. Recent advancements in artificial intelligence (AI) are greatly enhancing these strategies. AI, which includes technologies such as machine learning (ML), deep learning, computer vision, etc., is becoming increasingly important in pest management (Rahman et al., 2022). Figure 7.3 shows the various applications of AI in pest management. AI detection systems improve pest identification and monitoring by using cameras and sensors to gather data, such as heat signatures, movement, and sounds (Toscano-Miranda et al., 2022). Machine learning algorithms analyze this data against large datasets, enabling accurate pest identification, early detection, and improved accuracy and suggesting effective treatment strategies. Traditional pest management methods can be labor-intensive and costly, especially for large farms or conventional production systems (Demirel et al., 2021). AI provides a promising alternative by automating and optimizing various aspects of pest control. This technology not only is transforming agricultural pest management but also has the potential to revolutionize household pest control. By applying these advanced techniques to residential settings, AI can significantly improve the management of pests at home. In a recent study by Mendoza et al. (2023), a basic insect detection system was developed using a manual-focus camera called Jetson Nano, which is a budget-friendly, low-power, single-board computer and a trained deep learning model. This system was tested with live visual feeds to detect, classify, and monitor insect pests in realtime. The camera captures insect images, which are then analyzed by the Jetson Nano running the deep learning model to determine the insect's presence and species. The system was assessed under three different lighting conditions: white LED, yellow LED, and no light. It effectively identified and classified adult cigarette beetles and warehouse beetles with good accuracy. The study concludes that this automated system is an effective and cost-efficient solution for insect detection, which can help reduce pest control expenses and save time and effort for producers while ensuring the quality of stored products.

7.6 CASE STUDIES ON THE MANAGEMENT OF COMMON HOUSEHOLD PESTS

Case studies provide a comprehensive understanding of the pest infestation, identify effective control methods, and inform evidence-based decision-making. By conducting in-depth examinations of specific pest management scenarios, researchers and practitioners can gather valuable insights into the complex interactions between pests, environments, and control strategies (Buczkowski et al., 2014). These studies also facilitate the evaluation of novel control methods, assessment of treatment efficacy, and identification of best practices for managing specific pest species while elucidating the social, economic, and environmental factors that influence pest management outcomes. Examples of related case studies from different countries are presented in Table 7.3.

TABLE 7.3
Case Studies on Various Household Pest Management Approaches Using Innovative Techniques

Insects	Country	Techniques Used	Findings	References
Periplaneta americana (Cockroach) (Blattodea)	China	RNAi	The study found that most tissues of *P. americana* are sensitive to extracellular dsRNA, making RNAi a powerful and effective tool for gene function studies in this species.	Li et al. (2021)
Drosophila suzukii (Fruit fly) (Diptera)	United Kingdom	SIT	The results showed that the release of sterile males led to suppression of the wild female *D. suzukii* population by up to 91% compared to control sites without sterile male releases.	Homem et al. (2022)
Trogoderma granarium (Khapra beetle) (Coleoptera)	India	Biological approaches: diatomaceous earth (DE), IGR (lufenuron and tebufenozide)	The results showed 84.9% mortality rate when DE was used alone; lufenuron alone registered 66.4% mortality rate, tebufenozide alone showed 41.5% mortality rate, andthe combined treatment of lufenuron and DE exhibited the highest mortality rate at 91.6%.	Sagheer et al. (2020)
Periplaneta americana (American cockroach) (Blattodea)	Iran	Biological approaches; biocontrol agent (parasitoid *Evania appendigaster*), along with physical method	This study investigated the use of four physical treatments (freezing, heating, UV irradiation, and gamma irradiation) to inhibit the embryogenesis of *Periplaneta americana* oothecae before releasing them parasitized by *Evania appendigaster* for biological control. The findings revealed that UV-treated oothecae allowed hatching, while the other treatments effectively inhibited it. Among the treatments, gamma irradiation had the least impact on the parasitism rate, emergence rate, and developmental time of *E. appendigaster*, making it the most suitable for preserving the parasitic wasp's efficacy. Freezing significantly reduced the emergence rate of *E. appendigaster*to below 20%. The results suggest that combining gamma-irradiated, parasitized oothecae with bait control could effectively manage *P. americana* populations in habitats like sewer networks.	Hwang et al. (2004)
Drosophila melanogaster (Fruit fly) (Diptera)	Turkey	Nanotechnology	The research found that various nanopesticides, including permethrin (20 ± 5 nm), copper (II) hydroxide [$Cu(OH)_2$] (15 ± 6 nm), acephate (55 ± 25 nm), and validamycin (1,177 ± 220 nm), exhibited significant toxic and genotoxic effects on *Drosophila melanogaster* at higher doses.	Demir et al. (2022)
Sitophilus oryzae (Rice weevil) (Coleoptera)	USA	Acoustics	The device captured distinct sound signals from the insects, which were recorded and analyzed. The sound pulses from the insects were found to have the highest energy between 500 and 1,500 Hz, with some additional energy in the 6–8 kHz range. This indicates that the TreeVibes system is capable of identifying insect activity in stored grain, which could help in monitoring and managing pest infestations.	Mankin et al. (2021)
Female *Cimex lectularius* (Bedbugs) (Hemiptera)	USA	Ultrasonic repellent devices	The study found that ultrasonic devices did not repel or attract bedbugs. Female bedbugs were equally likely to be found in both the sound and no-sound areas, indicating that these devices are ineffective in repelling bedbugs. Interestingly, more bedbugs preferred to stay in the middle corridor between the test and control areas when the ultrasonic devices were active, but overall, the results suggest that commercial ultrasonic devices are not a viable tool for bedbug control.	Yturralde et al. (2012)
Blattella germanica (German cockroach) (Blattodea)	Iran	Ultrasonic waves	Fisher's test indicated that the highest repellency in both male and female German cockroaches occurred at frequencies of 35 and 40 kHz. Regression test results revealed that the most lethal effects were observed at frequencies of 40 and 75 kHz. The ultrasonic frequencies studied in this research show potential for repelling and killing German cockroaches, which pose risks to human health and the environment.	Khan-Ahmadi et al. (2023)

7.7 CONCLUSION

Managing pests in homes can be challenging, often complicated by decisions driven more by personal opinions and up-front costs rather than informed judgment. The first step in effective pest management is to identify the pests and understand the potential damage they might cause, then choose the most suitable control methods. Non-chemical approaches, such as maintaining cleanliness and regular maintenance, can be very effective in preventing infestations. Integrated pest management (IPM) is a sustainable method that can be applied in households but can be difficult to implement because pest management is often reactive rather than proactive. Many people lack the necessary knowledge of pest biology, identification skills, and understanding of various control methods required for effective pest management. Additionally, trusted sources of information, such as pest management professionals, insecticide salespeople, family, and friends, may be biased or lack expertise, which can hinder the adoption of alternative methods, like biological control. While chemical insecticides can quickly eliminate pests, the benefits of preventive measures, like habitat modification, are less obvious. Therefore, it is essential to shift perceptions, improve knowledge among non-professionals, and make informed pest management decisions. The advent of advanced techniques, such as gene editing, artificial intelligence (AI), and the sterile insect technique (SIT), has introduced new, environmentally friendly methods for pest management, offering improved precision, effectiveness, and sustainability. AI facilitates real-time monitoring and early detection, gene editing provides targeted pest control, and SIT reduces pest populations without chemicals. However, these methods also present challenges, including high costs, ethical concerns, the need for specialized knowledge, and potential unintended ecological impacts. Despite their significant potential to revolutionize pest management, these techniques must be carefully evaluated for their limitations and overall impact.

REFERENCES

Alam, A., Abbas, S., Abbas, A., Abbas, M., Hafeez, F., Shakeel, M., and Zhao, C. R. (2023). Emerging trends in insect sex pheromones and traps for sustainable management of key agricultural pests in Asia: Beyond insecticides – A comprehensive review. *International Journal of Tropical Insect Science*, 43(6): 1867–1882.

Ando, T., Inomata, S. I., and Yamamoto, M. (2004). Lepidopteran sex pheromones. In: S. Schulz (ed.), *The chemistry of pheromones and other semiochemicals I. Topics in current chemistry* (Vol. 239). Berlin, Heidelberg: Springer. https://doi.org/10.1007/b95449

Asaoka, T., Almagro, J., Ehrhardt, C., Tsai, I., Schleiffer, A., Deszcz, L., Junttila, S., Ringrose, L., Mechtler, K., Kavirayani, A., Gyenesei, A., Hofmann, K., Duchek, P., Rittinger, K., and Ikeda, F. (2016). Linear ubiquitination by LUBEL has a role in *Drosophila* heat stress response. *EMBO Reports*, 17(11): 1624–1640. https://doi.org/10.15252/embr.201642378

Athanassiou, C. G., Kavallieratos, N. G., Benelli, G., Losic, D., UshaRani, P., and Desneux, N. (2018). Nanoparticles for pest control: Current status and future perspectives. *Journal of Pest Science*, 91: 1–15.

Balatsos, G., Karras, V., Puggioli, A., Balestrino, F., Bellini, R., Papachristos, D. P., and Michaelakis, A. (2024). Sterile Insect Technique (SIT) field trial targeting the suppression of *Aedes albopictus* in Greece. *Parasite*, 31.

Bale, J. S., VanLenteren, J. C., and Bigler, F. (2008). Biological control and sustainable food production. *Philosophical Transactions of the Royal Society B: Biological Sciences*, 363(1492): 761–776.

Barrangou, R., and Doudna, J. A. (2016). Applications of CRISPR technologies in research and beyond. *Nature Biotechnology*, 34(9): 933–941.

Bell, C. H. (2014). Pest control of stored food products: Insects and mites. In: *Hygiene in food processing* (pp. 494–538). Woodhead Publishing.

Benedict, M. Q. (2021). Sterile insect technique: Lessons from the past. *Journal of Medical Entomology*, 58(5): 1974–1979.

Blore, K., Baldwin, R., Batich, C. D., Koehler, P., Pereira, R., Jack, C. J., and Xue, R. D. (2022). Efficacy of metal nanoparticles as a control tool against adult mosquito vectors: A review. *Frontiers in Tropical Diseases*, 3: 969299.

Bourtzis, K., and Vreysen, M. J. (2021). Sterile insect technique (SIT) and its applications. *Insects*, 12(7): 638.

Buczkowski, G., Roper, E., and Chin, D. (2014). Polyacrylamide hydrogels: An effective tool for delivering liquid baits to pest ants (Hymenoptera: Formicidae). *Journal of Economic Entomology*. 107(2): 748–757. https://doi.org/10.1603/EC13508

Byrne, D. N., Carpenter, E. H., Thoms, E. M., andCotty, S. T. (1984). Public attitudes toward urban arthropods. *Bulletin of the ESA*, 30(2): 40–44.

Chitra, G. A., Kaur, P., Bhatnagar, T., Manickam, P., and Murhekar, M. V. (2013). High prevalence of household pesticides and their unsafe use in rural South India. *International Journal of Occupational Medicine and Environmental Health*, 26: 275–282.

Christiaens, O., Whyard, S., Vélez, A. M., and Smagghe, G. (2020). Double-stranded RNA technology to control insect pests: Current status and challenges. *Frontiers in Plant Science*, 11: 451.

Cranshaw, W. (2011). A review of nuisance invader household pests of the United States. *American Entomologist*, 57(3): 165–169.

Daniels, J. L., Olshan, A. F., Teschke, K., Hertz-Picciotto, I., Savitz, D. A., Blatt, J., and Castleberry, R. P. (2001). Residential pesticide exposure and neuroblastoma. *Epidemiology*, 12(1): 20–27.

Demir, E., Kansız, S., Doğan, M., Topel, Ö., Akkoyunlu, G., Kandur, M. Y., and Turna Demir, F. (2022). Hazard assessment of the effects of acute and chronic exposure to permethrin, copper hydroxide, acephate, and validamycin nanopesticides on the physiology of *Drosophila*: Novel insights into the cellular internalization and biological effects. *International Journal of Molecular Sciences*, 23(16): 9121.

Demirel, M., and Kumral, N. A. (2021). Artificial intelligence in integrated pest management. In: *Artificial intelligence and IoT-based technologies for sustainable farming and smart agriculture* (pp. 289–313). IGI Global.

Dermauw, W., Jonckheere, W., Riga, M., Livadaras, I., Vontas, J., and VanLeeuwen, T. (2020). Targeted mutagenesis using CRISPR-Cas9 in the chelicerate herbivore *Tetranychus urticae*. *Insect Biochemistry and Molecular Biology*, 120: 103347.

DuBois, K. P. (1958). Insecticides, rodenticides, herbicides; household hazards. *Postgraduate Medicine*, 24(3): 278–288.

Dyck, V. A., Hendrichs, J., and Robinson, A. S. (2021). Sterile insect technique: Principles and practice in area-wide integrated pest management (p. 1216). Taylor and Francis.

Ehler, L. E. (2006). Integrated pest management (IPM): Definition, historical development and implementation, and the other IPM. *Pest Management Science*, 62(9): 787–789.

Fields, P. G., and Muir, W. E. (2018). Physical control. In: *Integrated management of insects in stored products* (pp. 195–221). CRC Press.

Frederick, Brandon. (2015). *Household insect pest guide* (p. 3). Distance Master of Science in Entomology Projects. http://digitalcommons.unl.edu/entodistmasters/3

Georghiou, G. P. (1972). The evolution of resistance to pesticides. *Annual Review of Ecology and Systematics*: 133–168.

Gilles, A. F., Schinko, J. B., and Averof, M. (2015). Efficient CRISPR-mediated gene targeting and transgene replacement in the beetle *Tribolium castaneum. Development*, 142(16): 2832–2839.

Graf, J. F. (1993). The role of insect growth regulators in arthropod control. *Parasitology Today*, 9(12): 471–474.

Gratz, S. J., Wildonger, J., Harrison, M. M., and O'Connor-Giles, K. M. (2013). CRISPR/Cas9-mediated genome engineering and the promise of designer flies on demand. *Fly*, 7(4): 249–255.

Grossman, J. (1995). What's hiding under the sink: Dangers of household pesticides. *Environmental Health Perspectives,* 103(6): 550–554. https://doi.org/10.1289/ehp.95103550.

Hajek, A. E., and Eilenberg, J. (2018). *Natural enemies: An introduction to biological control*. Cambridge University Press.

Hendrichs, J., Kenmore, P., Robinson, A. S., and Vreysen, M. J. (2007). Area-wide integrated pest management (AW-IPM): Principles, practice and prospects. In: M. J. B. Vreysen, A. S. Robinson, and J. Hendrichs (eds.), *Area-wide control of insect pests: From research to field implementation* (pp. 3–33). Dordrecht, The Netherlands: Springer.

Homem, R. A., Mateos-Fierro, Z., Jones, R., Gilbert, D., Mckemey, A. R., Slade, G., and Fountain, M. T. (2022). Field suppression of spotted wing drosophila (SWD) (*Drosophila suzukii* Matsumura) using the sterile insect technique (SIT). *Insects*, 13(4): 328.

Hwang, S. Y., and Chen, L. M. (2004). Effects of four physical treatments of oothecae of *Periplaneta americana* on parasitism and development of parasitic wasp, *Evania appendigaster. Environmental Entomology*, 33(5): 1321–1326.

i5K Consortium. (2013). The i5K Initiative: Advancing arthropod genomics for knowledge, human health, agriculture, and the environment. *Journal of Heredity*, 104(5): 595–600.

Khan-Ahmadi,A., Vatandoost, H., Akhavan, A.A., Baniardalani, M., Khalifeh-Soltani, K., Azarm, A., and Zahraei-Ramazani, A. (2023). Evaluation of repellency and lethal effects of ultrasonic waves on the *Blattella germanica* (Blattodea: Blattellidae). *Journal of Arthropod Borne Diseases*, 17(1): 83–93. https://10.18502/jad.v17i1.13204 PMID: 37609562; PMCID: PMC10440496.

Kogan, M. (1998). Integrated pest management: Historical perspectives and contemporary developments. *Annual Review of Entomology*, 43(1): 243–270.

Komal, J., Desai, H. R., Samal, I., Mastinu, A., Patel, R. D., Kumar, P. D., and Bhoi, T. K. (2023). Unveiling the genetic symphony: Harnessing CRISPR-Cas genome editing for effective insect pest management. *Plants*, 12(23): 3961.

Kyrou, K., Hammond, A. M., Galizi, R., Kranjc, N., Burt, A., Beaghton, A. K., and Crisanti, A. (2018). A CRISPR – Cas9 gene drive targeting double sex causes complete population suppression in caged Anopheles gambiae mosquitoes. *Nature Biotechnology*, 36(11): 1062–1066.

Leong, W. H., Teh, S. Y., Hossain, M. M., Nadarajaw, T., Zabidi-Hussin, Z., Chin, S. Y., and Lim, S. H. E. (2020). Application, monitoring and adverse effects in pesticide use: The importance of reinforcement of Good Agricultural Practices (GAPs). *Journal of Environmental Management*, 260: 109987.

Li, F., and Scott, M. J. (2016). CRISPR/Cas9-mediated mutagenesis of the white and Sex lethal loci in the invasive pest, *Drosophila suzukii. Biochemical and Biophysical Research Communications*, 469(4): 911–916.

Li, L., Jing, A., Xie, M., Li, S., and Ren, C. (2021). Applications of RNA interference in American cockroach. *Journal of Visualized Experiments*, (178): e63380.

Ligonniere, S., Bantz, A., Raymond, V., and Goven, D. (2024). Using RNA interference targeting a nicotinic acetylcholine receptor subunit to counteract insecticide accommodation mechanisms: Example of the β1 subunit in the imidacloprid-accommodated American cockroach, *Periplaneta americana. Journal of Pesticide Science*, 49(1): 58–64.

Mankin, R., Hagstrum, D., Guo, M., Eliopoulos, P., and Njoroge, A. (2021). Automated applications of acoustics for stored product insect detection, monitoring, and management. *Insects*, 12(3): 259.

Mankin, R. W. (2012). Applications of acoustics in insect pest management. *CABI Reviews*, 1–7.

Mawthoh, A. B. T., Seram, D., and Watt, H. J. (2023a). Enhancing crop resilience through biotechnology. In: B. Koul, P. K. Prabhakar, and E. Goutam (eds.), *Smart and sustainable agricultural technology* (p. 177). Lovely Professional University Publications. ISBN 978-81-19334-75-9.

Mawthoh, A. B. T., Seram, D., and Watt, H. J. (2023b). Green synthesized plant-based nanotechnology: Cutting edge innovation fostering sustainability and revolutionizing agriculture. In: *E3S web of conferences* (Vol. 453, p. 01018). EDP Sciences. https://doi.org/10.1051/e3sconf/202345301018

Mendoza, Q. A., Pordesimo, L., Neilsen, M., Armstrong, P., Campbell, J., and Mendoza, P. T. (2023). Application of machine learning for insect monitoring in grain facilities. *AI*, 4(1): 348–360.

Metcalf, R. L. (1989). Insect resistance to insecticides. *Pesticide Science*, 26(4): 333–358.

Molly,M. G. (2016). *Control household pests without scary poisons*. www.nrdc.org/stories/control-household-pests-without-scary-poisons

Mumford, J. D. (2012). Science, regulation, and precedent for genetically modified insects. *PLoS Neglected Tropical Diseases*, 6(1): e1504.

Murugan, K., Benelli, G., Panneerselvam, C., Subramaniam, J., Jeyalalitha, T., Dinesh, D., Nicoletti, M., Hwang, J.S., Suresh, U., and Madhiyazhagan, P. (2015). *Cymbopogon citratus*-synthesized gold nanoparticles boost the predation efficiency of copepod Mesocyclops aspericornis against malaria and dengue mosquitoes. *Experimental Parasitology*, 153: 129–138. https://doi.org/10.1016/j.exppara.2015.03.017.

Panthawong, A., Doggett, S.L., and Chareonviriyaphap, T. (2021). The efficacy of ultrasonic pest repellent devices against the Australian paralysis tick, *Ixodes holocyclus* (Acari: Ixodidae). *Insects*, Apr 30; 12(5): 400. https://10.3390/insects12050400 PMID: 33946185; PMCID: PMC8145533.

Rabiee, M. H., Mahmoudi, A., Siahsarvie, R., Kryštufek, B., and Mostafavi, E. (2018). Rodent-borne diseases and their public health importance in Iran. *PLoS Neglected Tropical Diseases*, 12(4): e0006256. https://doi.org/10.1371/journal.

Rahman, S. M., and Ravi, G. (2022). Role of artificial intelligence in pest management. *Current Topics in Agricultural Sciences*, 7: 64–81.

Reeves, R. G., Denton, J. A., Santucci, F., Bryk, J., and Reed, F. A. (2012). Scientific standards and the regulation of genetically modified insects. *PLoS Neglected Tropical Diseases*, 6(1): e1502.

Sagheer, M., Asif, M. U., Ali, Q., and Zaman, S. (2020). Combined effect of diatomaceous earth and insect growth regulators against *Trogoderma granarium* (Coleoptera: Dermestidae). *Scence Letters*, 8: 55–60.

Schairer, C. E., Najera, J., James, A. A., Akbari, O. S., and Bloss, C. S. (2021). Oxitec and Mosquito Mate in the United States: Lessons for the future of gene drive mosquito control. *Pathogens and Global Health*, 115(6): 365–376.

Schoelitsz, B., Meerburg, B. G., and Takken, W. (2019). Influence of the public's perception, attitudes, and knowledge on the implementation of integrated pest management for household insect pests. *Entomologia Experimentalis et Applicata*, 167(1): 14–26.

Schwindenhammer, S. (2020). The rise, regulation and risks of genetically modified insect technology in global agriculture. *Science, Technology and Society*, 25(1): 124–141

Srivani,M. V. K. (2019). Green pest management practices for sustainable buildings: Critical review. *Science Progress*, 102(2): 141–152.

Sudakin, D. L. (2003). Biopesticides. *Toxicological Reviews*, 22: 83–90.

Tiwari, D. K., and Ansari, M. A. (2016). Electronic pest repellent: A review. In *International conference on innovations in information embedded and communication systems* (ICIIECS'16).

Toscano-Miranda, R., Toro, M., Aguilar, J., Caro, M., Marulanda, A., and Trebilcok, A. (2022). Artificial-intelligence and sensing techniques for the management of insect pests and diseases in cotton: A systematic literature review. *The Journal of Agricultural Science*, 160(1–2): 16–31.

Tunaz, H., and Uygun, N. (2004). Insect growth regulators for insect pest control. *Turkish Journal of Agriculture and Forestry*, 28(6): 377–387.

Urban, J. A., Doherty, C. A., Jordan, W. T., Bliss, J. E., Feng, J., Soruco, M. M., and Larschan, E. N. (2017). The essential Drosophila CLAMP protein deferentially regulates non-coding roX RNAs in male and females. *Chromosome Research*, 25: 101–113.

Walker, T. J. (1996). Acoustic methods of monitoring and manipulating insect pests and their natural enemies. In: *Pest management in the subtropics: Integrated pest management-a Florida perspective* (pp. 245–257). Andover, UK: Intercept.

Weintraub, P. G. (2009). Physical control: An important tool in pest management programs. *Biorational Control of Arthropod Pests: Application and Resistance Management*, 317–324.

William, F. D. (1978). Electromagnetic repellers – Fact or fiction? *Proceedings of the 8th Vertebrate Pest Conference*, 17. https://digitalcommons.unl.edu/vpc8/17

Witzgall, P., Kirsch, P., and Cork, A. (2010). Sex pheromones and their impact on pest management. *Journal of Chemical Ecology*, 36: 80–100.

Wynant, N., Santos, D., VanWielendaele, P., and VandenBroeck, J. (2014). Scavenger receptor-mediated endocytosis facilitates RNA interference in the desert locust, *Schistocerca gregaria. Insect Molecular Biology*, 23(3): 320–329.

Young, S. L. (2017). A systematic review of the literature reveals trends and gaps in integrated pest management studies conducted in the United States. *Pest Management Science*, 73(8): 1553–1558.

Yturralde, K. M., and Hofstetter, R. W. (2012). Efficacy of commercially available ultrasonic pest repellent devices to affect behavior of bed bugs (Hemiptera: Cimicidae). *Journal of Economic Entomology*, 105(6): 2107–2114, https://doi.org/10.1603/EC12166

Zalucki, M. P., Adamson, D., and Furlong, M. J. (2009). The future of IPM: Whither or wither? *Australian Journal of Entomology*, 48(2): 85–96.

8 Recent Advancements to Tackle Grasshopper/Locust

Devina Seram[1*], *Aquiny Befairlyne T. Mawthoh*[1], *Yashika*[2], *Jayesh Gulia*[3], *and Haobijam James Watt*[4]

[1]Department of Entomology, School of Agriculture, Lovely Professional University, Phagwara 144411, Punjab, India

[2]Department of Genetics and Plant Breeding, Acharya Narendra Deva University of Agriculture and Technology, Kumarganj 224229, Ayodhya, UP, India

[3]School of Agriculture and Environment Science, University of Western Australia, Perth, Australia

[4]Department of Agricultural Economics and Extension, School of Agriculture, Lovely Professional University, Phagwara 144411, Punjab, India

[*]Corresponding author

8.1 INTRODUCTION

In many parts of the world, locusts and grasshoppers are considered one of the most detrimental agricultural pests that can result in significant financial losses (Lecoq, 2001; Zhang and Hunter, 2017). Throughout history, locust plagues, floods, and droughts have been regarded as the three most devastating natural calamities (Matthews, 2021). Controlling grasshoppers and locusts is thus crucial in ensuring global food security and environmental well-being. These highly destructive insect pests have the ability to cause extensive crop damages, resulting in substantial reductions in crop yields and food availability. This can worsen issues of hunger, malnutrition, and food security, while also imposing significant economic burdens on farmers and the agricultural sector. There is also a risk of disrupting global food systems and international trade due to their impact. Swarms of grasshoppers and locusts have the potential to disturb ecosystems, resulting in alterations in the composition of plants and a decrease in biodiversity. Additionally, they can contribute to soil degradation by intensifying erosion and depleting nutrients, which can have lasting impacts on the overall health of the ecosystem. While the current management strategies for grasshopper and locust populations are focused on the application of toxic pesticides, it can lead to water source and soil contaminations, thus posing a great threat to aquatic organisms, human beings, and the environment. For example, broad-spectrum neurotoxins were applied to 13 million hectares across 22 nations and three continents to eliminate locust swarms (Belayneh, 2005). One of the primary reasons traditional methods have become less effective is the growing resistance among locust populations to commonly used chemical pesticides. Overreliance on these chemicals has led to the evolution of resistant strains, rendering different pesticides ineffective (Ahmad et al., 2020). Furthermore, climate change can intensify locust infestations, hence increasing the importance of implementing effective control approaches. Efficient management strategies are crucial in reducing the effects of these pests, safeguarding vulnerable commodities and guaranteeing a sustainable food supply. Failing to take action can have catastrophic consequences, such as food shortages, heightened poverty rates, environmental deterioration, and pollution. This emphasizes the necessity for proactive, comprehensive, and long-term approaches in controlling grasshoppers and locusts, which consider both human requirements and environmental sustainability. Globally, various countries have implemented cutting-edge technologies to manage locust outbreaks more effectively. In China, for instance, drone technology has been utilized to monitor and spray affected areas with precision, reducing the need for large-scale chemical application. Similarly, in Africa, where locust swarms have devastated crops, satellite-based monitoring systems have been employed to predict and track the movement of swarms, enabling early intervention and reducing crop damage.

8.2 HISTORICAL IMPACT OF LOCUST PLAGUES ON CIVILIZATIONS

The historical influence of locust outbreaks on civilizations has been significant and extensive, with countless accounts of catastrophic infestations that have devastated agricultural societies, resulted in widespread hunger, and even contributed to the downfall of empires (Deguine et al., 2021). Locust plagues have been a recurring phenomenon throughout history, occurring in various regions, including ancient Egypt, medieval Europe, Africa, Asia, and the Middle East (Lecoq and Cease, 2022). These plagues are often described

DOI: 10.1201/9781003516538-8

TABLE 8.1
Common Species of Locust and Grasshopper Found in Different Regions of the World (Orthoptera: Acrididae)

Common Name	Scientific Name	Distribution
Desert locust	*Schistocerca gregaria* (Forskål)	Africa, Near East Asia, Southwest Asia
Migratory locust	*Locusta migratoria* (Linnaeus)	Africa, Asia, Australia, East Indies, New Zealand
Red locust	*Nomadacris septemfasciata* Audinet-Serville	Africa (Malawi, Mozambique, Tanzania and Zambia)
Brown locust	*Locustana pardalina* (Walker)	Southern Africa
Australian plague locust	*Chortoicetes terminifera* (Walker)	Australia
Rocky mountain locust	*Melanoplus spretus* (Walsh)	Western half of USA and Canada (now extinct)
Italian locust	*Calliptamus italicus* (Linnaeus)	Western Europe, Central Asia
South American locust	*Schistocerca cancellata* (Serville)	South America
Moroccan locust	*Dociostaurus maroccanus* (Thunberg)	Morocco, Egypt, Libya, Europe, Kazakhstan
Desert grasshopper	*Sphingonotus rubescens* (Walker)	North Africa, Middle East
American grasshopper	*Schistocerca Americana* (Drury)	Florida, Mexico, Bahamas
Oriental migratory locust	*Locusta migratoria manilensis* (Franz Meyen)	Southeastern Asia (Philippines, Malaysia, Eastern China); outbreaks in Thailand, Java, Sulawesi, and southern Japan

in biblical and apocalyptic terms, as the insects devour entire crops, leaving behind barren landscapes and causing widespread starvation (Martin, 2015). The repercussions have been severe and widespread, as locust infestations have been associated with the downfall of historical civilizations, such as the Akkadian Empire in Mesopotamia, and the demise of Pharaonic Egypt (Kenawy and Abdel-Hamid, 2015). Locust swarms, including species like *Schistocerca gregaria, Locusta migratoria, Calliptamus italicus, Nomadacris septemfasciata, Chortoicetes terminifera, Dociostaurus maroccanus*, etc., regularly occur on every continent except Antarctica, impacting the lives of approximately 10% of the global population (Latchininsky et al., 2011) (Table 8.1). In 2003–2005, the desert locust (*S. gregaria*) outbreak in Africa caused devastating agricultural losses ranging from 80 to 100% for nearly 8 million people (Brader et al., 2005). Additionally, they have worsened social and economic pressures, resulting in migration, violence, and even the transmission of various dreadful diseases. Locusts have long been regarded as a forewarning of catastrophe, a divine retribution, or an indication of imminent calamity in numerous societies (Price, 2023). This underscores the profound fear and reverence that these insects have evoked throughout human history, emphasizing the necessity for efficient management approaches to alleviate the consequences of these insect swarms on susceptible communities and guarantee food security (Deguine et al., 2021). By using preventive measures, such epidemics are now more effectively managed and have decreased in frequency and magnitude compared to the past. However, these invasions continue to occur in severe forms at specific locations of the world.

8.3 BIOLOGY AND BEHAVIOR OF GRASSHOPPERS AND LOCUSTS

8.3.1 Description of Life Cycle

Locusts are a type of grasshopper that exhibits swarming behavior; hence, their life cycle is similar to that of grasshoppers. However, locusts tend to have a more rapid development and reproduction rate, which allow them to quickly colonize new areas and form large swarms (Pener and Simpson, 2009). The life cycle of grasshoppers and locusts consists of three distinct stages: egg, nymph, and adult (hemimetabolous insects, hence no pupal stage). It begins with eggs laid in moist soil or vegetation, which hatch into pronymphs after 2–4 weeks depending on temperature and humidity (Karpakakunjaram et al., 2002). Eggs are typically white, yellow, or brown and range from 0.5 to 2 mm in length. Females can lay up to 50 eggs at a time and may lay multiple batches (Symmons and Cressman, 2001). The developing nymphs, also known as hoppers, look like small wingless adults and undergo a series of molts as they grow by feeding on crops, shedding their skin five to six times (5–6 molts) over several weeks before reaching adulthood (Karpakakunjaram et al., 2002). Finally, the adult stage is reached, characterized by fully developed wings and reproductive organs, enabling flight, reproduction, and migration in search of food or mates. The adults can live for several weeks to a few months, depending on species and environmental factors. This cycle is similar for both grasshoppers and locusts, although locusts tend to develop and reproduce more rapidly, facilitating their swarming behavior.

8.3.2 Factors Triggering Locust Swarming and Transition from Solitary to Gregarious Phase

Locust swarming is a complex phenomenon where solitary locusts undergo a remarkable transformation into a gregarious phase, causing widespread destruction and posing serious challenges to food security and ecosystems across the globe (Anstey et al., 2009). During the transitory phase, locusts undergo significant changes in behavior, physiology, and morphology, including heightened mobility, feeding, and breeding, which ultimately result in the formation of large swarms that can migrate from one place to another (Pener and Simpson, 2009). The phase transition is brought about by a mix of various factors, such as the following.

8.3.2.1 Environmental Factors

Locust swarming is triggered by a combination of environmental factors, including rainfall, temperature, and vegetation (Veran et al., 2015). Heavy rainfall and subsequent drought create an ideal breeding ground for locusts, allowing them to multiply rapidly (Wang et al., 2023). Warm temperatures and abundant vegetation provide the necessary resources for locusts to grow and develop. When these conditions coincide, locust populations can explode, leading to the formation of swarms and plagues.

8.3.2.2 Density and Crowding

As locust populations grow, so does the density of individuals in a given area. When locusts are forced into close proximity, they undergo a series of physiological and behavioral changes, triggering the transition from the solitary to the gregarious phase (Du et al., 2021). This phase change is characterized by changes in color, behavior, and morphology, preparing locusts for the highly social and mobile lifestyle of a swarm.

8.3.2.3 Serotonin and Hormonal Changes

The transition from solitary to gregarious phase is also influenced by hormonal changes, particularly the increase of serotonin levels. Serotonin promotes the development of gregarious traits, such as increased mobility, aggression, and attraction to other locusts (Anstey et al., 2009). As locusts interact with one another, their serotonin level increases, which strengthens their social behavior and drives the formation of swarms (Pener, 1991).

8.3.2.4 Visual and Chemical Cues

Locusts use visual and chemical cues to communicate and coordinate their behavior, facilitating the formation of swarms. They detect visual cues, such as movement and color, to orient themselves toward other locusts and swarms. Chemical cues, like pheromones, are released by locusts to signal their presence and attract other individuals. As locusts aggregate, these cues amplify, creating a positive feedback loop that drives the growth and movement of the swarm (Despland, 2002).

8.3.2.5 Genetic Predisposition

Some locust species, such as desert locust, migratory locust, brown locust, Australian plague locust, etc., are more prone to swarming due to their genetic makeup. These species have evolved to respond to environmental cues and undergo phase changes more readily, allowing them to capitalize on favorable conditions and exploit new resources. Genetic factors can influence the likelihood and intensity of swarming, making some locust populations more susceptible to outbreaks (Topaz et al., 2012).

8.4 TRADITIONAL METHODS OF LOCUST/GRASSHOPPER CONTROL

8.4.1 Early Use of Chemical Insecticides

The use of insecticides to control locust and grasshopper populations dates back to ancient times, with early civilizations employing natural substances, like sulfur, arsenic, and plant extracts, to ward off these pests. However, it was not until the mid-20th century that synthetic insecticides became widely used for locust control. Organochlorine (OC) insecticides, such as DDT and dieldrin, were initially effective but soon raised concerns due to their persistence nature, bioaccumulation in the environment, and toxicity to non-target organisms. In the 1960s and the 1970s, organophosphate (OP) and carbamate insecticides became popular for locust control due to their relatively shorter persistence and lower toxicity to mammals. However, these chemicals still posed risks to beneficial insects, birds, and other wildlife. Moreover, locust populations developed resistance to these insecticides, reducing their effectiveness and necessitating increased application rates. The introduction of pyrethroid insecticides (deltamethrin, cypermethrin, permethrin, etc.) in the 1980s offered a more targeted approach to locust control, with lower toxicity to mammals and birds. However, pyrethroids are still harmful to aquatic life and beneficial insects (MacCuaig et al., 1979). Modern insecticides, like neonicotinoids (imidacloprid, acetamiprid, thiamethoxam, etc.) and fipronil, have also been used for locust control, but their environmental impacts are still being assessed and debated (Parkinson et al., 2017; Simon-Delso et al., 2015).

8.4.2 Integrated Management Approach

Combining various available control methods is crucial for sustainable and effective management of locusts and grasshoppers. An IPM approach considers the ecosystem as a whole and aims to minimize harm to non-target organisms and the environment. Cultural control involves modifying agricultural practices to prevent locust and grasshopper infestations. This includes crop rotation, sanitation, and irrigation management. For example, planting locust-resistant crop varieties or using trap crops can help reduce locust populations. Traditional mechanical control for grasshoppers and locusts includes techniques like hand-picking, trapping, and using physical barriers. Handpicking is effective for small-scale infestations but labor-intensive and time-consuming. Trapping using pitfall traps or sticky traps and flamethrower can capture or kill significant numbers but are not effective for large-scale infestations. Physical barriers, like fine mesh screens, or fine netting, and trenching, can prevent locusts from reaching crops but may not be practical in larger areas (Han, 1997). Biological control methods involve using natural predators, parasites (parasitoids), or pathogens to control grasshopper and locust populations, involving the introduction, augmentation, and conservation of biocontrol agents that feed or parasitize all the stages of locusts and grasshoppers. Predatory insects, like ants, ground beetles, and spiders, feed on locusts, while parasitic wasps and flies lay eggs inside locust bodies. Entomopathogenic pathogens, like fungi, bacteria, and viruses, can also infect and kill locusts (Lomer et al., 2001). These methods are environmentally friendly and can be cost-effective but may be slow-acting, require precise timing, and are affected by environmental factors. Moreover, these eco-friendly control strategies can be effective for small-scale infestations with low to moderate locust densities in localized areas following organic farming (Wang et al., 2023). They may not be effective for large-scale infestations and may not provide immediate results since precise timing and environmental conditions are vital, which can be affected by weather, terrain, and other factors. Monitoring and forecasting are critical components of IPM. Regular monitoring of locust and grasshopper populations helps identify areas requiring control measures, whereas forecasting models can predict locust and grasshopper outbreaks, enabling proactive control measures. Alternatively, the most common locust control method is the application of chemical insecticides using various methods, like spraying, dusting, or baiting. However, this approach should be used judiciously and in combination with other methods, as overuse can lead to pesticide resistance development and environmental pollution. To overcome these limitations, IPM strategies combining all the available control methods can improve the effectiveness, efficiency, and sustainability in managing grasshopper and locust populations. IPM aims to minimize environmental impacts, reduce chemical use, and optimize locust control. For example, using cultural control in combination with biological control can reduce locust populations while minimizing harm to non-target organisms. Similarly, combining chemical with physical control can improve their efficacy while reducing environmental impact, ensuring effectiveness and sustainability.

8.5 RECENT ADVANCEMENTS IN LOCUST OR GRASSHOPPER CONTROL

8.5.1 Genetic Research Advances

Recent advances in genetic research have enabled scientists to better understand the genetic makeup of grasshoppers and locusts, allowing for the development of targeted control measures. Genome sequencing and gene editing technologies like CRISPR/Cas9 have facilitated the identification of key genes controlling locust behavior, development, and reproduction (Zhang et al., 2023). This knowledge enables researchers to design precise genetic interventions to disrupt locust populations. One approach is to use RNA interference (RNAi) to silence specific genes essential for locust survival or reproduction. This can be achieved through genetic engineering of crops to produce RNAi molecules or by spraying RNAi-based pesticides. Another approach is to exploit genetic differences between locust species or populations to develop species-specific control measures. For example, genetic markers can be used to identify and target specific locust species or strains. Another strategy is the gene drive technology, which allows researchers to spread genetic modifications through locust populations, potentially leading to population suppression or elimination. This technology harnesses the power of genetic elements that can drive the inheritance of a particular trait, ensuring its spread throughout the population. Gene drive can be used to introduce traits that reduce locust fertility, disrupt mating behavior, or increase susceptibility to pesticides.

Genetic research also enables precision agriculture approaches, where genetic markers are used to monitor locust populations, track their movement, and identify areas requiring control measures. This information can be integrated with geographic information systems (GIS) and remote sensing technologies to optimize control strategies and reduce pesticide use (Ahmad et al., 2022). The advances in genetic research offer a promising avenue for sustainable pest management, reducing the reliance on broad-spectrum pesticides and minimizing environmental impacts. Targeted control measures can be designed to minimize harm to non-target organisms, preserving ecosystem balance and promoting biodiversity. By leveraging genetic knowledge, researchers can develop innovative solutions for managing grasshopper and locust populations, ensuring food security and environmental sustainability.

8.5.2 CRISPR/Cas9 Applications in Combating Locust Attack

CRISPR/Cas9 technology offers a wide range of applications, including targeted gene editing, gene knockout, and gene replacement (Zhang et al., 2023). With this technology, precise gene editing is made possible, ensuring that only the desired traits are introduced. With this level of precision, the potential for off-target effects is greatly minimized, resulting in a safer and more effective method for enhancing crops and managing insect populations. Scientists have the ability to create guide RNAs (gRNAs) that can be used to specifically target certain genes. This allows for precise editing of the genome without any unintended effects on other parts of the genetic material. This technique has the ability to target and disable certain genes that play a crucial role in the fertility of grasshoppers and locusts, resulting in the production of sterile individuals. This method guarantees that the insects that are set free cannot breed, which leads to a decrease in population numbers and lowers the chances of pest infestations (Yan et al., 2023). By utilizing modified versions of certain genes, it has the ability to render grasshoppers and locusts infertile. This method allows scientists to incorporate favorable characteristics while preserving the genetic makeup of the insect population. Nevertheless, the potential of CRISPR technology is undeniable. However, it is crucial to establish strict regulations to guarantee the safety and effectiveness of its use in agriculture and insect control. Efforts are underway to establish regulatory frameworks that will effectively govern the application of CRISPR in crop improvement and SITs. The aim is to ensure responsible use of these technologies while minimizing any potential environmental impact.

8.5.2.1 Development of Locust-Resistant Crop Varieties

CRISPR technology has emerged as a powerful tool for developing crop varieties resistant to grasshoppers and locusts. By editing specific genes, scientists can introduce traits that confer resistance to these pests. For instance, CRISPR can be used to introduce genes that produce insecticidal proteins, such as *Bacillus thuringiensis* (Bt) toxins, which are toxic to grasshoppers and locusts (Komal et al., 2023). Alternatively, CRISPR can be used to modify plant architecture, making it less accessible to these pests. This approach will enable farmers to reduce pesticide usage, minimizing environmental impacts and promoting sustainable agriculture.

8.5.2.2 Utilization in Sterile Insect Technique

CRISPR can also be applied to develop sterile insect techniques (SITs) for managing grasshoppers and locusts. SITs involve releasing genetically modified insects that are unable to reproduce, thereby reducing population sizes. CRISPR enables researchers to precisely edit genes responsible for insect fertility, creating sterile individuals. When released into the wild, these sterile insects mate with wild individuals but fail to produce offspring, leading to population decline (Kandul et al., 2019). This approach offers a promising alternative to traditional pesticides, reducing environmental impacts and minimizing harm to non-target organisms.

8.5.3 RNAi Methods Targeting Specific Genes to Disrupt Locust Development or Reproduction

RNA interference (RNAi) has emerged as a promising approach for controlling locust and grasshopper populations by targeting specific genes involved in their development and reproduction, such as those involved in molting, wing formation, metamorphosis, egg production in female, sperm development in male, etc. (Santos et al., 2014). Researchers can disrupt normal locust life cycles, preventing them from reaching adulthood or reproducing. For instance, targeting and silencing genes involved in wing development can prevent locusts from flying, reducing their ability to migrate and form swarms. Also, RNAi technique can reduce locust fertility, preventing population growth and minimizing the risk of outbreaks. This technology involves introducing double-stranded RNA (dsRNA) molecules that are complementary to the target gene's messenger RNA (mRNA), leading to its degradation and silencing (Figure 8.1). The external RNA molecules can be delivered through various methods, including microinjection, which involves injecting dsRNA molecules directly into locust embryos or tissues; oral delivery, that is, feeding locusts with dsRNA-coated food or water; topical application of dsRNA molecules directly to locust bodies; and nanoparticle-based, using specific nanoparticles to deliver dsRNA molecules to specific target tissues (He et al., 2022). Two types of RNAi can be implemented for locust control, namely, systemic RNAi and topical RNAi. Systemic RNAi involves introducing dsRNA molecules into the locust body, where they are transported to target tissues and cells, allowing the silencing of genes in specific tissues or organs, reducing off-target effects and increasing efficacy. This can be achieved through microinjection, oral delivery, or topical application of dsRNA molecules. On the other hand, topical RNAi involves applying dsRNA molecules directly to the locust body surface, where they are absorbed and transported to target tissues. This approach is particularly useful for controlling locust populations in definite and precise areas, such as agricultural fields, and can be achieved through spraying or dusting dsRNA-coated particles onto locusts or their habitats.

RNAi methods have shown high efficacy in disrupting locust development and reproduction, with minimal off-target effects (Wynant et al., 2014). However, further research is needed to ensure safety and efficacy in field applications. Additionally, RNAi methods can be combined with other control strategies, such as insecticides or biological control agents, to enhance their effectiveness. While sys-RNAi has been observed across many insect species, the sensitivity to dsRNA varies significantly. Many economically important insect species are resistant to systemically delivered dsRNA. However, locusts are particularly sensitive to intra-abdominally injected dsRNA, though naked dsRNA delivered through feeding has not been effective in inducing efficient RNAi responses. Locusts have been utilized as model organisms in various fields of insect biology, including endocrinology, neurophysiology, biochemistry, behavior, and ecology. Their robust RNAi response, particularly to intra-abdominal injection, highlights their utility in studying complex biological processes and developing novel pest control strategies.

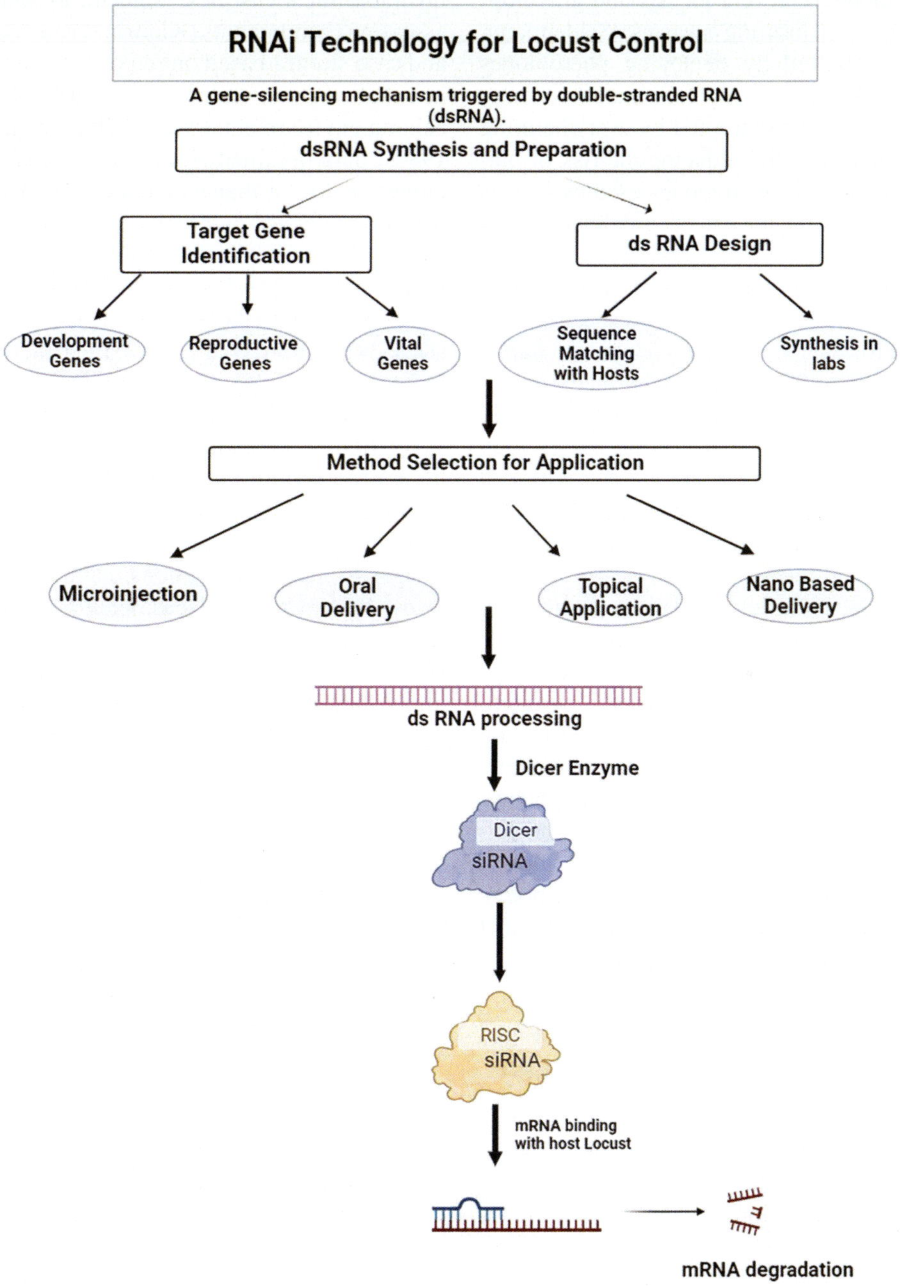

FIGURE 8.1 Introduction of double-stranded RNA (dsRNA) molecules that are complementary to the target gene's messenger RNA (mRNA), leading to its degradation and silencing; one of the advanced technologies which can be applied for locust management.

8.5.4 Development and Use of Pheromones to Disrupt Locust Mating and Swarming Behaviors

Pheromones are intra-specific chemical signals that play a crucial role in insect communication, mating, locating food, defense, etc. (El-Ghany, 2020). Locusts depend on pheromones to communicate and synchronize their activities, including mating, foraging, and migration. Pheromone-based management is an innovative and promising approach for controlling locust and grasshopper populations that involves the synthesis and deployment of artificial pheromones, thereby confusing, manipulating, and interfering with their behavior and stopping infestations (Ferenz and Seidelmann, 2003). For example, synthetic sex pheromones can be released to overwhelm males, making it difficult for them to locate females, a process known as mating disruption, thereby reducing mating success and population growth, and it can be achieved through various methods, like spraying pheromone-coated particles or using pheromone-impregnated traps or baits (Guo et al., 2020). Another application is swarming disruption, which involves using pheromone to prevent locusts and grasshoppers from aggregating and forming swarms (Seidelmann, 2006). This can be achieved by deploying pheromones that mimic the signals used by insects to initiate swarming behavior, and swarming can be disrupted by overwhelming locusts with these signals, thereby reducing the risk of outbreaks. Identifying the specific pheromones used by locusts and grasshoppers is crucial for the effective deliberation of this approach (Guo et al., 2020). For the isolation and characterization of pheromones, researchers use various techniques, such as gas chromatography and mass spectrometry (GC-MS), high-performance liquid chromatography (HPLC), etc. (Wadhams, 1990). Once identified, pheromones can be synthesized and tested for their effectiveness in disrupting both mating and swarming. Developing effective pheromone delivery systems is another essential consideration for successful pheromone-based control. This may involve the utilization of specialized equipment, such as sprayers or diffusers, to release pheromones in targeted areas. However, multi-location field trials are essential for evaluation, for researchers to conduct large-scale field trials in order to assess the impact of pheromones on locust and grasshopper populations, which can be measured by monitoring population densities, mating success, and swarming behavior. This approach provides a focused and ecologically conscious substitute for conventional pesticides, which have the potential to negatively impact beneficial insects and pollute soil and water. Utilizing this control offers a hopeful solution for sustainable locust management since it has potential to significantly mitigate the destructive consequences of locust infestations on agriculture and food security.

8.5.5 Use of Satellite Imagery and Drones for Real-Time Monitoring of Locust Movements

Despite advancements in technology and monitoring, locust outbreaks continue to cause significant devastation and food insecurity. One reason for this is ineffective monitoring, management, or population control in some locust-prone areas, often due to a lack of resources and technology. Environmental shifts, such as changes in land use, and variability in weather patterns within locust habitats can create optimal conditions for breeding. It is crucial to recognize these changes and act promptly to prevent locust populations from growing and transitioning from their solitary phase to the gregarious phase, which can trigger an outbreak. Continuous monitoring during the solitary phase is therefore essential (Zhang et al., 2019). Managing locusts is a complex task that necessitates a multidisciplinary approach, including entomology, biology, and ecology, along with aspects like spatial distribution modeling, climate analysis, weather forecasting, the use of chemical insecticides or bioagents, and remote sensing applications. Geospatial and remote sensing data have emerged as key resources in various locust-related research and management applications (Latchininsky et al., 2010). Remote sensing plays a vital role in forecasting and monitoring ecological conditions favorable for locust development, as well as mapping and assessing their habitats (Klein and Kuenzer, 2021). Research and case studies based on remote sensing have been critical in enhancing our understanding of locust-relevant ecological and environmental factors (Piou et al., 2023). Since the 1980s, data obtained from remote sensing have supported numerous locust management efforts, leading to improved control of locust outbreaks and plagues worldwide. The use of remote sensing data to study locust outbreaks was first introduced by Pedgley and Hielkema and later incorporated into the FAO's Desert Locust Information Service (DLIS) (Meynard et al., 2020). In 2018, the FAO began testing a long-range, fixed-wing drone to survey regions where locusts might be present by mapping areas of green annual vegetation. Unmanned aerial vehicles (UAVs), or drones, offer a complementary approach to satellite imagery for monitoring locust populations. Equipped with high-resolution cameras and sensors, drones can capture detailed images of locust populations, habitats, and movement patterns. They can also release pheromones or insecticides in targeted areas, enhancing control efforts, but they require skilled operators. For drones to be useful in locust-affected areas, they must be easy to operate both manually and automatically, robust, affordable, and simple to maintain. After three years of testing, the eLocust drone, developed by HEMAV, was introduced for use in locust-prone countries in 2020. These drones process imagery during flight; hence, once the survey is completed, the results are instantly available and are passed on to field teams using the eLocust3 tablet for recording and transmitting real-time data on survey and control efforts (Klein et al., 2023). Using drones for surveillance with the right detection equipment allows for larger areas to be monitored compared to ground teams, which are often limited by terrain (Figure 8.2).

Two examples of countries using drones for locust control include Kenya and India. In Kenya, drones were tested in 2020 to evaluate their efficiency in controlling locusts, specifically targeting hoppers and roosting locusts (Matthews, 2021). The goal was to develop standard operating procedures (SOPs) for optimal drone use. A DJI Agras T16 drone, equipped with a 16 L tank and six rotors, sprayed a 6.5 m swath at a rate of 4.8 L per minute, treating a total of 10 ha. In India, 15 drones, each with a 10 L tank, were used to spray insecticides over locust-infested areas (Figure 8.3). Each drone treated 1 ha in 15 minutes and covered 12 to 15 ha daily. Over 60 days, 1,753 L of insecticide were applied across 7,017 ha. The drones operated at heights between 12.2 and 13.7 m, with speeds of 10–20 km/h, depending on locust position, and achieved a locust mortality rate of 50–90% (Matthews, 2021).

8.5.6 Integration of AI, Machine Learning, and Deep Learning to Exploit Locust Outbreaks

Locust and grasshopper infestations have long been a threat to crops and human livelihoods. The most recent locust outbreak, which started in June 2019, has continued into

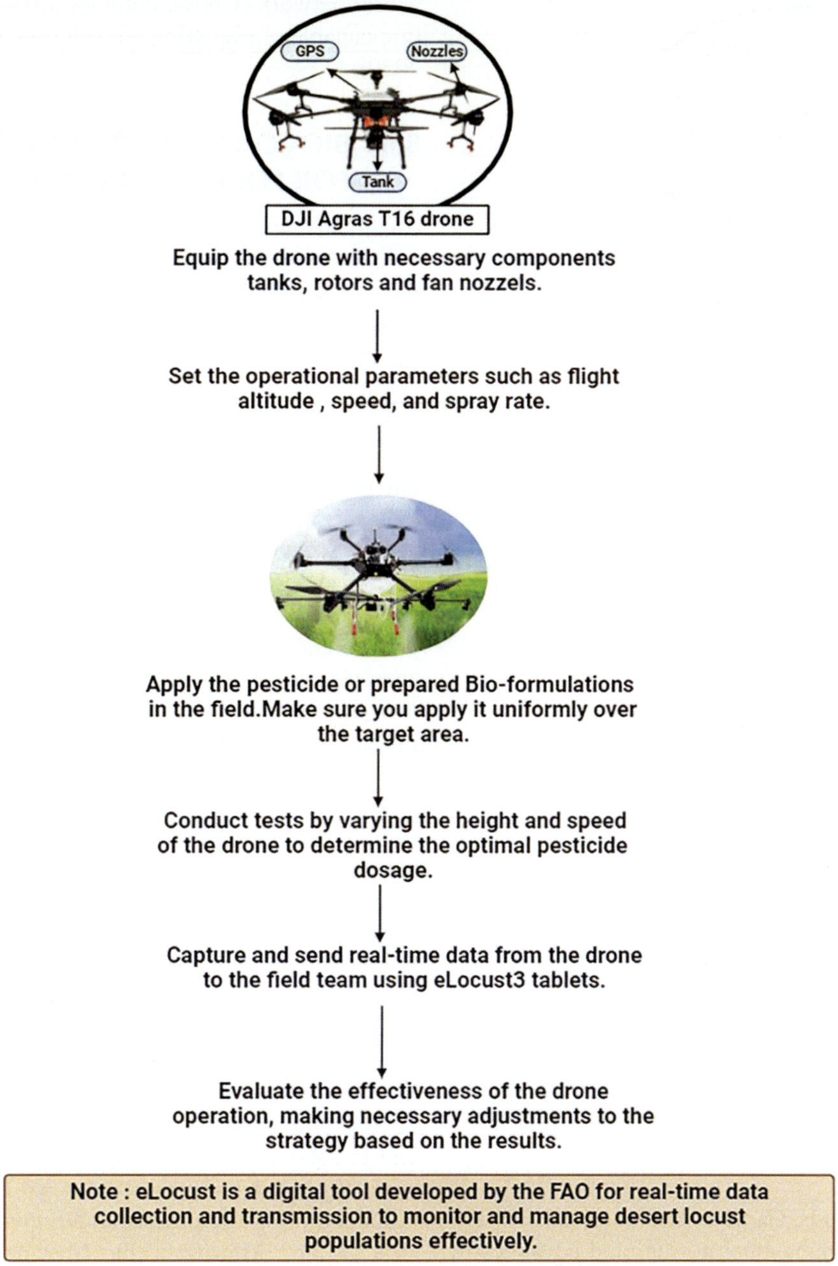

FIGURE 8.2 Important steps required to be undertaken for operating a digital tool (eLocust), from data collection to evaluation of data, for effective management of dessert locust.

2020 and is the worst in the last 70 years, affecting regions in Middle Africa, the Middle East, South Asia, and South America. Considering these challenges, countries are taking measures to prevent another economic disaster, especially after the COVID-19 pandemic. To effectively address the consequences of locust swarms, smart and efficient solutions are needed. One approach is to monitor key agricultural factors, such as soil moisture, temperature, and humidity, using sensors. These sensors can provide real-time data to farmers, alerting them to potential locust infestations via mobile devices (Salim et al., 2021). Manual insect detection, however, is time-consuming, labor-intensive, and costly. To overcome these challenges, researchers in agriculture have developed a perception-based computerized image analysis system that leverages AI, machine learning (ML), and deep learning for precise insect classification. AI plays a crucial role in pest management by supporting decision-making in areas such as monitoring and control (Maran, 2022). Applications of AI in pest management include pest identification, pest counting, and predicting the spread of infestations. Insect detection through AI offers speed, timeliness, and reliability, significantly improving

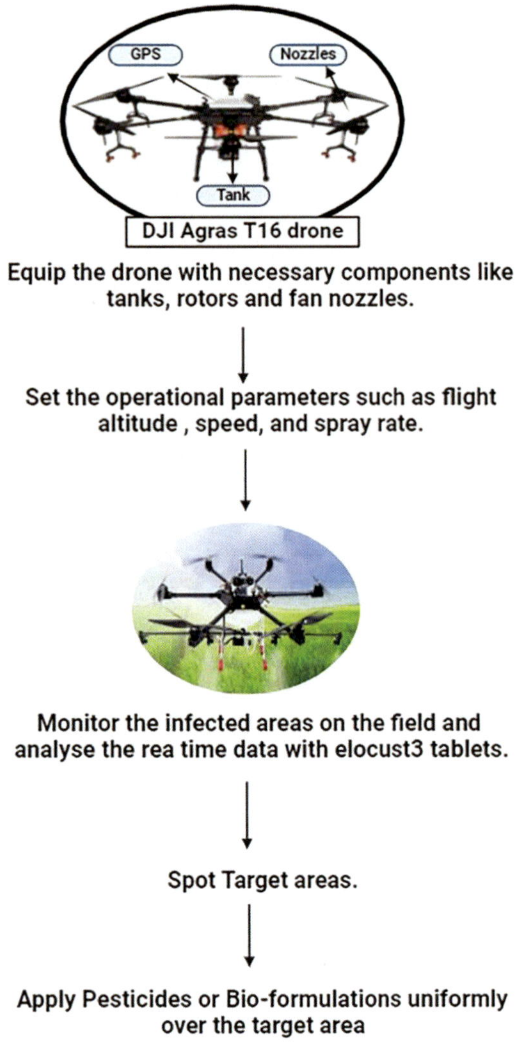

FIGURE 8.3 Visual representation of insecticide spray application over locust-infested areas using eLocust drone equipped with a 10 L tank in India.

pest management methods. Cameras and non-invasive sensors can perform continuous monitoring, and advances in computer vision, combined with deep learning algorithms, make AI-based insect identification possible (Demirel et al., 2021). Training deep learning models with images enables the estimation of insect abundance, biomass, and diversity. Among modern techniques for non-destructive, real-time, and rapid insect detection, near-infrared (NIR) spectroscopy and machine vision stand out. While NIR spectroscopy is reliable and accurate, it struggles to detect low levels of infestation or distinguish between live and dead insects. However, integrating these sensor outputs into deep learning models can further enhance estimates of insect populations (Du et al., 2024). Various algorithms and systems have been developed to improve throughput, reduce costs, allow real-time monitoring, and provide easier data access. Machine vision, when combined with machine learning, has proven to enhance the accuracy of insect detection and classification. A widely used technique, the convolutional neural network (CNN), employs a complex network structure capable of executing convolution operations (HC et al., 2020).

8.6 BIOCONTROL APPROACHES FOR LOCUST CONTROL

Chemical methods for killing insects and pathogens can lead to significant side effects and toxin buildup in the human body. Therefore, alternative methods are urgently needed. Biological approaches which make use of natural enemies, such as insect predators, parasitoids, and pathogens, offer a viable solution (Lacey et al., 2015). The introduction of chemical pesticides has also paved the way for the use of biopesticides as an alternative. Although biocontrol methods have potential for managing locusts and grasshoppers, their use has been limited in the past due to a reliance on chemical controls. Most grasshoppers and locusts are native to their environments, making classical biological control seem less promising. Additionally, their rapid population growth and movement often outpace that of their natural enemies, and pathogens typically have a minimal impact on their populations. However, a closer examination of these limitations reveals some opportunities (Li et al., 2023). Greathead documented initial observations regarding the influence of egg parasitoids on desert locust populations. Building on this, Shah et al. (1998) carried out an extensive four-season study in Malanville, northern Benin, to investigate the natural enemies affecting egg pods. Their research revealed that Meloid beetle predators, including *Epicauta* sp., *Mylabris* (four species), and *Psalydolytta* (three species), caused up to 50% mortality in egg pods. In contrast, parasitism by *Scelio africanus* Risbec, *S. mauritanicus* Risbec, and three additional *Scelio* species resulted in a maximum mortality rate of 3.3% (Lomer et al., 1999). Several entomopathogenic microbes for controlling locusts and grasshoppers have been investigated. The earliest commercially available product was No-Lo®, a bait formulation containing the protozoan *Nosema locustae* (Canning) (Microspora: Microsporidia: Nosematidae), developed by Evans Biocontrol. There are also more aggressive microsporidian species, including *Nosema cuneata*. Two fungal biopesticides have been developed and made commercially available for locust control. The first, Mycotrol®, produced by Mycotech, contains spores from a virulent strain of *Beauveria bassiana* (Balsamo) Vuillemin strain GHA (Hyphomycetes: Moniliales). Field trials in Canada demonstrated its effectiveness against grasshoppers, although later studies identified a limitation related to temperature when grasshoppers were able to regulate their body heat. Despite this, the product has proven highly effective against other insect species and is available for commercial use. The second fungal product for locust control is *Metarhizium anisopliae* (¯avoviride) var. *acridum* strain IMI 330189, which was developed by the LUBILOSA project, and this product is marketed as Green Muscle®. This biopesticide is effective against major acridid pests, with initial signs of mortality typically observed 7 to 10 days after application and complete effects evident within 14 to 18 days. Field studies have shown that the half-life of the fungal spores varies, averaging 6.8 days in north Benin, but ranging from 2 to 12 days in different environments (Lomer et al., 2001).

Botanicals have also been investigated for locust management. Extracts from seeds of Meliaceae family trees, such as neem (*Azadirachta indica*) and *Melia* (*Melia volkensii* GuÈrke), have been explored. Neem extracts have been studied extensively. Their effects vary by locust species; for instance, aqueous neem extract applied to foliage resulted in high mortality in *L. migratoria migratorioides*, while it had a strong deterrent effect on *S. gregaria*, *N. septemfasciata*, and *Z. variegatus*. Neem oil formulations, when applied topically and enriched with azadirachtin, resulted in locust mortality, developmental abnormalities (especially during molting), and behavioral changes (notably, disruption of gregarious behavior). Consequently, neem could be used in integrated pest management (IPM) strategies either to repel locusts from crops or to increase mortality through topical application. Research indicates that oils from orange peel, caraway, and wintergreen are effective against locusts when incorporated into a linseed oil and bicarbonate emulsion. This mixture achieved a mean mortality rate of 80% for desert locusts and 100% for migratory locusts within 24 hours (Githae and Kuria, 2021). However, there is a lack of sufficient field trials to fully assess its effectiveness. There are additional botanical plants with insecticidal properties that have not yet been explored for locust management, suggesting a need for further research in this area.

8.7 CASE STUDIES CONDUCTED ON LOCUST CONTROL

8.7.1 Case Study 1: Australia

In Australia, locusts and grasshoppers are major pests that can severely damage crops and pastures. A notable example is the 1984 plague of the Australian plague locust, which could have caused $103 million in damage. However, control measures costing $3.4 million reduced the damage to $5 million, showing a great cost–benefit ratio. During the late 1980s in Africa, widespread chemical use led to the creation of the LUBILOSA program, which focused on biological controls. This program successfully used a fungus called *Metarhizium anisopliae* var. *acridum* to kill locusts. Inspired by this success, Australia began similar research in 1992. Their pest management system aims to treat 30–60% of locust populations each generation to prevent them from reaching plague levels. This strategy relies on accurate forecasting, quick surveys, and targeted treatments. To meet the demand for eco-friendly alternatives, Australia developed a mycopesticide called Green Guard, which contains the same fungus. This approach combines traditional chemical methods with modern, environmentally friendly solutions.

8.7.2 Case Study 2: Kenya

During the severe desert locust outbreak in Kenya in 2019–2020, authorities used advanced technology to control the pests and protect agriculture. They moved away from harmful insecticides like dieldrin and adopted safer methods. Key to their strategy was the use of drones, specifically the DJI Agras T16. These drones, with a 16 L tank and six rotors, allowed precise and efficient pesticide application while minimizing environmental harm. The DJI Agras T16 drones were carefully set up to operate up to 3 km away and apply the right amount of pesticide at the right speed. They also used biopesticides like Green Muscle® (*Metarhizium acridum*) to control locusts sustainably without affecting other species. Kenya improved monitoring and early detection by using satellite images from NASA's Terra satellite. This data helped track vegetation health and predict locust activity. Combined with drone surveillance, it enabled timely and effective responses. These approaches reduced pesticide use and effectively targeted locusts, helping protect crops and uphold environmental standards. Kenya's use of these innovative, technology-driven methods highlighted the importance of combining new techniques with sustainability in pest control.

8.7.3 Case Study 3: Egypt

The management of desert locust (*Schistocerca gregaria*) was carried out using nanoparticles in Egypt. Nanotechnology is an emerging interdisciplinary research field, offering eco-friendly and efficient alternatives for pest management in agriculture. The study by Ramadan et al. (2020) explored the development of safe, effective, and cost-efficient insecticides using zinc oxide (ZnO) and copper oxide (CuO) nanoparticles to manage *S. gregaria*. These nanoparticles were synthesized in the lab, and their nanostructures were verified with transmission electron microscopy (TEM). The lethal time for 50% mortality (LT50) was determined to be 10.994 days for ZnO and 21.148 days for CuO. The study also revealed that ZnO nanoparticles exerted stronger latent biological effects on the desert locust than CuO nanoparticles. Additionally, both treatments significantly inhibited the activity of the enzyme acetylcholinesterase (AChE).

8.7.4 Case Study 4: Mauritania

Many previous studies have analyzed certain environmental predictors, such as NDVI or soil moisture, individually in an effort to detect suitable areas. Gómez et al. (2019) introduced new variables (surface temperature, LAI, and soil moisture root zone) from the SMAP satellite and applied different machine learning methods in their species distribution model in order to identify desert locust presence, and they obtained highly satisfactory model results (KAPPA and TSS = 0.901 and ROC = 0.986) to detect the probability of presence and, hence, likely breeding areas based on environmental factors. The most relevant variables were surface temperature, NDVI, and soil moisture at root zone under different time scenarios. This study also confirmed the potential of the SMAP satellite to retrieve critical temperatures due to its time pass, in addition to reinforcing the NDVI product from MODIS as a reliable environmental predictor.

8.7.5 Case Study 5: Belgium

Insects rely on juvenile hormone (JH) and ecdysteroids to regulate their development. The gene krüppel-homolog 1 (Kr-h1) helps keep insects in a juvenile state during their larval stages, while the ecdysteroid-induced gene E93 is crucial for transforming them into adults. Gijbels et al. (2020) investigated the role of the E93 gene in the development of the desert locust, and they utilized RNAi to knock down E93. The result showed that injecting locust nymphs with RNA targeting E93 stopped them from undergoing normal metamorphosis. Instead, these locusts went through extra nymphal stages and retained juvenile traits like their color and body shape, even though they grew to or exceeded adult size. Interestingly, the fifth nymphal stage was shorter than usual, and there were changes in ecdysteroid levels and related gene expressions, such as Spook (SgSpo), which affects ecdysteroid production. The levels of ecdysone receptor (SgEcR) and retinoid X receptor (SgRXR) transcripts also changed, showing that silencing E93 affects both ecdysteroid production and signaling. They also saw a significant increase in Kr-h1 transcript levels in the head and fat body of these locusts, but no major changes in the genes responsible for the final steps of juvenile hormone production. Moreover, the molting process was disrupted in the extra nymphal stages, leading to high mortality rates. This research could help in developing new, more targeted, and sustainable methods for controlling locust swarms. Other case studies utilizing advanced control methods have been conducted in different countries and are presented in Table 8.2.

TABLE 8.2
Use of Advanced Techniques for the Management of Different Species of Locust

Species	Target Gene/Model/Chemical	Technique Used for Control	References
Schistocerca gregaria (desert locust)	SPOOK (SPO)	RNAi	Sugahara et al. (2017)
Schistocerca gregaria (desert locust)	Ecdysone receptor (EcR) and retinoid-X-receptor (RXR)	RNAi	Lenaerts et al. (2016)
Locusta migratoria (migratory locust)	LmdsRNase1 and LmdsRNase4	RNAi	Song et al. (2019)
Locusta migratoria (migratory locust)	LmCHS2 (midgut gene) and LmCHS1 (cuticle gene)	RNAi	Lu et al. (2022)
Locusta migratoria (migratory locust)	Krüppel homologue 1 (Kr-h1), chitin metabolism [vermiform (Verm); chitin synthase 1 (CHS1)], (V-ATPase subunit A and E, V-ATPase A and E), [G-protein-coupled receptor (GPCR)], and cell structure (α- and β-tubulin)	RNAi	Luo et al. (2013)
Locusta migratoria manilensis (Oriental migratory locust)	Lm-ATP5A	RNAi	Hu et al. (2016)
Locusta migratoria (migratory locust)	LmDicer1	RNAi	Wang et al. (2013)
Locusta migratoria (migratory locust)	Odorant receptor co-receptor (Orco)	CRISPR/Cas9	Li et al. (2016)
Locusta migratoria (migratory locust)	LmigOr35 (olfactory receptor gene)	CRISPR/Cas9	Chen et al. (2018)
Locusta migratoria manilensis (Oriental migratory locust)	DEAD-box DDX10	CRISPR-Cas	Zhang et al. (2023)
Calliptamus italicus, CIT	Ecological niche modelling (ENM) and sentinel-2 (time series analyses of high–spatial resolution remote sensing data) and fused HSI model	Remote sensing	Klein et al. (2022)
Calliptamus italicus	H3-hexagonal hierarchical geospatial indexing system	Geospatial and remote sensing	Klein et al. (2023)
Locusta migratoria (migratory locust)	Silica nanoparticles	Nanotechnology	Rahal et al. (2021)
Locusta migratoria (migratory locust)	Chitosan nanoparticles (CNPs)	Nanotechnology	Alfy et al. (2020)

(*Continued*)

TABLE 8.2 (*Continued*)
Use of Advanced Techniques for the Management of Different Species of Locust

Species	Target Gene/Model/Chemical	Technique Used for Control	References
Schistocerca gregaria (migratory locust)	Nano-destruxin	Nanotechnology	Sabbour et al. (2014)
Schistocerca cancellata		Band movement and thermoregulation-examining the behavior of the South American locust in comparison to the desert locust (*Schistocerca gregaria*) to determine if their behaviors are similar, where gaining insights into these behaviors could improve strategies for managing locust outbreaks	Piou et al. (2022)
Locusta migratoria manilensis	*Metarhizium anisopliae* (entomopathogenic fungus)	Biopesticide	Finch et al. (2024)
Locusta migratoria migratorioides (nymphal stage) *Schistocerca gregaria* (nymphal stage)	*Nosema locustae* (entomopathogenic fungus)	Microbial agent	Zhang et al. (2021)
Schistocerca gregaria (desert locust)	Lambda cyhalothrin 2.5EC, malathion 57EC, deltamethrin 2.5EC, chlorpyrifos 40EC, pyriproxyfen 10.8EC, Metarhizium 100g/l, Buprofezin 25WP	Biopesticide	Ahmad et al. (2020)
Schistocerca gregaria Forskål (desert locust)	Green Muscle, Green Guard, *Metarhizium anisopliae*, and an isolate of *Beauveria bassiana*	Entomopathogenic fungal (EPF) formulations	Wakil et al. (2022)
Locusta migratoria (migratory locust)	*Yersinia* sp. (MH96 strain)	Entomopathogenic bacterium	McNeill et al. (2008)

8.8 CONCLUSION

The continuous progress in developing eco-friendly, efficient, and precise methods for managing locusts and grasshoppers could provide a blueprint for improving crop protection. With growing limitations on the use of large-scale chemical pesticides, there is an increasing need to explore new pest control strategies. One promising alternative is the use of biological approaches. These biological approaches include methods such as biopesticides, natural predators, microorganisms, and plant-based extracts that target pests without harming the environment. These could be useful for controlling locusts and should be studied more. New ways of delivering these agents, such as using nanocarriers, might improve the effectiveness of both current and new biopesticides. While commercial biopesticides like *Metarhizium* are currently available, enhancing their effectiveness through genetic modification and selection for traits such as increased virulence and stress resistance could be beneficial. Despite progress in understanding locust behavior, more clarity is needed on the effective use of pheromones and other semiochemicals for monitoring and control. One of the key steps in locust management is the need for quicker methods to detect and control locust and grasshopper populations. Remote sensing, GIS, GPS, and unmanned aerial vehicles (UAVs) can play a vital role in achieving this by providing efficient surveillance and targeted treatment over large areas, especially in hard-to-reach regions. Drones can efficiently survey large areas and apply pesticides in remote or difficult-to-access locations. Advances in functional genome sequencing could identify crucial molecular targets for pest management, including those related to RNA interference (RNAi) and CRISPR/Cas9 technologies. Genetic engineering with CRISPR could potentially target genes critical for locust development or reproduction, offering new avenues for pest control. RNAi could be used to silence key genes that affect locust survival, providing a more sustainable, targeted approach to pest management. Combining these genomic tools with new insights from sequencing could significantly enhance pest control strategies. Integrating modern technologies, such as artificial intelligence (AI), can greatly transform locust control efforts. AI can process extensive data from monitoring systems to enhance pest prediction and reduce response times. Additionally, it is important to understand how human activities contribute to the rise in locust and grasshopper populations. These innovations, combined with traditional and biological methods, can significantly improve the precision and sustainability of locust and grasshopper control efforts.

REFERENCES

Ahmad, K. J., Aslam, A., Munir, M., Ali, Q., Hussain, D., Malik, H., and Zubair, M. (2020). Toxicological impact of different insecticides on the desert locust (*Schistocerca gregaria* Forsk.) (Acrididae). *Life Science Journal*, 17(8): 6–10.

Ahmad, M. N., Shao, Z., and Altan, O. (2022). Effect of locust invasion and mitigation using remote sensing techniques: A case study of North Sindh Pakistan. *Photogrammetric Engineering and Remote Sensing*, 88(1): 47–53.

Alfy, H., Ghareeb, R. Y., Soltan, E. L. S. A. Y. E. D., and Farag, D. A. (2020). Impact of chitosan nanoparticles as insecticide and nematicide against *Spodoptera littoralis*, *Locusta migratoria*, and *Meloidogyne incognita*. *Plant Cell Biotechnology Molecular Biology*, 21: 126–140.

Anstey, M. L., Rogers, S. M., Ott, S. R., Burrows, M., and Simpson, S. J. (2009). Serotonin mediates behavioral gregarization underlying swarm formation in desert locusts. *Science*, 323(5914):627–630.https://10.1126/science.1165939PMID: 19179529.

Belayneh, Y. T. (2005). Acridid pest management in the developing world: A challenge to the rural population, a dilemma to the international community. *Journal of Orthoptera Research*, 14(2): 187–195.

Brader, L., Djibo, H., and Faye, F. G. (2005). Towards a more effective response to desert locusts and their impacts on food insecurity, livelihoods and poverty. *Independent Multilateral Evaluation of the 2003–2005 Desert Locust Campaign*, FAO, Rome, Italy.

Chen, D., Tang, J. X., Li, B., Hou, L., Wang, X., and Kang, L. (2018). CRISPR/Cas9-mediated genome editing induces exon skipping by complete or stochastic altering splicing in the migratory locust. *BMC Biotechnology*, 18: 1–9.

Deguine, J. P., Aubertot, J. N., Flor, R.J., Lescourret, F., Wyckhuys, K. A. J., and Ratnadass, A. (2021). Integrated pest management: Good intentions, hard realities. A review. *Agronomy for Sustainable Development*, 41: 38. https://doi.org/10.1007/s13593-021-00689-w

Demirel, M., and Kumral, N. A. (2021). Artificial intelligence in integrated pest management. In: *Artificial intelligence and IoT-based technologies for sustainable farming and smart agriculture* (pp. 289–313). IGI Global.

Despland, E. (2002). Role of olfactory and visual cues in the attraction/repulsion responses to conspecifics by gregarious and solitarious desert locusts. *Journal of Insect Behavior*, 14(1): 35–46. https://0892-7553/01/0100-0035

Du, B., Ding, D., Ma, C., and Kang, L. (2021). Locust density shapes energy metabolism and oxidative stress resulting in divergence of flight traits. *PNAS*, 119(1): e2115753118. https://doi.org/10.1073/pnas.2115753118

Du, Q., Wang, Z., Huang, P., Zhai, Y., Yang, X., and Ma, S. (2024). Remote sensing monitoring of grassland locust density based on machine learning. *Sensors*, 24(10): 3121.

El-Ghany, N. M. A. (2020). Pheromones and chemical communication in insects. In: D. Kontogiannatos, A. Kourti, and K. F. Mendes (Eds.), *Pests, weeds and diseases in agricultural crop and animal husbandry production*. IntechOpen. http://dx.doi.org/10.5772/intechopen.92384

Ferenz, H. J., and Seidelmann, K. (2003). Pheromones in relation to aggregation and reproduction in desert locusts. *Physiological Entomology*, *28*(1): 11–18. https://doi.org/10.1046/j.1365-3032.2003.00318.x

Finch, E. A., Li, H., Cornelius, A., Styles, J., Beeken, J., Cheng, Y., and Luke, B. (2024). An updated and validated model for predicting the performance of a biological control agent for the oriental migratory locust. *Pest Management Science*, 80(2): 442–451.

Gijbels, M., Marchal, E., Verdonckt, T. W., Bruyninckx, E., and Vanden Broeck, J. (2020). RNAi-mediated knockdown of transcription factor E93 in nymphs of the desert locust (*Schistocerca gregaria*) inhibits adult morphogenesis and results in supernumerary juvenile stages. *International Journal of Molecular Sciences*, 21(20): 7518.

Githae, E. W., and Kuria, E. K. (2021). Biological control of desert locust (*Schistocerca gregaria* Forskål). *CABI Reviews*, 16(013): 1–8. https://doi.org/10.1079/PAVSNNR202116013

Gómez, D., Salvador, P., Sanz, J., Casanova, C., Taratiel, D., and Casanova, J. L. (2019). Desert locust detection using Earth observation satellite data in Mauritania. *Journal of Arid Environments*, 164: 29–37.

Guo, X., Yu, Q., Chen, D., Wei, J., Yang, P., Yu, J., Wang, X., and Kang, L. (2020). 4-Vinylanisole is an aggregation pheromone in locusts. *Nature*, 584(7822): 584–588. https://doi.org/10.1038/s41586-020-2610-4.

Han, S. H. (1997). Mechanical control of the desert locust *Schistocerca gregaria* (Forskål, 1775). In: *New strategies in locust control*. Birkhäuser Basel. https://doi.org/10.1007/978-3-0348-9202-5_40

He, L., Huang, Y., and Tang, X. (2022). RNAi-based pest control: Production, application and the fate of dsRNA. *Frontiers in Bioengineering and Biotechnology*, 10: 1080576. https://doi.org/10.3389/fbioe.2022.1080576. PMID: 36524052; PMCID: PMC9744970.

Hu, J., and Xia, Y. (2016). F1-ATP synthase α-subunit: A potential target for RNAi–mediated pest management of *Locusta migratoriamanilensis*. *Pest Management Science*, 72(7): 1433–1439.

Kandul, N. P., Liu, J., Sanchez, C. H. M., Wu, S. L., Marshall, J. M., and Akbari, O. S. (2019). Transforming insect population control with precision guided sterile males with demonstration in flies. *Nature Communications*, 10(1): 84. https://doi.org/10.1038/s41467-018-07964-7. PMID: 30622266; PMCID: PMC6325135.

Karpakakunjaram, V., Kolatkar, M. D., and Muralirangan, M. C. (2002). Effects of abiotic factors on the population of an acridid grasshopper, *Diabolocatantops pinguis* (Orthoptera: Acrididae) at two sites in southern India: A three-year study. *Journal of Orthoptera Research*, 11(1): 55–62. https://doi.org/10.1665/1082-6467(2002)011[0055:EOAFOT]2.0.CO;2

Kenawy, M. A., and Abdel-Hamid, T. M. (2015). Insects in ancient (Pharaonic) Egypt: A review of fauna, their mythological and religious significance and associated diseases. *Egyptian Academic Journal of Biological Sciences*, 8(1): 15–32.

Klein, I., Oppelt, N., and Kuenzer, C. (2021). Application of remote sensing data for locust research and management – A review. *Insects*, 12(3): 233.

Klein, I., Uereyen, S., Eisfelder, C., Pankov, V., Oppelt, N., and Kuenzer, C. (2023). Application of geospatial and remote sensing data to support locust management. International *Journal of Applied Earth Observation and Geoinformation*, 117: 103212.

Klein, I., van der Woude, S., Schwarzenbacher, F., Muratova, N., Slagter, B., Malakhov, D., and Kuenzer, C. (2022). Predicting suitable breeding areas for different locust species – A multi-scale approach accounting for environmental conditions and current land cover situation. *International Journal of Applied Earth Observation and Geoinformation*, 107: 102672.

Komal, J., Desai, H. R., Samal, I., Mastinu, A., Patel, R. D., Kumar, P. V. D., Majhi, P. K., Mahanta, D. K., and Bhoi, T. K. (2023).

Unveiling the genetic symphony: Harnessing CRISPR-Cas genome editing for effective insect pest management. *Plants (Basel)*, 12(23): 3961. https://doi.org/10.3390/plants12233961. PMID: 38068598; PMCID: PMC10708123.

Lacey, L. A., Grzywacz, D., Shapiro-Ilan, D. I., Frutos, R., Brownbridge, M., and Goettel, M. S. (2015). Insect pathogens as biological control agents: Back to the future. *Journal of Invertebrate Pathology*, 132: 1–41.

Latchininsky, A., Sword, G., Sergeev, M., Cigliano, M. M., and Lecoq, M. (2011). Locusts and grasshoppers: Behavior, ecology, and biogeography. *Psyche*. ID 578327, 4. https://doi:10.1155/2011/578327

Latchininsky, A. V., and Sivanpillai, R. (2010). Locust habitat monitoring and risk assessment using remote sensing and GIS technologies. In: *Integrated management of arthropod pests and insect borne diseases* (pp. 163–188). Dordrecht: Springer Netherlands.

Lecoq, M. (2001). Recent progress in Desert and Migratory Locust management in Africa. Are preventative actions possible? *Journal of Orthoptera Research*, 10(2): 277–291. https://doi.org/10.1665/1082-6467

Lecoq, M., and Cease, A. (2022). What have we learned after millennia of locust invasions? *Agronomy*, 12: 472. https://doi.org/10.3390/agronomy12020472

Lenaerts, C., Van Wielendaele, P., Peeters, P., Broeck, J. V., and Marchal, E. (2016). Ecdysteroid signalling components in metamorphosis and development of the desert locust, *Schistocerca gregaria*. *Insect Biochemistry and Molecular Biology*, 75: 10–23.

Li, H., Kadzamira, M. A., Ogunmodede, A., Finch, E., Zhu, J., Romney, D., and Luke, B. (2023). Lessons learned and challenges of biopesticide usage for locust management – The case of China. *Sustainability*, 15(7): 6193.

Li, Y., Zhang, J., Chen, D., Yang, P., Jiang, F., Wang, X., and Kang, L. (2016). CRISPR/Cas9 in locusts: Successful establishment of an olfactory deficiency line by targeting the mutagenesis of an odorant receptor co-receptor (Orco). *Insect Biochemistry and Molecular Biology*, 79: 27–35.

Lomer, C. J., Bateman, R. P., Dent, D., De Groote, H., Douro-Kpindou, O. K., Kooyman, C., and Thomas, M. (1999). Development of strategies for the incorporation of biological pesticides into the integrated management of locusts and grasshoppers. *Agricultural and Forest Entomology*, 1(2).

Lomer, C. J., Bateman, R. P., Johnson, D. L., Langewald, J., and Thomas, M. (2001). Biological control of locusts and grasshoppers. *Annual Review of Entomology*, 46: 667–702. https://10.1146/annurev.ento.46.1.667. PMID: 11112183

Lu, Q., Cui, H., Li, W., Liu, T., Chen, Q., and Yang, Q. (2022). Synthetic nanoscale RNAi constructs as pesticides for the control of *Locusta migratoria*. *Journal of Agricultural and Food Chemistry*, 70(35): 10762–10770.

Luo, Y., Wang, X., Wang, X., Yu, D., Chen, B., and Kang, L. (2013). Differential responses of migratory locusts to systemic RNA interference via double-stranded RNA injection and feeding. *Insect Molecular Biology*, 22(5): 574–583.

MacCuaig, R. D., Le Berre, R., Thompson, A. T., and Castel, J. M. (1979). Pesticides for locust control [and discussion]. *Philosophical Transactions of the Royal Society of London. Series B, Biological Sciences*, 287(1022): 447–455. www.jstor.org/stable/2418193

Martin, A. E. (2015). The interacting effects of the historic landscape structure, human landscape change, and species mobility on species extinction risk in human-altered landscapes: An evolutionary perspective (Ph.D. thesis). Carleton University, Ottawa, Ontario.

Maran, S. P. M. (2022). Application of Artificial Intelligence (AI) tools in integrated pest management (IPM) – An insect-plant interaction perspective. *Acta Scientific Agriculture*, 6(1). ISSN: 2581-365X

Matthews, G. A. (2021). New technology for desert locust control. *Agronomy*, 11: 1052. https://doi.org/10.3390/agronomy11061052

McNeill, M. R., and Hurst, M. R. H. (2008). Yersinia sp (MH96) a potential biopesticide of migratory locust *Locusta migratoria* L. *New Zealand Plant Protection*, 61: 236–242.

Meynard, C. N., Lecoq, M., Chapuis, M. P., and Piou, C. (2020). On the relative role of climate change and management in the current desert locust outbreak in East Africa. *Global Change Biology*, 26(7): 3753–3755.

Neha, H. C., and Munavalli, J. R. (2020). Automated real-time locust management using artificial intelligence. *International Journal of Engineering Applied Sciences and Technology*, 5(4): 133–138. ISSN: 2455-2143

Parkinson, R. H., Little, J. M., and Gray, J. R. (2017). A sublethal dose of a neonicotinoid insecticide disrupts visual processing and collision avoidance behaviour in *Locusta migratoria*. *Scientific Reports*, 7(1): 936. https://doi.org/10.1038/s41598-017-01039-1. PMID: 28428563; PMCID: PMC5430526.

Pener, M. P. (1991). Locust phase polymorphism and its endocrine relations. In: P. D. Evans (Ed.), *Advances in insect physiology* (pp. 1–79). London: Elsevier. https://doi.org/10.1016/S0065-2806(08)60091-0

Pener, M. P., and Simpson, S. J. (2009). Locust phase polyphenism: An update. *Advances in Insect Physiology*, 36: 1–272. https://doi.org/10.1016/S0065-2806(08)36001-9

Piou, C., and Marescot, L. (2023). Spatiotemporal risk forecasting to improve locust management. *Current Opinion in Insect Science*, 56: 101024.

Piou, C., Zagaglia, G., Medina, H. E., Trumper, E., Brizuela, X. R., and Maeno, K. O. (2022). Band movement and thermoregulation in *Schistocerca cancellata*. *Journal of Insect Physiology*, 136: 104328.

Price, R. E. (2023). Invasions and local outbreaks of four species of plague locusts in South Africa: A historical review of outbreak dynamics and patterns. *Insects*, 14(11): 846. https://doi.org/10.3390/insects14110846

Rahal, M. A., Saidi-Touati, M., Zidane, A., and El-Samahy, M. F. (2021). Impact of rice husk derived silica nanoparticles on the migratory locust (*Locusta migratoria*, Acrididae). *Annals of Oradea University, Biology Fascicle/Analele Universităţii din Oradea, Fascicula Biologie*, 28(2).

Ramadan, A. F., El-Shafiey, S. N., Abol-Noor, K. M. A., Aboghalia, A., and Elshiekh, A. (2020). Responses of desert locust *Schistocerca gregaria* (Orthoptera: Acrididae) to treatment with chemically synthesized zinc and copper oxides nanoparticles. *Egyptian Journal of Plant Protection Research Institute*, 3(1): 339–345

Sabbour, M. (2014). Evaluating toxicity of extracted nano-Destruxin against the desert locust *Schistocerca gregaria* in Egypt. *Journal of Egyptian Academic Society for Environmental Development. D, Environmental Studies*, 15(2): 9–17.

Salim, S. A., Amin, M. R., Rahman, M. S., Arafat, M. Y., and Khan, R. (2021). An IoT-based smart agriculture system with locust prevention and data prediction. In: *8th international conference on information technology, computer and electrical engineering* (ICITACEE) (pp. 201–206). IEEE.

Santos, D., Broeck, J. V., and Wynant, N. (2014). Systemic RNA interference in locusts: Reverse genetics and possibilities for locust pest control. *Current Opinion in Insect Science*, 6: 9–14. https://doi.org/10.1016/j.cois.2014.09.013

Seidelmann, K. (2006). The courtship-inhibiting pheromone is ignored by female-deprived gregarious desert locust males. *Biology Letters*, 2: 525–527.

Shah, P. A., Gbongboui, C., Godonou, I., Houssou, A., and Lomer, C. J. (1998). Survival and mortality of grasshopper egg pods from semi-arid cereal cropping areas of northern Benin. *Bulletin of the Entomological Research*, 88: 451 ± 459.

Simon-Delso, N., Amaral-Rogers, V., and Belzunces, L. P. (2015). Systemic insecticides (neonicotinoids and fipronil): Trends, uses, mode of action and metabolites. *Environmenatl Science and Pollution Research*, 22: 5–34. https://doi.org/10.1007/s11356-014-3470-y

Song, H., Fan, Y., Zhang, J., Cooper, A. M., Silver, K., Li, D., and Zhang, J. (2019). Contributions of dsRNases to differential RNAi efficiencies between the injection and oral delivery of dsRNA in *Locusta migratoria. Pest Management Science*, 75(6): 1707–1717.

Sugahara, R., Tanaka, S., and Shiotsuki, T. (2017). RNAi-mediated knockdown of SPOOK reduces ecdysteroid titers and causes precocious metamorphosis in the desert locust *Schistocerca gregaria. Developmental Biology*, 429(1): 71–80.

Symmons, P. M., and Cressman, K. (2001). *Desert locust guidelines: Biology and behavior*. Rome: Food and Agriculture Organization of the United Nations.

Topaz, C. M., D'Orsogna, M. R., Edelstein-Keshet, L., and Bernoff, A. J. (2012). Locust dynamics: Behavioral phase change and swarming. *PLoS Computational Biology*, 8(8): e1002642. https://10.1371/journal.pcbi.1002642 PMID: 22916003; PMCID: PMC3420939.

Veran, S., Simpson, S. J., Sword, G. A., Deveson, E., Piry, S., Hines, J. E., and Berthier, K. (2015). Modeling spatiotemporal dynamics of outbreaking species: Influence of environment and migration in a locust. *Ecology*, 96(3): 737–748.

Wadhams, L. J. (1990). The use of coupled gas chromatography: Electrophysiological techniques in the identification of insect pheromones. In: A. R. McCaffery and I. D. Wilson (Eds.), *Chromatography and isolation of insect hormones and pheromones*. Chromatographic Society Symposium Series. New York, NY: Springer. https://doi.org/10.1007/978-1-4684-8062-7_28

Wakil, W., Ghazanfar, M. U., Usman, M., Hunter, D., and Shi, W. (2022). Fungal-based biopesticide formulations to control nymphs and adults of the desert locust, *Schistocerca gregaria* Forskål (Orthoptera: Acrididae): A laboratory and field cage study. *Agronomy*, 12(5): 1160.

Wang, X., Li, G., Wang, S., Feng, C., Xu, W., Nie, Q., and Liu, Q. (2023). The effect of environmental changes on locust outbreak dynamics in the downstream area of the Yellow River during the Ming and Qing Dynasties. *Science of the Total Environment*, 877: 162921. ISSN: 0048-9697. https://doi.org/10.1016/j.scitotenv.2023.162921

Wang, Y. L., Yang, M. L., Jiang, F., Zhang, J. Z., and Kang, L. (2013). MicroRNA-dependent development revealed by RNA interference-mediated gene silencing of LmDicer1 in the migratory locust. *Insect Science*, 20(1): 53–60.

Wynant, N., Santos, D., Van Wielendaele, P., and Vanden Broeck, J. (2014). Scavenger receptor-mediated endocytosis facilitates RNA interference in the desert locust, *Schistocerca gregaria. Insect Biochemistry and Molecular Biology*, 23(3): 320–329. https://doi.org/10.1111/imb.12083

Yan, Y., Aumann, R. A., Hacker, I., and Schetelig, M. F. (2023). CRISPR-based genetic control strategies for insect pests, *Journal of Integrative Agriculture*, 22(3): 651–668. https://doi.org/10.1016/j.jia.2022.11.003

Zhang, L., and Hunter, D. M. (2017). Management of locusts and grasshoppers in China. *Journal of Orthoptera Research*, 26(2): 155–159. https://doi.org/10.3897/jor.26.20119

Zhang, L., and Lecoq, M. (2021). *Nosema locustae* (Protozoa, Microsporidia), a biological agent for locust and grasshopper control. *Agronomy*, 11(4): 711.

Zhang, L., Lecoq, M., Latchininsky, A., and Hunter, D. (2019). Locust and grasshopper management. *Annual Review of Entomology*, 64(1): 15–34.

Zhang, L., Yang, L., Li, H., Yu, Y., and Zhai, J. (2023). Rapid detection of insect pest with CRISPR-Cas13a-based lateral flow strip: Using *Locusta migratoriamanilensis. Journal of Applied Entomology*, 147(6): 416–422.

9 Threats to Pollinators and Agricultural Systems

Penuballi Swathi[1*], *Nagarjuna Reddy K. V.*[2], *Desavath Gouthami*[1], *and M. Siri Chandana*[1]
[1]Department of Agricultural Entomology, College of Agriculture, Kerala Agricultural University, Thrissur, Kerala 680656, India
[2]School of Agriculture, Lovely Professional University, Phagwara, Punjab 144411, India
*Corresponding author

9.1 INTRODUCTION

Pollination services are crucial for global agriculture and food security, contributing significantly to the economy. Pollinators, such as bees, butterflies, and other insects, are responsible for the pollination of approximately 75% of the world's food crops, which translates to about 35% of global crop production by volume (Van der Sluijs *et al.*, 2016). The economic value of these services is substantial, with recent estimates suggesting that the global value of crop pollination services ranges from $195 billion to ~$387 billion annually (Porto *et al.*, 2020). Animal pollinators play a crucial role in the production of 87 global crops, such as cocoa, kiwi, passion fruit, and watermelon, across 200 countries. The primary categories of insect-pollinated crops include vegetables and fruits, each generating approximately EUR 50 billion, followed by oilseed crops, nuts, and spices. The value of 1 t of crop production not reliant on insect pollination is approximately EUR 151, in contrast to an average of EUR 761 for crops that depend on pollinators (Gallai *et al.*, 2009). Insect pollination is essential for the production of numerous agricultural products globally, including aromatic and medicinal plants. Throughout the year, honeybees, native bees, and flies pollinate 48 crops of the world's most valuable commodities, significantly impacting the global economy (Klein *et al.*, 2007). In the United States, pollination contributes USD 16 billion annually, with USD 12 billion directly linked to the presence of honeybees (Calderone, 2012; Rader *et al.*, 2016).

9.1.1 Pesticide Types and Their Mechanisms

The global agriculture sector is the predominant consumer of pesticides, utilizing about 3.69 Mt annually (FAO, 2023). Pesticides have been utilized extensively for decades, significantly enhancing food output. The utilization of pesticides is indispensable in contemporary agriculture, as they are intended to optimize crop yields and curb insect populations. They are classified as insecticides, herbicides, and fungicides, each with distinct mechanisms and effects on both target and non-target organisms.

Insecticides, like organophosphates and neonicotinoids, are designed to exterminate insect pests by interfering with their nervous systems, resulting in paralysis and mortality. Chlorpyrifos and imidacloprid are prevalent examples. Herbicides, such as glyphosate, eliminate undesirable plants by obstructing vital physiological functions. Glyphosate inhibits the enzyme 5-enolpyruvylshikimate-3-phosphate synthase (EPSPS), which is essential for the production of aromatic amino acids in plants. Fungicides, including azoles and strobilurins, are employed to manage fungal diseases by interfering with essential cellular mechanisms, such as sterol production or mitochondrial respiration. Common examples include pyraclostrobin and tebuconazole.

Pesticides have been widely employed worldwide, especially in underdeveloped countries. Due to farmers' limited knowledge and insufficient training in developing nations, few can comply with pesticide label instructions, impairing their capacity to manage pesticides effectively. As a result, farmers frequently misuse pesticides by applying either excessive or inadequate amounts or by using pesticides that have been prohibited or banned by the government. Pesticides are effective in managing pest populations; however, their unregulated and indiscriminate use can adversely affect human health, various life-forms, and ecosystems, dependent upon the sensitivity of organisms and the toxicity of the pesticides (Colosio *et al.*, 2023).

This is due to the fact that insecticides can be absorbed by plants and transported to nectar and pollen, exposing pollinators to detrimental sublethal levels. Bees are particularly susceptible to these risks. Although herbicides are typically less harmful to insects, their widespread application can result in habitat fragmentation and a decline in floral resources, so indirectly impacting pollinator populations. The reduction in plant diversity and abundance resulting from pesticide use decreases the availability of nectar and pollen, essential for the survival and reproduction of pollinators. While fungicides are not inherently hazardous to insects, their residues can synergistically combine with other pesticides, intensifying the toxic effects on pollinators (Cullen *et al.*, 2019). Furthermore, fungicides can modify

DOI: 10.1201/9781003516538-9

the microbial populations within the gastrointestinal tract of pollinators, impacting their digestive processes and immunological responses.

9.1.2 Direct Impacts of Pesticides on Pollinators

The acute toxicity and life-threatening consequences of pesticides on pollinators have become a major issue in recent years, with multiple studies emphasizing the harmful implications of various chemical agents. Particularly well-known for their extreme toxicity to pollinators are neonicotinoids, such as clothianidin and imidacloprid. Imidacloprid has been demonstrated to induce mortality in honeybees (*Apis mellifera*) at dosages as minimal as 0.081 μg/bee (de Castro Lippi *et al.*, 2024). This pesticide targets nicotinic acetylcholine receptors in the central nervous system, resulting in paralysis and death. Clothianidin has been documented to cause fatal consequences at doses of 1.80 ng/bee (Yalçin *et al.*, 2021). Neonicotinoids are systemic insecticides that are absorbed by plants, potentially contaminating pollen and nectar, which exposes pollinators to fatal dosages while foraging. Some pesticides and their LC50 values that affect honeybees are presented in Table 9.1.

Biopesticides have been employed worldwide since their inception in the late 19th century. Presently, they represent a modest portion of the global pesticide business, estimated at around US$3 billion, accounting for only 5% of the whole crop protection market (Arthurs and Dara, 2019). Numerous advantages of employing biopesticides in agriculture have been recognized. These biopesticides appear to leave no detrimental residues, addressing a significant issue for consumers, and their efficacy can be comparable to that of conventional pesticides. Comparative laboratory and field research suggest that biopesticides are not only safer for vertebrates and less detrimental to non-target arthropods but also more successful for the long-term management of certain pests (Rasoul *et al.*, 2013). For example, mycoinsecticides derived from *Metarhizium anisopliae* have been effectively utilized in conjunction with synthetic organophosphates to manage locusts and grasshoppers in areas including Africa, Australia, and China. These biopesticides have comparable efficacy to chemical pesticides, resulting in an approximate 75% decrease in pest populations, while not endangering non-target arthropods.

The possible effects of entomopathogenic fungi (EPF) on honeybees have been evaluated in both immature and adult stages. Multiple investigations have indicated that *Beauveria bassiana* and *Metarhizium anisopliae* do not adversely affect honeybees. Nonetheless, there are cases where exposure to this fungus has demonstrated effects on individual survival, development, and physiology. Larval exposure to these fungi can alter acetylcholinesterase (AchE) activity (Rasoul *et al.*, 2013), and adult workers exposed during the pupal stage have demonstrated considerable weight reduction and heightened expression of immune-related genes (Hamiduzzaman *et al.*, 2012). Moreover, these EPF have been observed to affect the behavioral characteristics of honeybees at both the individual and colony levels. Cappa et al. (2019) demonstrated that exposure to *B. bassiana* substantially alters the cuticular hydrocarbon (CHC) profiles of foraging bees, thus impairing the colony-level identification system. Field bioassays demonstrated that bees infected with this pathogen are more readily accepted by guard bees at the entrances of unfamiliar colonies. This heightened acceptance may promote the movement of foragers among colonies, thereby augmenting the dissemination of parasites and diseases.

Recent research has highlighted the capacity of *Bacillus thuringiensis* to alter the intestinal microbiota of honeybees (Steinigeweg *et al.*, 2023). Spinosad, a bioinsecticide, has been shown to possess significant toxicity to honeybees. Moreover, spinosad affects foraging behavior and triggers transcriptional alterations in metabolic genes in the brains of adult bees (Araújo *et al.*, 2023). Exposed workers exhibited considerable downregulation of genes that encode enzymes related to oxidative phosphorylation and metabolism. Moreover, a reduction in the activity of acetylcholinesterase (AChE) and adenosine triphosphatase (ATPase) in several organs of exposed workers was noted. This biopesticide generated structural alterations in the midgut and Malpighian tubules, as well as epithelial disorganization (Cappa *et al.*, 2022). Exposure to the biopesticide azadirachtin has been demonstrated to negatively impact honeybees. Significantly, over 90% of prepupae and pupae

TABLE 9.1
Different Pesticides and Their LC50 Values That Affect Honeybees

Pesticide	Chemical Class	LC_{50} (μg/bee)	Effects on Pollinator Activity	Reference
Imidacloprid	Neonicotinoid	3	Affects foraging and navigation	(Abbassy *et al.*, 2020)
Clothianidin	Neonicotinoid	0.003	Similar effects as imidacloprid; increases viral replication in bees	(Di Prisco *et al.*, 2013)
Chlorpyrifos	Organophosphate	0.5	Affects the nervous system; has mixed effects on viral susceptibility	(Johnson *et al.*, 2010)
Fipronil	Phenyl pyrazole	0.004	Neurotoxic effects; can lead to colony collapse	(Johnson *et al.*, 2010)
Glyphosate	Herbicide	>200	Indirect effects on nutrition and gut microbiota	(Xue *et al.*, 2024)
Atrazine	Herbicide	>100	Primarily affects floral resource availability; indirect toxicity	(Boily *et al.*, 2013)

subjected to treatment during the larval stage had ontogenetic anomalies, such as premature and irregular coloring on their mouthparts and other appendages (Peng *et al.*, 2000). Moreover, azadirachtin has been shown to disrupt foraging behavior, diminishing the visitation frequency of foragers in treated agricultural fields. While azadirachtin may not have immediate impacts on brood growth, it can influence colony overwintering, indicating possible long-term detrimental effects on bees exposed during the brood stage. This overview emphasizes the sublethal impacts of biopesticides on different species of pollinating insects, as evidenced by the examined research (Cappa *et al.*, 2022).

Recent years have witnessed an increasing emphasis on global pollination groups, prompted by the confirmed fall in bee and lepidopteran numbers. Research has underscored the substantial roles of certain insects, especially specific categories of flies and beetles, within pollination networks. These insects are essential for promoting and sustaining floral diversity. This shift in focus emphasizes the significance of a more diverse array of pollinators in the persistence of ecosystems. These investigations frequently yield contentious results. Research has suggested minimal negative impacts on non-target populations, or even beneficial effects. In contrast, additional research indicates that microbial and botanical pesticides, EPF, entomopathogenic nematodes (EPNs), and transgenic *Bt* crops can markedly diminish populations of non-target beneficial insects, including hoverflies and parasitoid wasps (Cappa *et al.*, 2022).

9.1.3 Sublethal Effects on Behavior and Physiology

Sublethal doses of pesticides, characterized as quantities that do not induce instantaneous mortality, have attracted considerable interest in contemporary entomological studies (Desneux *et al.*, 2007). Although these doses are not fatal, they can elicit several physiological and behavioral alterations in honeybees, impacting their foraging efficiency, navigation, and overall colony vitality. Comprehending the sublethal impacts of pesticides is essential, given that bees are integral to pollination, which is critical for sustaining biodiversity and agricultural output. Examining these effects aids in recognizing potential hazards associated with pesticide exposure, therefore guiding safer pest management strategies and safeguarding the sustainability of pollinator populations. This is crucial for formulating measures to alleviate the detrimental effects of pesticides, so aiding in the preservation of these key pollinators.

This collection of case studies explores the sublethal effects of pesticides on pollinators, offering a thorough analysis of the impact of different chemical agents on these vital organisms. These case studies elucidate the intricate and frequently complex connections between pollinators and their habitats through the examination of particular instances of pesticide exposure. This investigation is essential for comprehending the wider ramifications of pesticide application and for formulating ways to alleviate their detrimental impacts, therefore safeguarding the sustainability of pollinator populations and the ecosystems they sustain.

Decourtye *et al.* (2004a) revealed that administering 24 µg/kg of imidacloprid in sugar water to honeybee workers can diminish their foraging activity and adversely affect their performance in olfactory learnt discriminating tasks. Yang *et al.* (2008) investigated the effects of sublethal imidacloprid dosages on honeybee foraging by assessing the time gap between trips to a feeding station. Typically, bees returned within 300 seconds; however, imidacloprid-treated bees exhibited prolonged return times, with delays correlating to concentration levels. The minimal effective dose was 50 µg/L, whereas concentrations over 1,200 µg/L resulted in notable abnormalities. Interference with waggle dance, essential for foraging, may adversely impact colony food resources and general fitness.

Henry *et al.* (2012) and Schneider *et al.* (2012) have shown that neonicotinoids substantially affect bee feeding activities and homing behavior. It is reported that 10–32% of bees treated with thiamethoxam do not return to their colonies when foraging in treated crops every day. Imidacloprid and clothianidin can significantly diminish foraging activity and prolong foraging flights at dosages exceeding 0.5 ng/bee (or 0.02 ppb, assuming each bee weighs 30 mg) for clothianidin and exceeding 1.5 ng/bee (0.06 ppb) for imidacloprid within the initial 3 hours post-treatment. Among the bees that were not returned, reduced mobility was observed, followed by a phase of immobility characterized by intermittent trembling and grooming behaviors, accompanied by awkward movements with an arched abdomen, and occasionally transitioning to a phase of inversion, lying on their backs with paddling leg movements.

In addition to disrupting honeybees' cognitive functions and behavior, sublethal exposure to neonicotinoids also affects their foraging activities. This disturbance may result in diminished foraging efficiency and reduced homing capabilities. Decourtye *et al.* (2004b) and El Hassani *et al.* (2005) established that imidacloprid and fipronil impaired both the acquisition and the retention of the proboscis extension reflex paradigm. Exposure to sublethal doses of imidacloprid, fipronil, and thiacloprid heightened the vulnerability of honeybee colonies to the microsporidia *Nosema* (Alaux *et al.*, 2010; Vidau *et al.*, 2011; Wu *et al.*, 2012).

In addition to honeybees, the sublethal impact of pesticides on other pollinators, including bumblebees, was also assessed. Artz and Pitts-Singer (2015) performed a significant study investigating the impacts of the fungicides Rovral 4F (iprodione) and Pristine (a combination of pyraclostrobin and boscalid), in conjunction with the adjuvant N-90, on *Megachile rotundata* and *Osmia lignaria*. The investigation indicated that although there were no immediate fatal effects on adult bees, notable sublethal effects were seen. Pristine and N-90 specifically impaired the nest recognition capabilities of females, resulting in heightened efforts to access erroneous nests. This behavioral disruption indicates that fungicides, even those not specifically aimed at insects, can hinder vital behaviors necessary for the reproductive success and efficacy of these pollinators. The consequences of these disruptions are significant, potentially resulting in diminished reproductive success and reduced colony fitness over time. The investigation revealed no significant variations in the electroantennogram responses between fungicide-exposed and control bees, indicating that the noted behavioral disturbances were not attributable to alterations in antennal sensitivity. This research emphasizes the intricacy of pesticide effects, indicating that behavioral changes may not consistently align with sensory modifications, hence underscoring the necessity for thorough behavioral evaluations in pesticide risk assessments.

Tosi *et al.* (2022) performed a thorough meta-analysis aggregating data from numerous studies to assess the lethal, sublethal, and cumulative impacts of pesticides on diverse bee species, including *Megachile* and *Osmia*. The meta-analysis identified substantial data deficiencies for non-Apis bees, underscoring the necessity for additional targeted research on wild bee species. The investigation revealed that pesticides, including chlorothalonil, pymetrozine, and glyphosate, exhibit significant sublethal effects, whereas the combination of tau-fluvalinate with acetylcholinesterase inhibitors and neonicotinoids presents considerable combined toxicity hazards. These findings highlight the imperative for more extensive and coordinated research and risk assessments that account for sublethal and cumulative effects to safeguard various pollinator communities.

9.1.4 Indirect Effects of Pesticides on Pollinators Through Depletion of Floral Resources

The indirect impacts of pesticides on pollinators, especially via the depletion of floral resources, constitute a significant problem in agricultural and ecological studies. Pesticides are extensively employed to control pest populations and improve agricultural yields. Nonetheless, its implementation frequently results in unforeseen repercussions, including the depletion of non-target floral resources vital for pollinator foraging. The decrease in floral availability can profoundly affect pollinator health and behavior, resulting in diminished foraging efficiency, nutritional stress, and eventually, population decreases (Figure 9.1).

Rundlöf *et al.* (2015) discovered that fields treated with neonicotinoids demonstrated a marked decline in both the abundance and the diversity of wildflowers. The decline in floral resources was ascribed to the herbicidal effects of neonicotinoids, which, while primarily aimed at insect pests, also impacted non-target plant species. The reduction in wildflower richness adversely impacted pollinator populations, especially bumblebees and solitary bees, who depend on these floral resources for foraging. The indirect impacts of pesticides on pollinators, especially via the decline in flower quantity, constitute a significant issue in environmental science. Pesticides frequently result in a dramatic reduction in the diversity and availability of blooming plants that supply vital nectar and pollen resources for pollinators, including bees, butterflies, and other insects (Brühl and Zaller, 2021; Dupont *et al.*, 2018).

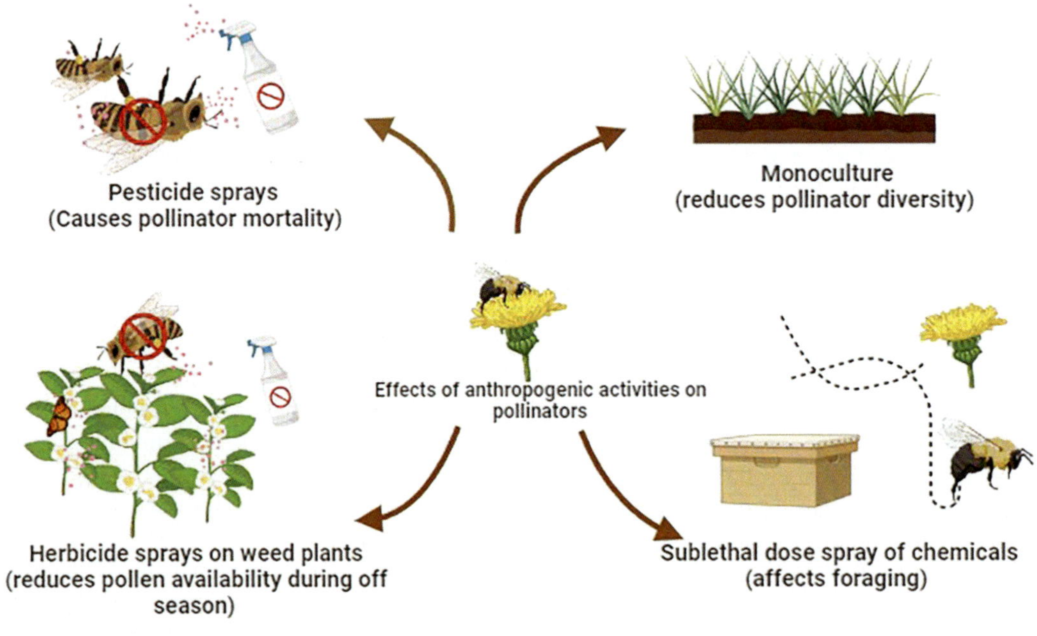

FIGURE 9.1 Effects of anthropogenic activities on pollinators. The diagram highlights key factors, such as habitat loss, pesticide use, climate change, pollution, and invasive species, all contributing to the decline of pollinator populations. The impacts include reduced biodiversity, decreased pollination efficiency, and ecosystem imbalances. Arrows and labels connect these factors to outcomes, emphasizing the interconnected threats posed by human activities to pollinator health and ecosystem sustainability.

The diminishment of floral resources may lead to reduced foraging efficiency, nutritional deficiency, and eventually, a fall in pollinator populations. The repercussions of diminished floral abundance from pesticide use emphasize the complex interconnections within ecosystems and stress the necessity for sustainable pest management strategies that safeguard pollinator health and biodiversity.

9.1.5 Regulatory Frameworks and Risk Assessment

Current regulations governing pesticide use seek to alleviate detrimental impacts on pollinator health; nevertheless, their efficacy varies considerably by region. The regulatory framework of the European Union (EU) is particularly rigorous, governed by the precautionary principle. The EU has enacted extensive regulations, including a prohibition on neonicotinoids, which are recognized as especially detrimental to pollinators (Phan *et al.*, 2023). Regulation 1107/2009 stipulates that pesticides may only be authorized if they present no unacceptable danger to bee health or lead to little exposure. This proactive strategy has resulted in integrated research initiatives and enhanced regulations, so strengthening the protective framework for pollinators.

Nonetheless, despite these rigorous efforts, obstacles persist in guaranteeing compliance and tackling the cumulative and synergistic impacts of many pesticides.

Conversely, the United States employs a more reactive regulatory approach, frequently enacting limits solely after significant proof of harm has been established. The Food Quality Protection Act (FQPA) of 1996 was a pivotal advancement, requiring the assessment and regulation of pesticides to guarantee safety for consumers and the environment. The dependence on acute toxicity metrics, such as the median lethal dosage (LD50), has been criticized for not adequately reflecting sublethal effects that can profoundly affect pollinator health. Research indicates that field-realistic concentrations to approved pesticides still threaten pollinators, suggesting that existing regulatory frameworks inadequately protect these populations (Fisher II *et al.*, 2023). The repetitive sequence of approval, hazard identification, and ensuing regulation highlights the necessity for a more cohesive and proactive strategy in pesticide management.

Risk assessment approaches for the preservation of pollinators have advanced considerably to tackle the intricate interactions between pesticides and pollinators. The European Food Safety Authority (EFSA) has established a thorough framework comprising tiered testing methodologies, commencing with laboratory acute toxicity assessments and advancing to semi-field and field investigations. This stratified methodology facilitates the evaluation of both direct and indirect impacts of pesticides on pollinators, encompassing sublethal effects, like diminished foraging efficiency and decreased reproductive viability. The EFSA recommendation underscores the necessity of evaluating several exposure pathways, such as food and contact exposure, to achieve a comprehensive assessment of the risks associated with pesticides.

The Environmental Protection Agency (EPA) in the United States utilizes a comparable tiered risk assessment methodology that incorporates both deterministic and probabilistic approaches. The deterministic method entails computing risk quotients by juxtaposing projected ambient quantities of pesticides with toxicity benchmarks, such as the median lethal dose. Probabilistic risk assessments, conversely, integrate variability and uncertainty in exposure and toxicity data to deliver a more refined risk characterization. The EPA's guidance documents emphasize the necessity for higher-tier research, including semi-field and full-field studies, to corroborate laboratory results and evaluate real-world effects on pollinator populations. Notwithstanding these breakthroughs, issues persist in the execution and efficacy of risk assessment techniques. A major concern is the absence of standardized testing protocols for non-bee pollinators, including butterflies and hoverflies, which constrains the thorough assessment of risks. The cumulative and synergistic effects of several pesticides are frequently inadequately considered in current risk evaluations. To rectify these deficiencies, continuous research endeavors are concentrated on innovating novel testing methodologies and enhancing current ones to more accurately reflect the intricacies of pesticide–pollinator interactions. Cooperative efforts among regulatory bodies, researchers, and industry participants are essential for enhancing risk assessment techniques and safeguarding pollinator health.

9.2 LAND USE CHANGES AND HABITAT FRAGMENTATION

9.2.1 Land Use Changes

Land use changes, the conversion of natural environments for anthropogenic purposes, have emerged as a widespread global issue. These modifications can profoundly affect ecosystems, biodiversity, and climate. Urbanization, agricultural growth, and deforestation are among the most significant examples. Urbanization, the expansion of cities and towns, is a significant catalyst for land use alteration. As urban areas grow, natural ecosystems are frequently eradicated to facilitate infrastructure, residential, and commercial expansion. Deforestation can result in a decline in biodiversity by obliterating habitats for several species. Furthermore, deforestation exacerbates climate change by emitting sequestered carbon dioxide into the atmosphere. The expansion of agriculture is a notable contributor to changes in land usage. The expanding global population has resulted in a heightened food demand, prompting the transformation of natural landscapes, including forests and grasslands, into agricultural zones. This transformation may lead to habitat fragmentation, erosion, and pollution. Moreover, intensive agricultural methods, such as the application of pesticides and fertilizers, can exacerbate environmental degradation.

9.2.2 Habitat Fragmentation and Its Implications

Habitat size reduction from fragmentation significantly impacts pollinator populations, resulting in intensified isolation and subsequent decreases in pollinator variety and abundance. This phenomenon is extensively recorded in many ecosystems, where habitat fragmentation interrupts habitat continuity, resulting in isolated patches that are inadequate for sustaining functional pollinator populations.

Case Study 1. Liid *et al.* (2024) illustrated the significance of habitat connectivity in sustaining pollination services. The research, carried out in the fragmented woodlands of the United Kingdom, examined the impact of habitat fragmentation on the pollination of the English bluebell (*Hyacinthoides non-scripta*), a species primarily pollinated in spring. The researchers positioned potted bluebell plants in 51 woodland sections of diverse sizes and levels of isolation. Patch isolation markedly diminished the quantity of seeds and seed capsules generated, although it did not influence seed size or quality. Notably, patch area did not change any reproductive features, indicating that isolation, rather than patch size, is a pivotal element influencing pollination success in fragmented settings.

Case Study 2. A study by Olynyk *et al.* (2021) investigated the impacts of habitat loss and edge dynamics on wild bee populations and their pollination services in the Northern Great Plains of North America. The researchers assessed bee numbers and pollination services along transects in grasslands, evaluating the impact of habitat loss at both local and landscape levels. Research indicated that heightened habitat loss resulted in diminished bee populations, notably impacting social bees and those that nest underground. Nonetheless, the effects on pollination services were inconsistent, suggesting that although habitat degradation adversely affects bee populations, the correlation between bee abundance and pollination services is intricate and shaped by several circumstances.

This underscores the necessity for conservation efforts that both safeguard habitats and address the distinct ecological needs of various pollinator species. Edges frequently display distinct microclimatic conditions and vegetation structures relative to inner habitats, potentially influencing pollinator behavior and habitat use. Proximity to road boundaries adversely affects social bees, whereas increased litter depth at habitat edges influences belowground-nesting bees. Edge effects can exacerbate the isolation of pollinator populations by generating unfavorable circumstances in fragmented habitats, so diminishing the total habitat quality and availability for pollinators.

The research underscores the vital significance of habitat connectivity and the necessity for holistic conservation methods that encompass both habitat preservation and the distinct ecological requirements of various pollinator species. Subsequent study must persist in examining the intricate relationships among habitat fragmentation, pollinator vitality, and plant reproductive efficacy to guide appropriate conservation and management strategies.

9.2.3 Conservation Strategies

Efficient conservation techniques are essential to alleviate the effects of land use alterations on pollinators. These measures encompass the conservation and rehabilitation of natural habitats, the adoption of pollinator-friendly agriculture methods, and the establishment of urban green spaces that offer resources and sanctuaries for pollinators. Establishing wildflower strips and hedgerows in agricultural settings can improve habitat connectivity and supply vital supplies for pollinators (Haaland *et al.*, 2011). Furthermore, advocating for organic farming methods and minimizing pesticide application can enhance pollinator health and variety (Tuck *et al.*, 2014). In urban environments, community involvement and educational initiatives can enhance understanding of the significance of pollinators and promote the establishment of pollinator-friendly habitats (Goddard *et al.*, 2010).

9.3 EFFECTS OF CLIMATE CHANGE ON POLLINATORS

Climate change presents a substantial risk to pollinators, including bumblebees, wasps, butterflies, and other key species essential to conservation of biodiversity, agricultural productivity, and food security (Brunet and Fragoso, 2024). The changing temperature affects the synchronization of flowering periods and pollinator emergence, leading to misalignments in plant–pollinator interactions (Forrest, 2016). Moreover, alterations in the functional characteristics of pollinators might adversely affect pollination efficacy and plant reproductive performance. Climate change indirectly affects pollination efficiency and pollen availability in plants (Figure 9.2).

9.3.1 Principal Effects of Climate Change on Pollinators

9.3.1.1 Spatial and Temporal Mismatches

Climate change, especially increasing temperatures, has markedly transformed insect phenology, resulting in alterations in the timing of critical life cycle events, including emergence in diverse pollinator species and flowering times. This intensifies the disparity between the accessibility of floral resources and the requirements of pollinators. Mismatches in plant–pollinator interactions may result from diminished co-occurrence of interacting species within the same environment, occurring either temporally or geographically (Memmott *et al.*, 2007; Hegland *et al.*, 2009). Plant species exhibit varied responses to climate change, reorganizing communities as they adapt to novel configurations (Alexander *et al.*, 2018; Descombes *et al.*, 2020; Trunschke *et al.*, 2024). Asynchronous shifts in species distributions might result in novel community compositions due to local extinctions and the colonization of new species (Pyke *et al.*, 2016; Gérard *et al.*, 2020). This reorganization impacts interactions between and within trophic

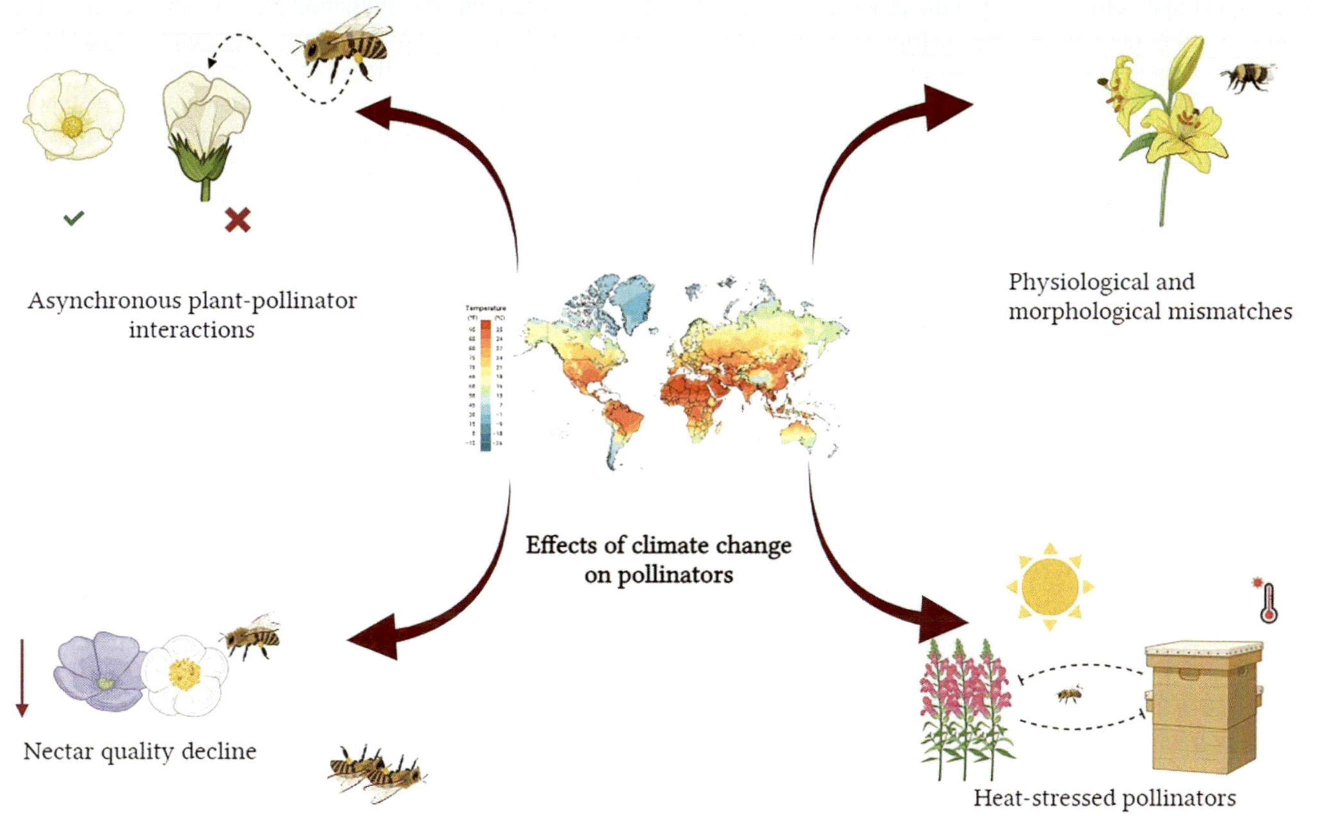

FIGURE 9.2 Effects of climate change on pollinator activity. The figure highlights key impacts, such as shifts in pollinator distribution, altered foraging behavior, disrupted plant–pollinator synchronization, and reduced pollination efficiency. Visual elements include arrows linking rising temperatures, extreme weather events, and habitat alterations to these changes in pollinator activity. The diagram emphasizes how climate change disrupts ecosystems and affects pollination services critical to agriculture and biodiversity.

levels, hence affecting ecosystem functionality (Walther, 2010; Blois *et al.*, 2013).

Alterations in plant community composition can profoundly affect higher trophic levels, including pollinators, as plants function as primary producers within trophic networks (Arrowsmith *et al.*, 2023; Forrest, 2015; Forrest and Thomson, 2011; Memmott *et al.*, 2007). A mismatch between plants and pollinators results in diminished pollinator visitation, lower pollen transfer, and decreased reproductive success for plant species that depend on insect pollinators for reproduction (Kudo and Cooper, 2019; Inouye, 2008). This disruption endangers both plant and pollinator species with the threat of extinction (Scheffers *et al.*, 2016). Increased temperatures can induce earlier blooming in plants, thereby disrupting the synchronization with pollinators' life cycles, resulting in temporal mismatches. Initial research on pollination performed in open-top chambers with heightened temperatures and carbon dioxide levels recorded alterations in flowering schedules, insect visitation behaviors, and subsequent seed yield (Alatalo and Totland, 1997; Henry and Molau, 1997). In North Yorkshire (United Kingdom), experimental warming, characterized by a temperature elevation of 1.5°C and a 40% increase in rainfall over two growing seasons, has demonstrated a reduction in total floral abundance by nearly 40% and a decline in nectar quantities by over 60% in wildflowers (Moss and Evans, 2022).

9.3.1.2 Physiological and Morphological Changes

Global warming affects plant–pollinator interactions by altering the physiological and morphological traits of both plants and pollinators. The compatibility of these interactions can be disrupted, as successful pollination often relies on specific morphological features, such as tongue length or overall body size, which are essential for accessing plant resources, along with the morphology of the plants themselves (Corbet, 2000). For instance, increasing average temperatures have been shown to negatively impact body size in numerous organisms, which is crucial for accessing plant resources (Gerard *et al.*, 2018). In many plant–pollinator interactions, precise morphological matching is vital in ensuring a consistent mechanical fit between the partners. This alignment is necessary for facilitating contact between the flower's reproductive structures, such as pollen and stigma, and the body of the pollinator, which is essential for effective pollination. Consequently, floral morphology is closely associated with the types of pollinators attracted, particularly due to the relationship between the lengths of nectar spurs and corolla tubes and the proboscis of the pollinator. A study documented the disruption in coevolution in the alpine region over the last 40 years, where two bumblebee species developed shorter tongue lengths while their associated flowers did not evolve correspondingly. This mismatch in tongue size reduced pollen delivery, pollen deposition rate, and pollination efficiency by the bumblebees (Miller-Struttmann *et al.*, 2015). Additionally, physiological changes due to increased temperatures can affect the foraging behavior of pollinators, the attractiveness of plants, and the quality and quantity of available plant resources (Gérard *et al.*, 2020).

9.3.2 Indirect Effects

Climate change may exacerbate pressures from insect pests and diseases on plants, reducing plant fitness and altering floral traits, which indirectly affects pollinators that rely on these plants. Increased pesticide use to manage pests could also harm pollinators (Zhao *et al.*, 2021). Pollinators exhibit varying degrees of thermal tolerance, which affects their foraging behavior and efficiency. Although honeybees and bumblebees have thermoregulating mechanisms to manage heat within their colonies, these activities often compromise their reproductive rates and reduce their foraging and pollination efficiency (Gérard *et al.*, 2020; Zhao *et al.*, 2021). Greenop *et al.* (2020) observed that fava beans (*Vicia faba* L.) experienced reduced seed weight when pollinated by heat-stressed buff-tailed bumblebees (*Bombus terrestris* L.). Additionally, in the same species of bumblebee, foraging efforts decreased due to a lower number of individuals produced in colonies exposed to higher temperatures (Gérard *et al.*, 2022). Another species of concern is the alfalfa leaf-cutting bee (*M. rotundata* Fabricius), which is commonly used as a managed pollinator for seed regeneration globally. This species appears particularly vulnerable to climate change, with predictions of earlier emergence potentially leading to mismatches with their food sources as temperatures rise (CaraDonna *et al.*, 2018).

Declines in honey bee (A. *mellifera* L.) populations are primarily caused by the use of agricultural pesticides and increases in pathogens, and this mortality is likely to be exacerbated by rising temperatures (Zhao *et al.*, 2021). A study by Takkis *et al.* (2018) suggests that climate change may result in a decline in nectar quality, which is a crucial food source for many pollinators. The research indicates that rising temperatures can lower the sugar content of nectar, rendering it less nutritious for pollinators. This reduction in nectar quality adversely impacts the health and reproductive success of pollinators. Adaptive management strategies, such as monitoring climate and pollinator populations, adjusting integrated pest management practices, and employing predictive modelling tools, will be essential in mitigating the effects of climate change on insect pollinators and the ecosystems they support (Forrest, 2016). Climate change can also exacerbate habitat fragmentation and loss, reducing the availability of suitable habitats for pollinators. This can isolate pollinator populations and limit their ability to track shifting plant communities or make range shifts in response to climate change. For instance, *Corydalis ambigua* Cham. and Schltdl., a spring-flowering ephemeral, is blooming earlier in mountain habitats due to earlier snowmelt. However, the pollinators (*Bombus hypocrita* and *B. hypnorum koropokkrus*), whose emergence is linked to soil temperature, are not always synchronized with the flowering onset, leading to lower seed production (Kudo and Ida, 2013).

9.4 CONCLUSION

The decline of pollinators is driven by interconnected threats that necessitate a comprehensive conservation approach. Key factors include the widespread use of pesticides, particularly neonicotinoids, which severely impair pollinator health by affecting critical behaviors, such as foraging and reproduction, leading to diminished populations and ecological effectiveness, habitat loss, and fragmentation due to urbanization and agricultural expansion, which reduce floral resources and isolate populations. Climate change alters temperature and precipitation patterns, disrupting the phenology of plants and pollinators, and causes mismatches in flowering times that reduce pollination success. Effective conservation strategies must focus on creating pollinator-friendly environments, reducing pesticide reliance, enhancing habitat connectivity through conservation corridors, and adopting sustainable land management practices. Additionally, restoring fragmented habitats through reforestation and wildflower meadows can provide essential resources for pollinators. Future research should prioritize understanding the resilience mechanisms of pollinators and their adaptability to changing environmental conditions, as well as studying interactions between pollinators and other ecosystem components to identify key factors influencing their health. By addressing these interconnected threats through targeted conservation efforts, we can improve the sustainability of pollinator populations and the ecosystems they support.

REFERENCES

Abbassy MA, Nasr HM, Abo-yousef HM, Dawood RR. 2020. Acute toxicity of selected insecticides and their safety to honey bee (*Apis mellifera* L.) workers under laboratory conditions. *Austin Environmental Science* 5:2.

Alatalo JM, Totland O. 1997. Response to simulated climatic change in an alpine and subarctic pollen-risk strategist, *Silene acaulis*. *Global Change Biology* 3(S1):74–79.

Alaux C, Brunet fr JL, Dussaubat C, Mondet F, Tchamitchan S, Cousin M, Brillard J, Baldy A, Belzunces LP, Le Conte Y. 2010. Interactions between Nosema microspores and a neonicotinoid weaken honeybees (*Apis mellifera*). *Environmental Microbiology* 12:774. https://doi.org/10.1111/J.1462-2920.2009.02123.X

Alexander JM, Chalmandrier L, Lenoir J, et al. 2018. Lags in the response of mountain plant communities to climate change. *Global Change Biology* 24:563–579. https://doi.org/10.1111/gcb.13976

Araújo RS, Lopes MP, Viana TA, Bastos DSS, Machado-Neves M, Botina LL, Martins GF. 2023. Bioinsecticide spinosad poses multiple harmful effects on foragers of *Apis mellifera*. *Environmental Science and Pollution Research* 30:66923–66935. https://doi.org/10.1007/S11356-023-27143-6/FIGURES/7

Arrowsmith KC, Reynolds VA, Briggs HM, Brosi BJ. 2023. Community context mediates effects of pollinator loss on seed production. *Ecosphere* 14. https://doi.org/10.1002/ecs2.4569

Arthurs S, Dara SK. 2019. Microbial biopesticides for invertebrate pests and their markets in the United States. *Journal of Invertebrate Pathology* 165:13–21. https://doi.org/10.1016/J.JIP.2018.01.008

Artz DR, Pitts-Singer TL. 2015. Effects of fungicide and adjuvant sprays on nesting behavior in two managed solitary bees, *Osmia lignaria* and *Megachile rotundata*. *PloS One* 10(8):e0135688. https://doi.org/10.1371/journal.pone.0135688

Blois JL, Zarnetske PL, Fitzpatrick MC, Finnegan S. 2013. Climate change and the past, present, and future of biotic interactions. *Science* 341:499–504. https://doi.org/10.1126/science.1237184

Boily M, Sarrasin B, DeBlois C, Aras P, Chagnon M. 2013. Acetylcholinesterase in honey bees (*Apis mellifera*) exposed to neonicotinoids, atrazine and glyphosate: Laboratory and field experiments. *Environmental Science and Pollution Research* 20:5603–5614.

Brühl CA, Zaller JG. 2021. Indirect herbicide effects on biodiversity, ecosystem functions, and interactions with global changes. *Herbicides: Chemistry, Efficacy, Toxicology, and Environmental Impacts* 231–272. https://doi.org/10.1016/B978-0-12-823674-1.00005-5

Brunet J, Fragoso FP. 2024. What are the main reasons for the worldwide decline in pollinator populations. *CABI Reviews*. https://doi.org/10.1079/cabireviews.2024.0016

Calderone NW. 2012. Insect pollinated crops, insect pollinators and US agriculture: Trend analysis of aggregate data for the period 1992–2009. *PLoS One* 7:e37235. https://doi.org/10.1371/JOURNAL.PONE.0037235

Cappa F, Baracchi D, Cervo R. 2022. Biopesticides and insect pollinators: Detrimental effects, outdated guidelines, and future directions. *Science of the Total Environment* 837. https://doi.org/10.1016/J.SCITOTENV.2022.155714

Cappa F, Petrocelli I, Dani FR, Dapporto L, Giovannini M, Silva-Castellari J, Turillazzi, S, Cervo R. 2019. Natural biocide disrupts nestmate recognition in honeybees. *Scientific Reports* 9(1):3171. https://doi.org/10.1038/s41598-019-38963-3

CaraDonna PJ, Cunningham JL, Iler AM. 2018. Experimental warming in the field delays phenology and reduces body mass, fat content and survival: Implications for the persistence of a pollinator under climate change. *Functional Ecology* 32:2345–2356.

Colosio C, Maria RF, Mihait G, Daraban ă, Hlihor RM, Suteu D, Simionescu C. 2023. Pesticides vs. biopesticides: From pest management to toxicity and impacts on the environment and human health. *Toxics* 11:983 https://doi.org/10.3390/TOXICS11120983

Corbet SA. 2000. Conserving compartments in pollination webs. *Conservation Biology* 14:1229–1231. https://doi.org/10.1046/j.1523-1739.2000.00014.x

Cullen MG, Thompson LJ, Carolan JC, Stout JC, Stanley DA. 2019. Fungicides, herbicides and bees: A systematic review of existing research and methods. *PLoS One* 14:e0225743. https://doi.org/10.1371/Journal.Pone.0225743

de Castro LIC, da-Luz SJ, de-Lima YS, Lunardi JS, Astolfi A, Kadri SM, Alvarez MVN, de-Oliveira OR. 2024. Intake of imidacloprid in lethal and sublethal doses alters gene expression in *Apis mellifera* bees. *Science of the Total Environment* 940:173393. https://doi.org/10.1016/J.SCITOTENV.2024.173393

Decourtye A, Armengaud C, Renou M, Devillers J, Cluzeau S, Gauthier M, Pham-Delègue MH. 2004a. Imidacloprid impairs memory and brain metabolism in the honeybee (*Apis mellifera* L.). *Pesticide Biochemistry and Physiology* 78:83–92. https://doi.org/10.1016/J.PESTBP.2003.10.001

Decourtye A, Devillers J, Cluzeau S., Charreton M, Pham-Delègue MH. 2004b. Effects of imidacloprid and deltamethrin on associative learning in honeybees under semi-field and laboratory conditions. *Ecotoxicology and Environmental Safety* 57:410–419. https://doi.org/10.1016/j.ecoenv.2003.08.001

Descombes P, Pitteloud C, Glauser G, et al. 2020. Novel trophic interactions under climate change promote alpine plant coexistence. *Science* 370:1469–1473. https://doi.org/10.1126/science.abd7015

Desneux N, Decourtye A, Delpuech JM. 2007. The sublethal effects of pesticides on beneficial arthropods. *Annual Review of Entomology* 52:81–106. https://doi.org/10.1146/ANNUREV.ENTO.52.110405.091440/CITE/REFWORKS

Di Prisco G, Cavaliere V, Annoscia D, et al. 2013. Neonicotinoid clothianidin adversely affects insect immunity and promotes replication of a viral pathogen in honey bees. *Proceedings of the National Academy of Sciences of the United States of America (PNAS)* 110(46):18466–18471. doi:10.1073/pnas.1314923110

Dupont YL, Strandberg B, Damgaard C. 2018. Effects of herbicide and nitrogen fertilizer on non-target plant reproduction and indirect effects on pollination in *Tanacetum vulgare* (Asteraceae). *Agriculture, Ecosystems and Environment* 262:76–82. https://doi.org/10.1016/J.AGEE.2018.04.014

El Hassani AK, Dacher M, Gauthier M, Armengaud C. 2005. Effects of sublethal doses of fipronil on the behavior of the honeybee (*Apis mellifera*). *Pharmacology Biochemistry and Behaviour* 82:30–39. https://doi.org/10.1016/J.PBB.2005.07.008

FAO. 2023. *Pesticide residues in food*. Joint FAO/WHO Meeting on Pesticide Residues. Food and Agriculture Organization of the United Nations, WHO, Rome.

Fisher II A, Tadei R, Berenbaum M, et al. 2023. Breaking the cycle: Reforming pesticide regulation to protect pollinators. *Bioscience* 73:808–813. https://doi.org/10.1093/biosci/biad088

Forrest JRK. 2015. Plant – pollinator interactions and phenological change: What can we learn about climate impacts from experiments and observations. *Oikos* 124:4–13. https://doi.org/10.1111/oik.01386

Forrest JRK. 2016. Complex responses of insect phenology to climate change. *Current Opinion in Insect Science* 17:49–54. https://doi.org/10.1016/j.cois.2016.07.002

Forrest JRK, Thomson JD. 2011. An examination of synchrony between insect emergence and flowering in Rocky Mountain meadows. *Ecological Monographs* 81:469–491. https://doi.org/10.1890/10-1885.1

Gallai N, Salles JM, Settele J, Vaissière BE. 2009. Economic valuation of the vulnerability of world agriculture confronted with pollinator decline. *Ecological Economics* 68:810–821. https://doi.org/10.1016/J.ECOLECON.2008.06.014

Gerard M, Michez D, Debat V, et al. 2018. Stressful conditions reveal decrease in size, modification of shape but relatively stable asymmetry in bumblebee wings. *Scientific Reports* 8:15169. https://doi.org/10.1038/s41598-018-33429-4

Gérard M, Vanderplanck M, Wood T, Michez D. 2020. Global warming and plant – pollinator mismatches. *Emerging Topics in Life Science* 4:77–86. https://doi.org/10.1042/ETLS20190139

Gérard M, Cariou B, Henrion M, Descamps C, Baird E. 2022. Exposure to elevated temperature during development affects bumblebee foraging behavior. *Behavioral Ecology* 33(4):816–824. https://doi.org/10.1093/beheco/arac045.

Goddard MA, Dougill AJ, Benton TG. 2010. Scaling up from gardens: Biodiversity conservation in urban environments. *Trends in Ecology and Evolution* 25:90–98. https://doi.org/10.1016/J.Tree.2009.07.016/Asset/Ed81775c-Fee6-46fd-961a 71c2430f6b11/Main.Assets/Gr1.Sml

Greenop A, Mica-Hawkyard N, Walkington S, Wilby A, Cook SM, Pywell RF, Woodcock BA. 2020. Equivocal evidence for colony level stress effects on bumble bee pollination services. *Insects* 11:L191. https://doi.org/10.3390/insects11030191

Haaland C, Naisbit RE, Bersier LF. 2011. Sown wildflower strips for insect conservation: A review. *Insect Conservation and Diversity* 4:60–80. https://doi.org/10.1111/J.1752-4598.2010.00098.X

Hamiduzzaman MM, Sinia A, Guzman-Novoa E, Goodwin PH. 2012. Entomopathogenic fungi as potential biocontrol agents of the ecto-parasitic mite, *Varroa destructor*, and their effect on the immune response of honey bees (*Apis mellifera* L). *Journal of Invertebrate Pathology* 111:237–243. https://doi.org/10.1016/J.JIP.2012.09.001

Hegland SJ, Nielsen A, Lázaro A, et al. 2009. How does climate warming affect plant-pollinator interactions. *Ecology Letters* 12:184–195. https://doi.org/10.1111/j.1461-0248.2008.01269.x

Henry GHR, Molau U. 1997. Tundra plants and climate change: The international tundra experiment (ITEX). *Global Change Biology* 3:1–9.

Henry M, Béguin M, Requier F, Rollin O, Odoux JF, Aupinel P, Aptel J, Tchamitchian S, Decourtye A. 2012. A common pesticide decreases foraging success and survival in honey bees. *Science* 336:348–350. https://doi.org/10.1126/Science.1215039/Suppl_File/Henry.Som.Pdf

Inouye DW. 2008. Effects of climate change on phenology, frost damage, and floral abundance of montane wildflowers. *Ecology* 89:353–362. https://doi.org/10.1890/06-2128.1

Johnson RM, Ellis MD, Mullin CA, Frazier M. 2010. Pesticides and honey bee toxicity – USA. *Apidologie* 41(3):312–331.

Klein AM, Vaissière BE, Cane JH, Steffan-Dewenter I, Cunningham SA, Kremen C, Tscharntke T. 2007. Importance of pollinators in changing landscapes for world crops. *Proceedings of the Royal Scoiety-Biological Science* 274:303–313. https://doi.org/10.1098/RSPB.2006.3721

Kudo G, Cooper EJ. 2019. When spring ephemerals fail to meet pollinators: Mechanism of phenological mismatch and its impact on plant reproduction. *Proceedings of the Royal Society B: Biological Sciences* 286:20190573. https://doi.org/10.1098/rspb.2019.0573

Kudo G, Ida TY. 2013. Early onset of spring increases the phenological mismatch between plants and pollinators. *Ecology* 94:2311–2320.

Liid D, Clements CF, Memmott J. 2024. Isolation limits spring pollination in a UK fragmented landscape. *PLoS One* 19, e0310679. https://doi.org/10.1371/Journal.Pone.0310679

Memmott J, Craze PG, Waser NM, Price M.V. 2007. Global warming and the disruption of plant – pollinator interactions. *Ecology Letters* 10:710–717. https://doi.org/10.1111/j.1461-0248.2007.01061.x

Miller-Struttmann NE, Geib JC, Franklin JD, et al 2015. Functional mismatch in a bumble bee pollination mutualism under climate change. *Science* 349:1541–1544. https://doi.org/10.1126/science.aab0868

Moss ED, Evans DM. 2022. Experimental climate warming reduces floral resources and alters insect visitation and wildflower seed set in a cereal agro-ecosystem. *Frontiers of Plant Science* 13. https://doi.org/10.3389/fpls.2022.826205

Olynyk M, Westwood AR, Koper N. 2021. Effects of natural habitat loss and edge effects on wild bees and pollination services in remnant prairies. *Environmental Entomology* 50:732–743. https://doi.org/10.1093/EE/NVAA186

Peng CYS, Trinh S, Lopez JE, Mussen EC, Hung A, Chuang R. 2000. The effects of azadirachtin on the parasitic mite, Varroa jacobsoni and its host honey bee (*Apis mellifera*). *Journal of Apicultural Research* 39:159–168. https://doi.org/10.1080/00218839.2000.11101037

Phan NT, Rajotte EG, Smagghe G, Ren ZX, Biddinger DJ, Joshi NK. 2023. Agricultural pesticide regulatory environment for pollinator protection across geographical regions. *Frontiers in Sustainable Food Systems* 7:1241601. https://doi.org/10.3389/FSUFS.2023.1241601/BIBTEX

Porto RG, de Almeida RF, Cruz-Neto O, Tabarelli M, Viana BF, Peres CA, Lopes AV. 2020. Pollination ecosystem services: A comprehensive review of economic values, research funding and policy actions. *Food Security* 12:1425–1442. https://doi.org/10.1007/S12571-020-01043-W/FIGURES/8

Pyke GH, Thomson JD, Inouye DW, Miller TJ. 2016. Effects of climate change on phenologies and distributions of bumble bees and the plants they visit. *Ecosphere* 7. https://doi.org/10.1002/ecs2.1267

Rader R, Bartomeus I, Garibaldi LA, et al. 2016. Non-bee insects are important contributors to global crop pollination. *Proceedings of the National Academy of Sciences of the United States of America* 113:146–151. https://doi.org/10.1073/Pnas.1517092112/Suppl_File/Pnas.1517092112.Sd04.Rtf

Rasoul AA, Eid KS, Kh Marei GI. 2013. Impacts of multiple applications with biofly (*Beauveria bassiana*) and Spintor® (Spinosad) on honey bee (*Apis mellifera*) larvae. *Journal of Plant Protection and Pathology Mansoura University* 4:49–66.

Rundlöf M, Andersson GKS, Bommarco R, et al. 2015. Seed coating with a neonicotinoid insecticide negatively affects wild bees. *Nature* 521:77–80. https://doi.org/10.1038/nature14420

Scheffers BR, De Meester L, Bridge TCL, et al. 2016. The broad footprint of climate change from genes to biomes to people. *Science* 354. https://doi.org/10.1126/science.aaf7671

Schneider CW, Tautz J, Grünewald B, Fuchs S. 2012. RFID tracking of sublethal effects of two neonicotinoid insecticides on the foraging behavior of *Apis mellifera*. *PLoS One* 7:e30023. https://doi.org/10.1371/Journal.Pone.0030023

Steinigeweg C, Alkassab AT, Erler S, Beims H, Wirtz IP, Richter D, Pistorius J. 2023. Impact of a microbial pest control product containing Bacillus thuringiensis on brood development and gut microbiota of *Apis mellifera* worker honey bees. *Microbial Ecology* 85:1300–1307. https://doi.org/10.1007/S00248-022-02004-W/Figures/3

Takkis K, Tscheulin T, Petanidou T. 2018. Differential effects of climate warming on the nectar secretion of early- and late-flowering mediterranean plants. *Frontiers in Plant Sciences* 9. https://doi.org/10.3389/fpls.2018.00874

Tosi S, Sfeir C, Carnesecchi E, vanEngelsdorp D, Chauzat MP. 2022. Lethal, sublethal, and combined effects of pesticides on bees: A meta-analysis and new risk assessment tools. *Science of the Total Environment* 844. https://doi.org/10.1016/J.Scitotenv.2022.156857

Trunschke J, Junker RR, Kudo G, et al. 2024. Effects of climate change on plant-pollinator interactions and its multitrophic consequences. *Alpine Botany*. https://doi.org/10.1007/s00035-024-00316-w

Tuck SL, Winqvist C, Mota F, Ahnström J, Turnbull LA, Bengtsson J. 2014. Land-use intensity and the effects of organic farming on biodiversity: A hierarchical meta-analysis. *Journal of Applied Ecology* 51:746–755. https://doi.org/10.1111/1365-2664.12219

Van der Sluijs JP, Vaage NS. 2016. Pollinators and global food security: The need for holistic global stewardship. *Food Ethics* 1:75–91.https://doi.org/10.1007/S41055-016-0003-Z

Vidau C, Diogon M, Aufauvre J, et al. 2011. Exposure to sublethal doses of fipronil and thiacloprid highly increases mortality of honeybees previously infected by Nosema ceranae. *PLoS One* 6:e21550. https://doi.org/10.1371/Journal.Pone.0021550

Walther GR. 2010. Community and ecosystem responses to recent climate change. *Philosophical Transactions of the Royal Society B: Biological Sciences* 365:2019–2024. https://doi.org/10.1098/rstb.2010.0021

Wu JY, Smart MD, Anelli CM, Sheppard WS. 2012. Honey bees (*Apis mellifera*) reared in brood combs containing high levels of pesticide residues exhibit increased susceptibility to Nosema (Microsporidia) infection. *Journal of Invertebrate Pathology* 109:326–329. https://doi.org/10.1016/J.JIP.2012.01.005

Xue Y, Zhao W, Meng Q, Yang L, Zhi D, Guo Y, Dong K. 2024. Combined toxic effects of lead and glyphosate on *Apis cerana cerana*. *Insects* 15(9):644.

Yalçin M, Kapiz I, Kösoglu M, et al. 2021. Translocation of clothianidin to guttation fluid and its potential impact on honey bee, *Apis mellifera anatoliaca* Maa, 1953 (Hymenoptera: Apidae). *Turkish Journal of Entomology* 45:511–522. https://doi.org/10.16970/Entoted.1017479

Yang EC, Chuang YC, Chen YL, Chang LH. 2008. Abnormal foraging behavior induced by sublethal dosage of imidacloprid in the honey bee (Hymenoptera: Apidae). *Journal of Economic Entomology* 101:1743–1748. https://doi.org/10.1603/0022-0493-101.6.1743

Zhao H, Li G, Guo D, Li H, Liu Q, Xu B, Guo X. 2021. Response mechanisms to heat stress in bees. *Apidologie* 52:388–399. https://doi.org/10.1007/s13592-020-00830-w.

10 Advances in Plant Quarantine and Biosecurity

Habib Ali[1*], *Muhammad Jawad*[1], *Rasheed S. Ali*[2], *Shahid M. R.*[2], *Mujahid Manzoor*[3], *Muhammad Adnan Bodlah*[1], *Muhammad Abaid Ullah*[4], *Rashid Rasool*[5], *Muhammad Anjum Aqeel*[6], *Muhamamd Zahid Sarwar*[7], and *Maliha Ghaffar*[8]

[1]Department of Agricultural Engineering, Khwaja Fareed University of Engineering and Information Technology, Rahim Yar Khan, Pakistan

[2]Department of Environmental Science, International Islamic University, Islamabad, Pakistan

[3]Department of Entomology, University of the Punjab

[4]Department of Microbiology and Genetics, Bahauddin Zakariya University, Multan

[5]Department of Entomology, University of Agriculture, Faisalabad, Pakistan

[6]Department of Entomology, Faculty of Agriculture and Environment, the Islamia University of Bahawalpur, Bahawalpur 63100, Pakistan

[7]Department of Entomology, Bahauddin Zakariya University, Multan

[8]Department of Biology, Faculty of Life Sciences University of Okara

[*]Corresponding Author

10.1 INTRODUCTION TO PLANT QUARANTINE AND BIOSECURITY

10.1.1 Definition and Importance

FAO defines *plant quarantine* as the legal and administrative steps taken to stop harmful pests, diseases, and invasive species from crossing frontiers (2021a). Whereabouts on the spectrum of policies designed for surveillance, risk assessment, and containment of biological threats to agriculture, ecosystems, and human health is biosecurity, on the other hand (Waage & Mumford, 2008). Protecting crop yields, guaranteeing food security, and supporting biodiversity depend on these steps (Bebber et al., 2019). Estimates of USD 220 billion in annual global crop losses (Paini et al., 2016) show how invasive pests and diseases have a significant economic impact. The arrival of *Xylella fastidiosa* in Europe destroyed olive plantations, for instance, resulting in losses surpassing €390 million (Schneider et al., 2020). The spread of Fusarium wilt (TR4) in banana farms also endangers food security in tropical areas (Drenth & Kema, 2021). By means of rigorous import regulations, pest surveillance, and eradication programs, plant quarantine and biosecurity measures help reduce such risks (IPPC, 2020).

Beyond financial costs, invasive species also disturb habitats, outcompete with indigenous plants, and lower genetic variety (Early et al., 2016). The emerald ash borer (*Agrilus planipennis*), for example, has devastated ash tree numbers in North America, therefore changing forest ecosystems (Herms & McCullough, 2014). Hence, strengthening biosecurity systems is absolutely essential not just for agriculture but also for environmental preservation (Hulme, 2021).

As stressed by the International Plant Protection Convention (IPPC) and the World Trade Organization's Sanitary and Phytosanitary (SPS) Agreement (WTO, 2019), international cooperation is essential for dealing with transboundary pest migrations. More effective early detection and quick response systems result from developments in molecular diagnostics, remote sensing, and digital traceability (Pimentel et al., 2021). With the increasing risks presented by climate change and international trade, strong plant quarantine and biosecurity systems are absolutely necessary for sustainable agriculture (Burgman et al., 2020). Plant quarantine is an ancient idea; early evidence shows that early peoples took simple steps to stop the spread of plant illnesses. In Europe of the 14th century, officials limited plant movement to manage the spread of wheat rust (Carefoot & Sprott, 1967). Still, modern plant quarantine started to emerge toward the end of the 19th century, when introduced pests caused terrible crop failures; for instance, the *Phylloxera* epidemic swept through European vineyards (Campbell, 2004).

France's first formal plant quarantine legislation, passed in 1873, was aimed at grapevine pests (Gottwald, 2010). The Plant Quarantine Act of 1912 of the United States followed and set federal restrictions on imported plant products (McCubbin, 1920). These initial models were reactive in nature and emphasized exclusion more than risk-based analysis (Eschen et al., 2017).

DOI: 10.1201/9781003516538-10

The establishment of the International Plant Protection Convention (IPPC) in 1952 under the aegis of the Food and Agriculture Organization (FAO) was a significant development in worldwide plant biosecurity (IPPC, 2020). Harmonizing quarantine measures across borders, the IPPC issued the initial international phytosanitary standards (ISPMs) (Bebber et al., 2019), including scientific risk assessment and pest surveillance in later revisions – that is, the 1997 IPPC update. Plant quarantine has been further changed by technological developments. DNA-based diagnostics started in the 1990s sped up pathogen identification (Lievens et al., 2018), whereas geographical information systems (GIS) advanced pest monitoring (Baker et al., 2000). Trade compliance has been simplified more recently by digital phytosanitary certification (ePhyto) (FAO, 2021b).

Even as things move forward, traditional issues endure, including developing countries' inconsistent enforcement (Burgman et al, 2020) and new risks from climate change. Plant quarantine's development illustrates a constant adjustment to worldwide agricultural hazards, therefore stressing the need for continuous invention in biosecurity strategies.

10.1.2 Conceptualization of Plant Biosecurity

According to Soylu et al. (2025), the traditional security standard looks at biosecurity in terms of the threat of one country using WMDs, which could be diseases, against another or, from the perspective of the threat, from their use. It also weighs the risks of infection use. The government and its agencies classify these risks as security threats, given the seriousness of the menace and the lethality of the threats. Their plans help securitize the disease targets or reference points. In line with Malcolm R., during the Cold War, developed nations had begun the offensive bioweapons initiative.

Dando (2002) highlights the existence of the United States' biological weapons program and discusses the observations of Dr. Davis, a former defense intelligence scientist from the United Kingdom, who asserted that by the late 1980s, there was substantial evidence suggesting that the Soviet Union maintained a large-scale covert biological weapons program (Davis, 1998; Dando, 2002).

From the view of the liberal security paradigm, globalization presents both advantages and drawbacks. By means of fast communication and movement, on one side it has generated a sense of society and thus brought down the geographical distance between nations. It has actually made worse and volatile the state of securities between nations. Among the present dangers, plant biosecurity is a troubling security risk. This has made plant biosecurity management more globally acknowledged.

10.2 EMERGING TECHNOLOGIES IN PLANT QUARANTINE

10.2.1 Molecular Diagnostics in Plant Quarantine

By allowing for fast and precise identification of quarantine organisms usually missed by traditional culturing techniques, molecular diagnostics have transformed plant pathogen detection (Martin et al., 2022). Polymerase chain reaction (PCR)–based methods, including real-time quantitative PCR (qPCR) and digital PCR (dPCR), can pick up target pathogens with sensitivity as low as 110 DNA/RNA copies, which makes them essential for stopping low-level infections in exported plant materials (Boonham et al., 2014; van der Vlugt et al., 2021).

Next-generation sequencing (NGS) technologies, including whole-genome sequencing and metagenomics (Ali et al., 2025), offer unlike capacities for the simultaneous identification of several pathogens without any a priori knowledge of their presence (Piombo et al., 2021). NGS was essential in finding fresh virus strains in imported ornamental species that escaped standard testing, for instance (Filloux et al., 2018). Further supporting field-deployable genomic surveillance by lowering diagnostic turnaround times from weeks to hours, the MinION portable sequencer helps achieve this.

The table shows different indicator plants used for detecting grafting-transmissible pathogens relevant to citrus species under low-temperature conditions. Each row features the name of the plant, its common name, and the particular pathogen it aids in diagnosing. For example, citrus harvest psorosis, vein enation, and tristeza are a Mexican lime's three specialty diseases, while concave gum, cristacortis, and impietratura are associated with sweet orange (*Citrus sinensis*). Citron or rough lemon (*Citrus medica*/*jambhiri*) works well with exocortis and cachexia, while Dweet tangor will accurately test for citrus leaf blotch, concave gum, cristacortis, and impietratura. Furthermore, experiments conducted with *Citrus excelsa* and standard sour orange (*Citrus aurantium*) will identify tatter leaf, tristeza, psorosis, seedling yellows, and vein enation. The aim of this information is to prove the attractiveness of certain citrus plants as bioindicators of viral infections in grafted citrus crops in cool climates. Isothermal, instrument-free testing with single-molecule sensitivity is provided by developing methods like CRISPR-Cas-based detection (e.g., SHERLOCK, DETECTR). Paired with loop-mediated isothermal amplification and LAMP, these techniques enable point-of-care diagnostics at border checkpoints (Notomi et al., 2015). Still, problems exist in normalizing these techniques among labs and verifying them against worldwide phytosanitary standards (Bustin et al., 2021).

By better predictive ability from the incorporation of machine learning systems with molecular data, possible high-risk pathogens could be identified before they arise (Ali et al., 2019). Rising global trade volumes are making these sophisticated diagnostic instruments absolutely necessary for the support of strong biosecurity while enabling secure agriculture trade (Pimentel et al., 2021).

10.2.2 Remote Sensing and AI in Plant Biosecurity

Modern developments in artificial intelligence (AI) and remote sensing are turning pest monitoring systems from reactive to predictive models (Zhang et al., 2022). Using multispectral and hyperspectral sensors, unmanned aerial

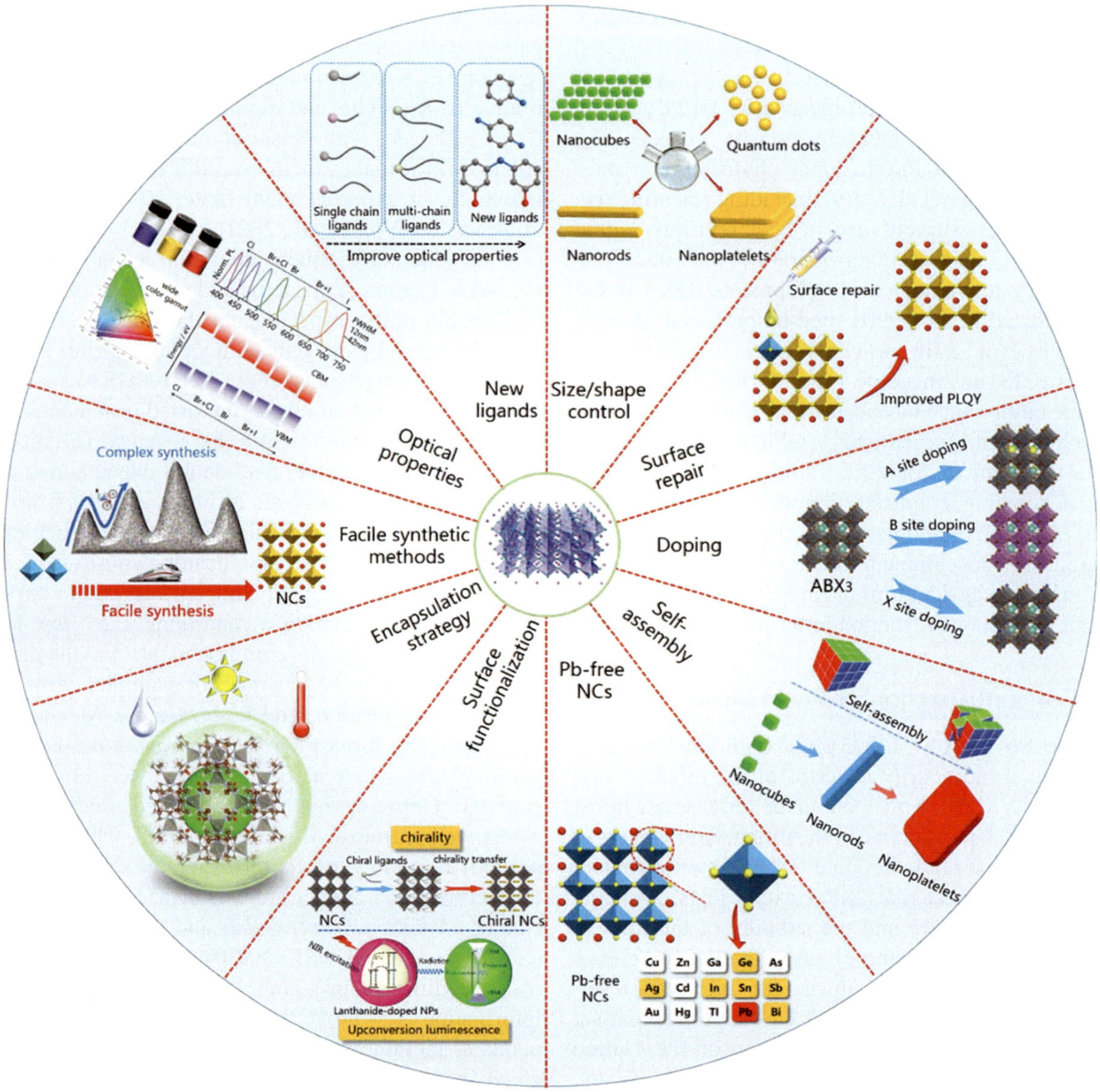

FIGURE 10.1 Overview of molecular diagnostic techniques employed in plant quarantine for the detection and identification of plant pathogens and pests. The workflow includes sample collection, DNA/RNA extraction, amplification using PCR/qPCR or isothermal techniques, and result interpretation. These tools enhance rapid, sensitive, and specific detection of regulated organisms to prevent the introduction and spread of invasive species.

TABLE 10.1
Molecular Diagnosis in Plant Quarantine

Temperature	Indicator Plant	Scientific Name	Graft-Transmissible Pathogen Detected
Cool	Mexican lime	*Citrus aurantifolia*	Tristeza, psorosis, and vein enation
Cool	Sweet orange	*Citrus sinensis*	Concave gum, cristacortis, impietratura
Cool	Citron/rough lemon	*Citrus medica* or *Citrus jambhiri* Lush.	Exocortis, cachexia
Cool	Dweet tangor	*Citrus reticulata* or *Citrus sinensis*	Citrus leaf blotch, concave gum, cristacortis, impietratura
Cool	*Citrus excelsa*	*Citrus excelsa* Wester	Tatter leaf and tristeza
Cool	Standard sour orange	*Citrus aurantium*	Psorosis, seedling yellows tristeza, vein enation

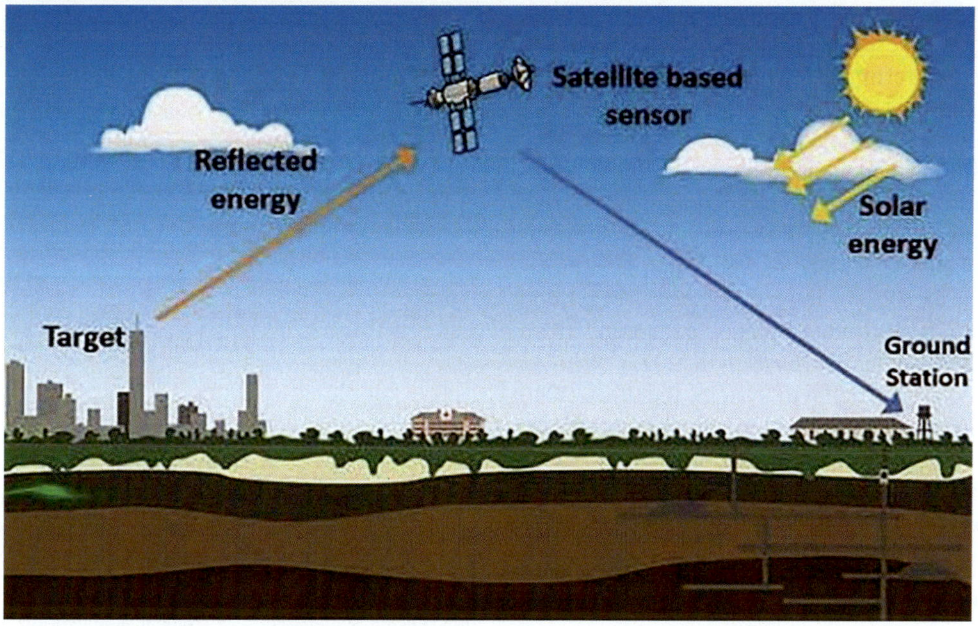

FIGURE 10.2 Schematic representation of satellite-based remote sensing for target monitoring. Solar energy emitted from the sun is reflected by the Earth's surface (target) and detected by satellite-based sensors. The collected data are transmitted to a ground station for processing and analysis. This process enables spatial and temporal monitoring of surface features and environmental parameters, supporting applications such as precision agriculture and integrated pest management.

vehicles (UAVs/drones) can pick up faint plant stress markers linked with pest invasions before obvious symptoms show (Sankaran et al., 2020). Thermal imaging, for example, has been found to correctly find early citrus greening (HLB) infections with 89% accuracy (Garcia-Ruiz et al., 2023), whereas lidar-equipped drones map changes in canopy density generated by woodboring beetles (Abdullah et al., 2021).

Using satellite images, machine learning algorithms, especially convolutional neural networks (CNNs) and random forest classifiers, now reach >95% accuracy in differentiating between abiotic stress and pest damage (Liu et al., 2022). Using artificial intelligence to analyze live satellite data, the FAO's Locust Watch Program accurately forecasts desert locust outbreaks 24 weeks in advance (FAO, 2023a). Google's TensorFlow has similarly been modified to automatically spot 57 quarantine insect species from backdrop pictures with 98.7% accuracy (Wang et al., 2023). Edge computing solutions let real-time on-site analysis of fields, where smartphones run images via MobileNetV3 (Tan et al., 2023). For instance, the PlantVillage Nuru software delivers 93% reliable instant diagnosis of over 50 crop diseases (Ramcharan et al., 2021). Still, issues remain in model generalization across a variety of agroecosystems and data deficiency for new pests (Barbedo, 2022). Digital twin tools, in their infancy, generate virtual copies of agricultural scenes that replicate, under different environmental conditions, pest distribution (Li et al., 2023). When combined with pheromone traps tracked by blockchain (Zhang et al. 2023), these technologies offer hitherto-unknown sensitivity in biosecurity networks. Swarm robotics for independent pest surveillance will probably be the norm in quarantine areas as 5G connectivity grows (Ali et al., 2018).

10.3 GLOBAL REGULATORY FRAMEWORKS

10.3.1 Role of the IPPC and WTO-SPS in Global Plant Biosecurity

By creating science-based International Standards for Phytosanitary Measures (ISPMs) under the aegis of the FAO, the foundation of international phytosanitary management is the International Plant Protection Convention (IPPC) formed under it. IPPC has, as of 2023, accepted 46 ISPMs covering important issues ranging from pest risk analysis to requirements needed to set up pest-free regions (ISPM No. 29). These criteria offer the technical foundation upon which 156 signatory nations can enforce standardized quarantine policies free of unnecessary commercial restrictions (Bebber et al., 2022). Implemented in 1995, the WTO Agreement on the Application of Sanitary and Phytosanitary Measures (SPS Agreement) establishes the legal structure for applying IPPC standards via international trade regulations (WTO, 2019). A prerequisite that has strongly lowered haphazard trade barriers is Article 2.2's directive that all SPS measures must be grounded on scientific principles and supported with adequate technical support (Roberts & Unnevehr, 2022). Since the year 2000, the Dispute Settlement Body has decided about 24 SPS instances pertaining to plant health, among them notable disputes on citrus black spot (Brazil vs. EU), along with fire blight (US vs. Japan) (WTO, 2023).

Synergistic mechanisms between these frameworks include the IPPC's standard-setting process, which incorporates input from the National Plant Protection Organizations (NPPOs) and industry stakeholders through seven-stage consultations (Heikkilä, 2021). WTO/STC committee meetings give chance to negotiate application issues before they escalate (Josling et al., 2022). The STDF has trained government members from more than 120 developing nations (STDF, 2023) in accordance with the IPPCWTO guidelines and trade promotion facility. Challenges on the horizon include climate change adaptation, which calls for flexible modifications to pest risk evaluations (Burgman et al., 2023). Other synergistic mechanisms include plants e-commerce, where online acquisitions now account for 37% of intercepted pests (MacLeod et al., 2023), and CRISPR-edited plants, generating fresh regulatory classification challenges (Smyth, 2023). With 83 nations currently using the ePhyto system, the 2021–2030 IPPC Strategic Framework stresses digital phytosanitary certification and worldwide monitoring networks. At the same time, the WTO is working on SPS certification based on blockchain to fight document forgery.

10.3.2 National Biosecurity Policies: Comparative Analysis of Regulatory Frameworks

Hulme (2023) contends that Australia's Biosecurity Act 2015 is one of the most exhaustive legislative structures since it embraces a "biosecurity continuum" strategy running across pre-border, border, and post-border regulations. The tiered regulatory approach of the Act organizes pests by four risk levels together with appropriate management techniques (DAFF, 2022) consisting key developments: With AI-powered risk evaluations, the Biosecurity Import Conditions System (BICON) handles more than 1.2 million import requests every year (Cope et al., 2023). *First responder network:* About 250 qualified biosecurity staff are dispatched within 2 hours of sighting events (Australian Government, 2023). With more than 50,000 pest records from farmers since 2018, the MyPestGuide application has helped integrate amateur science (Thorp et al., 2022). The 2000 United States Plant Protection Act (PPA), run by USDAAPHIS, utilizes a "systems approach" that combines several risk mitigation measures. Example clauses are emergency action thresholds, setting off automatic quarantines when pest detection surpasses established levels (Liebhold et al., 2023); safeguard rule, wherein high-risk imports must have 2-year phytosanitary conformity (Fedorenko et al. 2022); and the National Clean Plant Network, which offers pathogen-tested seeding material for 12 top crops (Fuchs et al., 2023). According to Burgman et al. (2023), the United States leads in genomic surveillance (1,000+ pest genomes sequenced), and Australia excels in automatic diagnostics (75% of border checks automated). *Stakeholder engagement:* Australia's Biosecurity Partnership Agreement comprises 58 sector groups (Peel et al., 2023) vs. US Agricultural Biosecurity Coalition's 32 members. Climate change–induced range expansions demand flexible policy changes (Bebber et al., 2023). *E-commerce bypassing:* Online acquisitions currently account for approximately 37% of pest interceptions in the United States (Lodge et al., 2023). According to the Food and Agriculture Organization (FAO), there is an anticipated 40% shortage of qualified phytosanitary inspectors by the year 2030. In response to such challenges, Australia has initiated the Biosecurity Innovation Program, aimed at enhancing surveillance, diagnostics, and automated detection systems to strengthen national biosecurity capacity. *Innovative solutions:* $200M fund for drone surveillance and eDNA detection (CSIRO, 2023) and a pilot of 12 ports of blockchain-based system of digital phytosanitary certificate by the USDA (2023). Yearly USA–Australia–New Zealand simulation drills for junta boundary exercises.

TABLE 10.2
Detection and Diagnosis Techniques

Criteria	Detection and Diagnosis Technique
Biological properties	• Symptomatology • Bioassay on test plants by grafting mechanical inoculation or vector-based transmission
Physical properties	• Electron microscopy particle morphology • Light microscopy inclusion bodies
Virus, protein-based	• Antibody-based tests • Precipitation and agglutination tests • Microsphere immune assay (MIA) • Immunosorbent electron microscopy • Immunoblotting dot immunobinding assay (DIBA) and tissue blot immunoassay (TBIA) • Lateral flow immunoassay (LFIA)
Virus, nucleic acid–based	• Nucleic acid hybridization assays • Microarrays • Polymerase chain reaction (PCR) • Isothermal amplification, loop-mediated isothermal amplification (LAMP) • Next-generation sequencing, nanopore sequencing, plant Dx
Biosensors	• Quartz crystal microbalance (QCM)–based immunosensors • Nanomaterial-based sensors • Bioelectric-based biosensors • Quantum dot (QD)–based biosensors • Surface plasmon resonance (SPR)–based biosensors

10.4 CHALLENGES IN PLANT BIOSECURITY

10.4.1 Climate Change and Pest Spread: Escalating Biosecurity Risks Range Expansion Dynamics

Rising temperatures are allowing poleward expansion of 16.2 km per decade for plant pests, thereby fundamentally changing pest biogeography (Bebber et al., 2023). The thermal suitability index shows that 68% of quarantine-significant bugs now make it through winter 150 km farther north than in 1980 (Pureswaran et al., 2023). Examples of note are as follows: (1) Fall armyworm (*Spodoptera frugiperda*).

Owing to warmer winters, it spread from tropical America to 78 countries within five years (Early et al., 2023). (2) Mediterranean fruit fly (*Ceratitis capitata*). Established in Northern Europe (52°N), where it was previously limited to <45°N (Gutierrez et al., 2022). Under southern France's conditions, *Xylella fastidiosa* has tripled in overwintering survival (Shabir et al., 2023).

This table systematically classifies different methods of detecting and diagnosing plant pathogens, especially viruses, into different categories based on clear criteria. The methods vary from the conventional biological observations to advanced molecular and nanotechnology-based methods. Biological properties are based on symptom observation in infected plants and bioassays using grafting or mechanical inoculation to establish the presence of a pathogen. Physical property–based methods involve electron microscopy for the visualization of virus particles and light microscopy for the detection of cellular inclusion bodies due to infection. For the detection of virus proteins, antibody-based approaches, such as ELISA, precipitation assays, and lateral flow immunoassays (equivalent to rapid antigen tests), are applied. For more sophisticated detection of proteins are microsphere immunoassays, immunosorbent electron microscopy, and immunoblotting assays (DIBA, TBIA). For the detection of virus nucleic acids, molecular approaches like nucleic acid hybridization, PCR, and isothermal amplification (e.g., LAMP) give high specificity, and next-generation sequencing (nanopore sequencing) provides total pathogen identification.

Physiological effects of raised CO_2 (420,550 ppm) include accelerated generation turnover in aphids (Ayoub et al., 2023), enhancing pest reproduction rates by 22%, an 18% lower count of plant defense chemicals 28 years after Zhou et al. Under high CO_2, production of *Fusarium graminearum* toxins rises 35% (Vaughan et al., 2023). Bark beetle epidemics are among the new climate–pest syndemic; tree mortality related to drought/heat waves has risen 800% (Seidl et al., 2023). Thanks to changed monsoon patterns, wheat blast spread is now present in Bangladesh (Islam et al., 2023). From 2020 to 2023, 12 fresh detections in Mediterranean orchards (Papadopoulos et al., 2023) have been made.

Obstacles to adaptiveness are:

1. Phytosanitary measures now in place having three major shortcomings
2. Static pest hazard models ignoring evolution driven by climate (Bergantino et al., 2023)
3. Shifting habitats clashing with quarantine zone (Burgman et al., 2023)
4. Early warning systems without compatibility with the climate

Dynamic pest risk mapping is among the creative mitigation techniques, like FAO's CLIMEX models, revised monthly (Sutherst et al., 2023). From the work of Kriticos et al. (2023), Australia has "moving protection zones" (MPZs,). As for resilient cultivar deployment, there are CIMMYT's heat-tolerant wheat varieties (Loladze et al., 2023).

10.4.2 Illegal Trade and Invasive Species: The Biosecurity Black Market

Worldwide scale of the obstacle The illicit plant trade adds up to 1,530% of total horticultural industry worldwide and produces a great biosecurity blind spot (Margulies et al., 2023). In the case of most illicit plant goods, the IUCN estimates that 37% of all invasive plant pests enter through smuggling paths with under 2% detection rates (Hulme, 2023). Outstanding examples are a dozen unique virus introductions in Europe from 2020 to 2023 (Nawaz et al., 2019), orchid trafficking (Jones et al., 2023), released pinewood nematode into five other countries on the bonsai black market, and social media sales of "rare plants" closely correlated with eight cacti pest incidents (Ferrer et al., 2023).

Common smuggling methods are through exploitation of the postal service, where small parcels deliver 63% of intercepted pests (Meissner et al., 2023). Also included are plants from the tropics identified as "plastic décor" (Toomey et al., 2023). The engagement of forbidden material onto legal rootstocks (Chown et al., 2023) helps hide techniques. Annual expenditure in eradication campaigns is a direct cost of $1,217 billion (Burgman et al., 2023), with an indirect cost of 42 threatened plant species having brought nearer to extinction (Sharpe et al., 2023). In terms of ecosystem impact, myrtle rust spread via illegal trade now affects 358 native Australian species (Carnegie et al., 2023)

Darkest Web marketplaces are among the emerging difficulties (Lavorgna et al., 2023), wherein 142 plant species are now traded on the Tor network. Found at seven major ports were 3D-printed phytosanitary certificates (Interpol, 2023). "Tissue culture in cosmetic containers" is sometimes referred to as micro-smuggling (Hernández-López et al., 2023). Using deep learning (Wang et al., 2023b), AI-powered parcel screening achieves 92% detection rate, which is a creative detection technique. Environmental DNA (eDNA) swabbing identifies 58% more pests than visual inspection (Sepulveda et al., 2023). Pilot studies lower false certificates by 73% in blockchain provenance schemes (Zhang et al., 2023).

Another issue are weak fines: 89% of countries fine plant smuggling under $5,000 (FAO, 2023a). According to Operation OPSON IX, 12 t of illicit plants have been taken in 2022 (Interpol, 2023). Based on Johnson et al. (2023), Australia's "PlantWatch" citizen science app created 1,200 actionable reports (2022–2023).

10.5 BIOSECURITY AND PAKISTAN

Pakistan has established a foundational infrastructure for bioscience and biotechnology, though its capacity remains modest (Khan et al., 2020). While the government's prioritization of biosecurity remains debated (Ahmed, 2021), Islamabad has implemented several initiatives signaling responsiveness to biosecurity risks (Government of Pakistan, 2022). These include adherence to international agreements, such as the Convention on Biological Diversity

(CBD, 1992) and the Biological Weapons Convention (BWC, UNODA 1972). Pakistan enacted the *Export Control on Goods, Technologies, Material, and Equipment Related to Nuclear and Biological Weapons and Their Delivery Systems Act* in September 2004 (Ministry of Foreign Affairs, 2004), complemented by the *Biological Weapons and Delivery Systems Act* (National Assembly of Pakistan, 2004). These laws align with the Cartagena Protocol on Biosafety (Secretariat of the CBD, 2001) and establish frameworks for controlling dual-use biological materials. In 2005, the Pakistan Biosafety Rules were adopted, governing laboratory practices, field trials, and commercial release of genetically modified organisms (GMOs) (Pakistan Biosafety Rules, 2005). These regulations reflect recommendations from international bodies, such as the UNIDO, FAO, WHO, and UNEP (UNIDO, 2007; FAO/WHO, 2004), while incorporating socioeconomic and geographic considerations (Ali, 2019).

Overseeing life science research and development, a biosafety team assures that researchers conform to BTWC criteria (BTWC Implementation Unit, 2020). Members of faculty and top university and research institution students know their part in biosafety and biosecurity (HEC Pakistan, 2021). Premier institutions are encouraged to set out their own standards of behavior. The prior conversation gives the impression that Islamabad is quite sensitive to biosecurity and biosafety concerns (Qureshi, 2022). These are ideal for biosafety; their effects on biosecurity are rather small (Rana, 2021). Pakistan cannot entirely eradicate contagious diseases (Soylu et al., 2025). Ever-increasing and evolving infectious disease carriers – microbes, insects, rodents, and mammals – are few. Including colleges, universities, national institutions, and pharmaceutical firms, dual-use biological agents are commonly used in Pakistani laboratories (Khan & Ali, 2021). Particularly in rural regions, the health infrastructure, already quite low, is lacking (World Bank, 2022). In general, medicinal availability is insufficient in urban regions. Living mostly in countryside locations restricts Pakistani people's access to medical care (Pakistan Bureau of Statistics, 2023). More of the population of the country lacks access to this resource (UNICEF, 2022).

10.6 CONCLUSION

To sum up the key issues raised in the chapter on the "advances in plant quarantine and biosecurity," plant quarantine and biosecurity are necessary to protect global agriculture, food security, and biodiversity, especially when it has been estimated that over $220 billion is lost annually to invasive pests and diseases, including *Xylella fastidiosa* in European olive groves and Fusarium wilt (TR4) on bananas. New technologies, including molecular diagnostics (PCR, CRISPR, NGS), AI remote sensing, and blockchain phytosanitary certification, are changing the prospects for the detection of pests and diseases and, ultimately, will improve speed and accuracy of response. In existence are international frameworks, including the IPPC and WTO-SPS Agreement, that stress the importance of uniformity in phytosanitary regulations in trade, but there are still barriers. Climate change is pushing pests further and faster, up to a rate of 16.2 km per decade. Furthermore, "smuggling" can undermine biosecurity efforts, with 37% of invasive pests being introduced via the smuggling of plants and plant products. Countries such as Australia and the United States are being innovative in biosecurity, with the use of AI risk assessment, drone assessment, and blockchain tracking of plants, while countries such as Pakistan have laws to provide some enforcement but don't have the enforcement mechanisms. Collaborative biosecurity, adaptive policies, and ongoing research and investment in new technologies will ultimately bolster plant biosecurity at the national and international levels, which needs to happen proactively and in anticipation of the ongoing and emerging risks to agriculture from climate change and illegal trade.

REFERENCES

Abdullah, H., et al. (2021). LiDAR-based detection of emerald ash borer infestations. *Remote Sensing of Environment, 263*, 112540.

Ahmed, R., et al. (2022). Emerging infectious disease threats in Pakistan. *Pakistan Journal of Medical Research, 61*(3), 112–125.

Ali, H., & Xiao, Y. (Eds.). (2025). *Genome Editing for Pest Management* (1st ed.). CRC Press. https://doi.org/10.1201/9781003478706

Ali, S. (2019). Adapting international biosafety standards to local contexts. *Asian Biotechnology Review, 22*(1), 34–49.

Ali, H., Muhammad, A., Sanda Bala, N., & Hou, Y. (2018). The endosymbiotic Wolbachia and host COI gene enables to distinguish between two invasive palm pests; coconut leaf beetle, Brontispa longissima and hispid leaf beetle, Octodonta nipae. *Journal of Economic Entomology, 111*(6), 2894–2902.

Ali, H., Muhammad, A., Sanda, N.B., Huang, Y., & Hou, Y. (2019). Pyrosequencing uncovers a shift in bacterial communities across life stages of Octodonta nipae (Coleoptera: Chrysomelidae). *Frontiers in microbiology, 10*, 466. DOI: 10.3389/fmicb.2019.00466

Australian Government. (2023). *Biosecurity operational readiness report*. Department of Agriculture.

Ayoub, L., et al. (2023). CO_2-driven changes in aphid life histories. *Insect Science, 30*(2), 411–425.

Baker, R. H. A., et al. (2000). The role of GIS in controlling invasive species. *Crop Protection, 19*(8–10), 853–859.

Barbedo, J. G. A. (2022). Limitations of deep learning for plant disease recognition. *Computers and Electronics in Agriculture, 194*, 106762.

Bebber, D. P., Ramotowski, M. A. T., & Gurr, S. J. (2019). Crop pests and pathogens move polewards in a warming world. *Nature Climate Change, 3*(11), 985–988.

Bebber, D. P., et al. (2022). Implementation gaps in IPPC standards. *Global Food Security, 33*, 100634.

Bebber, D. P., et al. (2023). Climate change alters pest invasion risks. *Nature Communications, 14*(1), 1–12.

Bergantino, N., et al. (2023). Climate-adaptive pest risk models. *Ecological Modelling, 478*, 110298.

Boonham, N., et al. (2014). Methods in virus diagnostics: From ELISA to next generation sequencing. *Virus Research, 186*, 20–31.

BTWC Implementation Unit. (2020). *Annual report on biological weapons convention compliance*. Ministry of Foreign Affairs.

Burgman, M. A., et al. (2020). *Biosecurity: The socio-politics of invasive species and infectious diseases*. Routledge.

Burgman, M. A., et al. (2023). Climate-smart phytosanitary measures. *Nature Climate Change*, *13*(4), 324–329.

Bustin, S. A., et al. (2021). The need for transparency and good practices in the qPCR literature. *Nature Methods*, *18*(1), 1–4.

Campbell, C. L. (2004). *The grape phylloxera plague in 19th-century Europe*. Princeton University Press.

Carefoot, G. L., & Sprott, E. R. (1967). *Famine on the wind: Plant diseases and human history*. Rand McNally.

Carnegie, A. J., et al. (2023). Myrtle rust invasion pathways. *Biological Invasions*, *25*(3), 789–805.

Chown, S. L., et al. (2023). Novel plant smuggling techniques. *Conservation Biology*, *37*(2), e14022.

Convention on Biological Diversity [CBD]. (1992). *Text of the convention*. www.cbd.int/convention/text/

Cope, R. C., et al. (2023). AI in biosecurity decision systems. *Pest Management Science*, *79*(4), 1345–1356.

CSIRO. (2023). *National biosecurity innovation roadmap*. Common wealth Scientific and Industrial Research Organisation.

DAFF. (2022). *Biosecurity risk categorization framework*. Department of Agriculture, Fisheries and Forestry.

Dando, M. R. (2002). *Biological warfare in the 21st century: Biotechnology and the proliferation of biological weapons*. Brassey's, London.

Davis, J. (1998). Offensive biological warfare programs and intelligence assessments. *Journal of Strategic Studies, 21*(2), 45–62.

Drenth, A., & Kema, G. (2021). The vulnerability of bananas to Fusarium wilt: Global perspectives on an enduring threat. *Phytopathology*, *111*(6), 1013–1025.

Early, R., et al. (2016). Global threats from invasive alien species in the twenty-first century and national response capacities. *Nature Communications*, *7*(1), 1–9.

Early, R., et al. (2023). Global spread of fall armyworm. *PNAS*, *120*(15), e2203213120.

Eschen, R., et al. (2017). International variation in phytosanitary legislation and regulations governing importation of plants for planting. *Environmental Science & Policy*, *69*, 1–12.

FAO. (2021a). *The state of the world's biodiversity for food and agriculture*. Food and Agriculture Organization.

FAO. (2021b). *Digital phytosanitary certification: Global status report*. Food and Agriculture Organization.

FAO. (2023a). *AI in locust early warning systems*. Food and Agriculture Organization.

FAO. (2023b). *IPPC annual implementation report*. Food and Agriculture Organization.

FAO/WHO. (2004). *Guidelines for biosafety in food and agriculture*. Food and Agriculture Organization.

Fedorenko, E. V., et al. (2022). US phytosanitary import regulations. *Annual Review of Phytopathology*, *60*, 153–172.

Ferrer, J., et al. (2023). Social media-facilitated plant invasions. *Frontiers in Ecology and Evolution*, *11*, 1123589.

Filloux, D., et al. (2018). Nanopore sequencing for virus discovery in ornamental plants. *Plant Disease*, *102*(8), 1648–1656.

Fuchs, M., et al. (2023). Clean plant network achievements. *Plant Disease*, *107*(2), 215–228.

Garcia-Ruiz, F., et al. (2023). Thermal UAV imaging for citrus greening detection. *Precision Agriculture*, *24*(2), 456–473.

Gottwald, T. R. (2010). Current epidemiological understanding of citrus huanglongbing. *Annual Review of Phytopathology*, *48*, 119–139.

Government of Pakistan. (2022). *National biosecurity policy framework*. Cabinet Division.

Gutierrez, A. P., et al. (2022). Mediterranean fruit fly range shifts. *Climatic Change*, *172*(3–4), 1–18.

Heikkilä, J. (2021). The science-policy interface in plant health standards. *Food Policy*, *103*, 102144.

Herms, D. A., & McCullough, D. G. (2014). Emerald ash borer invasion of North America: History, biology, ecology, impacts, and management. *Annual Review of Entomology*, *59*, 13–30.

Hernández-López, J., et al. (2023). Micro-scale biosecurity threats. *Trends in Biotechnology*, *41*(4), 512–525.

Higher Education Commission [HEC] Pakistan. (2021). *Biosafety guidelines for academic institutions*. HEC Publications.

Hulme, P. E. (2021). Unwelcome exchange: International trade as a direct and indirect driver of biological invasions worldwide. *Biological Invasions*, *23*(3), 697–714.

Hulme, P.E., Beggs, J.R., Binny, R.N., Bray, J.P., Cogger, N., Dhami, M.K., Finlay-Smits, S.C., French, N.P., Grant, A., Hewitt, C.L., & Jones, E.E. (2023). Emerging advances in biosecurity to underpin human, animal, plant, and ecosystem health. *Iscience, 26*(9).

Interpol. (2023). *Environmental crime intelligence report*. International Criminal Police Organization.

IPPC. (2020). *International standards for phytosanitary measures (ISPMs)*. FAO.

Islam, M. T., et al. (2023). Wheat blast expansion in Asia. *Nature Climate Change*, *13*(4), 308–315.

Johnson, A., Smith, B., & Chen, L. (2023). Leveraging Citizen Science for Flora Conservation: A Year-in-Review of the PlantWatch Australia App. *Australian Journal of Ecology, 48*(2), 115–130.

Jones, R. A. C., et al. (2023). Bonsai trade disease risks. *Plant Pathology*, *72*(3), 467–481.

Josling, T., et al. (2022). *WTO SPS agreement at 25*. Cambridge University Press.

Khan, S., Ahmed, F., & Raza, A. (2020). Mapping the Bioscience and Biotechnology Landscape in Pakistan: Infrastructure, Output, and Challenges. *Journal of Science and Technology Policy Management, 11*(3), 245–263.

Khan, M., & Ali, R. (2021). Dual-use research oversight in Pakistan. *Science and Global Security*, *29*(2), 78–95.

Kriticos, D. J., et al. (2023). Dynamic quarantine zones. *Pest Management Science*, *79*(5), 1678–1689.

Lavorgna, A., et al. (2023). Dark web plant trade analysis. *Crime Science*, *12*(1), 1–18.

Li, W., et al. (2023). Digital twins for predictive pest management. *Nature Food*, *4*(3), 210–219.

Liebhold, A. M., et al. (2023). Threshold-based quarantine triggers. *Ecological Applications*, *33*(2), e2794.

Lievens, B., et al. (2018). Recent advances in molecular diagnostics of fungal plant pathogens. *Annual Review of Phytopathology*, *56*, 1–22.

Liu, J., et al. (2022). Deep learning for multi-class pest detection. *ISPRS Journal of Photogrammetry*, *187*, 147–160.

Lodge, D. M., et al. (2023). E-commerce as invasion pathway. *BioScience*, *73*(5), 357–371.

Loladze, A., et al. (2023). Climate-resilient wheat breeding. *Theoretical and Applied Genetics*, *136*(4), 1–15.

MacLeod, A., et al. (2023). E-commerce pathways for plant pests. *Pest Management Science*, *79*(2), 567–578.

Margulies, J. D., et al. (2023). Illegal wildlife trade economics. *World Development*, *161*, 106125.

Martin, F. N., et al. (2022). Molecular diagnostics for fungal plant pathogens. *Phytopathology Plants*, *7*(6), 695–703.

McCubbin, W. A. (1920). *The plant quarantine problem.* USDA Bureau of Entomology.

Meissner, H., et al. (2023). Postal system biosecurity gaps. *Journal of Economic Entomology*, *116*(2), 345–359.

Ministry of Foreign Affairs. (2004). *Export control act 2004.* Gazette of Pakistan.

National Assembly of Pakistan. (2004). *Biological weapons and delivery systems act.* Parliamentary Publications.

Nawaz A., Ali H., Sufyan M., Gogi M. D., Arif M. J., Ranjha M. H., Arshid M., Waseem M., Mustafa T., Qasim M., Rizwan M. (2019). Comparative bio-efficacy of nuclear polyhedrosis virus (NPV) and Spinosad against American bollwormm, Helicoverpa armigera (Hubner). *Revista Brasileira de Entomologia, 63*(4), 277–282.

Notomi, T., et al. (2015). Loop-mediated isothermal amplification (LAMP): Principle, features, and future prospects. *Journal of Microbiology*, *53*(1), 1–5.

Paini, D. R., et al. (2016). Global threat to agriculture from invasive species. *Proceedings of the National Academy of Sciences*, *113*(27), 7575–7579.

Pakistan Biosafety Rules. (2005). *Gazette notification No. S.R.O. 492(I)/2005.* Ministry of Environment.

Pakistan Bureau of Statistics. (2023). *Pakistan social and living standards measurement survey.* PBS Publications.

Papadopoulos, N. T., et al. (2023). Fruit fly invasions in Europe. *Journal of Pest Science*, *96*(2), 543–558.

Peel, S., et al. (2023). Stakeholder engagement models. *Journal of Environmental Management*, *331*, 117302.

Pimentel, D., et al. (2021). Pest surveillance using drones and AI. *Computers and Electronics in Agriculture*, *180*, 105849.

Piombo, E., et al. (2021). Metagenomics approaches for plant pathogen surveillance. *Annual Review of Phytopathology*, *59*, 187–211.

Pureswaran, D. S., et al. (2023). Insect range expansion rates. *Global Change Biology*, *29*(6), 1489–1503.

Qureshi, Z. (2022). Evaluating Pakistan's biosecurity framework. *Strategic Studies*, *42*(3), 67–89.

Ramcharan, A., et al. (2021). Mobile AI for smallholder farmers. *AI for Social Good*, *3*(1), 45–59.

Rana, W. (2021). *Gaps in Pakistan's biosecurity implementation.* Islamabad Policy Research Institute.

Roberts, D., & Unnevehr, L. (2022). Science and the SPS agreement. *Annual Review of Resource Economics*, *14*, 301–320.

Sankaran, S., et al. (2020). Low-altitude remote sensing for plant health monitoring. *Remote Sensing*, *12*(3), 598.

Schneider, K., et al. (2020). Impact of Xylella fastidiosa subspecies pauca in European olives. *Proceedings of the National Academy of Sciences*, *117*(17), 9250–9259.

Secretariat of the CBD. (2001). *Cartagena protocol on biosafety.* https://bch.cbd.int/protocol/

Seidl, R., et al. (2023). Climate-amplified bark beetle outbreaks. *Science*, *379*(6638), eabq1022.

Sepulveda, A. J., et al. (2023). eDNA for border biosecurity. *Methods in Ecology and Evolution*, *14*(5), 1234–1248.

Shabir, A., Sarwar, Z.M., & Ali, H. (2023). Eco-friendly approaches of zinc oxide and silver nitrate nanoparticles along with plant extracts against Spodoptera litura (Fabricius) under laboratory conditions. *Science progress, 106*(4), 00368504231219171.

Sharpe, D. M., et al. (2023). Extinction risks from plant trade. *Conservation Letters*, *16*(3), e12958.

Sheikh, N., et al. (2020). Infectious disease challenges in Pakistan. *Lancet Regional Health – Southeast Asia*, *1*(100003).

Smyth, S. J. (2023). Regulatory challenges for gene-edited crops. *GM Crops & Food*, *14*(1), 1–12.

Soylu, S., Bozkurt, A., Ali, H. et al. (2025). Evaluation of resistance of bean genotypes to common bean bacterial blight disease agents and characterisation of plant defence mechanisms involved in resistance. *Eur J Plant Pathol, 173*, 387–411. https://doi.org/10.1007/s10658-025-03066-5

STDF. (2023). *Global phytosanitary capacity report.* Standards and Trade Development Facility.

Sutherst, R. W., et al. (2023). Next-generation CLIMEX models. *Ecological Modelling*, *482*, 110401.

Tan, M., et al. (2023). EfficientNetV2: Smaller models and faster training. *ICML 2023.*

Thorp, J. R., et al. (2022). Citizen science for pest surveillance. *Biological Invasions*, *24*(5), 1467–1483.

Toomey, A. H., et al. (2023). Deceptive trade practices. *AMBIO*, *52*(4), 789–802.

UNIDO. (2007). *Capacity building for biosafety in Pakistan.* United Nations Industrial Development Organization.

United Nations Office for Disarmament Affairs [UNODA]. (1972). *Biological weapons convention.* www.un.org/disarmament/

USDA. (2023). *Digital phytosanitary certification pilot report.* Animal and Plant Health Inspection Service.

van der Vlugt, R. A., et al. (2021). Validation of real-time PCR methods for quarantine pathogens. *EPPO Bulletin*, *51*(1), 1–12.

Vaughan, M. M., et al. (2023). CO_2-enhanced mycotoxin production. *Toxins*, *15*(3), 189.

Waage, J. K., & Mumford, J. D. (2008). Agricultural biosecurity. *Philosophical Transactions of the Royal Society B*, *363*(1492), 863–876.

Wang, J., Chitsaz, F., Derbyshire, M.K., Gonzales, N.R., Gwadz, M., Lu, S., Marchler, G.H., Song, J.S., Thanki, N., Yamashita, R.A., & Yang, M. (2023). The conserved domain database in 2023. *Nucleic acids research*, 6;51(D1):D384–8.

Wang, C., et al. (2023a). Automated insect identification using deep learning. *Scientific Reports*, *13*(1), 6789.

Wang, C., et al. (2023b). Deep learning for parcel screening. *Pattern Recognition*, *137*, 109245.

World Bank. (2022). *Pakistan health sector review.* World Bank Reports.

WTO. (2019). *Agreement on the application of sanitary and phytosanitary measures (SPS).* World Trade Organization.

WTO. (2023). *SPS information management system.* World Trade Organization.

Zhang, L., et al. (2022). Blockchain for pest surveillance data integrity. *Computers and Electronics in Agriculture*, *198*, 107035.

Zhang, Y., et al. (2023). 5G-enabled swarm robotics in agriculture. *Nature Machine Intelligence*, *5*(4), 312–324.

Zhou, S., et al. (2023). Climate-altered plant defenses. *New Phytologist*, *238*(4), 1565–1579.

11 Future Directions and Opportunities for IPM Research and Development

Arzu Öztürk Kesebir
Vocational School of Patnos, Department of Property Protection and Security, Agri Ibrahim Cecen University, Agri, Patnos, Turkey

11.1 INTRODUCTION

Considering the depletion of fossil fuels over time and the increasing environmental pollution, making global energy consumption more reliable has become one of the world's notable issues (Abraham *et al.*, 2020). The visible effects of climate change and the increase in the world population together with the awareness have led scientists to work on renewable and low-carbon energy production (Wang *et al.*, 2023). The world population was 3.7 billion in 1970 and increased to 7.9 billion in 2021. It is estimated that this number will increase to 9 billion in 2050, and 11 billion in 2100 (Koop and van Leeuwen, 2017). It is thought that the rapidly increasing human population will bring with it challenges in food security (Godfray *et al.*, 2010). To meet the nutritional needs of millions of people, there has been an increase in both livestock and crop production; thus, the amount of crop residue (AW) has also increased at the same rate (Boserup, 1975; Tripathi *et al.*, 2019). Over the past century, China, Africa, and India have experienced both rapid population growth and economic growth, while the AW rate has also increased considerably (Wang *et al.*, 2018).

The country that produces the most agricultural waste after China is India, which produces over 130 Mts of paddy straw every year, and only half of it is used as animal feed, the remaining half thrown away (Singh and Sidhu, 2014). In addition, the practice of burning rice residues, called parali, in the northwestern region of India both creates air pollution and negatively affects public health (Shyamsundar *et al.*, 2019). The disposal of crop residues via inappropriate methods causes the formation of greenhouse gases (GHGs), such as carbon dioxide (CO_2), methane (CH_4), and nitrous oxide (N_2O), which pose a significant threat to human health and cause air pollution (Searchinger, *et al.*, 2008; Kaab *et al.*, 2019).

AW categories that threaten the sustainability of agricultural regimes and therefore have become a prominent issue in recent years are product residues (leaves, stems, seed husks, straw, fruit peels, and weeds), livestock waste (feathers, spilled feed, bedding materials, and droppings), slaughterhouse residues (meat, blood, hair, hide, etc.), agricultural industry wastes (fruit peels, pulp, molasses), fruit pulp, aquaculture wastes (leftover feed and feces), and oilseed cakes (soybean, coconut, ground peanuts, mustard, palm kernel meal, etc.), for example (Figure 11.1).

Among these wastes, dry plant mass (LCB) is thought to be carbon-neutral/negative as it is intended to capture and even store the carbon element. It is estimated that the annual global amount of LCB is approximately 200 billion tons, and the bioenergy that can be produced from LCB is approximately four times the current demand. It is expected that by 2050, as a result of the development of sustainable methods, 38% of the global fuel supply and 17% of the global electricity need will be obtained from AWs (Guo *et al.*, 2015; Shrivastava and Sharma, 2022).

Almost all the mentioned AWs can be separated by simple methods. In addition to being used as fertilizer for plants, the products help the soil become porous and aerate, thus increasing its water retention ability. In this way, we can transform agricultural waste and by-products into valuable resources, create green markets and employment opportunities, and lay the groundwork for clean, safe, and sustainable agriculture by reducing greenhouse gas production and fossil fuel necessity (Mohanty *et al.*, 2002; Scarlat *et al.*, 2018).

11.2 AGRICULTURAL WASTE SOURCES (AW)

In order to develop in situ or ex situ strategies for the management and reuse of AWs, we need to have information about the amount, type, and source of AW produced (Figure 11.2). AW resources can be examined under four main headings:

1. Crop residues
2. Industrial-agricultural wastes
3. Animal waste
4. Aquaculture waste

(Koul *et al.*, 2022)

11.2.1 Crop Residue

Corn stovers, plant leaves, and straw and seed pods from grains, such as barley, wheat, oats, and rice, can be considered crop waste. It is estimated that crop waste potential

DOI: 10.1201/9781003516538-11

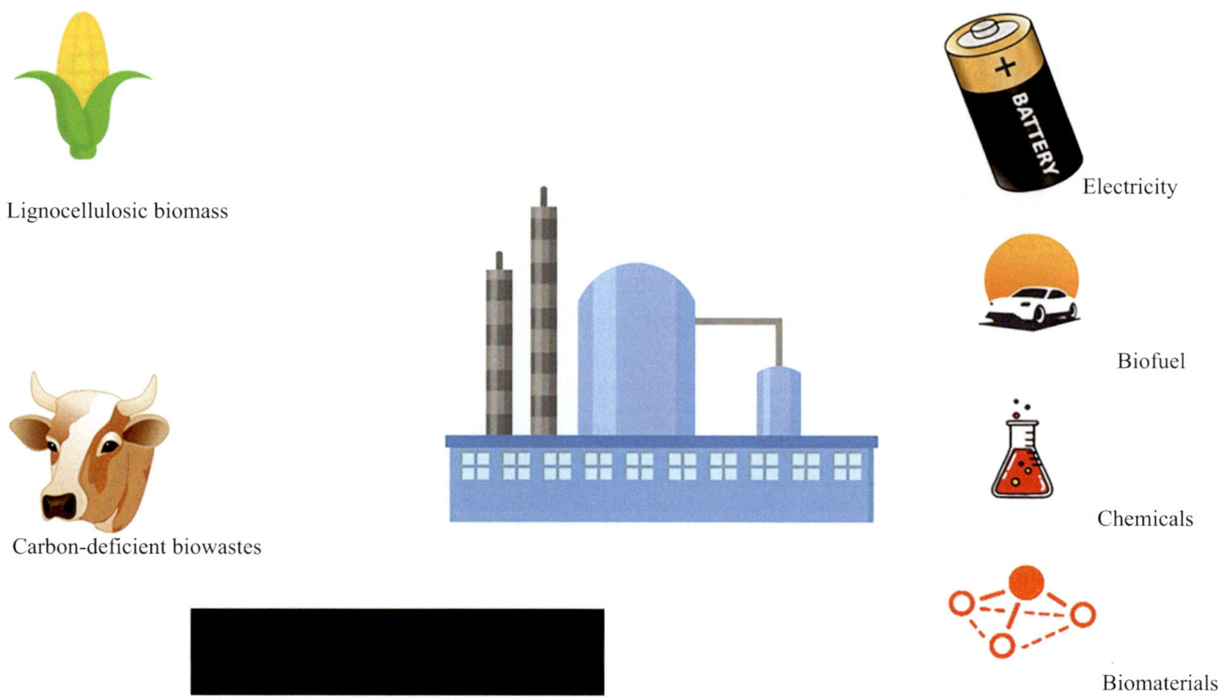

FIGURE 11.1 Biorefinery processes of lignocellulosic biomass.

Source: Wang et al. (2023).

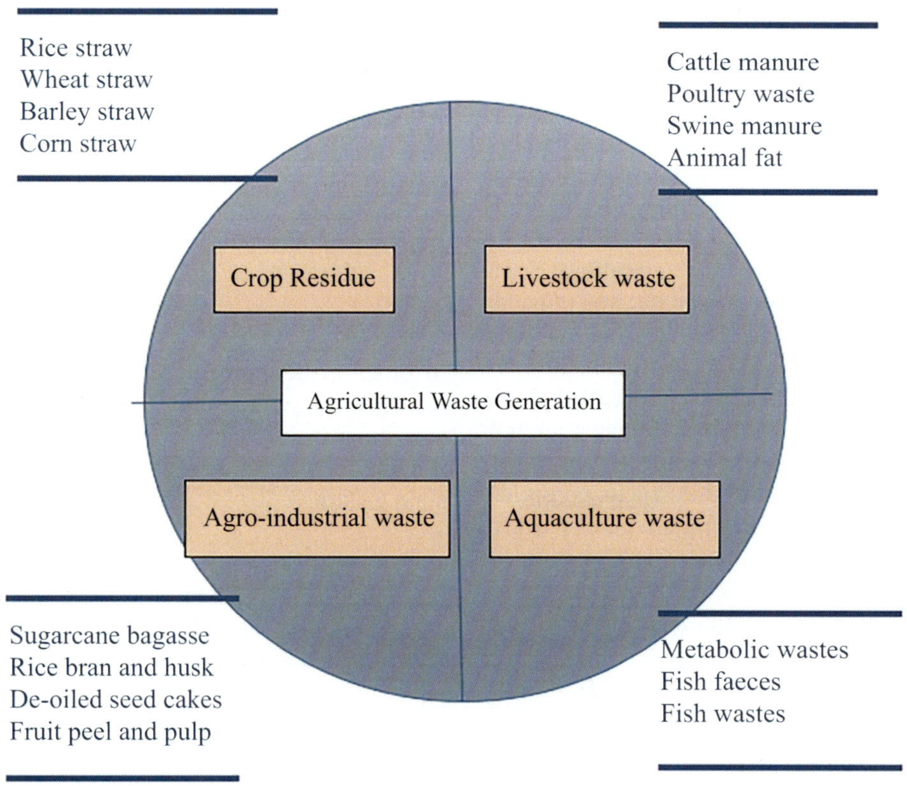

FIGURE 11.2 Various sources of agricultural waste.

Source: Koul et al. (2022).

will be 2,802 Mt annually (Zabed *et al.*, 2016). A very small portion of waste from corn and rice, which are among the most produced products in the world, is used as animal feed or in producing bioethanol. A large part of it is thrown away as garbage or randomly burned. It is thought that this situation will have serious negative effects on human health and the environment and also delays the sustainability of agriculture (Kumar and Singh, 2021).

It is known that 731 Mt of rice are produced annually all over the world, and approximately 92% of this amount is produced in Asia (Sarkar *et al.*, 2012). It is thus predicted that, by 2030, India will produce 221.8 Mt of rice straw annually (Kumar *et al.*, 2018a, 2018b). While the annual production of wheat straw is 354.34 Mt, the production of corn stovers (cob, leaf, husk, and stalk) is 128.02 Mt (Kim and Dale, 2003; Sarkar *et al.*, 2012).

11.2.2 Livestock Waste

Water waste, solid manure, and liquid manure (urine) from farms are animal waste. Annual animal manure production in Europe is around 1,500 Mt. Approximately 300 Mt of this is pig manure, and the remaining part is cattle manure. Untreated manure pollutes both air and water. The cause of surface water pollution and contamination from pathogens is the leakage of rich solid and liquid fertilizers. Fertilizers cause environmental pollution through the greenhouse effect by releasing carbon dioxide (18%) and methane (57%) gases (Holm-Nielsen *et al.*, 2009).

11.2.3 Agro-Industrial Waste

Molasses produced as by-products in the food processing industry, vegetables and fruits and their peels and pulps, bagasse, oilseed pulp, eggshells, starch residues, and poultry and farm animal skins can be considered agro-industrial waste. The pulp that remains after the juice is extracted from sugarcane is a major industrial waste. More than 180 Mt of sugarcane bagasse are produced worldwide every year, and this figure is expected to increase (Saini *et al.*, 2015; Pattanaik *et al.*, 2019). Approximately 86 Mt of dates are used in palm oil production, and over 35 Mt of waste such as bagasse and seeds are produced (Shuit *et al.*, 2009; Sukiran *et al.*, 2017). If we also take into consideration the fruit residues that arise from the production of ketchup, fruit wine, jam, pickle, jelly, fruit juice, and sauces, we see that a tremendous amount of waste is generated worldwide. When only apples are examined, the pulp of 770 Mt apples constitutes 25–30% of the total biomass (Cherubin *et al.*, 2018).

11.2.4 Aquaculture Wastes

The increase in the world population has led to an increase in fish consumption. Aquaculture has been developed to meet the demand, and it is thought that it will meet approximately 1/3 of the demand by 2030, with this rate gradually increasing. The care and production of freshwater fish, marine fish, shellfish, marine plants, and marine creatures in danger of extinction is becoming a growing sector (Cherubin *et al.*, 2018).

In addition to its benefits, aquaculture, called the blue revolution, also causes water and environmental pollution due to the residues it produces. This ceases to be a sustainable practice (Martins *et al.*, 2010; Henriksson *et al.*, 2018). Since the type of feed varies depending on the type of living creature produced, they have become the most important waste source of aquaculture, along with fish feces (Olusegun *et al.*, 2016).

Solid waste is dangerous because it clogs the gills of fish, so it must be removed immediately (Olusegun *et al.*, 2016). Another important type of waste in aquaculture are metabolic wastes and feed residues of fish. Remains of fish feed rich in N and P, which are important and necessary for the protein needs of fish, also dissolve in water and create waste. Fish can retain up to 49% nitrogen and 40% phosphorus. For this reason, they may not be able to benefit from P and N in feed, and it creates water pollution (Stephens and Farris, 2004). It can cause turbidity in water and undesirable microorganism growth, and if released into large water bodies, it can harm aquatic creatures and microorganisms and even cause eutrophication.

Ammonia is the form of nitrogen released into water and can break down into nitrite and nitrate. Ammonia in the form of NH_3 (non-ionized) and NH_4^+ (ionized) is in equilibrium in water; this equilibrium state depends on the temperature and pH value of the water. The unionized form of ammonia is more dangerous (Dauda *et al.*, 2019). Nitrite is toxic but does not pose a danger because it is very easily oxidized to nitrate. Nitrate does not have a toxic effect up to 200 mg per liter, but together with phosphates, it causes water aging. Depending on the amount of phosphates released into the water by fish feces and their concentration in the water, it may cause eutrophication. For all these reasons, there has been an increasing interest in aquaculture to minimize water pollution and protect and increase aquatic vitality (Lafferty *et al.*, 2015; Ahmed *et al.*, 2019; Dauda *et al.*, 2019).

11.3 POTENTIAL

In order to utilize agricultural residues, the amount, type, and content of the residue produced are important. This content varies from region to region, as soil content and type also have an impact on the product obtained. For this purpose, it will be necessary to examine agricultural residues, their potential, and their impact on sustainability in various regions of the world.

A study of 28 member states of the European Union investigated the use of agricultural feedstocks for bioenergy up to 2017 and their future potential. It reported that biodiesel is mostly produced for road transportation in Europe, with 80% of this production coming from rapeseed oil and the rest from bioethanol from sugarcane and wheat. There is potential for straw, which it defines as primary agricultural waste, but only a small proportion is used for bioenergy. In EU-28, over 4 million hectares of land are used to grow energy crops, with over 40 million hectares expected to be used by 2030. It is thought that 15–40% of the arable agricultural area in the EU will be allocated for this purpose. However, since European Union countries are located in a wide geography, the location, content, and product information of the land cannot be clarified (Kluts *et al.*, 2017).

Figure 11.3 provides an overview of the growth potential in agricultural products for 2020–2030. It can be seen that first- and second-generation crops will increase until 2030. The calculation of these potentials is based on the assumption that the total available land is cultivated with crops from only one crop group. There are sustainability constraints for food security in the EU. Due to these constraints, the potential for energy crop production is underestimated. These constraints are that less land is allocated to agriculture in order to preserve the natural landscape and that not all types of crops can be used to produce energy in line with GHG mitigation requirements. It is not yet clear whether these criteria will also apply to the production of biochemicals or plastics using biomass feedstocks from biorefineries, but it is expected that certain criteria will be introduced (European Commission, 2009).

The type of energy crops grown on a site also affects the potential, as it will change the energy content and dry matter content. However, trees and grasses are expected to have significant potential for sustainable bioenergy. In De Wit and Faaij's study, since the calculation was made by considering that a single type of product was planted on all available lands, it is clearly seen how much the type of product affects the amount of energy to be produced. Grass crops have the highest energy potential, with woody plants coming in second. Since they are subjected to more comprehensive agricultural practices compared to other products, efficiency increases and costs decrease (Wit and Faaij, 2010).

The European Environment Agency (EEA) has predicted a shift from oil and starch crops to grasses and woody plants. They first focused on the demands for the amount of bioenergy they targeted in their National Renewable Energy Action Plan in 2020. Considering the bioenergy content and the methods of obtaining it, they predicted that 7–17 million hectares of agricultural land would be needed. In addition, the type of raw materials to be processed when establishing a biorefinery and the equipment and requirements that may be needed during production will vary depending on the land types and the bioenergy plants produced (EEA, 2013).

It has been reported that between 2003 and 2013, the rate of agricultural waste increased by 33% compared to previous years and reached 5 billion Mg: 47% of these residues were produced in Asia, 29% in the Americas, 16% in Europe, 6% in Africa, and 2% in Oceania. Brazil produced 432 million Mg of agricultural residue, an increase of 41% in these 10 years. It corresponds to 30% of the total production of the American continent and 9% of the world.

Corn (*Zea mays* L.) and sugarcane (*Saccharum spp. L.*) produce the two crop residues with the highest energy content, respectively. Since wheat (*Triticum aestivum* L.) and

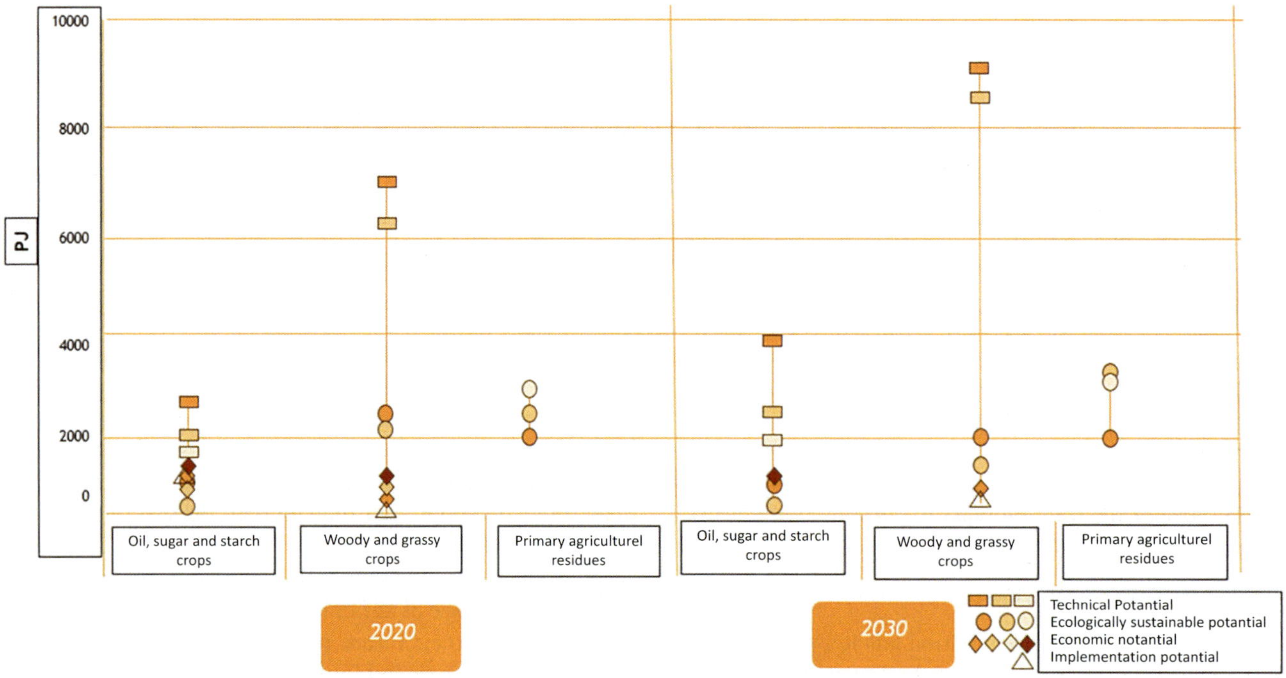

FIGURE 11.3 Estimated bioenergy potentials from energy crops and agricultural residues in 2020 and 2030 in EU-27.

Source: Based on Kluts et al. (2017).

rice (*Oryza sativa* L.) production accounts for 44% of total production, crop waste amounts are also very high, but they play a minor role in energy production. Cotton (*Gossypium hirsutum* L.) and soybeans (*Glycine max* (L.) Merr.) are not collected to produce energy due to their low production intensity and rapid decomposition (~663 million Mg) (Lal, 2005).

In Brazil, the reduction and termination of the burning of sugarcane straw since the early 2000s have led to an increase in the amount of residue in the soil. Therefore, a large amount of straw was obtained for energy production (Leal *et al.*, 2013).

In addition to electricity and heat production from sugarcane straw, cellulosic ethanol production is also very important. Straw constitutes 9–19% of the dry weight of sugarcane and has a structure of lignin, cellulose, and hemicellulose (Andrade *et al.*, 2013; Aguiar *et al.*, 2021). Straw is the part of sugarcane with the highest energy potential. If the energy production potential as a result of fermentation is listed, after the straw come the upper parts, and lastly the bagasse (Pereira *et al.*, 2015).

In 2015, it was reported that bioelectricity production from sugarcane met 4% of Brazil's entire energy consumption (EPE, 2015). Although corn is produced in a wide geography, America's potential to produce ethanol from corn grains is quite high (Lal, 2005).

After the grain is harvested, 10 Mg ha^{-1} residue remains in the soil (Lal, 2005), which forms a large biomass to produce cellulous ethanol. Corn has the potential to provide approximately 1.7 times more carbon than other grains, depending on its production level. Although the carbohydrate rate varies greatly in different parts of the corn plant, it is reported that this variability is not very effective in energy production, and the remaining mass volume is important (Mourtzinis *et al.*, 2016).

The United States is limited in the amount of space it can devote to planting corn, so it has instead adopted more effective plant breeding and biomass management on available areas (Bonner *et al.*, 2014). It is estimated that by 2040, 53–161 Mt of the annual 588–936 Mt of dry material will be obtained from corn (USDE, 2016).

Although not yet produced in large quantities, paper, cellulose, and wood industry residues are thought to be promising for energy production (Koizumi, 2015). In order for trees to be used in bioenergy production, their content must be suitable; they must be easily dried and harvested within the year (Long *et al.*, 2015). The best example that has these features and can be harvested in short rotations is currently eucalyptus (*Eucalyptus* spp.) (Seebaluck and Leal, 2015).

In a study conducted in Iran in 2016, It has been stated that there are other products with high potential for producing bioenergy in Iran, including industrial products, such as sugarcane, sugar beet, cotton, soybeans, rapeseed, sesame seeds, olives, sunflowers, safflower, almonds, walnuts, and hazelnuts, and horticultural products, such as apples, grapes, and dates. It has been stated as well that many research and projects have been carried out on bioenergy in Iran after 2010 and that the government is supportive in this regard. It has been reported that 11.33 Mt of the harvest residues remaining in the soil can be collected to produce biofuel, and bioethanol (3.84 GI), biodiesel (0.35 GI), and biobutanol (1.07 GI) can be produced from these residues and oilseeds (Alavijeh, and Yaghmaei, 2016).

In a study conducted on agricultural products in Greece, it was stated that there is potential for biodiesel and bioethanol production, and the available resources were reported to be oilseeds, such as sunflower seeds, soybeans, rapeseed, and waste oils. As a result of pyrolysis of sunflower shells, a maximum gas yield of 53% at 500°C and an oil yield of 21% at 400°C were reported. Again, a simple first-order kinetic model was applied to sunflower shells, and an energy value of 78.15 kJ/mol was calculated (Zabaniotou *et al.*, 2008).

When numerous studies conducted in various parts of the world are examined, it is a fact that harvest residues and production facility residues remaining in agricultural areas have a very serious green energy source potential. Governments are also aware of this potential, and various projects are being developed. The point that needs to be taken into consideration is that different products are cultivated in each geography, so special plans should be made for the product and residues in the region. It is also an important issue that green energy generation is not sustainable.

11.4 SUSTAINABLE RECYCLING OF PLANT RESIDUES

The process of producing energy from crop residues, which is thought to have a good potential to be an alternative to fossil fuels, which are harmful to the environment and human health, and to minimize greenhouse gas (GHG) emissions, also promises hope for reducing climate change (Edenhofer *et al.*, 2011). For this reason, these fuels, called second-generation biofuels, are very advantageous compared to fossil fuels (Gnansounou, 2010; Glithero *et al.*, 2013; Lindorfer *et al.*, 2014). The significant amount of bioethanol derived from straw in Europe is highlighted (Hamelin *et al.*, 2019). Although the amount of straw varies in various parts of the world, it has been seen in previous sections that it has a huge potential. Despite its advantages, there may also be some obstacles regarding environmental impacts, logistics, marketing, and conversion of biomass (IEA Bioenergy, 2017). The biggest concern about removing high amounts of biomass from the soil is the decrease in the amount of soil organic carbon (SOC). This decrease has possible negative consequences on soil functioning, productivity, and the environment in the long term (Lal, 2005; Carvalho *et al.*, 2017; Cherubin *et al.*, 2018).

While combating climate change, it is important to process crop residues with the correct methods and quantities so that sustainable methods are not abandoned and agricultural productivity is not affected. Sustainable intensification (SI) includes local practices, such as targeted fertilization for residue removal, to increase both productivity and environmental sustainability (IEA Bioenergy, 2017). These measures will affect environmental impacts, sustainability, amount of residue extraction, crop yield, land use, and amount of emissions (Hilst *et al.*, 2012; Zhao *et al.*, 2015; Haase *et al.*, 2016). Many studies have been published in the literature on the sustainable use of biomass to produce bioenergy using land-specific needs and modeling (Brandão *et al.*, 2011; Monforti *et al.*, 2015).

11.5 EFFECT OF CROP RESIDUE HARVESTING ON SOIL CHEMICAL PROPERTIES

Harvesting crop residues has changed the chemical composition of the soil (Adler *et al.*, 2015; Tan and Liu, 2015). Crop residues provide nutrients in the soil for subsequent crops. Therefore, harvesting residues for energy production may increase the need for soil fertilization (Trivelin *et al.*, 2013; Karlen *et al.*, 2015). Of course, these effects depend on the amount of crop residue harvested (Franco *et al.*, 2013; Khanal *et al.*, 2014; Karlen *et al.*, 2015).

While the residue affects the amount of K and P in the soil in the short term, it has an effect on micronutrients and N, Ca, Mg, S values in the long term (Jeschke, 2012). Crop residues with a high C/N ratio may cause soil N immobilization through microorganisms, increasing the need for N fertilizer in the short term (Trivelin *et al.*, 2013; Karlen *et al.*, 2015; Ferreira *et al.*, 2016; Meier and Thorburn, 2016). It was observed that 16% of the initial amount of N in sugarcane straw was recovered after three crop cycles. It means contribution to crop nutrition in the short term (Ferreira *et al.*, 2016). Leaving sugarcane straw in the soil in the long term will increase the amount of N accumulated in the soil, thus reducing the need for N fertilization (Robertson and Thorburn, 2007; Trivelin et al., 2013; Costa *et al.*, 2014; Ferreira et al., 2016; Meier and Thorburn, 2016).

It has been reported that corn cultivation in Lowa, USA, reduces the amount of N in the soil. In line with this information, harvesting corn straw in soils with high N content will prevent N leakage, and therefore water pollution (Khanal *et al.*, 2014).

These may be the reasons for phosphorus deficiency in the soil after the harvest of crop residues:

1. Less P released during the mineralization process
2. Low P adsorption influenced by soil organic matter (SOM) in soils where C concentration is high due to crop residue in the soil

(Reddy *et al.*, 2014)

Corn kernels contain five times more P than residues remaining in the soil (Avila-Segura *et al.*, 2011). However, when 1 t of corn is harvested, it leaves 0.76 kg/ha P in the field, which is not an amount to be ignored (Karlen *et al.*, 2015). In sugarcane, this rate is 0.41 kg/ha, 40% of the amount of P in the crop (Oliveira *et al.*, 2010).

Since K in the plant structure does not participate in any structure, it is easily released into the soil from the crop residue (Calonego *et al.*, 2012). Since crop residues also contain high amounts of K, allowing these residues to remain in the soil increases the K content in the soil (Sawyer and Mallarino, 2007; Calonego *et al.*, 2012, Blanco-Canqui, 2013). Corn residues contain approximately 73% of the K in the plant (Karlen *et al.*, 2015), while sugarcane straw contains 42% (Oliveira *et al.*, 2010).

Harvesting crop residues, both in the short term and in the long term, negatively affects the amount of minerals and nutrients in the soil, their storage, and therefore, their use by plants (Adler *et al.*, 2015; Tan and Liu, 2015; Karlen *et al.*, 2011). In order to establish a sustainable system, the minerals lost from the soil must be compensated (Adler *et al.*, 2015; Karlen *et al.*, 2011). Another perspective is to harvest the parts of the plant that are poorest in terms of nutrients and to leave the remaining parts in the soil (Cherubin *et al.*, 2018).

11.6 EFFECTS OF CROP RESIDUE HARVESTING ON THE PHYSICAL STRUCTURE OF THE SOIL

It is thought that the pressure created by machine traffic on the soil during crop residue harvest, the deterioration in the soil structure, and the decrease in C stocks in the soil cause the structural properties of the soil to deteriorate (Blanco-Canqui, 2013; Tormena *et al.*, 2017). Since the compaction in the soil as a result of pressure reduces macroporosity (Osborne *et al.*, 2014), it prevents the roots from advancing easily and the plant from reaching water (Tormena *et al.*, 2017).

It is also known that crop residues in the soil act as a barrier and protect the soil from the destructive effects of rain and wind (Hammerbeck *et al.*, 2012; Franzluebbers, 2015; Johnson *et al.*, 2016). It has also been reported that it stores water in the soil and prevents its easy evaporation by approximately 50% (Xiukang *et al.*, 2015; Tormena *et al.*, 2017).

The water storage effect of crop residues in the soil will be useful in reducing drought against climate change and will also be useful against erosion in case of excessive rainfall. Additionally, crop residues can provide thermal insulation between the soil and the air, and this effect decreases as crop residue rots (Wilhelm, 2004).

11.7 EFFECTS OF CROP RESIDUE HARVESTING ON THE BIOLOGICAL STRUCTURE OF THE SOIL

No-till corn harvest has a significant effect on the fungal/bacterial ratio in the upper soil layer. The elapsed time, soil type, and climate factors affect the microbial structure in the soil (Lehman *et al.*, 2006). Soil retention of maize residues affects the structure of bacterial and fungal communities in the presence of denitrification (napA) and N fixation (njfH) genes (Wakelin *et al.*, 2007). Studies show that crop residues remaining in the soil provide beneficial results in preventing the formation of microorganisms and diseases that damage crops. In fact, it has been observed that the compounds released during the decomposition of leaves and stems during corn harvest inhibit disease-causing *Rhizoctonia* and *Bipolaris* fungi (Qi *et al.*, 2015).

The retention of corn crop residues in the soil has led to an increase in the energy supply of soil microorganisms, which has led to the increased development of antagonists and predators and a decrease in the population of phytoparasitic nematodes. It was determined that the collection of residues of different bioenergy plants had no effect on the total number of microorganisms in the soil or on the microbial groups in annual/perennial systems (Govaerts *et al.*, 2007). Restructuring the soil as a result of the deterioration of the microbial structure in the soil and balancing the amounts of organic substances entering the soil is a very long and gradual process (Orr *et al.*, 2015).

In studies conducted in various regions around the world, it has been observed that sugarcane straw causes an increase in both microbial biomass (Graham and Haynes, 2006; Souza *et al.*, 2012; Junior *et al.*, 2015) and microbial diversity (Rachid *et al.*, 2016). The environment with the highest diversity of microorganisms in the soil was natural vegetation, followed by pastures, and relatively least was sugarcane straw (Franco *et al.*, 2016). Straw residues serve as a source of nutrients in the soil, support the microbiota, support the improvement of microclimate conditions (Portilho *et al.*, 2011; Benazzi *et al.*, 2013), and serve as a shelter for the microbiota. A significant decrease in the biomass and diversity of epigeic invertebrates was observed in soils where sugarcane straw was completely cleared (Portilho *et al.*, 2011). Although it is a fact that sugarcane crop residues support the macrofauna of the soil, increasing its amount in the soil did not change the result (Abreu *et al.*, 2014). Moreover, although straw in the soil supports beneficial microorganisms, it can also support the growth of harmful microorganisms, such as the root spittlebug [*Mahanarva fimbriolata* (Hemiptera; Cercopidae)] and sugarcane borer (*Sfenophora levis* Vaurie (Coleoptera: Curculionidae) (Dinardo-Miranda and Fracasso, 2013), and cause a decrease in sugarcane yield.

11.8 EFFECT OF CROP RESIDUE COLLECTION ON SOIL C STOCKS

C stocks in the soil depend on the C released as a result of SOM (soil organic materials) decomposition and the amount of C entering the soil (Poeplau and Don, 2015). Preserving crop residues by not tilling the soil is one of the strategies implemented to increase soil C stocks. It has been observed that in uncultivated soils in Brazil, there is an annual change of approximately 0.37 Mg/ha at 0–0.3 m, and this rate reaches approximately 0.57 Mg/ha annually below 1 m (Cherubin *et al.*, 2018).

West and Post examined 276 studies conducted around the world and reported that soil C stocks increased as a result of preserving crop residues and not tilling the soil (West and Post, 2002). In Brazil, it has been confirmed that the amount of soil C increases as a result of preserving crop residues, whether the soil is cultivated conventionally or not at all (Campos *et al.*, 2011).

Using the DayCent prediction model, changes in soil C content as a result of soil tillage and corn removal practices were estimated in the Nebraska region of the USA, and it was reported that soil C stocks decreased by 1 Mg/ha annually by 75–100% in soils cultivated with conventional methods. It was estimated that if the soil is not cultivated, soil C stocks will increase even if corn is not planted in the soil, and this increase is related to hay care (Carvalho *et al.*, 2017).

The literature came to the same conclusions that sugarcane maintenance is the main reason for soil C increase in sugarcane crops (Cerri *et al.*, 2011; Meier and Thorburn, 2016; Carvalho *et al.*, 2017; Oliveira *et al.*, 2017). Burning sugarcane crops used to be common in Brazil, but as a result of this practice being phased out, sugarcane straw remained in the soil and rotted over time. Some of the C formed as a result of decomposition was mixed into the air as CO_2, and the rest remained in the soil (Meier and Thorburn, 2016; Carvalho *et al.*, 2017; Oliveira *et al.*, 2017). After removal by burning, the C change in soil in sugarcane crops was 0.81 Mg/ha for 0–0.3 m and 0.62 Mg/ha for 0–0.1 m.

Again, in two studies in which 30-year soil C changes were estimated in Brazil using the DayCent model, it was concluded that when sugarcane straw was not removed from the soil, there was an increase in C stocks in the range of approximately 0.1–0.2 Mg/ha annually, and that the removal of straw negatively affected soil C stocks (Meier and Thorburn, 2016; Oliveira *et al.*, 2017).

There are various studies reporting that soil C stocks increase (Lima *et al.*, 2006; Zinn *et al.*, 2011), balance (Cook *et al.*, 2014; Fialho *et al.*, 2014), and decrease (Soares *et al.*, 2013) in different forested areas in Brazil, but the common point of these studies is that the harvest of crop residues reduces soil C stocks (Kumaraswamy *et al.*, 2014; Versini *et al.*, 2014; Epron *et al.*, 2015). Additionally, protecting forest residues has a positive effect on forest productivity (Leonardo *et al.*, 2013). Although crop residue harvesting causes a decrease in C stocks in the soil, it is necessary to transition to a low-C economy and utilize biomass for this purpose (Sanchez *et al.*, 2015). Since it is thought that the use of bioenergy will reduce greenhouse gas emissions, it is necessary to develop policies to protect soil C stocks.

11.9 EFFECTS OF CROP RESIDUE HARVESTING ON GREENHOUSE GAS EMISSIONS

The largest greenhouse gas emissions are thought to occur when large amounts of material are left on the soil surface (Carmo *et al.*, 2013; Diana *et al.*, 2014; Jin *et al.*, 2014). Harvesting 14–21 Mg/ha of sugarcane straw remaining on the soil surface has been shown to reduce greenhouse gas emissions by approximately 2.3 times (Carmo *et al.*, 2013). Similar results were obtained in a similar study conducted for corn harvest. In nine regions of the USA, corn harvesting has been observed to increase CO_2 emissions by 4% compared to no corn harvesting (Diana *et al.*, 2014).

In agriculture, CO_2 formed as a result of the decomposition of residues is not included in greenhouse gas emissions because it passes from soil to soil and into the atmosphere, creating a cycle, so it is not included in the CO_2 emitted by fossil fuels (Pourhashem *et al.*, 2013). However, increasing the temperature and decreasing the amount of soil moisture can support CO_2 emissions (Corradi *et al.*, 2013; Guzman *et al.*, 2015). Remaining crop residues in the soil will also cause an increase in N_2O emissions (Carmo *et al.*, 2013; Jin *et al.*, 2014). Considering the long term, it is thought that mineralized N_2O obtained from straw can reverse this situation (Meier and Thorburn, 2016). If this is the case, the dependence of N_2O needs on synthetic N fertilizers in agriculture will be reduced (Ferreira *et al.*, 2016; Meier and Thorburn, 2016).

Harvesting residues can also increase soil temperature and increase N_2O emissions by increasing microbial activity and N crystallization (Pourhashem *et al.*, 2013).

Various studies have examined CH_4 emission in straw-covered soils and suggested that it did not cause a change (Carmo et al., 2013; Jin et al., 2014). CH_4 emission has not been taken into account by some researchers because it is 1% lower than CO_2 emission and 10% lower than N_2O emission (Guzman *et al.*, 2015). In addition to using crop residues to produce bioenergy, it will also have benefits in reducing the need for land for raw materials. Most of the uncertainty about the sustainable production of biofuels is due to the fact that they cause greenhouse gas emissions as a result of changes in land use (Searchinger *et al.*, 2008; Lapola *et al.*, 2010). Cellulosic ethanol production could be a solution to these problems, as it would prevent further land use. Harvesting 50% of straw could contribute to a 35% increase in cellulosic ethanol production. This helps avoid the need to expand the agricultural area and prevent greenhouse gas emissions.

As a result, harvesting the remaining crop residues in the field will reduce greenhouse gas emissions directly or indirectly and contribute to the sustainability of industrial use. In fact, due to the economic value of cellulosic ethanol and bioelectricity production, it is not possible to preserve large amounts of crop residue in the field. The aim should be to determine the optimal amount of residue harvest without increasing greenhouse gas emissions. This amount will depend on the climate, soil type, and crop.

11.10 EFFECTS OF HARVESTING CROP RESIDUES ON PLANT GROWTH AND PRODUCTIVITY

Crop residues left in the soil directly or indirectly affect crop yield (Pittelkow *et al.*, 2015). It has been observed that the maintenance of sugarcane crop residue remaining in the soil in tropical regions affects crop yield in different ways (Campos *et al.*, 2010; Tavares *et al.*, 2010). It has been observed that leaving straw in the soil has positive effects on the plant population, while burning straw has the opposite effect. However, in cases where all the straw remained in the soil, a decrease in the growth and productivity of the plants was observed (Campos *et al.*, 2010)

In temperate climates, leaving the entire crop residue on the soil reduced the productivity of sugarcane, and burning the straw before harvest increased the amount of waste in the short term and had a negative effect in the long term (Wiedenfeld, 2009; Viator and Wang, 2011; Arceneaux and Selim, 2012).

The reason for the negative consequences that occur when straw remains in the soil is that the amount of biochemicals formed as a result of the decomposition of straw has a toxic effect on the soil. These biochemicals are organic acids, alcohols, phenols, aldehydes, ketones and similar functional groups (Villegas *et al.*, 2007).

It has been reported that harvesting forestry residues for bioenergy production will negatively affect the nutrition of eucalyptus trees in the long term, and harvesting all crop residues will initially reduce eucalyptus wood yield by 40% (Rocha *et al.*, 2016). It is thought that harvesting residue in northern forests will not affect tree growth and productivity (Kaarakka *et al.*, 2014).

As a result of the study conducted on corn for four years, they suggested that soil and crop fertility and soil C stocks should be constantly controlled in order to create a sustainable harvest management. They concluded that the reason it was not possible to agree on the amount of hay crop without reducing crop productivity varied according to climate, location, soil type, and topography (Blanco-Canqui, 2013; Birrell *et al.*, 2014).

11.11 COMBINATION OF OPTIMAL METHODS FOR SUSTAINABLE CROP RESIDUE

As can be seen, crop residues affect the ecosystem, plant growth and health, and soil quality. Uncontrolled harvesting of crop residues remaining in the soil to produce bioenergy may result in loss of these benefits. Therefore, a strategy should be determined by combining best practices. One of the best methods for this purpose is to replace the amount of carbon and nutrients lost in the soil with fertilizers, to perform crop rotation without tillage, and to include both annual and perennial plants in these practices (Blanco-Canqui, 2013; Paustian *et al.*, 2016). Maintaining soil C stocks will increase the amount of crop residue that can be used for bioenergy (Carvalho *et al.*, 2017).

Annual plants are harvested every year, so SOC loss is higher. Predictions made using the DayCent model showed that there would be an annual SOC loss of 1 Mg/ha when comparing conventional tillage with no-till farming in the corn belt region of the USA (Campbell et al., 2014; Carvalho *et al.*, 2017). Planting no-till legumes and semi-perennial crops supports crop residue harvesting by increasing SOC stocks. It is also thought that the inclusion of cover crops in the agricultural system will support this issue (Wegner *et al.*, 2015). No-till cover can retain 0.1–1 Mg/ha C annually, depending on plant type, soil type, and climate (Blanco-Canqui, 2013). A simulation was made in Brazil to use sugarcane straw as a sustainable bioenergy source. It was observed that the C rate in the soil decreased from 0.19 Mg/ha to 0.11 Mg/ha with the traditional method without tillage. It has been stated that if an alternative C source is used, annual C stocks can increase to 25 Mg/ha.

Studies examine the multiple effects of crop residue harvesting on soil properties. This will positively affect the structural degradation of the soil, water storage amounts, nutrient cycling, and biological activity ((Blanco-Canqui, 2013; Franzluebbers, 2015; Stavi *et al.*, 2016). It will be possible to minimize the negative environmental effects caused by planting a single type of crop in the cultivation areas. For example, minimizing crop residue harvesting in soils that may be subject to erosion (Cherubin *et al.*, 2018).

We can say that it is inevitable to use crop residues for bioenergy production. It has become important to develop sustainable methods without the need for new agricultural areas and without posing a threat to the functioning of the ecosystem and the future of food production.

Studies in the literature agree that harvesting crop residues has a negative impact on nutrient cycling and soil C stocks. Removing crop residues from the soil means a decrease in the nutrients of the soil, and this deficiency must be compensated with mineral fertilizers. This means extra costs. In addition, crop residue harvesting poses a threat to microorganisms in the soil. Crop residues in the soil also provide good protection against erosion by protecting the soil from the corrosive effect of rain. It retains temperature and water by acting as a temperature insulator for the soil. Residues remaining in the soil make the soil more sensitive, causing compaction in the soil, which reduces the infiltration and storage of water; as a result, the roots of the plants cannot be nourished and cannot grow sufficiently. On the other hand, leaving crop residue in the field causes pests to increase in sugarcane fields.

Harvesting crop residues is thought to have positive effects on reducing greenhouse gas emissions because CO_2 and N_2O gases are released as a result of decay. However, it is thought to not be effective on CH_4 emissions.

It is seen in the literature that plants in different regions respond differently to crop residue harvesting. In addition to edaphoclimatic conditions, such as light, water, temperature, altitude, and soil properties, crop management strategies are also effective. Therefore, planning that integrates the best crop management practices, such as conservation tillage, crop rotation and cover crops, nutrient management, and/or organic residues, can partially compensate for the negative consequences of crop residue harvesting and create a more sustainable strategy. It is also important to inform countries with high crop residue potential, such as Brazil, about bioenergy production and to provide the necessary support and studies by governments.

11.12 BIOETHANOL PRODUCTION PROCESS

Bioethanol is considered an alternative to gasoline, with limited reserves, because it causes less greenhouse gas emissions. Commercially produced bioethanol is generally obtained from corn and sugarcane, which is a costly process. Second- and third-generation (3G) sources have been turned to produce bioethanol at lower costs. Obtaining the desired compounds by increasing the ability of algae to capture CO_2 from the air is called fourth-generation bioethanol. Even though food is not used in its production, it is important to reduce the cost of the bioethanol production process (Edeh, 2020).

11.12.1 Pretreatment

Pretreatment is an expensive process, costing $0.30 per gallon. Different pretreatment processes have been developed for high cellulose reactivity and sugar yield. Conventional pretreatment processes can be biological, physical, chemical, and physicochemical. The chemical pretreatment process is the most effective and preferred method (Bensah and Mensah, 2013). Advanced pretreatment processes can be ionic-based or acid-mediated fractionation (Maurya *et al.*, 2015).

The purpose of this process is to disrupt the lignocellulosic structure, releasing the cellulose embedded in the matrix containing hemicellulose and lignin. If cellulose is not exposed in this way, enzymes cannot reach it, and enzymatic hydrolysis will not occur. In addition, deterioration of the structure of pentose sugars is prevented by pretreatment, and factors that may reduce the yield by disrupting the sugar structure are eliminated (Edeh, 2020).

11.12.1.1 Traditional Pretreatments

This method is categorized as follows:

1. *Physical pretreatment.* It is the breaking down of the size of the lignocellulosic biomass by methods such as grinding, extrusion, and irradiation. Thus, as both the surface area and the pores of the biomass increase, it becomes more suitable for enzymatic hydrolysis. Physical pretreatments can be supported by chemical pretreatments to sufficiently degrade the lignocellulosic structure.
2. *Chemical pretreatment.* Includes organic acid (organosolvation), oxidative delignification, alkali and acid methods. They select, degrade, and release structures such as lignin and hemicellulose in the polymer matrix. Chemical pretreatment methods are effective but require harsh operating conditions that may have a negative impact on subsequent steps. It is also necessary to destroy by-products (Brodeur *et al.*, 2011).

3. *Physicochemical pretreatment.* It involves a combination of physical and chemical methods, such as steam explosion, CO_2 explosion, liquid hot water, and microwave irradiation (Egüés *et al.*, 2012).
4. *Biological pretreatment.* It is the method in which microorganisms are used when more enzymatic hydrolysis is aimed. These microorganisms can be brown and white rot fungi, bacteria, and soft rot fungi (Hassan *et al.*, 2018).

11.12.2 Advanced Pretreatment Methods for Lignocellulose

It is a preliminary process that targets the production of cellulosic ethanol as a result of the fractionation of lignocellulose by producing value-added by-products at atmospheric pressure and 50°C (Zhang *et al.*, 2007). In order to make cellulose more accessible, solvents are used to decompose cellulose, hemicellulose, and lignin. The method, also known as cellulose solvent-based lignocellulose fractionation (CSLF), favors the use of less enzyme in subsequent enzymatic hydrolysis (Sathitsuksanoh *et al.*, 2013). CSLF is done in two ways, namely, acid-mediated fractionation and ionic liquid–based fractionation (ILF).

Acid-mediated fractionation. It is a method applied in atmospheric acid and at 50°C using organic solvents, such as acetone and ethanol, or cellulosic solvents, such as phospholic acid, to decompose the lignocellulosic biomass. The effectiveness of the separation method is directly proportional to the amount of dissolution of cellulose in cellulose solvent, hemicellulose in organic solvent, and lignin in pure water. High separation of hemicellulose and lignin from cellulose fractions is important to prevent competitive binding sites, undesirable degradation of sugar structures, cost of inhibitors, and reduction of production (Zhang *et al.*, 2007). It is a suitable method for lignocellulose varieties, such as bamboo, sugarcane, switchgrass, corn stover, and elephant grass (Sathitsuksanoh *et al.*, 2011).

Ionic liquid–based fractionation. Solutions that are liquid at room temperature and contain high amounts of organic cations and inorganic anions are called ionic liquids (IL). It is used for lignocellulosic fractionation to release purified and polymeric feedstocks that are intact value-added by-products. This method has the advantage of being easier and better separating important components by consuming less energy compared to the traditional method (Zhang et al., 2007). ILs are included in the class of green solvents because their vapor pressure is low and they are stable at high thermal temperatures. In addition, since the hydrophobic properties, polarities, and solvent power of ILs can be changed for a purpose, they are included in the tunable solutions group (da Costa *et al.*, 2013).

The most preferred pretreatment method is steam explosion. The reasons this method is preferred are that the system is cheap to install, works with high efficiency, does not require much hazardous chemicals, and produces less waste for the environment and 100% sugar recovery (Alvira *et al.*, 2010).

11.12.3 Hydrolysis

The purpose of this step is to produce sugar monomers, because the enzymes used in the fermentation stage can only act on monomeric sugars. Hydrolysis can be catalyzed by acid or enzyme, of which acid-catalyzed methods are preferred. Concentrated or dilute solutions of H_2SO_4 or HCL may be preferred. Methods involving high concentrations of acid are carried out at low temperature, and 90% sugar is recovered (Joshi *et al.*, 2011). However, the cost of excessive use, recycling, and disposal of acid is high (Azhar *et al.*, 2017). High acidity can disrupt the structure of sugar monomers in the environment. For these reasons, hydrolysis with dilute acid requires high temperature and low acid concentration. The disadvantage of this method is the formation of inhibitors. The most preferred method is dilute acid hydrolysis.

$$(C_6H_{10}O_5)_n + H_3O+ \longrightarrow H_2O + (C_6H_{12}O_6)_n \qquad E_1$$

The acid hydrolysis of lignocellulosic structures has two stages. In the first stage, hemicellulose is hydrolyzed with dilute acid. In the second stage, cellulose is hydrolyzed with concentrated acid (Joshi *et al.*, 2011).

Enzymatic hydrolysis is performed using enzymes that break down polymeric hydrocarbons into monomers in an environment where the temperature is 45–50°C and pH is around 5. It is an effective method, with which high efficiency is achieved without the risk of inhibitor formation or corrosion. Three types of enzymes can be used, namely, cellobiohydrolases, endo-1,4-β-glucanases, and β-glucosidases. They are expensive because they are widely used in the paper, food, and textile industries (Sharma *et al.*, 2020). Due to the cost of these enzymes, microorganisms that secrete cellulolytic enzymes are preferred. The most commonly used among these microorganisms are *Trichoderma* species (Imran *et al.*, 2016). In microbial enzymes, problems may occur in the inhibition of substrate or product, enzyme stability, and catalytic efficiency. Such problems are tried to be solved with recombinant enzyme technologies. Additionally, by adding polyethylene glycol (PEG) or Tween 20 to the medium, cellulosic saccharification can be increased and the absorption of cellulose onto lignin can be prevented (Joshi *et al.*, 2011).

The hydrolysis steps of cellulose to glucose are as follows:

1. The β-1,4 bond of cellulose is catalyzed by endoglucanase and binds to the water molecule. The result is cellodextrin, with shorter-chain free ends (Figure 11.4).
2. Exoglucanase enzyme breaks down cellodextrin into two-unit glucose (cellobiose) by catalyzing the reducing and non-reducing ends (Figure 11.5) (MetaCyc, 2014b).
3. Breakdown of cellobioses into glucose by β-glucosidases (Figure 11.6) (Brenda, 2015).

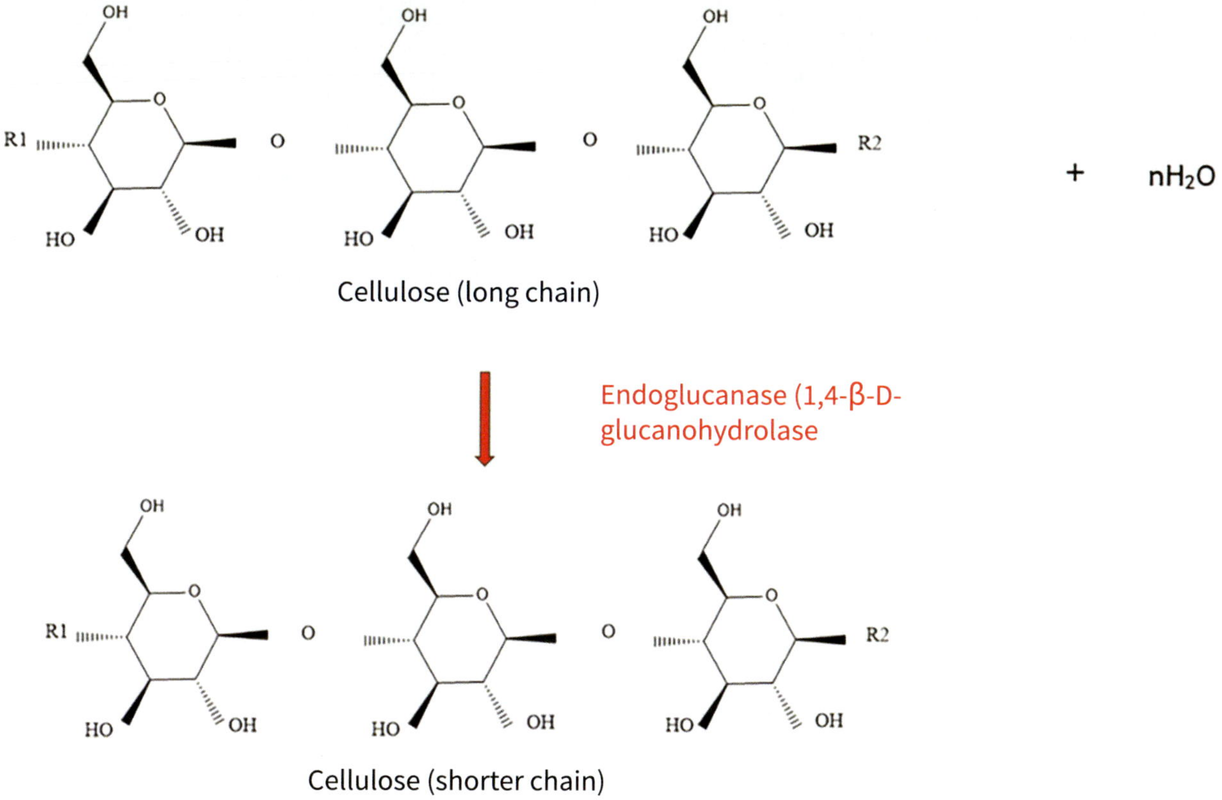

FIGURE 11.4 Degradation reaction of long-chain cellulose into cellodextrin.

Source: MetaCyc (2014a).

Since only monomers can be used in the fermentation step, it is important to break down cellulose to glucose.

Since hemicellulose has a more amorphous structure than cellulose, its hydrolysis is easier. It contains xylan in its structure, and xylan has both internal and external chains. The inner chain is cleaved by endo-β-1,4-xylanase (EC 3.2.1.8), and the outer chain by β-xylosidase (EC 3.2.1.37). The main chain of xylan is reduced to oligosaccharides by endo-p-1,4-xylanase, and oligosaccharides are reduced to xyropyranoses by p-xylosidase (Cherubin *et al.*, 2018).

11.12.4 Fermentation

It is the stage where sugar monomers released by hydrolysis are converted into acid, ethanol, and gas by yeast, bacteria, and fungi (Sharma *et al.*, 2020; Senatore *et al.*, 2020). *Saccharomyces cerevisiae*–type yeast is the most preferred microorganism because it produces ethanol with high yield and is resistant to reaction conditions (Surendhiran and Sirajunnisa, 2019). Just as the monomers of cellulose can convert glucose, mannose, and fructose into ethanol, the monomer of hemicellulose can convert xylan into xylose. Microorganisms used in fermentation and product yields are given in Table 11.1 (Cherubin *et al.*, 2018).

Fermentation technologies. The most commonly used fermentation technologies are separate hydrolysis and fermentation, simultaneous saccharification and fermentation (SSF), and simultaneous saccharification and co-fermentation (SSCF). Apart from these, the technologies used are non-isothermal simultaneous saccharification and fermentation, simultaneous saccharification, filtration and fermentation, and consolidated bioprocessing (CBP) (Azhar *et al.*, 2017).

Separate hydrolysis and fermentation (SHF). It is a method that allows the fermentation step to take place at medium temperatures after the enzymatic hydrolysis step takes place at high temperatures, separate from the fermentation (Azhar *et al.*, 2017). Since both processes are carried out at optimum temperatures, high efficiency is achieved.

The disadvantages of this method are the need for separate reactors for two separate reaction steps, the long processing time, and the possibility that cellulose activity may be limited due to the sugars released during hydrolysis (Figure 11.7) (Tavva *et al.*, 2016).

Simultaneous saccharification and fermentation (SSF). the saccharification step of celluloses and the fermentation of sugar monomers occur simultaneously in the same reactor. Since hydrolysis products are used simultaneously in fermentation, cellulase activities are preserved (Sakimoto et al., 2017). Its disadvantage is that since the temperatures required for hydrolysis and fermentation are different, there is a risk that microorganisms will be affected by the high temperature in the hydrolysis step and their numbers will decrease (Figure 11.8) (Rastogi and Shrivastava, 2017).

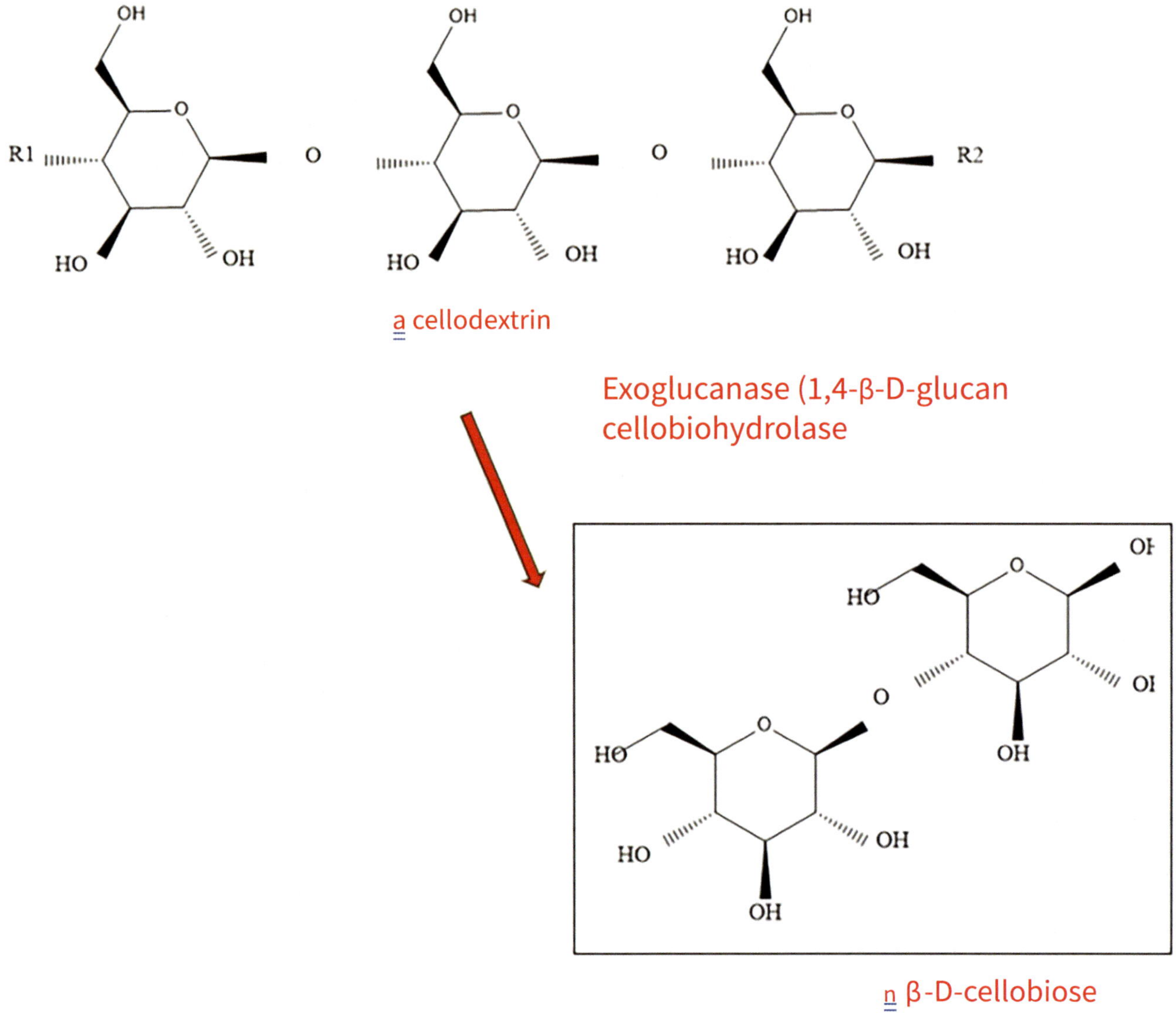

FIGURE 11.5 Hydrolysis reaction of cellodextrin to cellobiose.

Source: MetaCyc (2014b).

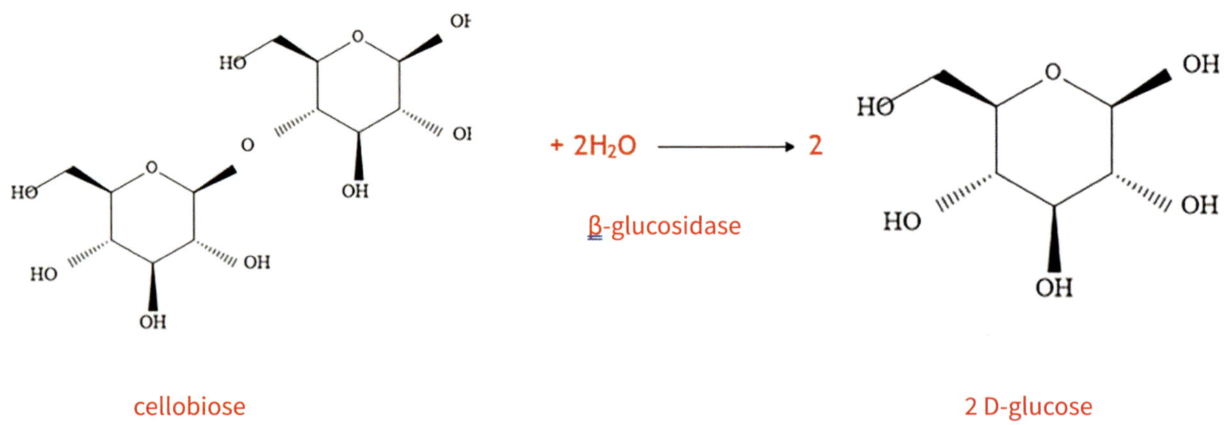

FIGURE 11.6 Hydrolysis reaction of each cellobiose into two D-glucose.

Source: Brenda (2015).

TABLE 11.1
Effects of Microorganisms and Reaction Medium on Ethanol Yield

Microorganism		Temperature (°C)	pH	Fermentation Time (h)	Sugar Concentration (g/L)	Ethanol Yield (g/L)	Ref.
Yeasts	*Saccharomyces cerevisiae* 3013	30	5.5	65	280	130.12	(Ji et al., 2012)
	Saccharomyces cerevisiae BY4742	35	5.0	96	80	39	(Zeng et al., 2014)
Bacteria	*Zymomonas mobilis* NRRL 806	30	6.5	18	117	30.4	(Letti et al., 2012; Ma et al., 2016)
	Zymomonas mobilis ZMA7–2	30	4.0	44	200	99.78	(Ma et al., 2016)
Fungi	*Aspergillus oryzae* 694	First aerobic step (30)	5.0	24	50	24.4	(Skory et al., 1997)
		Second anaerobic step (30)	5.0	144			
	Rhizobium javanicus 2871	First aerobic step (30)	5.0	24	100	33	(Skory et al., 1997)
		Second anaerobic step (30)	5.0	72			

Source: Cherubin *et al.* (2018).

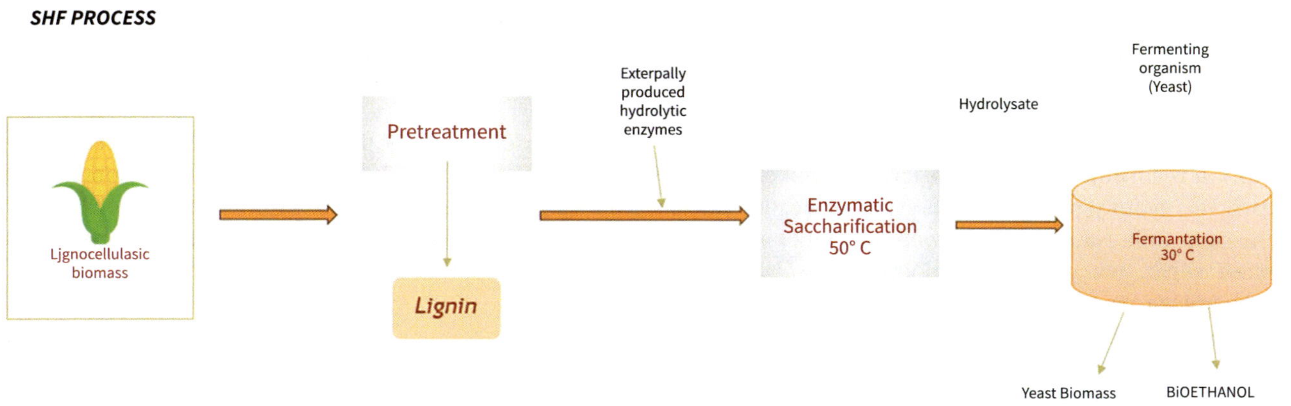

FIGURE 11.7 Separate hydrolysis and fermentation (SHF) steps.

Source: Choudhary et al. (2016).

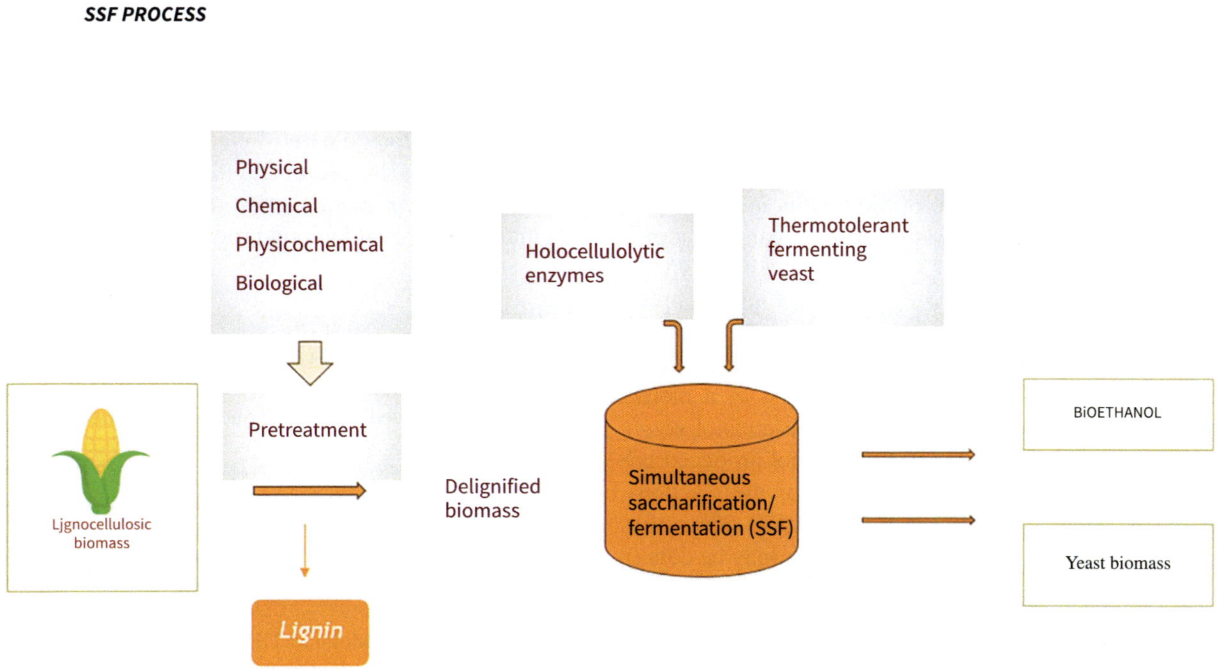

FIGURE 11.8 Schematic representation of the Simultaneous Saccharification and Fermentation (SSF) process for bioethanol production from lignocellulosic biomass. The process involves pretreatment (physical, chemical, physicochemical, or biological) to remove lignin and produce delignified biomass, followed by enzymatic hydrolysis using holocellulolytic enzymes and fermentation with thermotolerant yeast to yield bioethanol and yeast biomass

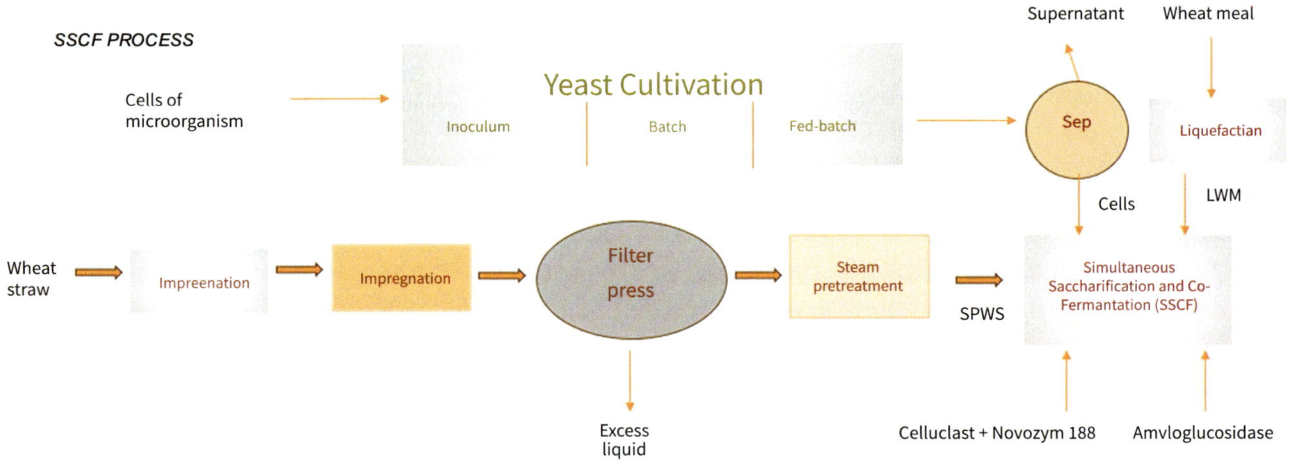

FIGURE 11.9 Simultaneous saccharification and co-fermentation (SSCF) steps.

Source: Choudhary et al. (2016).

Simultaneous saccharification and co-fermentation (SSCF). It is a method in which the fermentation of pentose sugars, the hydrolysis step, and the saccharification occur in the same unit. Natural *Saccharomyces* cannot hydrolyze pentose sugar, but genetically modified *Saccharomyces* can hydrolyze pentose. This method is relatively cheaper than the previously described methods, its ethanol yield is higher, and the processing time is shorter (Bondesson and Galbe, 2016). In addition, sugar-induced inhibition is minimized with this method, and since microorganisms generally feed on xylose, it increases the transition from xylose to glucose (Figure 11.9).

11.12.5 Optimum Fermentation Conditions

Bioethanol production is affected by the temperature and pH of the environment, sugar concentration, duration of the fermentation process, agitation rate in the reaction vessel, and inoculum size (Zabed *et al.*, 2014). The optimum fermentation temperature of biomass is in the range of 25–30°C, and high-temperature values can denature enzymes (Liu and Shen, 2008). Optimum bioethanol yield is achieved at a concentration of 150 g/L. The optimum pH value for fermentation with *Saccharomyces cerevisiae* is between 4 and 5; hence, it is necessary to extend the fermentation time in more acidic environments. In a more basic environment, a significant decrease in ethanol concentration is observed. The amount of ethanol produced is directly proportional to the agitation rate of the reaction vessel; the optimum agitation rate is around 150–200 rpm. If excessive shaking is done, microorganisms may be damaged (Zabed *et al.*, 2014).

11.12.6 Ethanol Recovery

The next step after fermentation is ethanol recovery. By reducing the water content to 0.5%, 99.5% anhydrous ethanol is formed. This process is done with different distillation methods by taking advantage of the difference in boiling points of ethanol and water. The most commonly used methods are azeotropic distillation, extractive distillation, and liquid–liquid extraction (Nitsche and Gbadamosi, 2017).

11.13 CONCLUSION

The world's population growth and technological advances have increased the demand for energy. Environmental health and the scarcity of fossil fuels have led to the search for new and sustainable energy sources. Biomass energy production is a promising alternative, but it is important that it be sustainable and harmless. To achieve this, regional studies are required. Harvesting crop residues can deplete soil nutrients and carbon reserves necessary for sustainable agriculture. Therefore, a careful plan should be developed that considers both short-term and long-term consequences to determine the optimum amount of crop residue harvest while maintaining healthy food production.

REFERENCES

Abraham A, Mathew AK, Park H, Choi O, Sindhu R, Parameswaran B, Pandey A, Park JH, Sang BI. 2020. Pretreatment strategies for enhanced biogas production from lignocellulosic biomass, *Bioresour Technol*, vol. 301, p. 122725.

Abreu RRL, de Lima SS, de Oliveira NCR, Leite LFC. 2014. Fauna edáfica sob diferentes níveis de palhada em cultivo de cana-de-açúcar, *Pesqui Agropecu Trop*, vol. 44, no. 4, pp. 409–416.

Adler PR, Rau BM, Roth GW. 2015. Sustainability of corn stover harvest strategies in Pennsylvania, *Bioenergy Res*, vol. 8, no. 3, pp. 1310–1320.

Aguiar A, Milessi TS, Mulinari DR, Lopes MS, da Costa SM, Candido RG. 2021. Sugarcane straw as a potential second generation feedstock for biorefinery and white biotechnology applications, *Biomass Bioenergy*, vol. 144, p. 105896, ISSN: 0961-9534.

Ahmed N, Thompson S, Glaser M. 2019. Global aquaculture productivity, environmental sustainability, and climate change adaptability, *Environ Manage*, vol. 63, no. 2, pp. 159–172.

Alavijeh MK, Yaghmaei S. 2016. Biochemical production of bioenergy from agricultural crops and residue in Iran, *Waste Manag*, vol. 52, pp. 375–394.

Alvira P, Tomás-Pejó E, Ballesteros M, Negro MJ. 2010. Pretreatment technologies for an efficient bioethanol production process based on enzymatic hydrolysis: A review, *Bioresour Technol*, vol. 101, no. 13, pp. 4851–4861.

Andrade LMG, Scarpari MS, Xavier MA, dos Anjos IA, Baptista AS, de Aguiar CL, da Silva DN, Bidóia MAP, Brancalião SR, Bressiani JA, de Campos MF, Miguel PEM, da Silva TN, da Silva VHP, Anjos LOS, Ogata BH. 2013. Residual biomass potential of commercial and pre-commercial sugarcane cultivars, *Scientia Agricola*, vol. 70, pp. 299–304.

Arceneaux AE, Selim HM. 2012. Mulch management strategies and sugarcane yield, *American Society of Sugar Cane Technologists (ASSCT)*, vol. 32, pp. 28–37.

Avila-Segura M, Barak P, Hedtcke JL, Posner JL. 2011. Nutrient and alkalinity removal by corn grain, stover and cob harvest in Upper Midwest USA, *Biomass Bioenergy*, vol. 35, no. 3, pp. 1190–1195.

Azhar SHM, Abdulla R, Jambo SA, Marbawi H, Gansau JA, Mohd Faik AA, Rodrigues KF. 2017. Yeasts in sustainable bioethanol production: A review, *Biochem Biophys Rep*, vol. 10, pp. 52–61.

Benazzi EDS, Bianchi MDO, Correia MEF, Lima E, Zonta E. 2013. Impactos dos métodos de colheita da cana-de-açúcar sobre a macrofauna do solo em área de produção no Espírito Santo – Brasil, *Semin Cienc Agrar*, vol. 34, no. 6, Supl 1, p. 3425.

Bensah EC, Mensah M. 2013. Chemical pretreatment methods for the production of cellulosic ethanol: Technologies and innovations, *Int J Chem Eng*, vol. 2013, pp. 1–21.

Birrell SJ, Karlen DL, Wirt A. 2014. Development of sustainable corn stover harvest strategies for cellulosic ethanol production, *Bioenergy Res*, vol. 7, no. 2, pp. 509–516.

Blanco-Canqui H. 2013. Crop residue removal for bioenergy reduces soil carbon pools: How can we offset carbon losses? *Bioenergy Res*, vol. 6, no. 1, pp. 358–371.

Bondesson PM, Galbe M. 2016. Process design of SSCF for ethanol production from steam-pretreated, acetic-acid-impregnated wheat straw, *Biotechnol Biofuels*, vol. 9, no. 1, p. 222,

Bonner IJ, Muth DJ, Koch JB, Karlen DL. 2014. Modeled impacts of cover crops and vegetative barriers on corn stover availability and soil quality, *Bioenergy Res*, vol. 7, no. 2, pp. 576–589.

Boserup E. 1975. The impact of population growth on agricultural output, *Q J Econ*, vol. 89, no. 2, p. 257.

Brandani CB, Abbruzzini TF, Williams S, Easter M, Cerri CEP, Paustian K. 2015. Simulation of management and soil interactions impacting SOC dynamics in sugarcane using the CENTURY model, *Global Change Biology Bioenergy*, vol. 7, pp. 646–657.

Brandão M, Milà Canals L, Clift R. 2011. Soil organic carbon changes in the cultivation of energy crops: Implications for GHG balances and soil quality for use in LCA, *Biomass Bioenergy*, vol. 35, no. 6, pp. 2323–2336.

Brodeur G, Yau E, Badal K, Collier J, Ramachandran KB, Ramakrishnan S. 2011. Chemical and physicochemical pretreatment of lignocellulosic biomass: A review, *Enzyme Res*, vol. 2011, pp. 1–17.

Calonego JC, Gil FC, Rocco VF, dos Santos EA. 2012. Persistence and nutrient release from maize, brachiaria and lablab straw., *Biosci J*, vol. 28, pp. 770–781.

Campbell EE, Johnson JMF, Jin VL, Lehman RM, Osborne SL, Varvel GE, Paustian K. 2014. Assessing the soil carbon, biomass production, and nitrous oxide emission impact of corn stover management for bioenergy feedstock production using DAYCENT, *Bioenergy Res*, vol. 7, no. 2, pp. 491–502.

Campos BHC, Amado TJC, Bayer C, Nicoloso RS, Fiorin JE. 2011. Carbon stock and its compartments in a subtropical oxisol under long-term tillage and crop rotation systems, *Rev Bras Cienc Solo*, vol. 35, no. 3, pp. 805–817.

Campos LHF, de Carvalho SJP, Christoffoleti PJ, Fortes C, da Silva JS. 2010. Sistemas de manejo da palhada influenciam acúmulo de biomassa e produtividade da cana-de-açúcar (var. RB855453), *Acta Sci Agron*, vol. 32, no. 2.

Carmo JB, Filoso S, Filoso S, Zotelli L, Zotelli L, de Sousa Neto ER, Pitombo LM, Duarte-Neto PJ, Vargas VP, Andrade CA, Gava GJC, Rossetto R, Cantarella H, Neto AE, Martinelli LA. 2013. Infield greenhouse gas emissions from sugarcane soils in Brazil: Effects from synthetic and organic fertilizer application and crop trash accumulation, *GCB Bioenergy*, vol. 5, no. 3, pp. 267–280.

Carvalho JLN, Hudiburg TW, Franco HCJ, DeLucia EH. 2017. Contribution of above- and belowground bioenergy crop residues to soil carbon, *GCB Bioenergy*, vol. 9, no. 8, pp. 1333–1343.

Cerri CC, Galdos MV, Maia SMF, Bernoux M, Feigl BJ, Powlson D, Cerri CEP. 2011. Effect of sugarcane harvesting systems on soil carbon stocks in Brazil: An examination of existing data, *Eur J Soil Sci*, vol. 62, no. 1, pp. 23–28.

Cherubin MR, da Silva Oliveira DM, Feigl BJ, Pimentel LG, Lisboa IP, Gmach MR, Varanda LL, Morais MC, Satiro LS, Popin GV, de Paiva SR, dos Santos AKB, de Vasconcelos ALS, de Melo PLA, Cerri CEP, Cerri CC. 2018. Crop residue harvest for bioenergy production and its implications on soil functioning and plant growth: A review, *Sci Agric*, vol. 75, no. 3, pp. 255–272.

Choudhary J, Singh S, Nain L. 2016. Thermotolerant fermenting yeasts for simultaneous saccharification fermentation of lignocellulosic biomass, *Electron J Biotechnol*, vol. 21, pp. 82–92.

Cook RL, Binkley D, Mendes JCT, Stape JL. 2014. Soil carbon stocks and forest biomass following conversion of pasture to broadleaf and conifer plantations in southeastern Brazil, *For Ecol Manage*, vol. 324, pp. 37–45.

Corradi MM, Panosso AR, Martins FMV, La Scala JN. 2013. Crop residues on short-term CO_2 emissions in sugarcane production areas, *Engenharia Agrícola*, vol. 33, no. 4, pp. 699–708.

Costa LAM., João KG, Morais ARC, Bogel-Łukasik E, Bogel-Łukasik R. 2013. Ionic liquids as a tool for lignocellulosic biomass fractionation, *Sustain Chem Process*, vol. 1, no. 1, p. 3.

Costa LG, Marin FR, Nassif DSP, Pinto HMS, Lopes-Assad MLRC. 2014. Simulação do efeito do manejo da palha e do nitrogênio na produtividade da cana-de-açúcar, *Rev Bras Eng Agrí Ambient*, vol. 18, no. 5, pp. 469–474.

Dauda AB, Ajadi A, Tola-Fabunmi AS, Akinwole AO. 2019. Waste production in aquaculture: Sources, components and managements in different culture systems, *Aquac Fish*, vol. 4, no. 3, pp. 81–88.

Diana S, Luísa LMP, Carlos EPC. 2014. Greenhouse gases emissions due to sugarcane trash on the soil, *Bragantia*, vol. 73, pp. 113–122.

Dinardo-Miranda LL, Fracasso JV. 2013. Sugarcane straw and the populations of pests and nematodes, *Sci Agric*, vol. 70, no. 5, pp. 305–310.

EEA. 2013. *EU bioenergy potential from a resource-efficiency perspective*, Copenhagen, Denmark: European Environment Agency (EEA), ISBN: 978-92-9213-397-9

Edeh I. 2020. Bioethanol production: An overview, in *Bioethanol technologies*, IntechOpen.

Edenhofer O, Pichs-Madruga, R, Sokona Y, Seyboth K, Matschoss P, Kadner S, Zwickel T, Eickemeier P, Hansen G, Schloemer S, von Stechow C. 2011. *Renewable energy sources and climate change mitigation: Special report of the intergovernmental panel on climate change*, Cambridge, UK: Cambridge University Press.

Egüés I, Sanchez C, Mondragon I, Labidi J. 2012. Effect of alkaline and autohydrolysis processes on the purity of obtained hemicelluloses from corn stalks, *Bioresour Technol*, vol. 103, no. 1, pp. 239–248.

Empresa de Pesquisa Energética [EPE]. 2015. *Analysis of the biofuels scenario: Análise de conjuntura dos biocombustíveis*, Empresa de Pesquisa Energética (EPE).

Epron D, Mouanda C, Mareschal L, Koutika LS. 2015. Impacts of organic residue management on the soil C dynamics in a tropical eucalypt plantation on a nutrient-poor sandy soil after three rotations, *Soil Biol Biochem*, vol. 85, pp. 183–189.

Erdei B, Galbe B, Zacchi G. 2013. Simultaneous saccharification and co-fermentation of whole wheat in integrated ethanol production, *Biomass Bioenergy*, vol. 56, pp. 506–514.

European Commission. 2009. Directive 2009/28/EC of the European Parliament and of the Council on the promotion of the use of energy from renewable sources and amending and subsequently repealing repealing Directives 2001/77/EC and 2003/30/EC. Off J Eur Union 160:16–62.

Ferreira DA, Franco HCJ, Otto R, Vitti AEC, Fortes C, Faroni CE, Garside AL, Trivelin PCO. 2016. Contribution of N from green harvest residues for sugarcane nutrition in Brazil, *GCB Bioenergy*, vol. 8, no. 5, pp. 859–866.

Fialho RC, Zinn YL. 2014. Changes in soil organic carbon under eucalyptus plantations in Brazil: A comparative analysis, *Land Degrad Dev*, vol. 25, no. 5, pp. 428–437.

Franco AL, Bartz ML, Cherubin MR, Baretta D, Cerri CE, Feigl BJ, Wall DH, Davies CA, Cerri CC. 2016. Loss of soil (macro)fauna due to the expansion of Brazilian sugarcane acreage, *Sci Total Environ*, vol. 563–564, pp. 160–168.

Franco HCJ, Pimenta MTB, Carvalho JLN, Magalhães PSG, Rossell CEV, Braunbeck OA, Vitti AC, Kölln OT, Neto JR. 2013. Assessment of sugarcane trash for agronomic and energy purposes in Brazil, *Sci Agric*, vol. 70, no. 5, pp. 305–312,

Franzluebbers AJ. 2015. Farming strategies to fuel bioenergy demands and facilitate essential soil services, *Geoderma*, vol. 259–260, pp. 251–258.

Glithero NJ, Ramsden SJ, Wilson P. 2013. Barriers and incentives to the production of bioethanol from cereal straw: A farm business perspective, *Energy Policy*, vol. 59, pp. 161–171.

Gnansounou E. 2010. Production and use of lignocellulosic bioethanol in Europe: Current situation and perspectives, *Bioresour Technol*, vol. 101, no. 13, pp. 4842–4850.

Godfray HC, Beddington JR, Crute IR, Haddad L, Lawrence D, Muir JF, Pretty J, Robinson S, Thomas SM, Toulmin C. 2010. Food security: The challenge of feeding 9 billion people, *Science (New York, N.Y.)*, vol. 327, no. 5967, pp. 812–818.

Govaerts B, Fuentes M, Mezzalama M, Nicol JM, Deckers JD, Etchevers JD, Figueroa-Sandoval B, Sayre KD. 2007. Infiltration, soil moisture, root rot and nematode populations after 12 years of different tillage, residue and crop rotation managements, *Soil Tillage Res*, vol. 94, no. 1, pp. 209–219,

Graham MH, Haynes RJ. 2006. Organic matter status and the size, activity and metabolic diversity of the soil microbial community in the row and inter-row of sugarcane under burning and trash retention, *Soil Biol Biochem*, vol. 38, no. 1, pp. 21–31.

Guo M, Song W, Buhain J. 2015. Bioenergy and biofuels: History, status, and perspective, *Renew Sustain Energy Rev*, vol. 42, pp. 712–725, Feb.

Guzman J, Al-Kaisi M, Parkin T. 2015. Greenhouse gas emissions dynamics as influenced by corn residue removal in continuous corn system, *Soil Sci Soc Am J*, vol. 79, no. 2, pp. 612–625.

Haase M, Rösch C, Ketzer D. 2016. GIS-based assessment of sustainable crop residue potentials in European regions, *Biomass Bioenergy*, vol. 86, pp. 156–171.

Hamelin L, Borzęcka M, Kozak M, Pudełko R. 2019. A spatial approach to bioeconomy: Quantifying the residual biomass potential in the EU-27, *Renew Sustain Energy Rev*, vol. 100, pp. 127–142.

Hammerbeck AL, Stetson SJ, Osborne SL, Schumacher TE, Pikul JL. 2012. Corn residue removal impact on soil aggregates in a no-till corn/soybean rotation, *Soil Sci Soc Am J*, vol. 76, no. 4, pp. 1390–1398.

Hassan SS, Williams GA, Jaiswal AK. 2018. Emerging technologies for the pretreatment of lignocellulosic biomass, *Bioresour Technol*, vol. 262, pp. 310–318.

Henriksson PJG, Belton B, Jahan KM, Rico A. 2018. Measuring the potential for sustainable intensification of aquaculture in Bangladesh using life cycle assessment, *Proc Natl Acad Sci*, vol. 115, no. 12, pp. 2958–2963.

Hilst F, Lesschen JP, van Dam JMC, Riksen M, Verweij PA, Sanders JPM, Faaij APC. 2012. Spatial variation of environmental impacts of regional biomass chains, *Renew Sustain Energy Rev*, vol. 16, no. 4, pp. 2053–2069,

Holm-Nielsen JB, Al Seadi T, Oleskowicz-Popiel P. 2009. The future of anaerobic digestion and biogas utilization, *Bioresour Technol*, vol. 100, no. 22, pp. 5478–5484.

IEA Bioenergy. 2017. Mobilization of agricultural residues for bioenergy and higher value bio-products: Resources, barriers and sustainability. IEA Bioenergy, IEA Bioenergy Task 43.

Imran M, Anwar Z, Irshad M, Asad MJ, Ashfaq H. 2016. Cellulase production from species of fungi and bacteria from agricultural wastes and its utilization in industry: A review, *Adv Enzyme Res*, vol. 04, no. 02, pp. 44–55.

Jeschke MHA. 2012. Sustainable corn stover harvest for biofuel production, *Crop Insights*, vol. 22, pp. 1–6.

Ji, H., Yu, J., Zhang, X., Tan, T. (2012). Characteristics of an Immobilized Yeast Cell System Using Very High Gravity for the Fermentation of Ethanol, *Appl Biochem Biotechnol*, vol. 168, no. 1, pp. 21–28, doi: 10.1007/s12010-011-9280-5.

Jin VL, Baker JM, Johnson JMF, Karlen DL, Lehman RM, Osborne SL, Sauer TJ, Stott DE, Varvel GE, Venterea RT, Schme, MR, Wienhold BJ. 2014. Soil greenhouse gas emissions in response to corn stover removal and tillage management across the US corn belt, *Bioenerg Res*, vol. 7, pp. 517–527.

Johnson JMF, Strock JS, Tallaksen JE, Reese M. 2016. Corn stover harvest changes soil hydrology and soil aggregation, *Soil Tillage Res*, vol. 161, pp. 106–115.

Joshi B, Bhatt MR, Sharma D, Joshi J, Malla R, Sreerama L. 2011. Lignocellulosic ethanol production: Current practicesand recent developments, *Biotechnol Mol Biol Rev*, vol. 6, no. 8, pp. 172–182.

Junior FP, Portilho IIR, Mercante FM. 2015. Atributos microbiológicos de um latossolo sob cultivo de cana-de-açúcar com e sem queima da palhada, *Semin Cienc Agrar*, vol. 36, no. 1, p. 151,

Kaab A, Sharifi M, Mobli H, Nabavi-Pelesaraei A, Chau KW. 2019. Combined life cycle assessment and artificial intelligence for prediction of output energy and environmental impacts of sugarcane production, *Sci Total Environ*, vol. 664, pp. 1005–1019.

Kaarakka L, Tamminen P, Saarsalmi A, Kukkola M, Helmisaari HS, Burton AJ. 2014. Effects of repeated whole-tree harvesting on soil properties and tree growth in a Norway spruce (Picea abies (L.) Karst.) stand, *For Ecol Manage*, vol. 313, pp. 180–187.

Karlen DL, Birell SJ, Hess JR. 2011. A five-year assessment of corn stover harvest in central Iowa, USA, *Soil Tillage Res*, vol. 115–116, pp. 47–55.

Karlen DL, Kovar J, Birrell S. 2015. Corn stover nutrient removal estimates for Central Iowa, USA, *Sustainability*, vol. 7, no. 7, pp. 8621–8634.

Khanal S, Anex RP, Gelder BK, Wolter C. 2014. Nitrogen balance in Iowa and the implications of corn-stover harvesting, *Agric Ecosyst Environ*, vol. 183, pp. 21–30.

Kim S, Dale BE. 2003. Global potential bioethanol production from wasted crops and crop residues, *Biomass Bioenergy*, vol. 26, no. 4, pp. 361–375.

Kluts IN, Brinkman MLJ, de Jong SA, Junginger HM. 2017. *Biomass resources: Agriculture*, Biorefineries, Advances in Biochemical Engineering/Biotechnology. Springer, 166, pp. 13–26. https://doi.org/10.1007/10_2016_66

Koizumi T. 2015. Biofuels and food security, *Renew Sustain Energy Rev*, vol. 52, pp. 829–841.

Koop SHA, van Leeuwen CJ. 2017. The challenges of water, waste and climate change in cities, *Environ Dev Sustain*, vol. 19, no. 2, pp. 385–418.

Koul B, Yakoob M, Shah MP. 2022. Agricultural waste management strategies for environmental sustainability, *Environ Res*, vol. 206, p. 112285.

Kumar G, Shobana S, Nagarajan D, Lee DJ, Lee KJ, Lin CY, Chen CJS. 2018a. Biomass based hydrogen production by dark fermentation – recent trends and opportunities for greener processes, *Curr Opin Biotechnol*, vol. 50, pp. 136–145.

Kumar P, Singh RK. 2021. Selection of sustainable solutions for crop residue burning: An environmental issue in northwestern states of India, *Environ Dev Sustain*, vol. 23, no. 3, pp. 3696–3730.

Kumar S, Paritosh K, Pareek N, Chawade A, Vivekanand V. 2018b. De-construction of major Indian cereal crop residues through chemical pretreatment for improved biogas production: An overview, *Renew Sustain Energy Rev*, vol. 90, pp. 160–170.

Kumaraswamy S, Mendham DS, Grove TS, O'Connell AM, Sankaran KV, Rance SJ. 2014. Harvest residue effects on soil organic matter, nutrients and microbial biomass in eucalypt plantations in Kerala, India, *For Ecol Manage*, vol. 328, pp. 140–149.

Lafferty KD, Harvell CD, Conrad JM, Friedman CS, Kent ML, Kuris AM, Powell EN, Rondeau D, Saksida SM. 2015. Infectious diseases affect marine fisheries and aquaculture economics, *Ann Rev Mar Sci*, vol. 7, pp. 471–496.

Lal R. 2005. World crop residues production and implications of its use as a biofuel, *Environ Int*, vol. 31, no. 4, pp. 575–584.

Lapola DM, Schaldach R, Alcamo J, Bondeau A, Koch J, Koelking C, Priess JA. 2010. Indirect land-use changes can overcome carbon savings from biofuels in Brazil, *Proc Natl Acad Sci USA*, vol. 107, no. 8, pp. 3388–3393.

Leal MRLV, Galdos MV, Scarpare FV, Seabra JEA, Walter A, Oliveira COF. 2013. Sugarcane straw availability, quality, recovery and energy use: A literature review, *Biomass Bioenergy*, vol. 53, pp. 11–19.

Lehman RM, Ducey TF, Jin VL, Acosta-Martinez V, Ahlschwede CM, Jeske ES, Drijber RA, Cantrell KB, Lima AMN, Silva IR, Neves JCL, Novais RF, Barros NF, Mendonça ES, Smyth TJ, Moreira MS, Leite FP. 2006. Soil organic carbon dynamics following afforestation of degraded pastures with eucalyptus in southeastern Brazil, *For Ecol Manage*, vol. 235, no. 1–3, pp. 219–231.

Leonardo MGJ, Clayton AA, Antonio RH, James S, Frosini BFS, Walter PL, Henrique SBP, Ayeska H, Jean-Pierre B, Jean-Paul L, Daniel E, Yann N. 2013. Integrating genetic and silvicultural strategies to minimize abiotic and biotic constraints in Brazilian eucalypt plantations, *For Ecol Manage*, vol. 301, pp. 6–27.

Letti, L.A.J., Karp, S.G., Woiciechowski, A.L., Soccol, C.R. (2012). Ethanol production from soybean molasses by Zymomonas mobilis, *Biomass Bioenergy*, vol. 44, pp. 80–86, doi: 10.1016/j.biombioe.2012.04.023.

Lima, A. M. N., Silva, I. R., Neves, J. C. L., Novais, R. F., Barros, N. F., Mendonça, E. S., Smyth, T. J., Moreira, M. S., & Leite, F. P. 2006. Soil organic carbon dynamics following afforestation of degraded pastures with Eucalyptus in southeastern Brazil. *Forest Ecology and Management*, vol. 235, pp. 219–231.

Lindorfer J, Fazeni K, Steinmüller H. 2014. Life cycle analysis and soil organic carbon balance as methods for assessing the ecological sustainability of 2nd generation biofuel feedstock, *Sustain Energy Technol Assess*, vol. 5, pp. 95–105.

Liu R, Shen F. 2008. Impacts of main factors on bioethanol fermentation from stalk juice of sweet sorghum by immobilized Saccharomyces cerevisiae (CICC 1308), *Bioresour Technol*, vol. 99, no. 4, pp. 847–854, doi: 10.1016/j.biortech.2007.01.009.

Long SP, Karp A, Buckeridge MS, Davis SC, Jaiswal D, Moore PH, Moose SP, Murphy D, Onwona-Agyeman S, Vonshak A. 2015. Crop feedstocks for biofuels and bioenergy, in GM Souza, V Victoria, C Joly, LM Verdade (eds.), *Bioenergy & sustainability: Bridging the gaps*, Paris, France: SCOPE, pp. 302–346.,

Ma, R., Castro-Dominguez, B., Mardilovich, I.P., Dixon, A.G., Ma, Y.H. (2016). Experimental and simulation studies of the production of renewable hydrogen through ethanol steam reforming in a large-scale catalytic membrane reactor, *Chemical Engineering Journal,* vol. 303, pp. 302–313, doi: 10.1016/j.cej.2016.06.021.

Martins CIM, Eding EH, Verdegem MCJ, Heinsbroek LTN, Schneide O, Blancheton JP, d'Orbcastel ER, Verreth JAJ. 2010. New developments in recirculating aquaculture systems in Europe: A perspective on environmental sustainability, *Aquac Eng*, vol. 43, no. 3, pp. 83–93, Nov. 2010.

Maurya DP, Singla A, Negi S. 2015. An overview of key pretreatment processes for biological conversion of lignocellulosic biomass to bioethanol, *3 Biotech*, vol. 5, no. 5, pp. 597–609.

Meier EA, Thorburn PJ. 2016. Long term sugarcane crop residue retention offers limited potential to reduce nitrogen fertilizer rates in Australian wet tropical environments, *Front Plant Sci*, vol. 7.

MetaCyc. 2014a. MetaCyc reaction: 3.2.1.4.

MetaCyc. 2014b. MetaCyc reaction: 3.2.1.91.

Mohanty AK, Misra M, Drzal LT. 2002. Sustainable Bio-Composites from renewable resources, opportunities and challenges in the green materials world, *J Polym Environ*, vol. 10, no. 1/2, pp. 19–26.

Monforti F, Lugato E, Motola V, Bodis K, Dallemand F. 2015. Optimal energy use of agricultural crop residues preserving soil organic carbon stocks in Europe, *Renew Sustain Energy Rev*, vol. 44, pp. 519–529.

Mourtzinis S, Cantrell KB, Arriaga FJ, Balkcom KS, Novak JM, Frederick JR, Douglas L, Karlen F. 2016. Carbohydrate and nutrient composition of corn stover from three southeastern USA locations, *Biomass Bioenergy*, vol. 85, pp. 153–158.

Nitsche M, Gbadamosi R. 2017. *Extractive and azeotropic distillation, practical column design guide*, Cham: Springer International Publishing, pp. 153–164.

Oliveira DMS, Williams S, Cerri CEP, Paustian K. 2017. Predicting soil C changes over sugarcane expansion in Brazil using the DayCent model, *GCB Bioenergy*, vol. 9, no. 9, pp. 1436–1446.

Oliveira ECA, Freire FJ, de Oliveira RI, dos Freire MBG, Simões Neto DE, da Silva SAM. 2010. Growth performance of African catfish (Clarias gariepinus) juveniles reared in wastewater treated with alum and Moringa oleifera seed, *J Aquac Res Dev*, vol. 34, no. 4, pp. 1343–1352

Olusegun AA, Dauda AB, Ololade OA. 2016. Haematological response of Clarias gariepinus Juveniles reared in treated wastewater after waste solids removal using alum or moringa oleifera seed powder, *International Journal of Aquaculture*, vol. 6, p. 11.

Orr MJ, Gray MB, Applegate B, Volenec JJ, Brouder SM, Turco RF. 2015. Transition to second generation cellulosic biofuel production systems reveals limited negative impacts on the soil microbial community structure, *Appl Soil Ecol*, vol. 95, pp. 62–72.

Osborne SL, Johnson JMF, Jin VL, Hammerbeck AL, Varvel GE, Schumacher TE. 2014. The impact of corn residue removal on soil aggregates and particulate organic matter, *Bioenergy Res*, vol. 7, no. 2, pp. 559–567.

Pattanaik L, Pattnaik F, Saxena DK, Naik SN. 2019. Biofuels from agricultural wastes, in *Second and third generation of feedstocks*, Elsevier, pp. 103–142.

Paustian K, Lehmann J, Ogle S, Reay D, Robertson GP, Smith P. 2016. Climate-smart soils, *Nature*, vol. 532, no. 7597, pp. 49–57.

Pereira SC, Maehara L, Machado CM, Farinas CS. 2015. 2G ethanol from the whole sugarcane lignocellulosic biomass, *Biotechnol Biofuels*, vol. 8, p. 44.

Pittelkow CM, Liang X, Linquist BA, van Groenigen KJ, Lee J, Lundy ME, van Gestel N, Six J, Venterea RT, van Kessel C. 2015. Productivity limits and potentials of the principles of conservation agriculture, *Nature*, vol. 517, no. 7534, pp. 365–368.

Poeplau C, Don A. 2015. Carbon sequestration in agricultural soils via cultivation of cover crops – A meta-analysis, *Agric Ecosyst Environ*, vol. 200, pp. 33–41.

Portilho IIR., Borges CD, Costa AR, Salton C, Mercante FM. 2011. Resíduos da cultura da cana-de-açúcar e seus efeitos sobre a fauna invertebrada epigéica, *Semin Cienc Agrar*, vol. 32, no. 3, pp. 959–970.

Pourhashem G, Adler PR, McAloon AJ, Spatari S. 2013. Cost and greenhouse gas emission tradeoffs of alternative uses of lignin for second generation ethanol, *Environ Res Lett*, vol. 8, no. 2, p. 025021.

Qi Y, Zhen W, Li H. 2015. Allelopathy of decomposed maize straw products on three soil-born diseases of wheat and the analysis by GC-MS, *J Integr Agric*, vol. 14, no. 1, pp. 88–97.

Rachid CT, Pires CA, Leite DC, Coutinho HL, Peixoto RS, Rosado AS, Salton J, Zanatta JA, Mercante FM, Angelini GA, Balieiro FC. 2016. Sugarcane trash levels in soil affects the fungi but not bacteria in a short-term field experiment, *Braz J Microbiol*, vol. 47, no. 2, pp. 322–326.

Rastogi M, Shrivastava S. 2017. Recent advances in second generation bioethanol production: An insight to pretreatment, saccharification and fermentation processes, *Renew Sustain Energy Rev*, vol. 80, pp. 330–340.

Reddy DD, Kushwaha S, Srivastava S, Khamparia RS. 2014. Long-term wheat residue management and supplementary nutrient input effects on phosphorus fractions and adsorption behavior in a vertisol, *Commun Soil Sci Plant Anal*, vol. 45, no. 4, pp. 541–554.

Rocha JHT, Gonçalves JLM, Gava JL, Godinho TO, Melo EASC, Bazani JH, Hubner A, Arthur Junior JC, Wichert MP. 2016. Forest residue maintenance increased the wood productivity of a Eucalyptus plantation over two short rotations, *For Ecol Manage*, vol. 379, pp. 1–10.

Robertson FA, Thorburn PJ. 2007. Management of sugarcane harvest residues: Consequences for soil carbon and nitrogen, *Soil Res*, vol. 45, no. 1, p. 13.

Saini JK, Saini R, Tewari L. 2015. Lignocellulosic agriculture wastes as biomass feedstocks for second-generation bioethanol production: Concepts and recent developments, *3 Biotech*, vol. 5, no. 4, pp. 337–353.

Sakimoto K, Kanna M, Matsumura Y. 2017. Kinetic model of cellulose degradation using simultaneous saccharification and fermentation, *Biomass Bioenergy*, vol. 99, pp. 116–121.

Sanchez DL, Nelson JH, Johnston J, Mileva A, Kammen DM. 2015. Biomass enables the transition to a carbon-negative power system across western North America, *Nat Clim Chang*, vol. 5, no. 3, pp. 230–234.

Sarkar N, Ghosh SK, Bannerjee S, Aikat K. 2012. Bioethanol production from agricultural wastes: An overview, *Renew Energy*, vol. 37, no. 1, pp. 19–27.

Sathitsuksanoh N, George A, Zhang YP. 2013. New lignocellulose pretreatments using cellulose solvents: A review, *J Chem Technol Biotechnol*, vol. 88, no. 2, pp. 169–180.

Sathitsuksanoh N, Zhu Z, Wi S, Zhang YHP. 2011. Cellulose solvent-based biomass pretreatment breaks highly ordered hydrogen bonds in cellulose fibers of switchgrass, *Biotechnol Bioeng*, vol. 108, no. 3, pp. 521–529, doi: 10.1002/bit.22964.

Sawyer JE, Mallarino AP. 2007. Nutrient removal when harvesting corn stover, Integrated Crop Management – IC-498(22), Iowa State University Extension, August 6, 2007.

Searchinger T, Heimlich R, Houghton RA, Dong F, Elobeid A, Fabiosa J, Tokgoz S, Hayes D, Yu TH. 2008. Use of U.S. croplands for biofuels increases greenhouse gases through emissions from land-use change, *Science (New York, N.Y.)*, vol. 319, no. 5867, pp. 1238–1240.

Seebaluck V, Leal MRLV. 2015. Feedstock supply chains, in GM Souza, R Victoria, C Joly, LM Verdade (eds.), *Bioenergy & sustainability: bridging the gaps*, Paris, France: SCOPE, pp. 347–373.

Senatore A, Dalena F, Basil A. 2020. Novel bioethanol production processes and purification technology using membranes, in *Catalysis, green chemistry and sustainable energy*, Elsevier – *Studies in Surface Science and Catalysis*, pp. 359–384. doi: 10.1016/B978-0-444-64337-7.00019-7.

Scarlat N, Dallemand JF, Fahl F. 2018. Biogas: Developments and perspectives in Europe, *Renew Energy*, vol. 129, pp. 457–472.

Sharma B, Larroche C, Dussap CG. 2020. Comprehensive assessment of 2G bioethanol production, *Bioresour Technol*, vol. 313, p. 123630.

Shrivastava A, Sharma RK. 2022. Lignocellulosic biomass based microbial fuel cells: Performance and applications, *J Clean Prod*, vol. 361, p. 132269.

Shuit SH, Tan KT, Lee KT, Kamaruddin AH. 2009. Oil palm biomass as a sustainable energy source: A Malaysian case study, *Energy*, vol. 34, no. 9, pp. 1225–1235.

Shyamsundar P, Springer NP, Tallis H, Polasky S, Jat ML, Sidhu HS, Krishnapriya PP, Skiba N, Ginn W, Ahuja V, Cummins

J, Datta I, Dholakia HH, Dixon J, Gerard B, Gupta R, Hellmann J, Jadhav A, Jat HS, Keil A, Somanathan R. 2019. Fields on fire: Alternatives to crop residue burning in India, *Science (New York, N.Y.)*, vol. 365, no. 6453, pp. 536–538.

Singh Y, Sidhu HS. 2014. Management of cereal crop residues for sustainable rice-wheat production system in the Indo-Gangetic plains of India, *Proc Indian Natl Sci Acad*, vol. 80, no. 1, p. 95.

Skory, C.D., Freer, S.N., Bothast, R.J. (1997). Screening for ethanol-producing filamentous fungi, *Biotechnol Lett*, vol. 19, pp. 203–206.

Soares EMB., Silva IR, de Novais RF, Hu YY, Schmidt-Rohr K. 2013. Alterations in molecular composition of humic substances from eucalypt plantation soils assessed by 13 C-NMR spectroscopy, *Soil Sci Soc Am J*, vol. 77, no. 1, pp. 293–306.

Souza RA, Telles TS, Machado W, Hungria M, Filho JT, Guimarães M. 2012. Effects of sugarcane harvesting with burning on the chemical and microbiological properties of the soil, *Agric Ecosyst Environ,* vol. 155, pp. 1–6.

Stavi I, Bel G, Zaady E. 2016. Soil functions and ecosystem services in conventional, conservation, and integrated agricultural systems. A review, *Agron Sustain Dev*, vol. 36, no. 2, p. 32.

Stephens WW, Farris JL. 2004. A biomonitoring approach to aquaculture effluent characterization in channel catfish fingerling production, *Aquaculture*, vol. 241, no. 1–4, pp. 319–330.

Sukiran MA, Abnisa F, Wan Daud WMA, Abu Bakar N, Loh SK. 2017. A review of torrefaction of oil palm solid wastes for biofuel production, *Energy Convers Manag*, vol. 149, pp. 101–120.

Surendhiran D, Sirajunnisa, AR. 2019. Role of Genetic Engineering in Bioethanol production from algae, in *Bioethanol production from food crops*, Elsevier, pp. 361–381.

Tan Z, Liu S. 2015. Corn Belt soil carbon and macronutrient budgets with projected sustainable stover harvest, *Agric Ecosyst Environ*, vol. 212, pp. 119–126.

Tavares OCH, Lima E, Zonta E. 2010. Sugarcane growth and productivity under different tillage and crop systems/ Crescimento e produtividade da cana planta cultivada em diferentes sistemas de preparo do solo e de colheita, *Acta Sci Agron*, vol. 32, no. 1.

Tavva SSMD, Deshpande A, Durbha SR, Palakollu VAR, Goparaju AU, Yechuri VR, Bandaru VR, Muktinutalapati VSR. 2016. Bioethanol production through separate hydrolysis and fermentation of Parthenium hysterophorus biomass, *Renew Energy*, vol. 86, pp. 1317–1323.

Tormena CA, Karlen DL, Logsdon S, Cherubin MR. 2017. Corn stover harvest and tillage impacts on near-surface soil physical quality, *Soil Tillage Res*, vol. 166, pp. 122–130.

Tripathi N, Hills CD, Singh RS, Atkinson CJ. 2019. Biomass waste utilisation in low-carbon products: Harnessing a major potential resource, *NPJ Clim Atmos Sci*, vol. 2, no. 1, p. 35.

Trivelin PCO, Franco HCJ, Otto R, Ferreira DA, Vitti AC, Fortes C, Faroni CE, Oliveira ECA, Cantarella H. 2013. Impact of sugarcane trash on fertilizer requirements for São Paulo, Brazil, *Sci Agric,* vol. 70, no. 5, pp. 345–352.

U.S. Department of Energy. 2016. Billion-ton report: Advancing domestic resources for a thriving bioeconomy, in *Studies in Surface Science and Catalysis. U.S. Department of Energy*, Elsevier, pp. 359–384.

Versini A, Mareschal L, Matsoumbou T, Zeller B, Ranger J, Laclau JP. 2014. Effects of litter manipulation in a tropical Eucalyptus plantation on leaching of mineral nutrients, dissolved organic nitrogen and dissolved organic carbon, *Geoderma*, vol. 232–234, pp. 426–436.

Viator H, Wang JJ. 2011. Effects of residue management on yield after three production cycles of a long-term sugarcane field trial in Louisiana, *Journal of the American Society of Sugar Cane Technologists*, vol. 31, pp. 15–25. ISSN: 1075-6302

Villegas F, Torres JS, Larrahondo JE, Restrepo DF. 2007. Allelopathic effects of sugarcane postharvest residue, *Journal of the American Society of Sugar Cane Technologists*, vol. 26, pp. 374–378. ISSN: 1075-6302

Wakelin SA, Colloff MJ, Harvey PR, Marschner P, Gregg AL, Rogers SL. 2007. The effects of stubble retention and nitrogen application on soil microbial community structure and functional gene abundance under irrigated maize, *FEMS Microbiol Ecol*, vol. 59, no. 3, pp. 661–670.

Wang F, Cheng Z, Reisner A, Liu Y. 2018. Compliance with household solid waste management in rural villages in developing countries, *J Clean Prod*, vol. 202, pp. 293–298.

Wang J, Lou Y, Ma D, Feng K, Chen C, Zhao L, Xing D. 2023. Co-treatment with free nitrous acid and calcium peroxide regulates microbiome and metabolic functions of acidogenesis and methanogenesis in sludge anaerobic digestion, *Sci Total Environ*, vol. 870, p. 161924.

Wegner BR, Kumar S, Osborne SL, Schumacher TE, Vahyala IE, Eynard A. 2015. Soil response to corn residue removal and cover crops in Eastern South Dakota, *Soil Sci Soc Am J*, vol. 79, no. 4, pp. 1179–1187.

West TO, Post WM. 2002. Soil organic carbon sequestration rates by tillage and crop rotation, *Soil Sci Soc Am J*, vol. 66, no. 6, pp. 1930–1946.

Wiedenfeld B. 2009. Effects of green harvesting vs burning on soil properties, growth and yield of sugarcane in South Texas, *Journal of the American Society of Sugar Cane Technologists*, vol. 29, pp. 102–109. ISSN: 1075-6302

Wilhelm W, Johnson JMF, Hatfield JL, Voorhees WB, Linden DR. 2004. Crop and soil productivity response to corn residue removal: A literature review, *Agron J,* vol. 96, pp. 1–17.

Wit M, Faaij A. 2010. European biomass resource potential and costs, *Biomass Bioenergy*, vol. 34, no. 2, pp. 188–202.

Xiukang W, Zhanbin L, Yingying X. 2015. Effects of mulching and nitrogen on soil temperature, water content, nitrate-N content and maize yield in the Loess Plateau of China, *Agric Water Manag*, vol. 161, pp. 53–64.

Zabaniotou AA, Kantarelis EK, Theodoropoulos DC. 2008. Sunflower shells utilization for energetic purposes in an integrated approach of energy crops: Laboratory study pyrolysis and kinetics, *Bioresour Technol*, vol. 99, no. 8, pp. 3174–3181.

Zabed H, Faruq G, Sahu JN, Azirun MS, Hashim R, Boyce AN. 2014. Bioethanol production from fermentable sugar juice, *Sci World J*, pp. 1–11.

Zabed H, Sahu JN, Boyce AN, Faruq G. 2016. Fuel ethanol production from lignocellulosic biomass: An overview on feedstocks and technological approaches, *Renew Sustain Energy Rev*, vol. 66, pp. 751–774.

Zeng, Y., Zhao, S., Yang, S., Ding, S.Y. (2014). Lignin plays a negative role in the biochemical process for producing lignocellulosic biofuels, *Curr Opin Biotechnol*, vol. 27, pp. 38–45, doi: 10.1016/j.copbio.2013.09.008.

Zhang YH, Ding SY, Mielenz JR, Cui JB, Elander RT, Laser M, Himmel ME, McMillan JR, Lynd LR. 2007. Fractionating recalcitrant lignocellulose at modest reaction conditions, *Biotechnol Bioeng*, vol. 97, no. 2, pp. 214–223.

Zhao G, Bryan BA, King D, Luo Z, Wang E, Yu Q. 2015. Sustainable limits to crop residue harvest for bioenergy: Maintaining soil carbon in Australia's agricultural lands, *GCB Bioenergy*, vol. 7, no. 3, pp. 479–487.

Zinn YL, Lal R, Resck DVS. 2011. Eucalypt plantation effects on organic carbon and aggregation of three different-textured soils in Brazil, *Soil Res,* vol. 49, no. 7, p. 614.

12 Integrated Approaches to Pest Management in Litchi Orchards

Innovations and Strategies for Sustainable Production

Deepika Sorahia[1], Ansh Raj[2], Bimal Kumar Sahoo[3], Ankit Alok[4], and Ipsita Samal[2]*

[1]Department of Entomology, School of Crop Protection, College of Post-Graduate Studies in Agricultural Sciences, Central Agricultural University (Imphal), Umiam, Meghalaya 793103

[2]ICAR–National Research Centre on Litchi, Muzaffarpur, Bihar, India

[3]Department of Agricultural Entomology, Centre for Plant Protection Studies, Tamil Nadu Agricultural University, Coimbatore 641003

[4]Department of Plant Pathology, School of Crop Protection, College of Post-Graduate Studies in Agricultural Sciences, Central Agricultural University (Imphal), Umiam, Meghalaya 793103

[*]Corresponding author

12.1 INTRODUCTION

Insect pests have several significant negative effects on food production and availability, such as in conventional orchard pest management practices, including the use of harmful agrochemicals, which have monetary, health, and ecological costs (Verma *et al.*, 2022). Some other approaches, like biological ones or items derived from biologically active compounds and pheromones utilized in mating disorientation, provide challenges when it comes to invasive insect pests and their application. Knowing the species and its life cycle is crucial for effective insect pest management. Insect distribution and ecology changes, possible damage levels, and economic losses can be detected with the help of pest management methods, which fruit growers should implement. As an additional goal, litchi farmers can promote sustainable farming practices by implementing these integrated pest management strategies, which aim to effectively control pests. Litchi growers can reduce their environmental impact while increasing crop yields by integrating cultural, biological, and chemical approaches. The most recent innovations in pest management for litchi orchards have focused on integrated pest management, a strategy that combines chemical, cultural, and biological approaches to eradicate pests in an eco-friendly manner.

12.2 MAJOR INSECT PESTS OF LITCHI

Insects are the primary factors that restrict litchi production (Priya *et al.*, 2025). Approximately 42 species of insects and mites have been identified as pests that attack litchi trees and fruits at various stages of growth and fruiting. Borers, bugs, leaf rollers, loopers, weevils, and mites are the significant groups of pests that impact litchi production (Sahni *et al.*, 2024) (Table 12.1).

12.2.1 Litchi Fruit Borer *Conopomorpha Sinensis* Bradley Gracillariidae: Lepidoptera

Earlier, the litchi fruit borer was known as the litchi stem-end borer in China and litchi fruit borer in Thailand and India (Srivastava *et al.*, 2016a). It appears to infest litchi the most frequently and has a significant economic effect. The biology comprises of egg, larva, pupa, and adult. *C. sinensis* females lay yellow flat eggs measuring approximately 0.4 × 0.2 mm long on the undersides of new leaves, on shoots, or near the calyx of litchi fruits (Lall and Sharma, 1978). Each female can produce between 30 and 49 eggs daily, with an average of 5–6 eggs. These eggs hatch within 4–5 days. Both litchi and longan fruits are susceptible to infestation. The larvae burrow into the fruit, causing them to often fall prematurely. Within 4–5 days, the eggs hatch and larvae start to pierce the fruit, leaf, or stem. First-instar larvae are slender and milky white, with a distinct light-brown head. Upon emerging, these larvae burrow into the fruits and consume the pulp. Larvae are yellowish and measure 8.0 ± 0.1 mm in length and 0.8 ± 0.01 mm in width when fully mature. There are five larval stages. Mature larvae are 6–10 mm long, appear whitish, and have a brownish head when completely formed. The larval period lasts 8–9 days. Pupae appear greenish on the dorsal side and greenish yellow on the ventral side. Larvae emerge from the fruit and pupate in an oval-shaped cocoon on the litchi surface. The pupal

DOI: 10.1201/9781003516538-12

TABLE 12.1
Major Insect Pests of Litchi

Common Name	Scientific Name	Order	Family	Damaging Stage
Litchi fruit and shoot borer	*Conopomorpha sinensis* Snellen	Lepidoptera	Gracillariidae	Larva
Litchi fruit and shoot borer	*C. litchiella* Bradley	Lepidoptera	Gracillariidae	Larva
Litchi nut borer	*Blastobasis* sp.	Lepidoptera	Blastobasidae	Larva
Bark-eating caterpillar	*Indarbela tetraonis* Moore	Lepidoptera	Metarbelidae	Larva
Leaf roller	*Pletypepius aprobola* Meyer	Lepidoptera	Tortricidae	Larva
Ash weevil	*Myllocerus undecimpustulatusn*	Coleoptera	Curculionidae	Grub and adult
Red weevil	*Apoderus blandus*	Coleoptera	Curculionidae	Grub and adult
Litchi looper	*Perixera illepidaria*	Lepidoptera	Geometridae	Larva
Litchi stink bug	*Tessaratoma javanica*	Hemiptera	Pentatomidae	Nymph and adult
Leaf folder	*Dudua aprobola* Meyrick	Lepidoptera	Tortricidae	Larva
Litchi mite	*Acerya litchi* Keifer	Acari	Eriophyidae	Nymph and adult

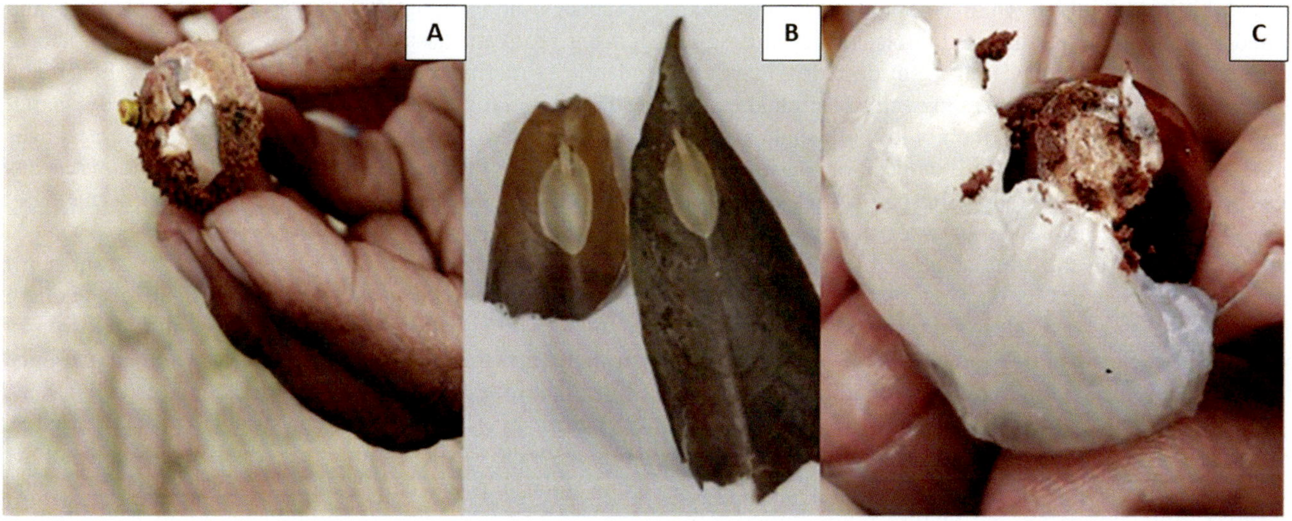

FIGURE 12.1 a. Litchi fruits infested with fruit borer, *Conopomorpha sinensis* b. *C. sinensis* pupa on litchi leaves c. Damage of *C. sinensis* in litchi after harvest.

phase continues for 6–7 days (Singh, 1992). *Conopomorpha sinensis* adults are smaller than other microlepidoptera, with a wingspan of 12 to 15 mm and a grayish-brown coloration. The body measures 6 to 7 mm long and exhibits zigzag patterns. The wing tips are golden brown, the hind wings are silver-gray, and the silvery antennae exceed the length of the wings. During fruiting season, the adult pest typically completes three overlapping generations (Srivastava *et al.*, 2019). The lifespan of adults varies from 3.12 to 6.84 days (Sharma and Agrawal, 1988).

12.2.1.1 Seasonal Incidence and Damage

In July, larvae cause indirect damage by creating mines in young leaves and shoots. During the winter months, leaf buds are the preferred sites for egg-laying. The larvae damage newly emerged shoots during September to October, preventing them from flowering. The highest pest population on shoots occurs in August. During April to May, larvae penetrate the peduncles of developing and maturing fruits, causing premature fruit drop and the presence of excreta when cut or opened after ripening. The extent of damage has been positively correlated with the amount of precipitation (Singh, 1992). Similarly, Hameed *et al.* (1992) observed maximum fruit damage during May to July and the highest pest population in August. They reported that the pest population was absent from orchards between October and March but reappeared in April. Young larvae create mines within the leaf blade and bore into the midrib of young leaves, eventually tunneling through them, causing branches to wither and fall (Figure 12.1). Infested fruits fail to reach their normal size and can be identified by the presence of a black spot near the stem. Larvae do not penetrate deeply into the pulp but feed beneath the calyx, approximately 15 mm deep. Fully developed larvae begin constructing cocoons, typically on old litchi leaves.

This pest is estimated to cause fruit loss and leaf infestation rates of 25–60% and 9–70%, respectively (Srivastava *et al.*, 2017). In Punjab and Himachal Pradesh, damage rates of 13.6–64.9% have been reported by Srivastav *et al.* (2017), with approximately 70% of fruits damaged in Bihar.

12.2.1.2 Erinose mite *Aceria litchi* (Keiffer) Acari: Eriophyidae

Female mites lay solitary eggs near the base of the erineum hairs on the leaf surface as they feed. The eggs are 0.032 mm in diameter, round, and translucent white. They hatch within 3–4 days, becoming a slow-moving protonymph that feeds on soft leaves. Immature stages undergo four developmental stages: two larval stages and two nymphal stages. The first nymphal stage, known as the protonymphal stage, lasts for approximately 2–3 days. The second nymphal stage, called the deutonymphal stage, typically lasts around 6 days. Immatures go through two larval and two nymphal instars. The two nymphal phases are protonymphal, which lasts 2–3 days, and deutonymphal, which lasts around 6 days. Nymphs are, on average, 80 µm long and molt to adults within five to seven days (Alam and Wadud, 1963). Adults are tiny, measuring approximately 0.13 mim long and exhibiting a pinkish-white coloration. Despite their limited number of legs, that is, four legs, they are agile and capable of readily relocating from decaying leaves to colonize fresh growth. Their feeding activity stimulates the formation of galls, providing them with both shelter and a source of nourishment. The pre-oviposition period averages for 1.5 days, and the adult lifespan ranges from 2 to 3 days. Observable differences between males and females are only apparent in the adult stage of these mites. Adult mites inhabit the soft, hairy growths, or erineum, that develop on leaves as a result of their feeding activity. The development cycle from egg to adult takes approximately 14 days, depending on environmental conditions (Jeppson *et al.*, 1975).

In India and China, 13–15 overlapping generations are produced annually (Prasad and Singh, 1981; Zhang, Z., 1997). Both immature and mature mites inflict damage upon leaves, flowers, and developing fruits. The adult population begins to increase rapidly in late March, reaching its maximum activity level approximately in July. The life cycle is completed in 8 to 20 days, with 10 to 12 generations occurring annually (Prasad and Singh, 1981). The mite population thrives and expands during periods of moderate heat and dryness, coinciding with the emergence of new plant growth.

12.2.1.2.1 Damage

Mites infest young leaves, stimulating the development of a fuzzy, mat-like growth (erineum) on the leaf undersides (Figure 12.2). These initially appear as small blisters but can eventually cover the entire leaf, leading to curling of leaves and chocolate-brown growth on the ventral surface of leaves at later stage (Singh and Raghuraman, 2011). In severe cases, entire terminals may become deformed.

Both nymphs and adults cause damage to leaves, inflorescences, and young developing fruits by piercing and tearing the plant tissues with their robust rostrum, subsequently extracting the cell sap (Lal and Rahman, 1975). The erineum initially has a silver-white color, gradually changing to light brown and then dark reddish brown as it ages (Nishida and Holdaway, 1945). The mite population reaches its highest density during the transitional (intermediate) stages of its life cycle.

If infestations become very severe, a significant number of leaves may fall. Mature trees are relatively resilient to heavy mite infestations, exhibiting minimal impact on their growth. In contrast, young trees may experience stunted development due to significant mite populations. Furthermore, a significant issue arises when mites move from leaves to the developing reproductive structures of the plant, such as flowers and fruit. Even after fruit has set and reached half its final size, mites can colonize it, producing erineum on the skin that detracts from its appearance

FIGURE 12.2 Effects of litchi erinose mite (*Aceria litchii*) infestation on litchi leaves: infested leaves displaying characteristic curling and distortion due to mite feeding; velvety growth on the underside of the leaves, resulting from the hyperplasia of leaf trichomes, indicative of severe mite activity.

and may render it unmarketable as a high-quality product (Waite and Hwang, 2002).

12.2.1.3 Bark-Eating Caterpillar, *Indarbela quadrinotata* (Walker) (Metarbelidae: Lepidoptera)

Eggs are laid in groups of 20 to 40 under loose bark, on twigs, in the crevices of older branches, or in areas where twigs have been broken or inadequately pruned. The eggs hatch approximately 8 to 10 days after being laid, and the newly emerged caterpillars begin feeding on the bark. Newly hatched larvae measure between 1.40 and 1.59 mm in length and 0.50 to 0.62 mm in breadth. They have light-brown bodies with dark-brown heads. Caterpillars consume the bark and excavate tunnels within the tree, creating elongated structures, where they seek refuge.

Fully grown larvae range from 48 to 57 mm in length and 4 to 5.5 mm in breadth and are pale brown to brown in color, have smooth skin, and possess dark chitinized patches on various abdominal segments. The total larval development period for *Indarbela* varies from 287 to 302 days under average temperature and relative humidity conditions of 26.72 ± 15.18°C and 58.50 ± 38.50%, respectively. Pupa formation began in April. The pupa possessed a series of spines or hooks on its abdominal sections, which it utilized to exit the tunnel it had created.

They are initially light yellowish-cream-colored pupae which later turn to light reddish within 20 to 24 hours. The pupae are elongated, sturdy, and cylindrical. Their length varies from 25 to 35 mm, and their breadth ranges from 4.10 to 5.20 mm. The pupal stage lasts from 30 to 37 days. Adult moths are most active during the beginning of the rainy season, usually in May or June. Male moths are whitish gray, compact, and robust, with brownish markings on their pale whitish forewings. Female moths are pale brown, with a row of dark reddish-brown spots on their forewings and a tuft of hairs at the tip of their abdomen, a feature that is not present in males. The lifespan of male moths varies from 7 to 10 days, while female moths live for 3 to 7 days (Patel and Patel, 2008).

12.2.1.3.1 Damage

Multiple caterpillars can simultaneously attack different sections of the same tree, causing severe damage to the bark and leading to the death of small branches (Figure 12.3). The holes left on the trunk can serve as entry points for other insects or plant pathogens (Kumar *et al.*, 2019). Caterpillars reside within these excavated tunnels during the day and venture out at night to consume the bark. Trees under attack exhibit the presence of winding tunnels and silken galleries filled with frass and fecal matter (Atwal and Dhaliwal, 2010). Their feeding activities on the trunk or main branches interfere with the flow of sap, and in instances of severe infestations, tree growth is inhibited. The tree's ability to produce fruit is adversely affected. Severe injuries weaken the stem, resulting in the drying of branches and, ultimately, the entire tree (Srivastava *et al.*, 2020). Additionally, trees that have been heavily infested may fracture at the points of attack.

12.2.1.4 Litchi Leaf Roller, *Platypeplusaprobola* (Meyrick) (Lepidoptera: Tortricidae)

The leaf roller is a widely distributed pest commonly found on numerous fruit crops across regions such as Hawaii, South Africa, Australia, and China. In Bihar and neighboring regions, it is recognized as a significant pest of litchi. Leaf roller infestations can be seen on litchi trees year-round, with a higher population noted between July and February (Singh, 1971a). Female moths deposit their eggs beneath the surface of newly emerged, delicate leaves, which hatch within a period of 2 to 8 days. The fully grown larvae construct a protective structure known as a larval

FIGURE 12.3 Damage caused by the bark-eating caterpillar (*Indarbela* spp.) on litchi trees: (a) Infested bark showing signs of feeding and tunneling, which can lead to tree stress and decline. (b) Close-up of caterpillar larvae on the bark, illustrating their destructive feeding habits.

clip, utilizing a small portion of the leaf margin both at the front and the back. They conceal themselves by folding and sealing the clipped leaf fragment. The mating and egg-laying activities of leaf folders take place during the night. The insect completes its life cycle in 21.6–31 days during July to August, and 46–46.5 days in February to March (Kulkarni *et al.*, 1967; Maiti and Sahoo, 1991).

12.2.1.4.1 Damage

Leaf roller infestations are reported from July to February (Singh, 1971a). The larval population reaches its highest point during the months of December to February, which precedes the flowering period of litchi trees (Figure 12.4). The breeding activity of the leaf roller on litchi leaves takes place from August to February, coinciding with the availability of new leaf growth. Breeding is limited during the off-season (March to July). Larval damage to leaves manifests as rolled tender leaves and internal feeding. As a result of larval injuries, infested twigs become distorted and wither (Srivastava *et al.*, 2016b). Litchi trees whose leaves are damaged by larvae experience minimal or, in the case of younger trees, no flowering during the flowering season. As a result, the yield of litchi fruit is substantially decreased.

According to Lall and Mallik (1976), leaf damage caused by the leaf roller ranges from 16.70 to 71.60%, while tree infestations vary from 12.88 to 53.54% during August to February.

12.2.1.5 Litchi Stink Bug, *Tessaratoma javanica* (Thunberg) (Hemiptera: Tessaratomidae)

Stink bugs are regular pests in all the litchi-growing areas in India, causing huge economic losses in quality and quantity of the litchi (Choudhary *et al.*, 2013). *T. javanica* is a significant pest species of stink bug in India. Adult *T. javanica* bugs lay round pale-pink eggs, but this process takes longer during their hibernation period (August to February), lasting approximately 180 days. Eggs are typically laid in clusters of 15 eggs arranged in 3–4 rows, 13 days after mating. While eggs are most commonly found on the undersides of leaves, they can also be located on flowers, on fruits, and on top of previously laid eggs. The average number of egg clusters laid by a female is 13.4. Immature stages are social. The bug undergoes five nymphal stages, which are sub-rectangular and dark brick red, except for the first instar, which is nearly sub-rectangular. Newly emerged nymphs are dirty white and have soft bodies, but their color changes to yellow-red within a few days (Choudhary *et al.*, 2013). The lifespan of each stage is 13 days for eggs, 60 days for nymphs, 60 days for male adults, and 75 days for female adults. The average duration of male and female adults is around 43–45 and 47–48 days, respectively (Parveen *et al.*, 2015).

12.2.1.5.1 Damage

The population of litchi stink bugs begins to increase with the emergence of inflorescences and reaches its peak during the full bloom stage of flowers, coinciding with pollinator activity (Choudhary *et al.*, 2013). Both adults and nymphs inflict direct damage by extracting sap from flowers and fruits, resulting in wilting, death, and ultimately, the shedding of fruit (Figure 12.5 and 12.6). When disturbed, stink bugs release an acidic substance that causes young leaves to wilt and brown spots to appear on older leaves and fruits, diminishing their commercial value. Severe infestations can lead to significant fruit loss and, in extreme cases, the complete destruction of the litchi crop. Zhang (1997) conducted an in-depth study on the physiology of stink bug stings and discovered that the damage they caused led to a higher incidence of immature fruit dropping from the tree compared to mature fruit. When bugs congregate on developing fruits, they cause the fruits to fall a couple of weeks later. Losses caused by these bugs can reach 70–90%. Liu and Lai (1998) reported that litchi stink bugs cause damage to 20–30% of the fruit, and chemical control methods have proven to be ineffective in achieving satisfactory outcomes.

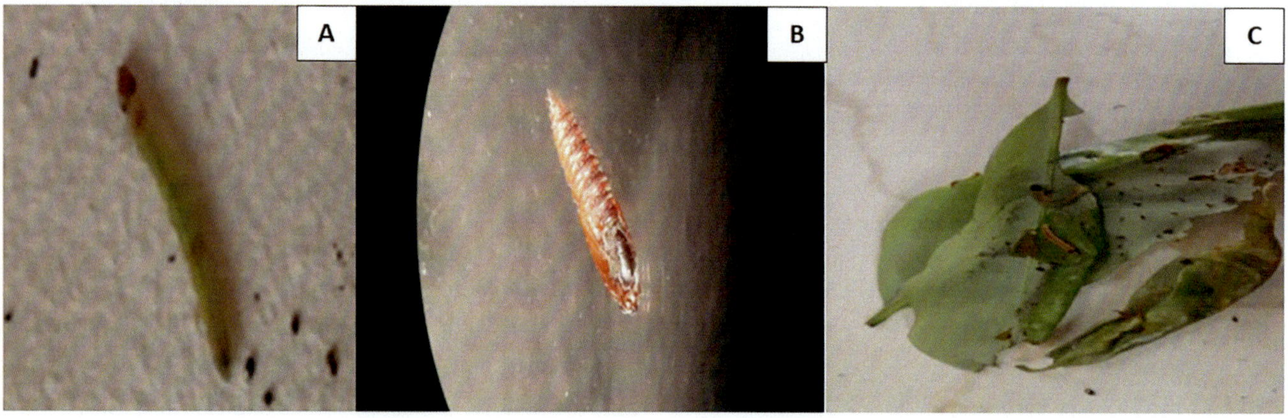

FIGURE 12.4 Damage caused by leaf rollers (*Statherotis* spp.) on litchi plants: (a) Pupa of the leaf roller, indicating the developmental stage within the rolled leaves. (b) Damaging symptoms on litchi leaves, characterized by rolling and distortion due to larval feeding. (c) Larvae of the leaf roller actively feeding on litchi leaves, contributing to significant foliage damage.

FIGURE 12.5 Litchi stink bug (*Tessaratoma javanica*) and associated damage: (a) Eggs of the litchi stink bug laid on the underside of leaves, indicating potential infestation. (b) Nymphs of the litchi stink bug feeding on tender plant parts, causing wilting and reduced vigor in the plant. (c) Adult litchi stink bugs, showcasing their characteristic appearance and potential for significant crop damage.

FIGURE 12.6 Biology of the litchi stink bug, *Tessaratoma javanica*. (a) Eggs laid in clusters on the underside of litchi leaves by female stink bugs. (b) Nymphs of the litchi stink bug feeding on tender plant parts, causing wilting and reduced vigor in the plant. (c) Life cycle of the litchi stink bug, showing the egg, nymphal, and adult stages. The bug completes one generation per year, with nymphs taking about 80 days to develop and adults living 203–371 days on average.

12.2.1.6 Gray weevil, *Myllocerus undecimpustulatus undatus* Marshall (Curculionidae: Coleoptera)

The female weevil can lay up to 360 eggs on organic material on the soil surface within 24 hours. These eggs are oval-shaped, measuring approximately 0.5 mm in length, and have a creamy white color. They are usually laid in groups. As hatching approaches, the creamy color of the eggs changes to a brownish hue (Atwal, 1976). The larva, characterized by its white body and brownish-red head, varies in size from 1 to 4 mm as it progresses from the first to the fourth instar. These larvae consume plant roots beneath the soil and eventually pupate within the soil (Butani, 1979). Pupation takes place in soil, lasting 5–7 days. Adult weevils of this species typically range in size from 6 to 8.5 mm, with females being approximately 1–2 mm larger than males. The weevils have dark, dotted to grayish elytra and a yellow-colored head. The female weevil's weight is about twice that of the male (George *et al.*, 2015). These weevils feed by consuming the leaf margins and creating small patches on the leaf lamina. The lifespan of the adults varies with the season, lasting 8 to 11 days in summer and 4 to 8 months in winter (Butani, 1979).

12.2.1.6.1 Damage

Adult weevils congregate on young leaves, where they create irregular holes by feeding. In some instances, they may consume the entire leaf, leaving only the midrib intact (Kumar *et al.*, 2019). Severe infestations by adults can lead to leaf notching, as these insects primarily feed on the edges or corners of leaves. In cases of heavy pest attack, complete defoliation may occur. Observations have shown that adults typically target the young leaves of plants, leading to plant decline and potentially stunting growth (Kiyanthy and Mikunthan, 2009) (Figure 12.7).

12.3 OVERVIEW OF PEST INTERACTIONS AND ECOSYSTEM DYNAMICS IN LITCHI

Litchi (*Litchi chinensis*) is susceptible to various pest interactions that significantly influence its ecosystem dynamics. Understanding these interactions is crucial for effective pest management and maintaining the health of litchi orchards. Litchi is affected by several insect pests, including litchi fruit and shoot borer (*Conopomorpha sinensis*), which causes significant damage during fruit development, necessitating timely insecticide applications. Litchi mite (*Aceria litchi*) attacks young leaves, leading to reduced vigor and potential yield loss. Mealybug (*Drosicha mangiferae*) sucks sap and secretes honeydew, promoting sooty mold growth, which can hinder photosynthesis. Natural enemies in litchi ecosystems, including parasitoid, predators, and pathogens, are provided in Table 12.2 and Figure 12.8, which follow.

FIGURE 12.7 Damage symptoms on litchi leaves caused by feeding activity, including irregular holes and leaf deformation, indicating the impact of this pest on litchi.

TABLE 12.2
Natural Enemies Associated with Litchi

Category	Organism	Target Pest	Reference
Parasitoid	*Anastatus dexingensis*	Litchi stink bug (*Tessaratoma papillosa*)	Wu *et al.* (2020)
Parasitoid	*Anastatus fulloi*	Litchi stink bug (*Tessaratoma papillosa*)	Wu *et al.* (2020)
Predator	*Chrysoperla zastrowi sillemi*	Various pests, including aphids and mites	NIPHM, Directorate of Plant Protection, Quarantine & Storage
Predator	*Amblyseius fallacis*	Mites	NIPHM, Directorate of Plant Protection, Quarantine & Storage
Predator	*Stethorus punctum*	Mites and other small pests	NIPHM, Directorate of Plant Protection, Quarantine & Storage
Predator	Lady beetles (Coccinellidae)	Various soft-bodied insect pests	Li et al. (2014)
Pathogen	Fungal pathogens (e.g., *Beauveria bassiana*)	Various insect pests	Li *et al.* (2014)

FIGURE 12.8 Interaction of beneficial insects in a litchi orchard.: The seven-spotted lady beetle (*Coccinella septempunctata*), a natural predator of aphids, actively foraging on litchi leaves, contributing to biological pest control. Red ant (*Oecophylla smaragdina*) observed in the litchi orchard, known for its role in protecting plants from various pests by preying on them and deterring herbivores.

12.3.1 Ecosystem Dynamics

12.3.1.1 Interactions with Companion Plants

The findings of another demonstrated that the joint planting of *Desmodium intortum* and litchi trees can enhance pest management. Its companion plant provides refuge and sustenance to the indigenous predators of pests that infest litchi, therefore resulting in a reduction in pest populations (Yuan *et al.*, 2023). Research findings indicate that the density of pests in a litchi–Desmodium complex is notably reduced compared to that of a litchi crop grown alone, particularly during the earlier stages of growth of both plants. Indeed, the predator/prey ratio is a vital factor that indicates the effectiveness of pest management strategies. In mixed environments, the ratio is seen to be notably greater throughout the flowering and fruit development phase of litchi. This fact demonstrates that companion plants contribute to the enhancement of the ecosystem, thereby increasing the likelihood of survival for other beneficial insects and decreasing the likelihood of pest infestation.

The IPM approach pursued within the AESA (agroecosystem analysis) also emphasizes the requirement of taking into account the interplay among the various components in the ecosystem (Mukta, 2020). These objectives are accomplished by evaluating the pest condition, employing natural predators, and using chemicals appropriately to minimize environmental damage while effectively managing the pest. Hence, the interconnections between litchi pests and their indigenous predators, as well as the implementation of intercropping with companion plants, are regarded as crucial elements of the natural ecology in litchi cultivation. These interactions can be enhanced by implementing integrated pest management techniques, therefore enhancing crop health and promoting sustainable farming.

12.4 INTEGRATED PEST MANAGEMENT (IPM) STRATEGIES

Different pest management strategies in litchi are highlighted in the following table (Table 12.3).

12.4.1 Cultural Practices

12.4.1.1 Pruning and Light Management

Litchi shoots and leaves affected by fruit and shoot borers, litchi mites, leaf rollers, litchi loopers, and weevils should be eliminated twice annually, in June and August. Litchi orchards should be tilled during the summer to destroy eggs and overwintering stages of soil insects by exposing them to sunlight, ants, and birds. Larvae of bark-eating caterpillars can be effectively destroyed by pouring kerosene oil into their holes and plugging them with clay or by using a mixture of ethylene glycol and kerosene (3:1) (Singh *et al.*, 2012). Using light traps and pheromone traps can reduce the population of moths, particularly fruit borers.

Light pruning and opening the center of litchi trees allow for adequate sunlight exposure, resulting in a reduced pest incidence. Encircling the tree trunk with a 300 mm wide alkathene sheet (400 gauge) prevents mealybug nymphs from ascending the tree.

12.4.1.2 Use of Seasonal Flowering Plants to Attract Natural Enemies

Plant flowering plants or compatible commercial crops along the orchard boundary, positioning shorter plants nearer to the main crop and taller plants toward the edge to attract natural enemies and discourage pest migration. Grow flowering plants on the internal embankments within the orchard. Select and grow appropriate companion plants that

TABLE 12.3
Different Pest Management Strategies in Litchi Orchard

Sl No.	Advancement	Description	Reference
1	Integrated pest management (IPM)	The implementation of IPM strategies combining biological control, cultural practices, and judicious pesticide use to manage litchi pests effectively	Regmi, 2022
2	AESA-based IPM	Adoption of agroecosystem analysis and surveillance approach (AESA) to monitor pest populations and enhance beneficial insect activity	NIPHM, Directorate of Plant Protection, Quarantine & Storage.
3	Biological control	The use of natural enemies, like parasitoids and predators, to reduce pest populations, especially the litchi fruit and shoot borer	Li *et al.*, 2014
4	Use of pheromone traps	Deployment of pheromone traps to monitor and control pest populations, particularly effective for litchi fruit borers	NIPHM, Directorate of Plant Protection, Quarantine & Storage.
5	Ecological engineering	Enhancing habitats for natural enemies through companion planting and habitat management to improve pest control	NIPHM, Directorate of Plant Protection, Quarantine & Storage.
6	Chemical control innovations	Development of new insecticides with lower toxicity to non-target organisms and improved efficacy against litchi pests	Upadhyay *et al.*, 2020
7	Cultural practices	Implementation of cultural practices, such as proper pruning, sanitation, and timing of planting, to reduce pest incidence	Singh *et al.*, 2012

can serve as trap crops and pest repellents. These plants can also attract natural enemies, as their flowers provide nectar and the plants offer a suitable environment. By increasing biodiversity through flowering plants, the abundance of parasitoids and predators (natural enemies) will also grow due to the availability of nectar, pollen, and insects. The flowering plants suggested for pest management strategies are known as attractant plants for the natural enemies of pests: French bean, marigold, carrot, sunflower, mustard, caraway, alfalfa, dill, cowpea, *Desmodium* species, chrysanthemum, and buckwheat.

12.4.1.3 Soil Management and Organic Amendments

Applying 4 kg of castor cake and 1 kg of neem cake to the root zone after the first monsoon shower has proven effective in reducing fruit borer, mites, leaf roller, and fruit borer infestations (Lal and Rahman, 1975). Alternate hosts, such as *kath*, *jamun*, and *chhotaamaltas*, which harbor the pest during the off-season, should be collected and destroyed to minimize the risk of severe leaf roller infestations in the following year. Dong *et al.* (2006) reported that spraying with azadirachtin against fruit borers reduced infestations by 89.4% and 87.5%, respectively. Sahoo *et al.* (2007) also found that azadirachtin was significantly superior to the untreated control.

12.4.2 Biological Control

12.4.2.1 Introduction of Natural Predators and Parasitoids

Releasing 10–15 grubs of the coccinellid predator *C. montrozieri* per tree is effective in controlling litchi mealybugs. To minimize the population of fruit borers, use *Trichogramma* at a rate of 20,000 eggs per acre. Natural enemies play a crucial role in reducing litchi pest populations. Hameed *et al.* (1992) identified eight species of predatory mites, including *Amblyseius caecineae* Gupta, *A. finlandicus* Oudemans, *A. largoensis* Muma, *A. paraacriallis* Muma, *A. purnigupta*, *Phytoseius intermedius* Evans and Macfarhane, *Typhlodromus fleschneri* Chant, and *T. homalii* Gupta, associated with litchi mites (Revynthi *et al.*, 2021). Populations of these predatory mites remain high during the summer months (April to September), peaking in June. With the onset of winter, the populations of predatory mites begin to decline. Three species of hymenopterous parasitoids belonging to the genera *Apanteles*, *Bracon*, and *Mesochorus* have been reported to parasitize the larvae of *Conopomorpha cramerella* during July to September (Hameed *et al.*, 1992). Hwang and Hsich (1989) recorded two species of parasitoids, *Phanerotoma* and *Apanteles*, from the pupae of *Acrocercops cramerella*. Two other parasitoids, *Tetrastichus* and *Elasmus*, were also found parasitizing *Acrocercops cramerella* larvae feeding on shoots or leaves.

12.4.2.2 Case Studies on Successful Biological Control Implementations

The classical biological control of litchi stink bugs using *Anastatus japonicus* (Hymenoptera: Eupelmidae) and *O. phongi* (Hymenoptera: Encyrtidae) has led to effective pest management in China, Hong Kong, and Thailand (Han *et al.*, 1999; Leksawasdi and Kumchu, 1991). Choudhary *et al.* (2013) also observed *Anastatus bangalorensis* Mani and Kurian (Hymenoptera: Eupelmidae) and *Ooencyrtus* sp. (Hymenoptera: Encyrtidae) in Ranchi. A predatory bug, *Eocanthecona furcellata*, was identified as a promising biocontrol agent in the litchi ecosystem at ICAR-NRCL Muzaffarpur. Two coccinellids, the seven-spotted beetle (*Coccinellaseptempunctata*) and *Cheilomenessexmaculata*, as well as praying mantises, have also been observed feeding on litchi insect pests in Muzaffarpur. Clearly, these findings will contribute to the development of effective integrated pest management strategies.

12.4.3 Chemical Control

12.4.3.1 Insect Growth Regulators (IGRs) and Other Pesticides

Insect growth regulators (IGRs) and avermectin formulations can be used to manage litchi fruit borers, providing an alternative to the adverse effects of chemical control measures. In this context, diflubenzuron 25 WP at 2 g/L, novaluron 10 EC at 1.5 mL/L, flubendiamide 39.35 SC at 1.5 mL/5 L, emamectin benzoate 5% SG at 2 mL/L, and spinosad 45 SC at 1.6 mL/5 L have proven highly effective against litchi fruit borers. Applying three sprays of any one of these molecules at 10- to 12-day intervals after the fruits reach pea size has successfully kept infestations below the threshold level. Under favorable weather conditions, such as rainy and humid conditions, the first and second sprays may be conducted with thiocloprid 21.7 SC at 0.7–1.0 mL/L for enhanced results.

12.4.3.2 Impact of Chemical Controls on Non-Target Species

While chemicals used for insect pest control have a strong effect, their drawbacks include the resurgence of pest populations after the decimation of natural enemies, the development of insecticide-resistant populations, and negative impacts on non-target organisms within and outside the crop system. Chemical pesticides have long-term detrimental effects, leading to environmental degradation and the elimination of natural parasitoids and predators. Several hundred insect and mite species have developed resistance to one or more chemical insecticides.

12.4.4 Monitoring and Surveillance Techniques

12.4.4.1 Scouting and Field Observations

12.4.4.1.1 Importance of Regular Monitoring for Pest Populations

Growers can conduct regular field scouting in their orchards to monitor major pest activity. Pest surveillance should begin immediately after tree establishment and continue on a weekly basis. For each tree, randomly select five branches to record insect presence according to standardized procedures specific to each insect species. Since fruit crops are perennial, it is essential that inspectors or scouts conducting surveillance be familiar with the crop's characteristics, growth stages, and development. Understanding these aspects is crucial for accurately identifying significant pests and diseases, as these organisms typically target specific stages or parts of the plant.

12.4.4.2 Techniques for Assessing Pest Damage and Thresholds

A range of sample methods are used in both agricultural crops and fruit plants, encompassing random and distributed approaches. The selection of the sample technique may differ based on the characteristics and goals of the research, such as the estimation of disease occurrence or levels of severity. Conventional sampling patterns in fruit crops include the zig-zag pattern, which is used to sample orchards without any apparent symptoms in order to evaluate the occurrence of diseases, such as viral and wilt diseases. An elliptical pattern is implemented within the drip line of trees and shrubs, specifically for the purpose of monitoring powdery mildew, downy mildew, and leaf spot diseases. A star pattern is employed for sampling in regions that have already shown signs of damage. Sampling frequency is determined by variables like the pest's generation interval, the annual number of generations, the possibility of population increase between generations, and the particular stage of crop–pathogen interaction.

To quantify insect pests such as mites, mealybugs, leaf rollers, shoot borers, leaf miners, and whiteflies, we tallied and documented the population of both nymphs and adults on five randomly chosen apical twigs from the upper, middle, and lower sections of each plant. For fruit borers, fruit-sucking moths, nut borers, and fruit flies, we sampled during the fruiting stage by enumerating the total number of fruits, damaged fruits caused by fruit borers, and the larval population on three randomly selected top, middle, and bottom branches of each plant.

12.4.4.3 Use of Pheromone Traps and Sticky Traps

Install fruit fly traps at a density of 4–5 traps per acre. Place traps for each species at least 75 ft apart, and attach them to a support pole at mid-canopy height. Replace the lures every 2–3 weeks, or as needed, based on their effectiveness. During weekly surveillance, count and record the number of flies caught per trap, and do this consistently throughout the year. After each recording, the trapped flies should be removed and destroyed. For monitoring pests like leaf miners, whiteflies, and mealybugs, set up yellow pan water traps on the ground or yellow sticky traps at mid-canopy level, using one trap per five trees. Locally available empty tins painted yellow and coated with grease, Vaseline, or castor oil on the outer surface can also serve as effective yellow sticky traps.

12.4.4.4 Role of Light Traps in Pest Management

Install light traps at a density of one trap per acre, positioning them at mid-canopy level to monitor and mass-trap moths of various pests, including fruit borers, leaf rollers, trunk borers, bagworms, fruit-sucking moths, bark-eating caterpillars, shoot borers, nut borers, and leaf miners. These light traps should be designed with an exit option for smaller natural enemies and should be operated during the evening hours, from 6:00 p.m. to 10:00 p.m. (Figure 12.9).

12.4.5 Advances in Research and Technology

12.4.5.1 Recent Developments in Pest Management

Insect pests pose a significant threat to agriculture, leading to substantial damage across many economically vital crops

FIGURE 12.9 Installation of light and delta traps in the litchi orchard at ICAR-NRCL, Muzaffarpur, Bihar. Light trap placed to attract nocturnal pests, enhancing monitoring efficiency in the orchard. Delta trap positioned among litchi trees, utilizing pheromone lures to capture target insect pests, providing critical data for pest management strategies.

(Isman, 2019). Research conducted by Mateos Fernández *et al.* (2022), and Kumari *et al.* (2022) estimates that insect pests contribute to a 25–30% reduction in agricultural yields. Traditional methods of pest control, such as chemical insecticides, have largely failed to prevent severe infestations (Dhananjayan *et al.*, 2020). Recent advancements in biological sciences have paved the way for novel pest control strategies. The field of genetic pest management is currently experiencing a period of rapid innovation, driven by the urgent need for new solutions to address the escalating problem of agricultural pests. Key biotechnological interventions, such as breeding for insecticide resistance and the transgenic introduction of new genes, are pivotal in pest management. Techniques like gene editing, RNA interference, and marker-assisted selection represent some of the cutting-edge methods in this area (Gandhi *et al.*, 2022 and Kumari *et al.*, 2022).

12.4.5.1.1 Gene Transformation and Genetic Engineering

To enhance crop resistance against insect pests, specific DNA segments or genes can be introduced into plants through gene transformation and genetic engineering. These inserted DNA fragments typically code for proteins with insecticidal properties (Latham *et al.*, 2017 and Rani *et al.*, 2022). The production of these insecticidal proteins within the plant provides protection against certain insect pests (Gatehouse, 2013). This approach has been effective in managing pests from the Lepidoptera, Coleoptera, and Diptera orders, among others (Nitnavare *et al.*, 2021 and Kamatham *et al.*, 2021). Gene editing, RNA interference (RNAi), and marker-assisted selection (MAS) are cutting-edge technologies that hold significant promise for pest management in litchi. These methods can enhance pest resistance and improve overall crop health. What follows is an overview of each approach, along with relevant references. Genome editing techniques can be used to enhance beneficial traits related to pest resistance, such as the litchi fruit and shoot borer (*Conopomorpha sinensis*). This methodology facilitates the development of litchi cultivars that exhibit reduced susceptibility to insect infestation (Li *et al.*, 2014).

RNA interference (RNAi) is a targeted gene-silencing method that disrupts the growth, development, and reproduction of target pest insects. Precision technology can be developed to selectively target certain genes of pests, such as the litchi stink bug (*Tessaratoma papillosa*), without causing harm to other species in the ecosystem (Christiaens *et al.*, 2020). RNA interference (RNAi) shows promise in managing various insect pests of litchi that are more environmentally benign than traditional chemical pesticides (Mamtaand Rajam, 2017). Marker-assisted selection is the use of molecular markers linked to important characteristics, like pest resistance, to expedite the genetic evolution of litchi cultivars. The screening and selection of plants with targeted resistance genes during early generation provide breeders with a valuable tool to enhance the efficiency of producing new litchi varieties with enhanced pest resistance (Upadhyay *et al.*, 2022).

The integration of gene editing, RNA interference, and marker-assisted selection represents a promising advancement in pest management strategies for litchi cultivation. These technologies not only enhance pest resistance but also contribute to sustainable agricultural practices, reducing reliance on chemical pesticides and promoting the development of resilient crop varieties. Continued research and field trials will be essential to fully realize the potential of these innovative approaches in litchi pest management.

12.4.5.1.2 Genome Editing for Litchi Stink Bug Management

Genome editing technologies have opened new avenues for sustainable management of insect pests, including the emerging threat of the litchi stink bug (*Tessarotoma javanica*). With the availability of advanced genomic tools, it is possible to identify and modify genes in pest populations that regulate essential biological processes such as reproduction, digestion, or resistance to insecticides (Molla *et al.*, 2021). By disrupting or silencing such genes, researchers can reduce pest survival or fertility, thereby curbing population growth (Vats *et al.*, 2019). For instance, gene-editing approaches could target *T. javanica* genes responsible for detoxification enzymes (such as cytochrome P450s or esterases), which often enable pests to tolerate chemical insecticides. Knockout of these genes would render the bugs more vulnerable to conventional or biopesticide treatments, reducing reliance on indiscriminate pesticide sprays in litchi orchards. Furthermore, editing genes linked to reproductive success, such as spermatogenesis- or oviposition-related genes, may suppress the pest population without affecting non-target insects, including pollinators that are crucial for litchi fruit set. Thus, genome editing represents a precision tool that, if applied responsibly, could become part of an integrated strategy for controlling *T. javanica* in litchi-growing regions.

12.4.5.1.3 RNA Interference (RNAi) for Litchi Pest Control

RNA interference (RNAi) is among the most promising next-generation tools for managing sucking insect pests like the litchi stink bug. The method works by silencing specific genes essential for pest survival, development, or reproduction, through the delivery of double-stranded RNA (dsRNA) (Silver *et al.*, 2021). In *T. javanica*, potential RNAi targets include genes associated with feeding (e.g., salivary gland proteins required for successful piercing and sucking of litchi sap), chitin biosynthesis (affecting cuticle formation), or hormonal regulation (affecting molting and reproduction). Silencing these genes would weaken or kill the insects (Munawar *et al.*, 2023), significantly reducing their capacity to cause damage to litchi flowers, young leaves, and fruits. RNAi offers the advantage of being highly species-specific, minimizing harm to non-target organisms such as bees and other beneficial pollinators in litchi orchards. RNAi delivery could be achieved through foliar sprays containing dsRNA, root uptake, or transgenic litchi rootstocks engineered to express dsRNA molecules against vital *T. javanica* genes. Although field application is still limited by delivery challenges and cost, the technology holds strong potential for integration into eco-friendly litchi pest management programs.

12.4.5.1.4 Gene Drive Approach for Litchi Stink Bug Suppression

The litchi stink bug poses a growing threat in India and other litchi-producing regions, with its rapid spread and ability to cause severe fruit drop and yield losses. Gene drive technology presents a futuristic approach to suppress or even eliminate such invasive and economically devastating pests (Books, 2019). In principle, a gene drive could be engineered in *T. javanica* to spread traits that reduce its fitness, such as sterility in females or skewed sex ratios. For example, releasing a small number of gene-drive-carrying individuals could bias inheritance patterns, ensuring that nearly all offspring inherit the engineered trait (Collins, 2018). Over successive generations, this could lead to population collapse or significant suppression of stink bug density in orchards. Moreover, gene drives could be used to overcome pesticide resistance by reintroducing susceptibility alleles into resistant *T. javanica* populations, thereby restoring the efficacy of chemical and biopesticides. However, ethical, ecological, and biosafety concerns remain paramount, especially since gene drives are self-propagating and can potentially spread beyond intended areas (Wedell *et al.*, 2019). Risk assessment and containment strategies would be essential before considering such applications in litchi ecosystems (Oberhofer *et al.*, 2019).

12.4.5.1.5 CRISPR-Cas9 Applications for Litchi Ecosystems

CRISPR-Cas9, the most widely adopted genome-editing tool, holds significant promise for both direct pest suppression and indirect enhancement of litchi plant resistance against the stink bug. From a pest management perspective, CRISPR-Cas9 could be employed to knock out *T. javanica* genes linked to reproduction, immune defense, or feeding behavior, creating strains with reduced survival and competitive ability (Gong, 2017). For example, disruption of chemosensory genes could impair the bug's ability to locate litchi hosts, thereby reducing infestation pressure. On the host plant side, CRISPR-Cas9 could be used to develop litchi varieties with enhanced insect resistance. Editing litchi genes to enhance production of natural defense compounds, structural barriers (like tougher epidermal cells), or volatiles that attract natural enemies of *T. javanica* could contribute to durable pest resistance. CRISPR-based tools could also be coupled with RNAi approaches, creating a two-pronged strategy that targets both the pest and its interaction with the host plant (Jaganathan *et al.*, 2018; Moon *et al.*, 2022). Importantly, these applications fit well within the principles of precision agriculture and sustainable pest management, offering environmentally sound alternatives to pesticide overuse in litchi cultivation.

12.4.5.1.6. Metagenomics Approaches in the Management of Litchi Stink Bug (Tessarotoma javanica)

Metagenomics, through high-throughput sequencing of microbial communities, offers novel avenues for eco-friendly management of *T. javanica*. By unraveling insect–microbe–plant interactions, it enables the identification of microbial partners, symbionts, and pathogens that can be exploited for targeted control strategies. The gut microbiota supports nutrition, detoxification, and pesticide resistance in *T. javanica*. Metagenomic profiling can identify key symbionts, which may be manipulated via microbiome disruption, paratransgenesis, or probiotic interventions to reduce pest survival. Endosymbionts essential for stink bug growth and reproduction can be disrupted using antibiotics, RNAi, or engineered microbial antagonists, reducing fecundity and survival. Metagenomics aids in detecting entomopathogenic microbes and their virulence factors. These can be developed into bioformulations or microbial pesticides against *T. javanica*. Microbiome-mediated detoxification contributes to resistance. Identifying these pathways can inform the design of inhibitors or novel insecticides to restore control efficacy. Exploring the phyllosphere and rhizosphere microbiomes of litchi reveals beneficial microbes that enhance plant defenses or produce metabolites antagonistic to stink bugs. Metagenomics enables ecosystem-based, species-specific, and sustainable strategies for litchi stink bug control. Integrated with RNAi, CRISPR, and precision biocontrol, it can significantly advance next-generation IPM in litchi orchards.

12.5 RESEARCH FINDINGS ON PEST RESISTANCE AND ADAPTATION

The litchi fruit borer is a highly prolific pest that has acquired resistance to several insecticides in recent years. Consequently, it is imperative to implement integrated pest control measures in order to mitigate the development of resistance. The researchers propose several approaches, including employing several categories of pesticides, adhering to economic thresholds, and utilizing biological control agents, like *Trichogramma* egg parasitoids (NIPHM, 2022). Moreover, studies indicate that the litchi stink bug (*T. papillosa*) has developed specific habits that enable it to evade traps and insecticides, therefore complicating its control. The efficacy of azadirachtin, a pesticide, in eradicating litchi stink bugs in northern Thailand has been demonstrated based on studies conducted by Schulte *et al.* (2006) and Regmi (2022). Litchi mites have a high propensity to overcome resistance to widely used acaricides. Therefore, it is advisable to employ alternative chemicals in the treatment of this pest (Butani, 1977). It has been documented that the mango scale, *Drosicha mangiferae*, has acquired a certain degree of resistance to pesticides, therefore evading chemical control measures. The recommended approach for controlling mealybugs is to employ the biological method, alter the cropping pattern, and utilize insecticides appropriately (Butani, 1977).

Molecular methodological approaches in integrated pest management for litchi involve utilizing high-throughput sequencing data to examine the genetic composition of the species present in the litchi production system, encompassing pests, beneficial insects, and microbes. This innovative methodology provides insights into the interactions between pests and natural enemies, as well as the overall health condition of the litchi ecosystem (Wu *et al.*, 2020). Further evidence has confirmed that there are significant changes in the bacterial microbiota based on the specific stage of development of the litchi stink bug. Analysis of microbiota was conducted by 16S rRNA amplicon sequencing, revealing that proteobacteria are the predominant microorganisms in the nymphal stages, whereas Firmicutes are the dominant microorganisms in the egg stage. A comprehensive understanding of these microbial communities can provide insights on the establishment and development of pests, as well as their capacity to adapt to novel environments. This knowledge has the potential to revolutionize pest management practices (Kumari *et al.*, 2021). The population dynamics of the invasive litchi stink bug were assessed in Taiwan, together with its egg parasitoids, and the genetic composition was examined by molecular DNA analysis. The results indicated population density variations similar to those observed in indigenous populations and the presence of many egg parasitoid species possible for biological control. The genetic information contains valuable data for effectively managing pest populations and implementing integrated pest management strategies (Chen *et al.*, 2020). Furthermore, integrating metagenomic data with traditional pest control techniques can enhance the effectiveness of integrated pest management (IPM) strategy for litchi. By leveraging the understanding of the litchi stink bug, its microbial partners, and its natural enemies, it becomes possible to devise more efficient and environmentally friendly control strategies that rely less on chemical pesticides (NIPHM, 2022). In conclusion, prominent litchi pests like fruit borer, stink bug, mite, and mealybug have demonstrated the ability to develop resistance and endure different strategies of treatment. Hence, it is crucial to embrace and implement balanced integrated pest management strategies to prevent resistance, reduce pesticide application, and improve sustainable litchi production.

12.6 CONCLUSION

Evidence suggests that the integration of novel pest control methods has demonstrated significant potential in enhancing litchi production with minimal environmental impact. Certain recommended methods, such as cultural practices, biological management, and safer pesticides, have been identified as highly effective means of pest control. Implementing these integrated pest management (IPM) techniques not only improves litchi yield and produces quality but also enhances the overall health of the trees and the orchard ecosystem.

Furthermore, the chapter underscores the need to do further studies and to adopt novel pest control treatments in response to emerging climate shifts and insect behaviors. By fostering cooperation among producers, researchers, and agricultural extension agencies, the litchi industry will enhance its ability to effectively address and overcome pest pressures.

In conclusion, the implementation of integrated pest management techniques in litchi orchards is a feasible strategy for achieving sustainable agricultural production, so ensuring the profitability of litchi crops for farmers. The ongoing emphasis on research and development, together with the sustainability of litchi production, will be crucial in advancing the fruit's farming in the future, given the increasingly difficult crop-producing environment.

REFERENCES

Alam, M. Z., and Wadud, M. A. (1963). On the biology of the litchi mite, *Acería litchi* Keifer (Eriophyidae, Acariña) in East Pakistan. *Revista Colombiana de Entomología*, 44(1).

Atwal, A. S. (1976). *Agricultural pests of India and Southeast Asia*. Kalyanri Publishers, Delhi, India.

Atwal, A. S., and Dhaliwal, G. S. (2010). *Agricultural pests of south asia and their management*. Kalyani Publishers, Ludhiana, 616 pp.

Books, A. (2019). *New biotechnological approaches to insect pest management and crop protection*. Gene Editing Approach (CRISPR-Cas system).

Butani, D. K. (1977). Pest of litchi in India and their control. *Fruits*, 32(4), 269–273.

Butani, D. K. (1979). *Insects and fruits*. Periodical Expert Book Agency, India.

Chen, Y., Hsu, Y., and Huang, Y. (2020). Population monitoring, egg parasitoids, and genetic structure of the invasive litchi stink bug, *Tessaratoma papillosa* in Taiwan. *Insects*, 11(10), 690. doi:10.3390/insects11100690.

Choudhary, J. S., Prabhakar, S. C., Moanaro, D. B., and Kumar, S. (2013). Litchi stink bug (*Tessaratoma javanica*) outbreak in Jharkhand, India, on litchi. *Phytoparasitica*, 41, 73–77.

Christiaens, O., Whyard, S., Vélez, A. M., and Smagghe, G. (2020). Double-stranded rna technology to control insect pests: Current status and challenges. *Frontiers in Plant Science*, 11, 451. doi:10.3389/fpls.2020.00451.

Collins, J. P. (2018, July). Gene drives in our future: Challenges of and opportunities for using a self-sustaining technology in pest and vector management. In *BMC proceedings* (Vol. 12, No. Suppl 8, p. 9). London: BioMed Central.

Dhananjayan, V., Jayakumar, S., and Ravichandran, B. (2020). Conventional methods of pesticide application in agricultural field and fate of the pesticides in the environment and human health. *Controlled Release of Pesticides for Sustainable Agriculture*, 1–39.

Dong, X. J., Wen, L. G., and Ling, Z. (2006). Suppressive effect of azadirachtin on litchi fruit borer *Conopomorpha sinensis* population. *Chinese Bulletin of Entomology*, 43(3), 327–330.

Gandhi Gracy, R., Mani, M., Swathi, R. S., Venkatesan, T., and Mohan, M. (2022). Biotechnological Applications in Horticultural Entomology. *Trends in Horticultural Entomology*, 185–209.

Gatehouse, J. A. (2013). *Genetic engineering of crops for insect resistance*. Springer.

George, J., Morse, W. C., and Lapointe, S. L. (2015). Morphology and sexual dimorphism of the weevil *Myllocerusundecimpustulatus* undatus (Coleoptera: Curculionidae). *Annals of the Entomological Society of America*, 108(3), 325–332.

Gong, P. (2017). Invasive Species Management on Military Lands: Clustered Regularly Interspaced Short Palindromic Repeat/CRISPR Associated Protein 9 (CRISPR/Cas9) Based Gene Drives (No. ERDCELSR172).

Hameed, S. F., Sharma, D. D., and Agarwal, M. L. (1992, May). Integrated pest management in litchi. In *National seminar on recent developments in litchi production held at Rajindra Agricultural University, Bihar, Pusa, May* (pp. 30–31, 38).

Han, S., Liu, W., and Chen, Q. (1999). Mass releasing *Anastatusjaponicasto* control *Tessaratomapapillosa* in Hong Kong. *Chinese Journal of Biological Control*, 15(2), 54–56.

Hwang, J. S., and Hsich, F. K. (1989). The bionomics of the cocoa pod borer in Taiwan. *Plant Protection Bulletin, Taipei*, 31(4), 387–395.

Isman, M. B. (2019). Challenges of pest management in the twenty first century: New tools and strategies to combat old and new foes alike. *Frontiers in Agronomy*, 1, 2.

Jaganathan, D., Ramasamy, K., Sellamuthu, G., Jayabalan, S., and Venkataraman, G. (2018). CRISPR for crop improvement: An update review. *Frontiers in Plant Science*, 9, 985.

Jeppson, L. R., Keifer, H. H., and Baker, E. W. (1975). *Mites injurious to economic plants*. University of California Press.

Kamatham, S., Munagapati, S., Manikanta, K. N., Vulchi, R., Chadipiralla, K., Indla, S. H., and Allam, U. S. (2021). Recent advances in engineering crop plants for resistance to insect pests. *Egyptian Journal of Biological Pest Control*, 31, 1–14.

Kiyanthy, S., & Mikunthan, G. (2009). Association of Insect Pests with Neem, Azadiracta indica with special reference to biology of ash weevils, Myllocerus sp in Jaffna, Sri Lanka. *American-Eurasian Journal of Scientific Research*, 4(4), 250–253.

Kulkarni, S. M., Malhotra, C. P., and Ghos, M. (1967). *Platypeplaaprobola* Meyer, a leaf roller on *Moghniamacrophylla* (Wild) O. Ketze. *Indian Forest*, 93, 111–119.

Kumar, S., Singh, S. K., Patel, R. K., Srivastava, K., and Pande, S. D. (2019). Litchi production through smart pest management. *Indian Horticulture*, 68(3), 53–57.

Kumari, A., Choudhary, J. S., Thakur, A. K., et al. (2021). Substantially altered bacterial diversity associated with developmental stages of litchi stink bug, Tessaratomajavanica. *Microorganisms*, 9(1), 1–15. doi:10.3390/microorganisms 9010001.

Kumari, P., Jasrotia, P., Kumar, D., Kashyap, P. L., Kumar, S., Mishra, C. N., and Singh, G. P. (2022). Biotechnological approaches for host plant resistance to insect pests. *Frontiers in Genetics*, 13, 914029.

Lall, B. S., and Mallik, S. N. (1976). Bionomics and control of litchi leaf roller (*Platypeplaaprobola* Meyrick) Lepidoptera: Tortricidae. *Proceedings of the National Academy of Sciences, India*, 22–23.

Lall, B. S., and Sharma, D. D. (1978). Studies on the bionomics and control of the cacao moth *Acrocercopscramerella* Snellen (Lepidoptera: Gracillariidae). *The Pan-Pacific Entomologist*, 93(4), 199–203.

Lall, B. S., and Rahman, M. F. (1975). Studies on the bionomics and control of the Erinose mite, *Eriophyes* litchi Keifer, Acarina: Eriophydae. *Pesticides*, 9(11), 49–54.

Latham, J. R., Love, M., and Hilbeck, A. (2017). The distinct properties of natural and GM cry insecticidal proteins. *Biotechnology and Genetic Engineering Reviews*, 33(1), 62–96.

Leksawasdi, P., and Kumchu, C. (1991). Mass rearing and releasing of the parasitoid *Anastatus* sp. *Kasetsart*Journal (*Natural Science*), 25, 47–53.

Li, D. S., Liao, C. Y., Zhang, B. X., and Song, Z. W. (2014). Biological control of insect pests in litchi orchards in China. *Biological Control*, 68, 23–36. doi:10.1016/j.biocontrol.2013.06.003.

Liu, X. D., and Lai, C. Q. (1998). Experiment of control of litchi stink bug by using *Anastatus japonicus* Ashmead. *South China Fruits* 27, 31.

Maiti, B., and Sahoo, A. K. (1991). Biological observations of *Platypepla aprobola* Meyer. on *Polyalthia longifolia*Var. Pendula in West Bengal. *Environment and Ecology*, 9(3), 647–649.

Mamta, B., and Rajam, M. V. (2017). *RNAi technology: A new platform for crop pest control*. NCBI. Retrieved from NCBI.

Mateos Fernández, R., Petek, M., Gerasymenko, I., Juteršek, M., Baebler, Š., Kallam, K., and Patron, N. J. (2022). Insect pest management in the age of synthetic biology. *Plant Biotechnology Journal*, 20(1), 25–36.

Molla, K. A., Sretenovic, S., Bansal, K. C., and Qi, Y. (2021). Precise plant genome editing using base editors and prime editors. *Nature Plants*, 7(9), 1166–1187.

Moon, T. T., Maliha, I. J., Khan, A. A. M., Chakraborty, M., Uddin, M. S., Amin, M. R., and Islam, T. (2022). CRISPR-Cas genome editing for insect pest stress management in crop plants. *Stresses 2022*, 2, 493–514.

Mukta, Z. N. (2020). *Understanding interactions and relationships in pest management innovation processes in Bangladesh* (Doctoral dissertation, Wageningen University and Research).

Munawar, I., Shamoon, F., Imran, M., Tariq, F., Ashraf, H., Nisar, T., and Munawar, N. (2023). RNAi for the management of insect pests: A review. *Agrobiological Records*, 11, 21–27.

National Institute of Plant Health Management (NIPHM). (2022). *Integrated pest management package for litchi*. Retrieved from NIPHM. https://niphm.gov.in/IPMPackages/Litchi.pdf

Nishida, T., and Holdaway, F. G. (1945). *The erinose mite of lychee*. Hawaii Agriculture Experiment Station Circular No. 48.

Nitnavare, R. B., Bhattacharya, J., Singh, S., Kour, A., Hawkesford, M. J., and Arora, N. (2021). Next generation dsRNA-based insect control: Success so far and challenges. *Frontiers in Plant Science*, 12, 673576.

Oberhofer, G., Ivy, T., and Hay, B. A. (2019). Cleave and Rescue, a novel selfish genetic element and general strategy for gene drive. *Proceedings of the National Academy of Sciences*, 116(13), 6250–6259.

Parveen, S., Choudhary, J. S., Thomas, A., and Ramamurthy, V. V. (2015). Biology, morphology and DNA barcodes of *Tessaratoma javanica* (Thunberg) (Hemiptera: Tessaratomidae). *Zootaxa*, 3936(2), 261–271.

Patel, A. N., and Patel, R. K. (2008). Biology of bark eating caterpillar, *Indarbela quadrinotata* (Walker). *Current Biotica*, 2(2).

Priya, M. K., Choudhary, J. S., Dhakar, M. K., Dey, D., S, R., Gurjar, M. S., . . . Das, A. (2025). Bacterial communities associated with litchi leaf roller, Statherotis leucaspis (Meyrick, 1902) (Lepidoptera: Tortricidae) reared on different litchi varieties. *Applied Fruit Science*, 67(4), 190.

Prasad, V. G., and Singh, R. K. (1981). Prevalence and control of litchi mite. *Aceria litchi* Keifer in Bihar. *Indian Journal of Entomology*, 43(1), 67–75.

Rani, D. D., Subhash, S., Gopalkrishna, H. R., and Chakravarthy, A. K. (2022). Newer genetic tools, techniques, vectors, promoters, and molecular markers for genetic engineering of herbivorous insects. In *Genetic methods and tools for managing crop pests* (pp. 3–25). Singapore: Springer Nature Singapore.

Regmi, N. R. (2022). A review on management options for litchi fruit and shoot borer, Conopomorpha Sinensis. *Big Data In Agriculture (BDA)*, 4(2), 58–61.

Revynthi, A. M., Crane, J. H., Wasielewski, J., Kendra, P. E., and Carrillo, D. (2021). The lychee erinose mite *Aceria litchii* (Keifer) (Acari: Eriophyidae): ENY2077/IN1347, 11/2021. *EDIS*, 2021(6).

Sahni, R., Kumari, S., Kumar, M., Kumar, A., Singh, S., and Kumar, M. (2024). Mechanization of Litchi (*Litchi chinensis S.*) cultivation: Current scenario and future needs of India. *Journal of Scientific Research and Reports*, 30(12), 786–797.

Sahoo, A. K., Roy, A., and Ray, S. K. (2007). Comparative efficacy of insecticides and neem formulations against Litchi fruit borer, *Conopomorpha cramerella* Snellen (Gracillariidae: Lepidoptera). Department of Agricultural Entomology, Bidhan Chandra KrishiViswavidtyalay, West Bengal, India. *ARC Journal*, 16, 1–3.

Schulte, M. J., Martin, K., and Sauerborn, J. (2006). Effects of azadirachtin injection in litchi trees (*Litchi chinensis* Sonn.) on the litchi stink bug (*Tessaratoma papillosa* Drury) in northern Thailand. *Journal of Pest Science*, 79, 241–250.

Sharma, D. D., and Agrawal, M. L. (1988). Studies on the biology and immature stages of litchi fruit borer *Conopomorpha cramerella* (Snellen)(Lepidoptera: Gracillariidae). *Journal of Research – Rajendra Agricultural University*, 6(1–2), 84–87.

Silver, K., Cooper, A. M., and Zhu, K. Y. (2021). Strategies for enhancing the efficiency of RNA interference in insects. *Pest Management Science*, 77(6), 2645–2658.

Singh, G. (1992). Loss assessment and biology of the litchi fruit borer. In *Proceedings of national seminar on recent developments in litchi production held at RAU, Pusa* (p. 40).

Singh, G., Nath, V., Pandery, S. D., Ray, P. K., and Singh, H. S. (2012). *The litchi (food and agriculture organization of the united Nations)*. New Delhi. *Open Access Library Journal*, 8(10).

Singh, J., and Raghuraman, M. (2011). Emerging scenario of important mite pests in north India. *Zoosymposia*, 6, 170–179.

Singh, M. P. (1971a). Occurrence of *Platypepla aprobola* Meyr. (Tortricidae: Lepidoptera) on litchi (*Litchi chinensis*) in North Bihar. *Indian Journal of Entomology*, 33(1), 98.

Srivastava, K., Pandey, S. D., Patel, R. K., Sharma, D., and Nath, V. (2016a). Insect pest management practices in litchi. *Insect Pests Management of Fruit Crops*, 127–149.

Srivastava, K., Patel, R. K., Gupta, A. K., and Sharma, D. (2020). Scenario of insect pests on litchi: Management options. innovative pest management approaches for the 21st century. *Harnessing Automated Unmanned Technologies*, 461–480.

Srivastava, K., Patel, R. K., Kumar, A., Pandey, S. D., Reddy, P. V. R., and Nath, V. (2017). Integrated management of litchi fruit and shoot borer (*Conomorpha sinensis*) using insect growth regulators under subtropics of Bihar. *Indian Journal of Agricultural Sciences*, 87(11), 1515–1518.

Srivastava, K., Patel, R. K., & Sharma, D. (2017). Insect pests management in litchi ecosystem. *Technological Innovations in Integrated Pest Management Biorational and Ecological Perspective*, 335.

Srivastava, K., Purbey, S. K., Patak, R. K., and Nath, V. (2016b). Managing fruit-borer for having healthy litchi. *Indian Horticulture*, 61(3).

Upadhyay, S. K., Aryal, S., Bhusal, B., and Chaudhary, B. (2020). Evaluation of insecticides for the management of litchi fruit and shoot borer. *Journal of Nepal Agricultural Research Council*, 6, 85–91.

Upadhyay, S. K., Aryal, S., Bhusal, B., and Chaudhary, B. (2022). Evaluation of insecticides for the management of litchi fruit and shoot borer. *Big Data in Agriculture*, 4(2), 58–61. Retrieved from Big Data in Agriculture.

Vats, S., Kumawat, S., Kumar, V., Patil, G. B., Joshi, T., Sonah, H., and Deshmukh, R. (2019). Genome editing in plants: Exploration of technological advancements and challenges. *Cells*, 8(11), 1386.

Verma, R. D., Patel, V. K., Thakur, M. V., and Patel, Y. (2022). *Impact and management strategy of fruit and shoot borer in litchi*. Vigyan Varta.

Waite, G. K., and Hwang, J. S. (2002). Pests of litchi and longan. In *Tropical fruit pests and pollinators: Biology, economic importance, natural enemies and control* (pp. 331–359). Wallingford, UK: CABI Publishing.

Wedell, N., Price, T. A. R., and Lindholm, A. K. (2019). Gene drive: Progress and prospects. *Proceedings of the Royal Society B*, 286(1917), 20192709.

Wu, Y.-H., Kamiyama, M. T., Chung, C.-C., and Tzeng, H.-Y. (2020). Population monitoring, egg parasitoids, and genetic structure of the invasive litchi stink bug, tessaratomapapillosa, in Taiwan. *Insects*, 11(3), 165. doi:10.3390/insects11030165

Yuan, B., Yu, D., Hu, A., Wang, Y., Sun, Y., and Li, C. (2023). Effects of green manure intercropping on soil nutrient content and bacterial community structure in litchi orchards in China. *Frontiers in Environmental Science*, 10, 1059800.

Zhang, D. P. (1997). A study of the pericarp of litchi fruit and the injury caused by litchi stinkbug, *Tessaritomapapillosa* (Hem.: Pentatomidae). *Wuyi Science Journal*, 13, 198–303.

Zhang, Z. W., Yuan, P. Y., Wang, B. Q., and Qui, Y. P. (1997). *Litchi pictorial narration of cultivation.* Guangzhou, China: Pomology Research Institute, Guangdong Academy of Agricultural Sciences.

13 Recent Advancements in Integrated Pest Management

Sumbal Nazir[1], *Sumaira Maqsood*[2*], *Habib Ali*[3], *Zain Mushtaq*[4], *Mujahid Manzoor*[5], *Hina Safdar*[6], *Naila Batool*[7], *Muhammad Naveed Riaz*[8], *Noor ul Ain*[9], *and Rafia Rashid Abbasi*[2]
[1]School of Zoology, Minhaj University, Lahore
[2]Department of Environmental Sciences, Kohsar University, Murree 47150, Pakistan
[3]Khwaja Fareed University of Engineering and Information Technology, Rahim Yar Khan, Pakistan
[4]Department of Soil Science, Punjab University, Lahore, Pakistan
[5]Department of Entomology, Punjab University, Lahore, Pakistan
[6]Department of Plant Pathology, University College of Agriculture, University of Sargodha, Sargodha
[7]Department of Psychology, Kohsar University, Murree 47150, Pakistan
[8]Department of Psychology, University of Sargodha, Sargodha
[9]Department of Entomology, Faculty of Agricultural Sciences, University of the Punjab, Lahore
*Corresponding author

13.1 INTRODUCTION

IPM is an ecosystem-based management strategy of combating pests that has been in existence for a long period of time. Now the time has come to change these old practices keeping in mind important factors, namely, economic viability, environmental security, and environmental safety. This will help in achieving pollution-free, sustainable food production and food security. This new IPM model encourages farmers and pest managers to revise agriculture strategies and techniques (Dara, 2020).

A large number of insects, weeds, and other pathogens are directly or indirectly responsible for the loss of approximately 1 billion individuals' food. Despite the use of pesticides, different methods of integrated pest management are used for pest control. IPM is a sustainable strategy for sustainable food security. IPM techniques are based on the relationships of insects with their associated environment and the life cycles of insects. IPM strategies can be applied in both agricultural and non-agricultural lands, like urban, forest, and natural areas, for pest control. IPM strategies limit the use of hazardous pesticide by using environment friendly approaches instead.

13.1.1 IPM Program Strategies

There are five steps for IPM:

1. Knowledge and identification of pest
2. Monitoring of pest population
3. Determination of pest intolerable points and generation of active threshold
4. Generation of IPM strategies and their implementation
5. Evaluation and observation of IPM strategic results

(www.epa.gov/safepestcontrol/integrated-pest-management-ipm-principles)

13.2 IPM ASSOCIATED WITH OTHER MANAGEMENT TOOLS FOR BETTER PEST MANAGEMENT

Different methods, namely, biological control, chemical control, mechanical or physical control, change in habitat, change in basic agricultural practices, and incorporation of resistant varieties, are associated with IPM to enhance its effectiveness (Figure 13.1).

13.2.1 Biological Control and IPM

Biological control techniques are basically the use of natural enemies against pests. So by utilizing these biological agents, the use of pesticides can be avoided. Biological control strategies are used by knowing the pest life cycle, its association with the host, and the climatic conditions. Three methods are used, including classical methods or importation, conservation biological control, and augmentation biological control. In classical methods or importation, exotic species (foreign species) are explored for the killing of the foreign pest because the

DOI: 10.1201/9781003516538-13

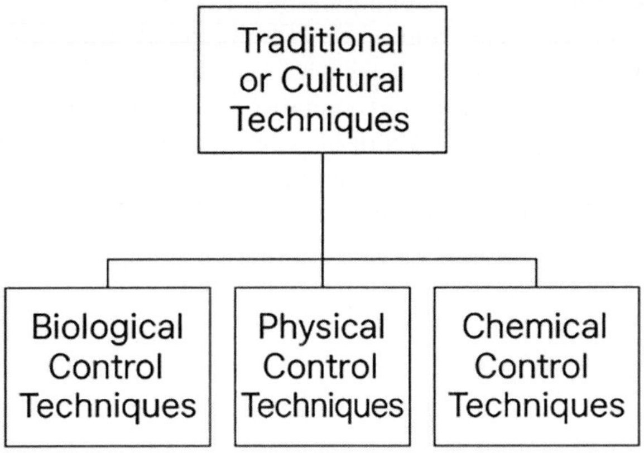

FIGURE 13.1 IPM strategies.

foreign pest has no predator in the region in which they have been introduced deliberately or accidentally but have predators in their native region. For example, decapitating flies are used as biological control against red imported fire ants, and a group of flea beetles, thrips, and stem borers are used against alligator weed. Alfalfa weevil, *Hypera Postica* (Gyllenhall), is a serious pest of weed alfalfa in Europe, and 12 parasitoid species were imported as a biological control of weevil pest (Landis and Orr, 2006; Jeffers and Chong, 2021). Mealybugs are common pests of different crops, so the eggs of these bugs are parasitized by parasitic wasps, hoverflies, and ladybird beetles. *Coccophagus*, *Acerophagus*, and *Leptomastix* are reported to be parasitic wasps of mealybugs (Le Vieux and Malan, 2013)

In the augmentation method of biological control, mass-rearing, genetic change, and colonization of biocontrol agents took place in the areas where pests were present to suppress the pest population. Mass-rearing of biocontrol agents takes place in insectaries. This method is widely used in nurseries, greenhouses, and crop fields by inoculative release (lesser number of biocontrol agents) and inductive release (larger number of biocontrol agents). Release of *Trichogramma* wasps to parasitize the pest eggs is an example of augmentative biocontrol technique used in agricultural practices.

In the conservation method or manipulative biocontrol agent, changes in habitat, plant production, and pest management approaches take place to control the pests. In this method, factors involved to limit the pest could be found and then modified to limit natural enemies. Biological control conservationists manipulate and modify the landscape and agriculture scape and then also reduce the use of pesticides, which also reduces pest diversity.

13.2.2 Physical Techniques of Integrated Pest Management

Change in temperature, humidity, and light; flooding; drying the seeds; and use of ionic radiations and sound waves are different physical strategies used in integrated pest management.

13.2.3 Modifications in Temperature, Light, and Humidity Variables

Variations in light, temperature, humidity, and day length can considerably influence the life cycle arrays of pests, persuading their activity levels. In the framework of stored grains, temperatures ranging from 13 to 35°C endorse pest activity, though temperatures below 13°C or beyond 35°C can defeat it. For illustration, investigation has shown that the egg period of *Helicoverpa armigera* persists for 7.9 days at 28°C, while on the other hand, it extends to 10.4 days at 25°C, representing temperature's effect on development rates. Furthermore, variations in temperature, such as increase from 10 to 27°C, destructively correlate with hatching degree days. Cold temperatures induce physiological, behavioral, and mechanical changes in insects, ultimately affecting their growth, reproduction, and survival (Khaliq et al., 2014). Employing freezing methods, like storing dry goods at temperatures between −13 and −18°C for 2–3 weeks, proves effective in disinfecting and is particularly beneficial in organic markets. For instance, freezing green coffee beans for 48 hours at −13.9 to −15.5°C has demonstrated efficacy against the coffee berry borer in Hawaii (Hollingsworth et al., 2013). Additionally, environmental factors, like humidity and soil moisture content, play a pivotal role in influencing insect pest dynamics. In regions like Asia, storing potatoes in cold storage helps safeguard them against the potato tuber moth (Omkar, 2018). Agricultural practices also employ light traps to attract and eliminate various pests, such as moths, termites, winged aphids, and whiteflies, with different pests being drawn to specific wavelengths of light (Shimoda and Honda et al., 2013). Ionizing radiations are utilized for pest control, employing techniques like the sterile insect technique and phytosanitary irradiation to hinder pest growth and reproduction. Flooding serves as a soil disinfectant strategy, particularly effective in destroying soilborne pests. When rice nursery beds are flooded, for instance, pests like the rice hispa are drowned, contributing to effective pest management strategies. Moreover, long-term summer soil flooding, with or without paddy rice culture, has been successful in reducing *Verticillium dahliae* populations, thereby controlling Verticillium wilt in subsequent cotton harvests and boosting lint production.

Light traps are widely used to destroy the insect pests of order Lepidoptera and Diptera. Light traps are also used to destroy aphids, termites, and whiteflies (Jonason et al., 2014).

Loud clapping and shouting is also used in getting rid of pests. Ultrasonic audio sounds of a frequency of more than 20,000 Hz are effectively used to get rid of pests. These sound waves create a hurdle in mating and the reproduction process of pests and act as repellant for pests. Ultrasonic waves of the frequency 50,000–240,000 Hz were experimentally used to kill pests in the fields by creating an effect on neurological processes and the brain (Rashid et al., 2017; Arvind et al., 2018).

Seed drying is another physical method used in stored grains to escape from pests. This method is used as protection and disinfectant (Beckett, 2011).

13.2.4 Mechanical Methods and IPM

Different mechanical approaches are used to control pests by restricting pest growth or changing their habitat. Hand netting and picking, grinding, beating, sieving, and perching are different approaches for the management of pests.

Handpicking method for pest management can be used in a small scale in kitchens, gardens, and nurseries. This method, however, is difficult on a large scale. This is a traditional old method of collection of insects and their eggs and destroys them. Sugarcane leafhoppers and cabbage butterflies are popular pests controlled using hand netting. Light traps are inefficient for trapping day pests.

Pheromone traps, bait traps, and suction traps are utilized for mass trapping and mating disruption to control insect pests. Pink bollworm, gypsy moth, cotton gray weevil, pine beetle, oriental fruit fly, melon fruit fly, and European chaffer monitoring, catching, mating disruption, and control are all being done successfully using pheromone traps. Different adhesive color traps are used for a variety of pest management (Ranamukhaarachchi and Wickramarachchi, 2007). Bait traps having odorless attractive pesticide were used to attract and kill the pest. GF-120 food bait is used to control fruit flies in cucurbit crops and fruit orchards.

Beating and shaking bushes and trees, especially during morning of the winter season, when insects are in dormant condition, then capturing and burying them in trenches, is effective, especially in cases of locusts, flies, and beetle pests.

In the banding method, various types of band are used on stems to prevent the crawling of insect pests toward leaves and fruits, where they die with starvation and entangled with a greasy band material. Jute gunny bags, polythene sheets, and adhesive bands on trees are used to avoid pests.

Water spray with pressure is used to remove insects like mites, aphids, and flies from crop plants, but this approach is only used on strong pants.

To attract predatory birds that feed on insect pests, wooden or metallic T or Y perches are randomly put in agricultural regions (Arif et al., 2019).

The sieving method is used to gather and kill pests in different life stages in stored grains in storage facilities. This approach is more common in stored wheat grain pests.

Trench digging and barrier techniques are used to ward off animals, low-flight birds, pests, and termites. Grasshopper and locust movement from one field to another can be prevented by this method (Adhikari, 2022).

13.2.5 Traditional or Cultural Control and IPM

In cultural pest control strategies, modifications in crop cultural and production strategies should take place to reduce or eliminate pest. These strategies include change in crop culturing time and dates, interplantating, crop rotation, weed management, sanitation, crop trapping, tillage and mulching practices, and other spatiotemporal strategies of the agroecosystem (Walgenbach, 2018; Frank et al., 2022). Cultural control strategies suppress insect outbreaks and prevent infestations. Cultural strategies create changes in irrigation system, because overwatering the crops can increase weeds and root diseases (All, 2004).

There are four main cultural strategies used in integrated pest management:

1. Pest habitat disruption or reduction
2. Adjustment of crop plantation and nutritional requirements
3. Diversion of pest population away from crop
4. Reduction in pest injury to avoid crop yield loss

13.2.6 Crop Rotation

Crop rotation is the most successful cultural method for pest management. This method distorts the normal life cycle of pests, having prolonged generation cycles. In this method, different types of crops are planted in a field every year. A longer rotation between the crops creates susceptibility against the same pests. The crop rotation method is used against tomato, soyabean, and corn diseases.

13.2.7 Crop Trapping

In this method, a small patch of area is planted as a trap, which attracts the pest and keeps it away from the crop. Pests can be destroyed when they get entangled in the crop trap. This method is used to control the striped cucumber beetle.

13.2.8 Change in Timing and Dates of Plantation

By changing the planting dates and times of host plant, the pest may not be available at that time for the destruction of crops. This method is used in controlling onion and cabbage maggots.

13.2.9 Tillage

In tillage, moving the soil surface by physical methods is done. Weeds and insect pests are controlled by the tillage method. Insect eggs and larvae also come on the soil surface by tillage method, where they become dead by drying or freezing method. The tillage method also destroys the habitat of insect pests. Allelopathy is also used for weed management. In allelopathy, competition between two plant species is reduced by adding toxic chemicals in soil.

13.2.10 Sanitation

Sanitation procedures are used to avoid pest infestations in stored products. Grain bins and warehouses should be cleaned. Sanitation strategies for pest control encompass various methods, including applying residual chemical treatments to cracks and crevices, routine sweeping and vacuuming to eliminate food dust and debris, sealing doors and windows, utilizing air curtains near entrances, installing screens, and implementing space treatments, such as aerosol fogging (Morrison et al., 2019).

13.2.11 Flooding

Flooding ascertains effectiveness in contrast to weed species that are at risk to extended periods underwater. It eliminates plants by depriving them of oxygen necessary for their growth. The efficacy of flooding hinges on the thorough submersion of weeds for extended durations. Flooding drowns insect pests and disrupts their life cycle (Skendžić et al., 2021. The rice water weevil, *Lissorhoptrus oryzophilus*, can be eliminated by shallow flooding (Tindall et al., 2013). This method is also used to inhibit pests of sugarcane crops and wireworms in vegetables.

13.2.12 Mulching

Mulching is the sustainable integrated pest management strategy which improves soil fertility and plant health. This increases crop yield and suppresses pest attack. Organic mulch, such as wood, plants, and crop residues, reduces fungal pathogen attack and inhibits soil splashing, which ultimately prevents fungal spores from reaching the leaves. For example, the purple blotch (*Alternaria porri*) fungal disease of green onions can be reduced through organic mulching (Quintanilla-Tornel et al., 2016). Synthetic mulches, for example, reflective plastic mulch and silver-colored plastic mulch, help in the reduction of insect pests, namely, aphids, beetles, spiders, thrips, and mites. The reflection of light repels and obscures these pests, which often enables controlling pests more efficiently than with insecticides (Summers et al., 2005). Insightful mulch raises the process of photosynthesis and keeps the plant and soil cool by reducing the temperature and moisture stresses (Kaufman and Kaufman, 2015). Solarization mulch, or clear man-made mulch, which safeguards the soil, traps solar heat to kill off the weed seed bank in the topsoil layer (Stapleton, 2000). This mulch also helps in controlling weed growth. Mulch obtained from cowpea has been confirmed to have successfully repressed insect pest inhabitants of pepper (Mochiah, 2012). Plastic mulch reduces the insect pest of okra and helps produce healthy plants (Ojiako et al., 2018). Mulching through organic constituent advantages promotes testing as part of the integrated management of sweet potato weevil, predominantly to defend emerging storage roots through dry periods when soil cracking can facilitate access by pests (Rehman et al., 2019). Paddy straw is able to work as mulch that preserves soil humidity and temperature and overturns weed progression (Paul et al., 2021); therefore, floras are raised well. Carbon-based materials, for example, cow urine and paddy straw mulch, retain the soil fertility and good soil texture and reduce pest attack (Pangaribuan et al., 2023).

13.2.13 Host Plant Defense Mechanisms

In cultural controls, plants have defense mechanisms to repel pest damage. Host plants maintain strong health so they can suppress pest pressure. Controlled temperature and ventilation methods should be used to store grains (Figure 13.4)

13.3 CHEMICAL CONTROLS AND IPM

The use of chemicals is not feasible in integrated pest management, but these chemicals are used in a limited manner and also in combination with other approaches. Chemical control is the dominant strategy used in IPM before World War II and the production of synthetic pesticides, horticulture oils, lead acetate, sulfur, lime, and nicotine (Brunner, 2014). A scouting program should be adopted before selecting any chemical to control pests. Pest identification, economic thresholds, and the crop/pest life stages also have a significant role in the management of pest outbreaks. Pesticides used in controlling the pest are applied on soil and foliage. A *pesticide* is the chemical or biological substance used for the repellence, destruction, and control of pests for better plant growth or crop yield. Pesticides play vital roles in plant processes:

1. Improving plant growth without using the nutrients
2. Preventing weed growth
3. Preserving plant products

(Bondareva and Fedorova, 2021)

Insecticides, fungicides, acaricides, herbicides, insect growth regulators, and pest attractant repellants are used in pest management. Biopesticides provide more environmentally friendly solutions of pest management as compared to synthetic pesticides.

13.4 INTEGRATED PEST MANAGEMENT AND MODERN TECHNOLOGY

Modern technology is also utilized for better pest management practices. Nanopesticides, GIS mapping, remote sensing, and artificial intelligence are now used in integrated pest management (Figure 13.2).

13.5 NANOPESTICIDES

Nanotechnology is an emerging field used for crop and environmental protection. Nanoparticle-based pesticides show more benefits compared to conventional pesticides. Nanopesticides have high solubility in water, highly stable formulation, and safer insecticidal activity than non-target organisms but increased activity against target organisms.

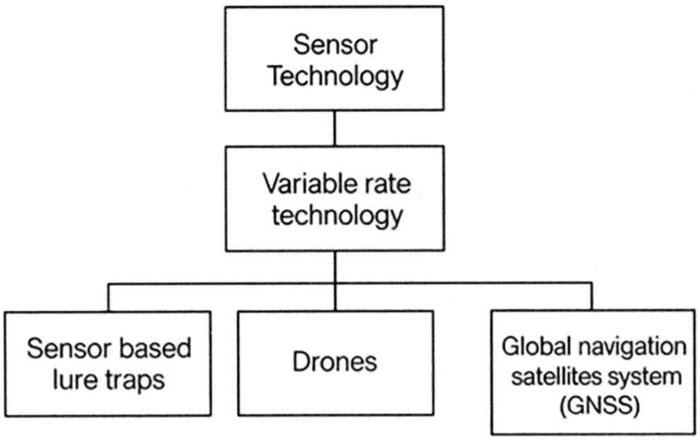

FIGURE 13.2 Modern technology and IPM.

Nanopesticides are environmentally friendly, have negligible residues, and do not accumulate in the food chain. This technology is used in IPM for the smart delivery of pesticides to their target organism and minimizes residual pollution. In insect pest management, nanotechnology is measured as an ecologically sustainable practice which is an outstanding insect controller tactic in green agriculture (Deka et al., 2021; Yousef et al., 2023). Toxicity of nanocides to non-target organisms is less than 43.1%, emphasizing the inadequate ecological damage (Wang et al., 2022).

Nanoparticles as biopesticides can be used in the form of nanospeheres, nanocapsules, nanogels, and micelles (Ragaei and Sabry, 2014; Harish et al., 2022).

1. *Nanospheres* are polymers with pesticide embedded in the matrix. A pesticide named imidacloprid is loaded to bowl-structured hollow porous silica nanospheres BHSNs (imi@BHSNs) by adsorption and entrapment mechanisms. This delivery mechanism provides accurate movement of the pesticide to its target organism (Nuruzzaman et al., 2020).
2. *Nanocapsules* are shell-like, with the pesticide embedded in the inner side. Nanocapsules containing two pesticides, thifluzamide and validamycin, were effectively synthesized using a double-emulsion technique in conjunction with high-pressure homogenization technology to combat rice sheath blight. This innovative approach is anticipated to address the growing concern of pesticide resistance by enhancing the efficiency of pesticide utilization and minimizing pesticide usage (Cui et al., 2020).
3. *Nanogel*s are hydrophilic polymers absorbing water mostly by cross-linking method. Nanogels are nontoxic and degradable carriers. By employing a temperature-dependent nanogel, researchers have devised a durable gel-like network that encases lambda-cyhalothrin (LC), resulting in the formation of a Pickering emulsion known as LC@TNPE. This pioneering formulation demonstrates efficacy against a range of pests, including *Plutella xylostella*, *Aphis gossypii*, and *Pieris rapae*, offering a holistic approach to pest management, and this nanogel remains safe for other, non-target organisms (Shen et al., 2020)
4. *Polymeric micelles* are an aggregate of hydrophilic and hydrophobic molecules in aqueous media. Micelles of poloxamers of different monoterpenes used against head lice gave good results, but the most effective result was seen by using linalool and thymol, which showed above 50% mortality of head lice (Lucia et al., 2017).

Cupper-based (Vignardi et al., 2020) and silver-based (Yin et al., 2020) nanopesticides are used in crop protection. These nanopesticides have antifungal, antibacterial, and antiviral activities against pests and have less activity against humans and ecosystems.

13.6 GIS MAPPING, REMOTE SENSING, AND IPM

New tools like GIS integration and remote sensing, data analytics, artificial intelligence, and biotechnology are introduced in integrated pest management for improved pest management approaches (Figure 11.3). Pest managers and farmers can make accurate data and target interventions which reduce dependency on pesticides. The implementation of these tactics nurtures sustainable agriculture, ensures biodiversity, and augments the long-term flexibility of our ecologies.

Continual investment in research and development is decisive to augment and optimize technical solutions for integrated pest management (IPM). By embracing and backing these advancements, we can ensure food security, environmental well-being, and a sustainable legacy for generations to come. Technology-enabled IPM presents a promising avenue toward fostering a balanced coexistence between agriculture and the natural world.

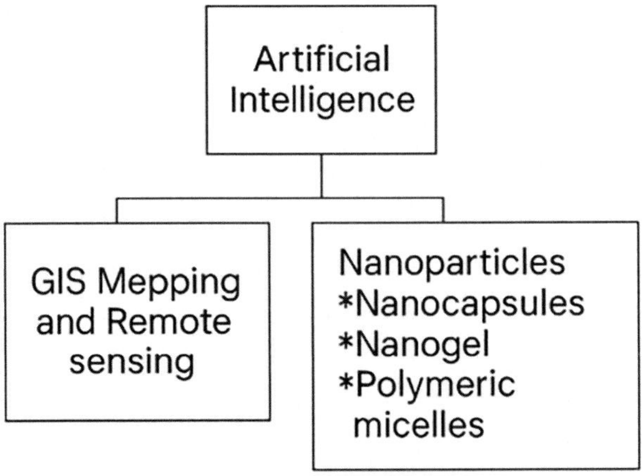

FIGURE 13.3 GIS mapping and remote sensing.

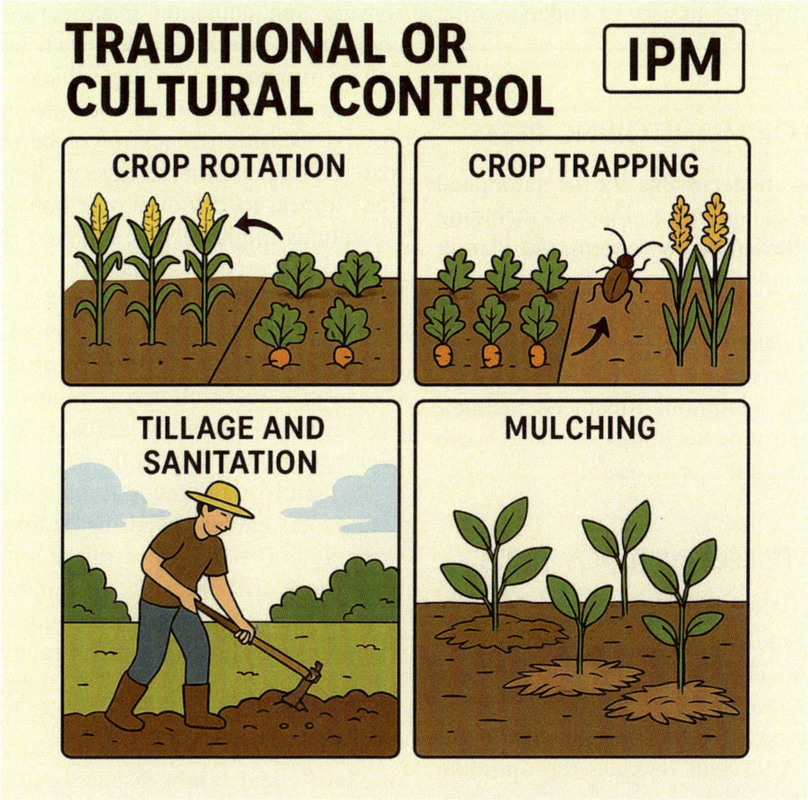

FIGURE 13.4 Traditional or cultural control methods in integrated pest management (IPM).

Geographical information system (GIS) and remote sensing approaches are used in integrated pest management because these technologies provide complete understanding of biotic and abiotic factors of crop fields by spatial modelling. Biochemical and biophysical monitoring of vegetation across the crop field has also be done with these latest technologies. The application of insecticides, fungicides, herbicides, and pesticides has also been done using remote sensing and global positioning system techniques. GPS with other technologies gives accurate spatial information of vegetation fields and crop fields (Kelly and Guo, 2007). Aerial photography by drones, thermal and multispectral cameras, and satellite multispectral scanning are used to detect and locate pest damage in crop fields. Drones are able of taking high-resolution photos and thermal data, shedding light on significant information about crop health, stress, and insect invasions, even at a plant-specific level. Desert locust attacks on crops in Rajasthan and Gujarat and whitefly infestation in cotton crops were detected, and pest control management strategies designed, by using the latest GIS mapping and remote sensing technologies (Willers et al., 2014; Yadav and Yadav, 2021). Citrus blackfly infestations in citrus orchards, harvester ant (*Pogonomyrmex barbatus* F. Smith) infestations on rangelands, and western pine beetle (*Dendroctonus brevicomis* LeConte) invasions in a forested area were also monitored by GIS mapping, aerial photography, and GPS technologies (Everitt et al., 2003)

13.7 PRECISION APPLICATION METHODS

A pest surveillance program is considered to be essential in detecting a wide range of pests on a significant scale. Contemporary tools employed for pest monitoring and administration include GPS, global navigation satellite system (GNSS), and artificial intelligence (AI). AI-based pest monitoring, leveraging its innovative competences, has become vital in undertaking abundant agricultural challenges. Deep learning (DL) models within AI play a dynamic part in ascertaining pests.

13.8 MONITORING PESTS THROUGH SENSOR TECHNOLOGY

Currently, there has been expanding usage of sensor-based expertise in field surveys aimed at different forms of pests. To illustrate, radar is engaged in monitoring the crusade patterns of pests throughout migration, thermal infrared imaging helps in revealing them, GNSS enables wildlife telemetry, video tools are utilized for observing flying insects, chemiluminescent tags are employed to track nocturnal insect activity, echo sounding detects larval movement, and habitat mapping assists in understanding pest habitats.

13.9 MOBILE APP FOR MONITORING PESTS

Various smartphone apps are accessible for the automated handling and regulation of agricultural pests, including popular options such as the Trapview System, the Plantix mobile app, Agri-App, and Kheti-Badi (Kanwal et al., 2022).

Outdated pest supervision approaches occasionally involve extensive pesticide actions, which can result in misuse and environmental pollution. Advanced technology-driven precision application techniques offer a more focused and effective strategy for pest control.

13.10 VARIABLE RATE TECHNOLOGY (VRT)

Variable rate technology (VRT) signifies a technical progression that regulates pesticide application proportions in response to real-time data. This method yields a complete methodology, taking into account aspects like insect population, crop strength, and ecological conditions. By investigating these variables, VRT can regulate the optimum quantity of pesticide need for application, endorsing embattled and exact usage. This not only recovers pest control but also lessens ecological impression and augments crop vigor (Campos et al., 2020).

13.10.1 Sensor-Based Lure Traps

Following pest populations is crucial in integrated pest management (IPM), and sensor-equipped traps have shown efficacy. These devices employ pheromones or substitute attractants to lure specific insect species by their sex, while an infrared (IR) sensor aids in detecting insects, simplifying the process of monitoring and evaluating pest populations (Flórián et al., 2023).

13.10.2 Drones

Due to a lack of steadfast and lucrative sampling methods, pest monitoring is time-consuming and considered to be the most vulnerable, so that we contend significant research as a challenge and focus on sustainable agriculture and modern pest management techniques for the betterment of crop vigor and yield. Biotic stress activates physiological defense responses in plants, like herbivory in case of arthropod pests, which cause significant changes in leaf reflectance. Modern imaging techniques are able to suspect these changes and can be used as the best crop inspection methods. Moreover, novel methods of treatment with precision are needed.

This analysis precisely focuses on the employment of small unmanned aerial robots, frequently attached to small drones, surrounded by agricultural outlines, where together sensing and actuation machineries can be deployed on equipment traversing fields (such as irrigation equipment) and on (un)manned driving vehicles.

Canopy reflectance data assembled and evaluated through sensing drones might be transferred as a digital map to direct additional class of drones, identified as actuation drones, to discover pest hotspots proficiently. These resolutions might include targeted releases of natural enemies or the precise application of pesticides via sprays. It is highlighted that the future of sustainable pest management in modern agriculture depends considerably on inventive technologies, emphasizing the demanding necessity for cooperative research across numerous disciplines, including agronomy, ecology, software development, and engineering (Filho et al., 2019).

Research on the recognition and classification of maize viruses has gathered significant attention. Some scientists presented a maize spectral recovery disease detection framework utilizing HSCNN+ and maize disease detector CNN, aiming for accurate and profitable detection. They identified an optimal spectral recovery model capable of transforming hyperspectral imaging (HSI) data from raw maize RGB data (Jun et al., 2022). A lightweight CNN model was proposed to categorize various severity stages of Maydis leaf blight disease, exhibiting superior performance compared to commonly used pretrained models in terms of classification accuracy (Haque et al., 2022). Similarly, EANet, an end-to-end learning CNN architecture for identifying multi-class maize diseases, has been introduced, integrating an attention mechanism to enhance disease recognition, especially in complex background settings. Additionally, scholars demonstrated pest and disease monitoring and recognition techniques for rice crops and jujube trees (Albahli et al., 2022). An intelligent monitoring system for detecting disease was developed, including pest lesions on rice canopies, utilizing an improved YOLO-DPD model to accurately capture rice disease and pest lesion features (Li et al., 2022). Hyperspectral remote sensing images from UAVs to detect and monitor spider mite infestations in jujube trees were employed to achieve promising results by incorporating spectral features and 3D information into their recognition model (Wu et al., 2023).

13.11 ARTIFICIAL INTELLIGENCE AND IPM

Nowadays, farmers are facing pest outbreak problems due to climatic conditions and worldwide exchange of pest-ridden materials. In order to overcome these problems, a sustainable system of pest control is required which ensures the principles of integrated pest management (IPM) are employed.

Integrated pest management (IPM) embodies an environmentally conscious tactic to pest control, encompassing various factors, such as natural predators, economic thresholds, plant resilience, breeding patterns, pest behavior, and prevailing climate conditions. Central to IPM is the pivotal role of knowledgeable professionals who contribute to system design, ecological monitoring, and decision-making processes. Sustainable pest management can be further enhanced by leveraging artificial intelligence (AI) to streamline routine tasks, like biological and environmental monitoring, as well as the strategic selection of optimal timing and methods. This chapter has delved into the application of AI within IPM, elucidating the utilization of algorithms, tools, and methodologies pertinent to artificial intelligence in pest management strategies (Tomar and Kaur, 2021)

The main barrier impeding the progress of integrated pest management (IPM) lies in the scarcity of reliable and up-to-date crop damage data. The absence of comprehensive information on insect pests and plant diseases hampers farm managers' ability to devise effective strategies for mitigating crop damage. This research aims to showcase how an integrated system has effectively guided farm managers toward implementing sustainable and data-driven IPM strategies. A system known as intelligent and integrated pest and disease management (I2 PDM) has been developed to address agricultural challenges. Specialized edge computing devices have been engineered to effectively identify and classify prominent greenhouse insect pests, like thrips (including *Frankliniella intonsa*, *Thrips hawaiiensis*, and *Thrips tabaci*) and whiteflies (such as *Bemisia argentifolii* and *Trialeurodes vaporariorum*), among others. These devices also monitor crucial environmental factors, like temperature, humidity, and light levels, transmitting this data to a centralized server. Implemented initially in greenhouses cultivating tomatoes and orchids, the system has gathered extensive spatial and temporal data on insect pest populations and environmental conditions over a period of up to 1,368 days. Analysis of the data revealed significant improvements in farm management practices, particularly in the context of integrated pest management (IPM). Notably, there was an annual reduction of up to 51.1% in insect pest populations on participating farms. The analysis revealed that employing a sophisticated integrated pest management (IPM) system can generate novel and efficient methods, thereby unlocking potential benefits that exceed those provided by conventional IPM programs. This research marks the first step toward developing an intelligent strategic framework for IPM, leveraging authentic long-term data collected automatically. The outcomes highlighted in this study seek to encourage cooperation among farm managers, researchers, specialists, and industries, facilitating the uptake of sustainable, data-driven IPM techniques (Rustia et al., 2022).

13.12 CONCLUSION

In the realm of agriculture, numerous challenges exist, ranging from crop diseases and improper soil treatment to insect pest attacks and inadequate drainage and irrigation, all of which can result in significant economic losses. Additionally, various climatic hazards arise due to the excessive use of chemicals, such as pesticides, weedicides, and herbicides. Agricultural experts have conducted numerous research endeavors to address specific issues within this domain. To confront the challenges of agriculture, artificial intelligence, with its adept learning capabilities, has emerged as a sought-after tool. Various automated and semi-automated systems have been developed to assist farmers and agricultural experts, offering improved solutions to agricultural problems worldwide.

Several factors contribute to the unpredictable outbreaks of insect pests, including meteorological changes, host plant phenological patterns, and the exchange of pest-infested materials internationally. Integrated pest management (IPM) programs emphasize the utilization of natural resources to control pest populations, relying on ecological principles and various biological and environmental variables. While the role of agricultural experts in IPM is crucial, manually monitoring pest populations is a time-consuming and labor-intensive process. Consequently, there is a pressing need to develop intelligent systems that consider all relevant factors, including different abiotic and biotic variables, to predict pest occurrences accurately for timely pesticide applications in the field.

REFERENCES

Adhikari, U. 2022. Insect pest management: Mechanical and physical techniques. *Reviews in Food and Agriculture (RFNA)* 3(1): 48–53. http://doi.org/10.26480/rfna.01.2022.48.53

Albahli, S., and Masood, M. 2022. Efficient attention-based CNN network (EANet) for multi-class maize crop disease classification. *Frontier Plant Science* 13: 1003152.

All, J. 2004. Cultural control of insect pests. In *Encyclopedia of Entomology*. Springer, Dordrecht. https://doi.org/10.1007/0-306-48380-7_1108

Arif, M. J., Gogi, M. D., Sufyan, M., Nawaz, A., and Sarfraz, R. M. 2019. Principles of insect pests management. In *Sustainable Insect Pest Management*, pp. 17–47.

Arvind, G., Athira, V. G., Haripriya, H., Rani, R. A., and Aravind, S. 2018. Automated irrigation with advanced seed germination and pest control. *Proceedings – 2017 IEEE Technological Innovations in ICT for Agriculture and Rural Development (TIAR) 7–8 April 2017*, pp. 64–67. https://doi.org/10.1109/TIAR.2017.8273687

Beckett, S. J. 2011. Insect and mite control by manipulating temperature and moisture before and during chemical-free storage. *Journal of Stored Products Research* 47(4): 284–292. https://doi.org/10.1016/j.jspr.2011.08.002

Bondareva, L., and Fedorova, N. 2021. Pesticides: Behavior in agricultural soil and plants. *Molecules* 26(17): 5370. https://doi.org/10.3390/molecules26175370. PMID: 34500803; PMCID: PMC8434383.

Brunner, J. F. 2014. Integrated pest management in tre e fruit crops. In *Encyclopedia of Agriculture and Food Systems, Chemical Control*.Elsevier. Academic Press.

Brunner, J. F. 2014. Integrated pest management in tree fruit crops. In *Encyclopedia of Agriculture and Food Systems, Chemical Control*.

Campos, J., Gallart, M., Llop, J., Ortega, P., Salcedo, R., and Gil, E. 2020. On-farm evaluation of prescription map-based variable rate application of pesticides in vineyards. *Agronomy* 10(1), Article 1. https://doi.org/10.3390/agronomy10010102

Cui, J., Sun, C., Wang, A., Wang, Y., Zhu, H., Shen, Y., Li, N., Zhao, X., Cui, B., Wang, C., Gao, F., Zeng, Z., and Cui, H. 2020. Dual-functionalized pesticide nanocapsule delivery system with improved spreading behavior and enhanced bioactivity. *Nanomaterials (Basel)* 10(2): 220. https://doi.org/10.3390/nano10020220. PMID: 32012747; PMCID: PMC7074971

Dara, S. K. 2020 Implementation of IPDM in strawberries and other berries. In M. L. Gullino, A. Albajes, P. Nicot, and J. C. van Lenteren (Eds.), *Pest and Disease Management in Greenhouse Crops*. Springer, pp. 597–624. https://doi.org/10.1007/978-3-030-22304-5_21

Deka, B., Babu, A., Baruah, C., and Barthakur, M. 2021. Nanopesticides: A systematic review of their prospects with special reference to tea pest management. *Frontier Nutrition* 8:686131. https://doi.org/10.3389/fnut.2021.686131. PMID: 34447773; PMCID: PMC8382848.

Everitt, J. H., Summy, K. R., Escobar, D. E., and Davis, M. R. 2003. An overview of aircraft remote sensing in integrated pest management. *Subtropical Plant Science* 55: 59–67.

Filho, F. H. I., Heldens, W. B., Kong, Z., and Lange, E. S. de. 2019. Drones: Innovative technology for use in precision pest management. *Journal of Economic Entomology* 113(1): 1–25. https://doi.org/10.1093/jee/toz268

Flórián, N., Jósvai, J. K., Tóth, Z., Gergócs, V., Sipőcz, L., Tóth, M., and Dombos, M. 2023. Automatic detection of moths (lepidoptera) with a funnel trap prototype. *Insects* 14(4), Article 4. https://doi.org/10.3390/insects14040381

Frank, S., Bradley, L. K., and Moore, K. A. 2022. Chapter 8 – Integrated pest management. In K. A. Moore and L. K. Bradley (Eds.), *North Carolina Extension Gardener Handbook*, 2nd ed. Raleigh, NC: NC State Extension.

Haque, M. A., Marwaha, S., Arora, A., Deb, C. K., Misra, T., Nigam, S., and Hooda, K. S. 2022. A lightweight convolutional neutral network for recognition of severity stages of maydis leaf blight disease of maize. *Frontier Plant Science* 13: 1077568.

Harish, V., Tewari, D., Gaur, M., Yadav, A. B., Swaroop, S., Bechelany, M., and Barhoum, A. 2022. Review on nanoparticles and nanostructured materials: Bioimaging, biosensing, drug delivery, tissue engineering, antimicrobial, and agro-food applications. *Nanomaterials* 12: 457. https://doi.org/10.3390/nano12030457

Hollingsworth, R. G., Jang, E. B., and Follett, P. A. 2013. Freezing as a treatment to prevent the spread of hypothenemus hampei (Coleoptera: Curculionidae), in coffee. *Journal of Economic Entomology* 106 (2): 653–660. https://doi.org/10.1603/EC12373

Jeffers, J.H., and Chong, JH, 2021. Biological control strategies in integrated pest management (ipm) programs. Clemson (SC): Clemson Cooperative Extension, Land-Grant Press by Clemson Extension; LGP 1111. http://lgpress.clemson.edu/publication/biological-control-strategies-in-integrated-pest-management-programs.1-9

Jonason, D., Franzén, M., and Ranius, T. 2014. Surveying moths using light traps: Effects of weather and time of year. *PLoS ONE*, 9(3). https://doi.org/10.1371/journal.pone.0092453

Jun, F., Jindai, L., Rongqiang, Z., Zhi, C., Yongliang, Q., and Dan, L. 2022. Maize disease detection based on spectral recovery from RGB image. *Frontier Plant Science* 13: 1–12.

Kanwal, S., Khan, M. A., Saleem, S., Tahir, M. N., Muntaha, S. T., Samreen, T., Javed, S., Nazir, M. Z., and Shahzad, B. 2022. Integration of precision agriculture techniques for pest management. *Environment Science Proceedings* 23: 19. https://doi.org/10.3390/environsciproc2022023019

Katan, J. 2000. Physical and cultural methods for the management of soilborne pathogens. *Crop Protection* 19(8–10): 725–731. https://doi.org/10.1016/S0261-2194(00)00096-X

Kaufman, D. W., and Kaufman, G. A. 2015. Plains harvest mice in north-central Kansas: Abundance, habitat association and individual attributes. *Transactions of the Kansas Academy of Science* 118(1–2).

Kelly, M., and Guo, Q. 2007. Integrated agricultural pest management through remote sensing and spatial analyses. In *General Concepts in Integrated Pest and Disease Management*. Springer, pp. 191–207. https://doi.org/10.1007/978-1-4020-6061-8_7

Khaliq, A., Javed, M., Sohail, M., and Sagheer, M. 2014. Environmental effects on insects and their population dynamics. *Journal of Entomology and Zoological Studies* 2: 1–7.

Landis, D. A., and Orr, D. B. 2006. *Biological Control: Approaches and Applications Radcliffe's IPM World Textbook*. Michigan State University.

Le Vieux, P. D., and Malan, A. P. 2013. An overview of the vine mealybug *(Planococcus ficus)* in South African vineyards and the use of entomopathogenic nematodes as potential biocontrol agent. *South African Journal of Enology and Viticulture* 34: 108–118. http://dx.doi.org/10.21548/34-1-108hypothalamus6

Li, S., Feng, Z., Yang, B., Li, H., Liao, F., Gao, Y., Liu, S., Tang, J., and Yao, Q. 2022. An intelligent monitoring system of diseases and pests on rice canopy. *Frontier Plant Science* 13: 972286.

Lucia, A., Toloza, A. C., Guzmán, E., Ortega, F., and Rubio, R. G. 2017. Novel polymeric micelles for insect pest control: Encapsulation of essential oil monoterpenes inside a triblock copolymer shell for head lice control. *Peer Journal* 5: e3171. https://doi.org/10.7717/peerj.3171. PMID: 28439460; PMCID: PMC5401625.

Mochiah, M. B., Baidoo, P. K., and Acheampong, G. 2012. Effects of mulching materials on agronomic characteristics, pests of pepper (Capsicum annuum L.) and their natural enemies' population. *American Journal of Agriculture Biology Science* 3(6): 253–261. http://doi.org/10.5251/abjna.2012.3.6.253.261.

Morrison, W.R., Bruce, A., Wilkins, R.V., Albin, C.E., and Arthur, F.H. 2019. Sanitation Improves Stored Product Insect Pest Management. *Insects*,1–20.

Nuruzzaman, M. D., Ren, J., Liu, Y., Rahman, M. M., Shon, H. K., and Naidu, R. 2020. Hollow porous silica nanosphere with single large pore opening for pesticide loading and delivery. *ACS Applied Nano Materials* 3(1): 105–113. https://doi.org/10.1021/acsanm.9b01769

Ojiako, F. O., Ibe, A. E., Ogu, E. C., and Okonkwo, C. C. 2018. Effect of varieties and mulch types on foliar insect pests of okra [*Abelmoschus Esculentus L.* (Moench)] in a humid tropical environment. *Agrosearch* 18(2): 38–58.

Omkar. 2018 Pests and their management. In *Pests and Their Management*.https://doi.org/10.1007/978-981-10-8687-8.2018

Pangaribuan, D. H., Widagdo, S., Hariri, A. M., Siregar, S., and Sardio, M. I. 2023. The effect of rice straw mulch and cow urine on growth, yield, quality on sweet corn and pest population density. *Asian Journal Agriculture Biology* (1): 202103123. https://doi.org/10.35495/ajab.2021.03.123

Paul, P. L. C., Bell, R. W., Barrett-Lennard, E. G., and Kabir, E. 2021. Impact of rice straw mulch on soil physical properties, sunflower root distribution and yield in a salt-affected clay-textured soil. *Agriculture*: 264.

Quintanilla-Tornel, M. A., Wang, K. H., Tavares, J., and Hooks, C. R. R. 2016. Effects of mulching on above and below ground pests and beneficials in a green onion agroecosystem. *Elsevier*: 1–36.

Ragaei, M., and Sabry, A. K. H. 2014. Nanotechnology for insect pest control. *International Journal of Science Environment Technology* 3(2): 528–545.

Ranamukhaarachchi, S. L., and Wickramarachchi, K. S. 2007. Color preference and sticky traps for field management of thrips Ceratothripoides claratris (Shumsher) (Thysanoptera: Thripdae) in tomato in Central Thailand. *International. Journal of Agriculture Biology* 9(6): 392–397.

Rashid, H., Ahmed, I. U., Reza, S. M. T., and Islam, M. A.2017. Solar powered smart ultrasonic insects' repellent with DTMF and manual control for agriculture. *IEEE International Conference on Imaging, Vision, and Pattern Recognition, IcIVPR*. https://doi.org/10.1109/ICIVPR.2017.7890869.2017

Rehman, M., Liu, J., Johnson, A. C., Dada, T. E., and Gurr, G. M. 2019. Organic mulches reduce crop attack by sweetpotato weevil (Cylas formicarius). *Scientific Reports* 9(1): 14860. https://doi.org/10.1038/s41598-019-50521-5. PMID: 31619690; PMCID: PMC6795849.

Rustia, D. J. A., Chiu, L. Y., Lum, C. Y., Wu, Y. F., Chen, S. K., Chung, J. Y., Hsu, J. C., and Lin, T. T. 2022. Towards intelligent and integrated pest management through an AIoT-based monitoring system. *Pest Management Science* 78(10): 4288–4302. http://doi.org/10.1002/ps.7048. Epub 2022 Jul 8. PMID: 35716088.

Shen, Y., An, C., Jiang, J., Huang, B., Li, N., Sun, C., Wang, C., Zhan, S., Li, X., Gao, F., Zhao, X., Cui, H., Gooneratne, R., and Wang, Y. 2020. Temperature-dependent nanogel for pesticide smart delivery with improved foliar dispersion and bioactivity for efficient control of multiple pests. *ACS Nano* 16(12): 20622–20632.

Shimoda, M., and Honda, K. I. 2013. Insect reactions to light and its applications to pest management. *Applied Entomology and Zoology* 48(4): 413–421. https://doi.org/10.1007/s13355-013-0219-x

Skendžić, S., Zovko, M., Živković, I. P., Lešić, V., and Lemić, D. 2021. The impact of climate change on agricultural insect pests. *Insects* 12: 440. https://doi.org/10.3390/insects12050440

Stapleton, J. J. 2000. Soil solarization in various agricultural production systems. *Crop Protection* 19: 837–841.

Summers, C., Mitchell, J., and Stapleton, J. 2005. Mulches reduce aphid-borne viruses and whiteflies in cantaloupe. *California Agriculture* 59(2): 90–94. https://doi.org/10.3733/ca.v059n02p90.

Tindall, K. V., Bernhardt, J. L., Stout, M. J., and Beighley, D. H. 2013. Effect of depth of flooding on the rice water weevil, *Lissorhoptrus oryzophilus*, and yield of rice. *Journal of Insect Science* 13: 62.

Tomar, P., and Kaur, G. 2021. *Artificial Intelligence and IoT-Base Technologies for Sustainable Farming and Smart*, p. 400. https://doi.org/10.4018/978-1-7998-1722-2

Vignardi, C. P., Muller, E. B., Tran, K., Couture, J. L., Means, J. C., Murray, J. L. S., Ortiz, C., Keller, A. A., Sanchez, S. N., and Lenihan, H. S. 2020. Conventional and nano-copper pesticides are equally toxic to the estuarine amphipod Leptocheirus plumulosus. *Aquatic Toxicology* 224: 105481.

Walgenbach, J. F. 2018. Chapter 16 – Integrated pest management strategies for field-grown tomatoes. In *Sustainable Management of Arthropod Pests of Tomato*. Academic Press, pp. 323–339.

Wang, D., Saleh, N. B., Byro, A., Zepp, R., Sahle-Demessie, E., Luxton, T. P., Ho, K. T., Burgess, R. M., Flury, M., White, J. C., and Su, C. 2022. Nano-enabled pesticides for sustainable agriculture and global food security. *Natural Nanotechnology* 17(4): 347–360. https://doi.org/10.1038/s41565-022-01082-8

Willers, J. L., DeFauw, S. L., English, P. J., and Jenkins, J. N. 2014. Chapter 3 – Application of remote sensing in integrated pest management. In *Integrated Pest Management, Current Concepts and Ecological Perspective*. pp. 37–58 https://doi.org/10.1016/B978-0-12-398529-3.00004-X

Wu, Y., Li, X., Zhang, Q., Zhou, X., Qiu, H., and Wang, P. 2023. Recognition of spider mite infestations in jujube trees based on spectral-spatial clustering of hyperspectral images from UAVs. *Frontier Plant Science* 14: 1078676.

Yadav, M., and Yadav, A. 2021. Remote sensing: A revolutionary approach in IPM. *Just Agricultue Newsletter* 2(2): 1–6.

Yin, I. X., Zhang, J., Zhao, I. S., Mei, M. L., Li, Q., and Chu, C. H. 2020. The antibacterial mechanism of silver nanoparticles and its application in Dentistry. *International Journal of Nanomedicine* 15: 2555.

Yousef, H. A., Fahmy, H. M., Arafa, F. N. et al. 2023. Nanotechnology in pest management: Advantages, applications, and challenges. *International Journal Tropical Insect Science* 43: 1387–1399. https://doi.org/10.1007/s42690-023-01053-z

14 Genome Advances in Pest Management

Riya Mishra[1*] *and Makhan Singh Karada*[2]
[1]Department of Genetics and Plant Breeding, Rajmata Vijayaraje Scindia Krishi Vishwa Vidyalaya, Gwalior, Madhya Pradesh 474002, India
[2]Department of Forestry, Jawaharlal Nehru Krishi Vishwa Vidyalaya, Jabalpur, Madhya Pradesh 482004, India
*Corresponding author

14.1 INTRODUCTION

Global food security, agricultural sustainability, and environmental health are being negatively impacted by the way different biotic and abiotic stressors are evolving due to climate change. In addition to changing ecosystems and decreasing soil fertility, rising temperatures, unpredictable rainfall patterns, and extreme weather events are hastening the spread of pests and diseases. In addition to increasing biotic pressures from emerging pathogens, these abiotic stresses – drought, salinity, and heat – are lowering crop yields and endangering sustainable agricultural methods (Mishra *et al.*, 2024a, 2024b). Meeting the expanding global food demand and preserving a robust agricultural system are significantly hampered by the ensuing fall in production and environmental deterioration.

Insects pose a serious threat to food security and crop productivity worldwide due to their ability to transmit diseases and damage the plants with their feeding habits (Moon *et al.*, 2022). Innovative pest management solutions are crucial to reduce the use for toxic synthetic insecticides while meeting the demand for increasing agricultural productivity (Komal *et al.*, 2023). The primary reason for choosing an alternative to conventional pesticides stems from the increasing development of resistance among pests (Borel, 2017), coupled with the environmental toxicity and associated health risks posed by these chemical agents. A variety of tactics are being evolved to enhance the insect pest resistance in crops incorporating approaches grounded in plant secondary metabolism, immunological responses, and microbiome science (Douglas, 2018). As sequencing technologies continue to progress, a growing number of insect genomes are being successfully sequenced (Li *et al.*, 2019). A recently developed breeding method called "genome editing" or "genome engineering" allows for the altering of genomes of microorganisms, humans, animals, and plants (Tyagi *et al.*, 2020). Ensuring the sustained advancement of novel pest management tactics requires the use of genome engineering of insects, which is crucial for functional gene analysis (Wang *et al.*, 2013). The growing interest in the precise application of gene editing technologies for genome alterations presents a plethora of opportunities that can influence important agronomic features, including improved resistance to biotic stressors (Komal *et al.*, 2023). Today, precise gene knockout, knock-in, and knockdown in other than model organisms are possible because of the development of genome editing technologies, which provide enhanced possibilities for targeted genetic alterations in a variety of biological systems (Xu *et al.*, 2019).

14.1.1 An Overview of Genome Editing

The exact altering of chromosomal DNA inside a biological genome by the introduction of targeted, particular sequence mutations is referred to as *genome editing* (Perez-Pinera *et al.*, 2012). Gene editing usually targets a single gene, but genome editing involves more extensive changes which make it different from gene editing (Khalil, 2020). Furthermore, genome editing is not the same as genetic engineering, which frequently involves introducing genetic material at random into host genome, whereas the precise and targeted insertion of genetic material into specific sections of the genome is made possible by genome editing techniques. It encompasses various methods, such as zinc finger nucleases (ZFNs), transcriptional activator-like effector nucleases (TALENs), and the more recently developed clustered regularly interspaced short palindromic repeats (CRISPR) in conjunction with CRISPR-associated nuclease 9 (Cas9) (Singh *et al.*, 2022). These technologies create double-strand breaks (DSBs) at particular target loci by using tailored endonucleases. Cellular systems then repair the ensuing DSBs. These mechanisms involve the two pathways, *viz.*, non-homologous end joining (NHEJ) pathway, which results in mutations, and homology-directed repair (HDR) pathway, which allows for sequence replacement (Zhang *et al.*, 2017). Basically, insect genome editing has primarily been used to create gene drives or overcome pesticide resistance. The capacity of genome editing technologies to accurately remove particular genomic areas while removing any extraneous vector sequences is a noteworthy benefit (Bisht *et al.*, 2019). Since the final genome-edited

DOI: 10.1201/9781003516538-14

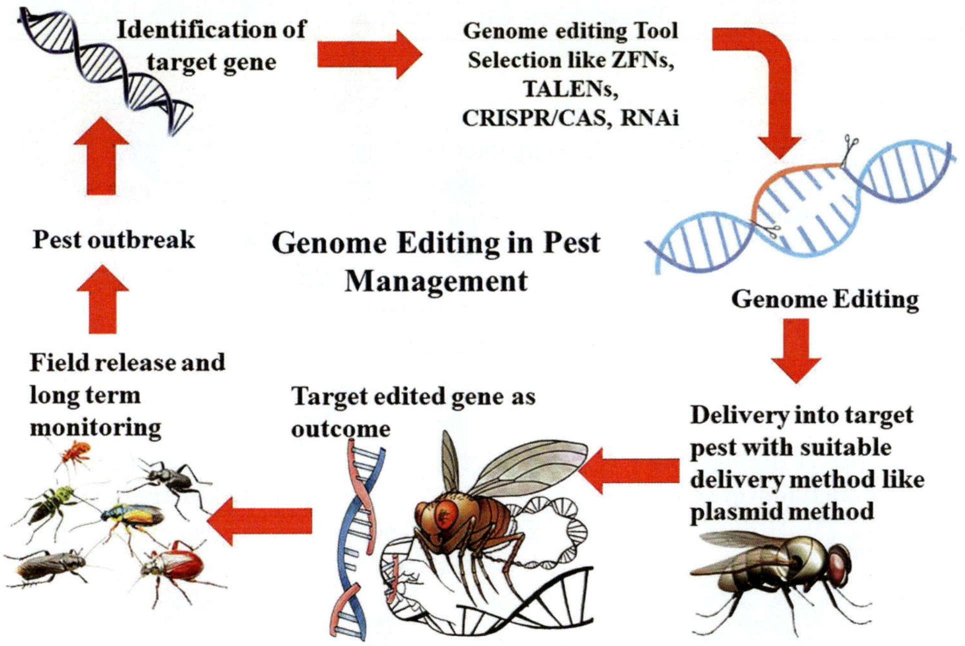

FIGURE 14.1 Genome editing process in pest management, outlining target gene identification, precise genetic modification, delivery to pests, field trials under monitoring, and the potential need for new cycles to address novel pest outbreaks.

organism (GEO) no longer contains any exogenous sequences, genome editing is now considered a reliable approach (Miglani, 2017).

14.1.2 Progress in Genome Editing

Pest control has advanced significantly as a result of genome editing, especially with the application of CRISPR/Cas9 technology, which enables precise and targeted genetic alterations (Ansori *et al.*, 2023). The invention of gene drives, which are designed to quickly spread particular genetic features through pest populations, is a significant breakthrough in this discipline (Hillary *et al.*, 2020). In particular, this technique has demonstrated great potential in controlling disease-carrying mosquitoes, which are the source of diseases including dengue, malaria, and the Zika virus (Weng *et al.*, 2024). Gene drives can be created to disseminate genes that limit these mosquitoes' ability to reproduce or prevent them from spreading infections (Tajudeen *et al.*, 2023). For instance, sterility or immunity to malaria has been introduced in Anopheles mosquitoes by the use of CRISPR-based gene drives, which has stopped the disease's spread (Nolan, 2021). This strategically offers a long-term, ecologically friendly alternative to chemical insecticides for managing disease vectors (van den Berg *et al.*, 2021).

Apart from gene drives, significant advancements in genome editing have enabled the modification of pest species' reproductive genes, providing a new approach to population control. Scientists can cause infertility in particular pest populations by focusing on critical genes that control fertility, which will eventually lead to suppression of pest population. For example, scientists have successfully manipulated the sex determination genes in certain insect species to skew the sex ratio in favor of males. Consequently, fewer females are available for reproduction, which leads to a progressive decline in pest populations (Compton *et al.*, 2022). In IPM programs, this strategy can be especially helpful when combined with more established approaches, such as the sterile insect strategy (SIT) (Marec & Vreysen, 2019; Bourtzis & Vreysen, 2021). Genome editing improves SIT by increasing sterilization efficiency and reducing the need for radiation or chemical treatments, which results in more successful field pest suppression (Alphey & Bonsall, 2018; Yan *et al.*, 2023).

Additionally, genome editing is being combined with cutting-edge breeding tools, such as quantitative trait loci (QTL mapping) and marker-assisted selection (MAS). With the use of these methods, scientists can pinpoint particular genetic markers linked to advantageous characteristics in crops, such as insect resistance. Scientists can more accurately and effectively introduce or improve resistance traits in crops by combining genome editing, MAS, and QTL mapping (Kushanov *et al.*, 2021; Naqvi *et al.*, 2022; Kamal *et al.*, 2023; Tyagi *et al.*, 2024). For example, genes linked to disease or insect resistance have been identified by QTL mapping, and these qualities can subsequently be improved in crop types through editing. By strengthening crop tolerance to pests, this integrated method also speeds up the breeding process, enabling the creation of crop types that are more resistant to pest attacks (Singh *et al.*, 2020; Kushanov *et al.*, 2021; Wani *et al.*, 2022; Manzoor *et al.*, 2024).

Functional genomics and insect biology are also essential to the advancement of genome editing for pest control. Scientists can find novel gene targets for pest control by comprehending the molecular pathways underpinning

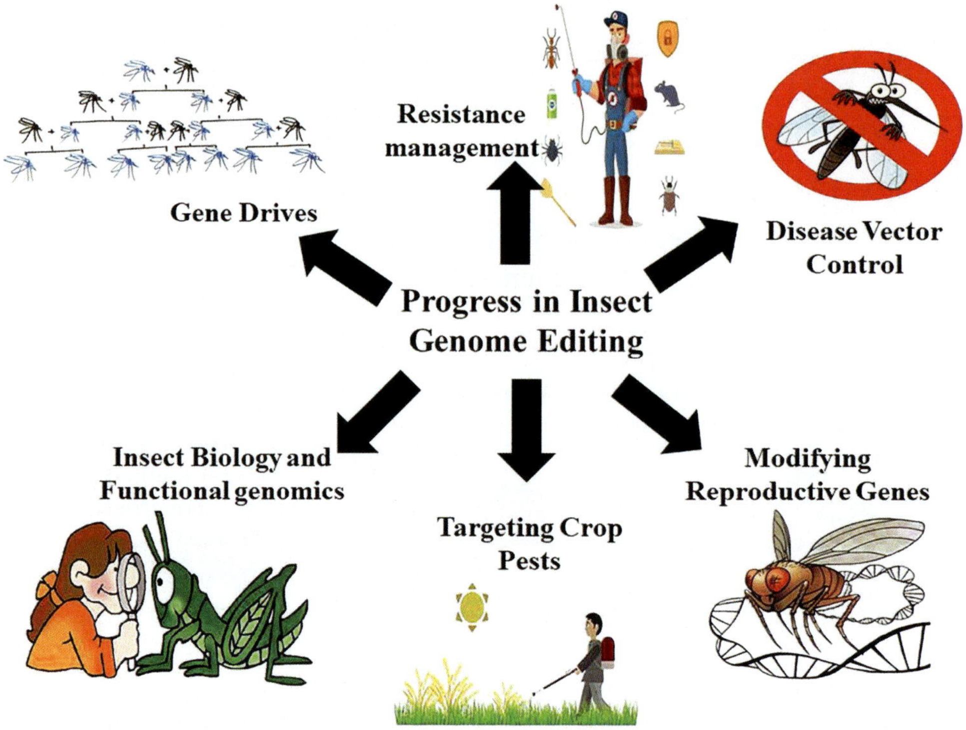

FIGURE 14.2 Diverse areas in which genome editing techniques have been successfully utilized for pest management, showcasing the progress made in controlling pest populations, enhancing resistance to insecticides, and implementing gene drives for sustainable pest control strategies.

insect development, metabolism, and reproduction (Gantz & Akbari, 2018; Panfilio *et al.*, 2019; Cagliari *et al.*, 2020). In addition to genome editing technologies, this improved understanding has resulted in the creation of biopesticides and other sustainable pest management techniques. The development of more efficient, eco-friendly pest management technologies that increase food security and agricultural output while reducing the environmental impact of conventional pesticide has advanced significantly as a result of these developments (Mishra *et al.*, 2020; Fenibo *et al.*, 2021; Zuma *et al.*, 2023; Ayilara *et al.*, 2023).

14.2 GENETICALLY MODIFIED CROPS

Recombinant DNA technology, or *genetic engineering*, is an umbrella term for a variety of methods used to modify genetic material. Incorporating the resulting hybrid DNA into organisms involves cutting and recombining DNA from various biological sources (Rosenberg, 2017). The middle of the 1990s saw the commercial introduction of the first transgenic crops resistant to insects. This led to the production of genetically modified maize, potatoes, and cotton that expressed the genes responsible for *Bacillus thuringiensis*'s insecticidal δ-endotoxin (Gatehouse *et al.*, 2011). Certain *Bacillus thuringiensis* (Bt) strains produce proteins called cry toxins, which consolidate to form crystalline formations during sporulation. The insecticidal activity of these toxins is particular, as they attack species belonging to different orders, such as Hymenoptera, Diptera, Lepidoptera, Nematoda, and Coleoptera (Abbas, 2018). As they increase crop yields and decrease the need for pesticides, insect-resistant crops have had a significant positive economic influence on the world (Talakayala *et al.*, 2020). For instance, transgenic cowpea lines expressing the Vip3Ba1 protein were created using vip genes from Australian *Bacillus thuringiensis* isolates that encode toxins effective against Maruca, providing total protection from the pest. Vip-transgenic cowpea combined with current cry-transgenic lines can help stop Maruca from becoming resistant to conventional treatments (Bett *et al.*, 2017). Additionally, in order to provide resistance against *Chilo infuscatellus*, Riaz *et al.* reported the creation of transgenic sugarcane lines that expressed the Vip3A toxin (Riaz *et al.*, 2020). In order to combat the Colorado potato beetle (CPB), Salim *et al.* created transgenic potato lines that expressed stacked insecticidal genes. These transgenic lines offer a viable approach to manage imidacloprid-resistant populations, as they displayed large reductions in the amount of foliage consumed by CPB (Salim *et al.*, 2021). Today the world is full of successful examples of genetically modified crops; however, the advent of CRISPR/Cas genome editing tool and RNAi technology has opened the door for novel developments in next-generation biotech crops. Without introducing toxin genes, these strategies allow the generation of plant genotypes with increased resistance to pests and diseases (Khabbazi *et al.*, 2024).

14.3 RNA INTERFERENCE (RNAI)

Insect science has greatly benefited from RNAi, which has been essential in identifying and comprehending the genes involved in a variety of physiological, developmental, and reproductive processes in insects (Palli, 2014). Over the past decade, the RNAi has gained much attention as a potentially effective method of controlling pests by causing detrimental effects (Table 14.1) (Christiaens *et al.*, 2020). Small RNA (sRNA) molecules can selectively inhibit gene expression at the transcriptional or post-transcriptional

TABLE 14.1
Application of RNAi for the Management of Different Insects/Pests

Insect/Pest	Target Gene	Function	Objective	Outcome	Reference
Tetranychus urticae	TuCht1, TuCht4, TuCht10	Chitinases, involved in molting process	Investigate the roles of TuCht genes in molting and identify targets for pest management	Suppression of TuCht4 and TuCht10 caused high mortality, molting abnormalities, and defects in chitinous layers, showing potential for RNAi-based pest control.	Liu *et al.*, 2024
Spodoptera litura	TLSP (trypsin-like serine protease)	Sperm activation, fertility regulation	Investigate the role of TLSP in sperm activation and explore its potential for pest control	TLSP deletion leads to prominent reduction in male fertility.	Yadav *et al.*, 2024
Henosepilachna vigintioctopunctata	Ecdysone receptor (EcR)	Regulates molting and metamorphosis	Evaluate the effect of dsRNA targeting EcR via dietary delivery and foliar spraying	Developmental arrest in larvae, abnormal pupae, reduced adult emergence.	Wu *et al.*, 2021
Diabrotica virgifera	Dicer-1, Dicer-2, Argonaute 1, Argonaute 2	Core components of RNAi pathway; involved in gene silencing and defense against dsRNA	Evaluate effects of dsRNA pathway on larval development and survivorship	Argonaute 1 reduced survivorship and delayed development.	Camargo *et al.*, 2018
Plutella xylostella	PxJHEH, PxEcR	Regulates metamorphosis and molting	Assess their role in DBM management	Gene silencing and mortality.	Chaitanya *et al.*, 2017
Tenebrio molitor	Tmol\Orco	Olfactory co-receptor	Investigate Tmol\Orco's role in mate-seeking behavior	Impaired mate-seeking behavior, decreased mate-finding, and increased search time.	Liu *et al.*, 2016
Drosophila	UB, L40	Involved in deltamethrin resistance	Investigate roles of UB and L40 in resistance	Confirms that both UB and L40 play roles in deltamethrin resistance.	Chi *et al.*, 2016
Aphis gossyppi	CYP6A2	Involved in spirotetramat resistance and alpha-cypermethrin cross-resistance	Investigate the role of CYP6A2 in resistance and cross-resistance	CYP6A2 associated with resistance and cross-resistance.	Peng *et al.*, 2016
Locusta migratoria	Syx1A	Regulates reproductive traits	Investigate the role of Syx1A in reproductive trait plasticity	Slower sexual maturation and smaller eggs.	Chen *et al.*, 2015
Leptinotarsa decemlineata	LdTorso, LdRas, LdRaf, LdERK	MAPK pathway components involved in ecdysteroidogenesis	Investigate the role of MAPK pathway components in larval development and ecdysteroidogenesis	Delayed development, increased pupal weight, impaired pupation and emergence, affected expression of ecdysteroidogenesis.	Zhu *et al.*, 2015
Pardosa pseudoannulata	Pp-AChE1, Pp-AchE2	Acetylcholinesterases, involved in insecticide sensitivity	Investigate the role of acetylcholinesterases in insecticide activity	Pp-AchE1 identified as major target for organophosphate and carbamate insecticides.	Meng *et al.*, 2015
Nilaparvata lugens	NlTOR	Regulates cell growth, nutrient-sensing lifespan, and reproduction	Study the function of TOR in male reproduction and mating behavior	NlTOR RNAi males produced no offsprings, exhibited poorly developed accessory glands and altered mating behavior.	Zhuo *et al.*, 2017
	NlPTTH	Regulates ecdysteroidogenesis and ecdysone response	Understand the role of NlPTTH in development	Nymphal development delayed; nymphal mortality increased; downregulation of ecdysteroidogenesis genes.	Chen *et al.*, 2017

stages through a biological process known as RNA interference (RNAi). This happens either by decreasing the stability and translation efficiency of the target mRNA or by directing restrictive chromatin changes (Liu *et al.*, 2020). This post-transcriptional gene silencing method is widely used in reverse genetics across a wide range of insect species because of its excellent selectivity (Santos *et al.*, 2014). Insects can experience stunted growth, aberrant developmental patterns, reduced ability to reproduce, and elevated mortality when they consume ds-RNAs intended to target vital genes (Baum & Roberts, 2014). More than 30 Hemiptera species from all suborders have benefitted from the successful application of RNAi-based technology, which has provided important new understandings of pesticide resistance, immunology, reproduction behavior, and metabolite synthesis (Jain *et al.*, 2020). Using RNAi, the white gene ortholog of the southern green stink bug (SGSB), *Nezara viridula*, was examined to determine whether it was a suitable marker for genetic transformation. White transcript knockdown created a different phenotype and had no effect on embryo development or viability (Souza *et al.*, 2022). Riga *et al.* optimized RNAi for southern green stink bug, demonstrating its effectiveness in gene knockdown and insecticide target discovery (Riga *et al.*, 2020). The major constraint of RNAi technique is the inability to produce stable genetic lines or null mutations, and their phenotypic results might vary greatly based on the timing and mode of dsRNA delivery (de Souza Pacheco *et al.*, 2022). Additionally, the temporary nature of RNAi makes it unsuitable for examining specific candidate genes, and it can differ between insect species, developmental stages, or tissues (Cagliari *et al.*, 2020).

14.4 GENE EDITING WITH CRISPR

Most prokaryotes (bacteria, archaea) contain repeated genomic sequences known as clustered regularly interspaced short palindromic repeats, which is abbreviated as CRISPR (Yadav *et al.*, 2023), and Cas-associated protein 9 as Cas9. The CRISPR/Cas system is widely utilized in the creation of genome editing tools due to its straightforward nature and ease of manipulation (Gu *et al.*, 2021; Mishra *et al.*, 2024c). CRISPR-Cas system of the type II is the foundation of this technology, which gives the bacteria adaptive defense mechanism against plasmids and viruses (Doudna & Charpentier, 2014). At particular genomic sites, CRISPR/Cas9 causes DSBs, which are primarily repaired by the erroneous NHEJ pathway. The targeted gene's function is frequently altered by this repair mechanism, which introduces different mutations at the specific location (Figure 14.3) (Asokan *et al.*, 2023). It is an endonuclease complex that induces DSBs in DNA at specific locations, guided by a complementary 20-base-pair sequence within the guide RNA (gRNA) (Lester *et al.*, 2020). A recent development in CRISPR technology called base editing allows point mutations to be precisely introduced into cellular DNA without causing DSBs (Kantor *et al.*, 2020). A highly effective alternative to previous techniques like ZFNs and

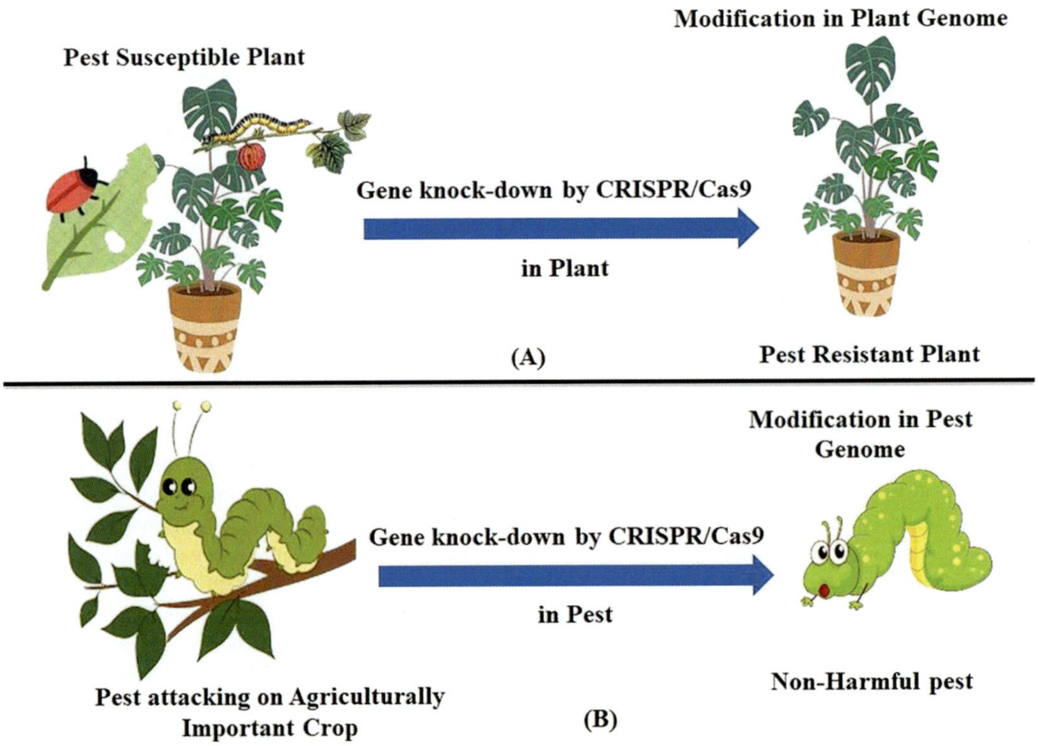

FIGURE 14.3 Utilization of CRISPR/Cas9 technology in sustainable pest management, showcasing its dual role in enhancing pest resistance in plants through targeted genome modifications and transforming pest genomes to render them non-harmful.

TALENs, which rely on protein-guided, sequence-specific DNA recognition and cleavage, the CRISPR system uses RNA-guided DNA cleavage for genome editing (Son & Park, 2022). It has sparked a great deal of advancement in the realm of functional genetics research and created new avenues for agricultural development, pest control, and disease vector control (Rouck *et al.*, 2024).

There are various instances reported in the application of CRISPR/Cas9 for controlling agriculturally important pests (Table 14.2), such as: In order to investigate the function of HaWnt1 gene, Fu *et al.* knocked out the HaWnt1 gene in *Helicoverpa armigera*, an agriculturally significant pest, and reported that, along with its importance in embryonic development, deletion of this gene can cause lethality, abnormal segmentation, defective appendages, and pigmentation in this pest (Fu *et al.*, 2022). To depict the potential of deleting Ser2 gene from *Plutella xylostella* for effective pest control strategy, Xu *et al.* used CRISPR/Cas9 in *Bombyx mori*, a model lepidopteran, to delete the BmSer2 gene, leading to the creation of male sterile individuals (Xu *et al.*, 2020). It has been effectively demonstrated that the CRISPR/Cas9 system can modify the Orco gene in pest *Spodoptera littorallis*. Orco knockdown confirmed its role in olfaction by interfering with the detection of sex

TABLE 14.2
Application of CRISPR/Cas9 in the Management of Different Insects/Pests

Insect/Pest	Target Gene	Function	Objective	Outcome	Reference
Helicoverpa armigera	HaCad	Receptor for Bt Cry1A toxins	Investigate the role of HaCad in resistance to Bt Cry1Ac toxin	SCD-Cad knockout line exhibited resistance to Cry1Ac compared to susceptible strain, confirming that HaCad is a functional receptor.	Wang *et al.*, 2016
	HaWnt1	Wnt signaling, embryonic development	HaWnt1 gene knockout to explore impacts on embryo and somatic development	High lethality, abnormal segmentation, defected appendages, pigmentation defects, crucial for embryonic development.	Fu *et al.*, 2022
Ectropis griescens	EgriBLOS2	Larval integument color formation	To understand the genetic and molecular basis of larval body color	EgriBLOS2 is responsible for larval integument color development.	Li *et al.*, 2022
Bombyx mori	BmSer2	Encodes a seminal fluid protein, male fertility	Disrupt Ser2 to induce male sterility with minimal effects on viability and mating competition	Dominant heritable male sterility; wild-type females mated with Ser2-deficient males laid non-hatching eggs; mutant males showed no change in other reproductive behaviors.	Xu *et al.*, 2020
Plutella xylostella	Ser2 orthologue	Male fertility	Target Ser2 gene for pest control strategy using engineered male sterility	Produced male sterility without affecting females; potential for pest control via gSIT with minimal environmental effect	Xu *et al.*, 2020
	PxRyR	Ryanodine receptor, resistance to diamides	Introduce 14790K mutation into PxRyR of susceptible strain to study resistance	Frequencies of 14790K allele significantly correlated with chlorantraniliprole resistance in field populations from China.	Jiang *et al.*, 2024
	Pxabd-A	Determines identity and functionality of abdominal segments	Target and mutagenize Pxabd-A to study its role in development	Abnormal prolegs, malformed segments, inherited defects, potential for pest management tools.	Huang *et al.*, 2016
Tetranychus urticae, Frankliniella occidentalis	Phytoene desaturase, CYP384A1	Genes related to pigmentation and development	Optimize a maternal injection protocol for CRISPR/Cas9 in *T. urticae* and other orthopods	SYNCAS formulation increased knockout efficiency (>20%) and facilitated both knockout and knock-in mutants in *T. urticae* and *F. occidentalis*	Rouck *et al.*, 2024
Locusta migratoria	Orco	Odorant receptor co-receptor	Disrupt Orco to study its role in the odorant receptor pathway	Impaired odor responses, body discoloration.	Li *et al.*, 2016
Drosophila suzukii	Beta-2-tubulin (β2t)	Spermatogenesis-specific promoter for fluorescent protein expression	Develop transgenic sperm-marking strain	Sperm-marking system established for monitoring pest control campaigns.	Ahmed *et al.*, 2019
Mythimna separata	NPC1b	Determines viability of eggs and post-embryonic development	Evaluate effects of injection timing	Cas9 protein had minimal impact on early-stage egg viability and high editing efficiency	Tang *et al.*, 2022
Nilaparvata lugens	NlCSAD	Melanin metabolism and body pigmentation; influences female oviposition and egg hatchability	Determine role in pigmentation, fecundity, and hatchability	Darker cuticle pigmentation and reduced fertility in females, leading to the use of this approach for pest management.	Chen *et al.*, 2021

pheromones and plant odor. This highlights the promise of CRISPR/Cas9 for research on gene functions and lepidopteran pest management techniques (Koutroumpa *et al.*, 2016).

14.5 GENE DRIVES FOR POPULATION CONTROL

Natural selection frequently includes genetic drive elements that display super-Mendelian inheritance, in which a particular genetic element is inherited by over half of the progeny. These elements include systems that reject offspring lacking the element, as well as sequences that can replicate within the genome, like jumping genes/transposons (Scott *et al.*, 2017). A phenomenon known as gene drive occurs when a certain heritable component skews inheritance in its favor over several generations, increasing the frequency of that element in the population. In this way, the gene is essentially driven to proliferate (Alphey *et al.*, 2020). Population suppression, which entails lowering the local insect population to lessen the harm done to livestock and crops, is one of the main goals of gene-based pest control (Scott *et al.*, 2017). Insect engineered gene drives offer substantial public advantages since they may address persistent issues with controlling disease vectors, agricultural pests, and invading species. They may also help with the conservation of endangered species (Devos *et al.*, 2021). The CRISPR/Cas9 enables a highly effective gene drive element through sexual reproduction, allowing it to rapidly disseminate throughout a population (Courtier-Orgogozo *et al.*, 2017). Gene drives are not expected to serve as a standalone solutions but are more likely to be incorporated into existing integrated pest management (IPM) strategies (Medina, 2018).

14.6 RISKS AND CHALLENGES

Over time, a greater number of crops have been cultivated using genome editing and transgenic technologies. Regulatory limitations that affect their marketing, cultivation, and application in human and animal nutrition have limited their uptake in many nations. In places where there is no particular legislation, people continue to argue and voice their opinions about the use of these in public domain (Marone *et al.*, 2023). An evaluation of the possible hazards to people, environment, and society posed by emerging technologies and their uses can be started with risk assessment (Lassoued *et al.*, 2019). While genome editing has a great deal of promise for plant breeding, it also raises serious biosafety issues due to unintended manipulations that need to be fully explored, such as potential for off-target consequences and the accidental spread of altered genes to other organisms (Saurabh, 2021; Movahedi *et al.*, 2023). A major risk is the establishment of resistant weeds (superweeds) and insects (superbugs). When there is enough selective pressure, resistance usually evolves. Large-scale planting of these cultivars may place the environment under intense selection, which could eventually result in the emergence of insect resistance within few years and reduce the efficacy of modified crop (Bawa & Anilakumar, 2013). Additionally, plants that have undergone genetic engineering can show decreased genetic stability at target or associated loci, which could compromise the efficacy of their increased resistance to pests. Thus, using a combination of molecular methods and phenotypic field assessments, plant breeders must evaluate and detect any pleiotropic effects, off-target effects, or stability concerns (Yıldırım *et al.*, 2023).

14.7 CONCLUSION

The integration of genomic advances in pest management marks a transformative era in agricultural practices, offering highly specific, sustainable, and efficient pest control strategies. By deciphering pest genomes, scientists can now target vulnerabilities at the genetic level, enabling precise interventions through tools like RNA interference (RNAi) and CRISPR-Cas9 gene editing. These approaches promise species-specific pest control, reducing reliance on traditional chemical pesticides and mitigating the negative effects on non-target species and the environment. Moreover, the application of genomics extends beyond pests themselves. The enhancement of biological control agents through genomic insights and the development of pest-resistant crops via marker-assisted breeding and genetic engineering represent holistic approaches to sustainable pest management. This not only improves crop resilience but also strengthens ecosystem health by promoting biodiversity and reducing chemical inputs. Genomic tools also play a pivotal role in resistance management, allowing for the early detection of resistance genes and supporting the design of dynamic pest control strategies. Furthermore, the integration of genomics with precision agriculture technologies facilitates real-time monitoring and targeted intervention, making pest control more efficient and minimizing environmental impacts. As research progresses, the potential for further innovations in genomic pest management is immense. This shift toward genetics-driven solutions represents a key step forward in achieving global food security, ensuring agricultural sustainability, and fostering environmentally sound practices. The continued development of genomic tools will be instrumental in creating a future where pest management is both scientifically informed and ecologically responsible.

REFERENCES

Abbas, Mohamed Samir Tawfik. 2018. "Genetically Engineered (Modified) Crops (*Bacillus thuringiensis* Crops) and the World Controversy on Their Safety." *Egyptian Journal of Biological Pest Control* 28 (1): 52. https://doi.org/10.1186/s41938-018-0051-2.

Ahmed, Hassan M. M., Luisa Hildebrand, and Ernst A. Wimmer. 2019. "Improvement and Use of CRISPR/Cas9 to Engineer a Sperm-Marking Strain for the Invasive Fruit Pest *Drosophila Suzukii*." *BMC Biotechnology* 19 (1): 85. https://doi.org/10.1186/s12896-019-0588-5.

Alphey, Luke S., Andrea Crisanti, Filippo (Fil) Randazzo, and Omar S. Akbari. 2020. "Standardizing the Definition of Gene Drive." *Proceedings of the National Academy of Sciences* 117 (49): 30864–67. https://doi.org/10.1073/pnas.2020417117.

Alphey, Nina, and Michael B. Bonsall. 2018. "Genetics-based Methods for Agricultural Insect Pest Management." *Agricultural and Forest Entomology* 20 (2): 131–40. https://doi.org/10.1111/afe.12241.

Ansori, Arif N. M., Yulanda Antonius, Raden J. K. Susilo, Suhaila Hayaza, Viol D. Kharisma, Arli A. Parikesit, Rahadian J. K. Zainul, et al. 2023. "Application of CRISPR-Cas9 Genome Editing Technology in Various Fields: A Review." *Narra J* 3 (2): e184. https://doi.org/10.52225/narra.v3i2.184.

Asokan, R, Anil Rai, Sangeetha Dash, Manamohan M, AshoK K, Bhargava C N, Rohan Wishard, Sanjay Kumar Pradhan, and Parvathy M S. 2023. "Application Of Genome Editing In Entomology." *Indian Journal of Entomology*, January, 96–103. https://doi.org/10.55446/IJE.2022.918.

Ayilara, Modupe S., Bartholomew S. Adeleke, Saheed A. Akinola, Chris A. Fayose, Uswat T. Adeyemi, Lanre A. Gbadegesin, Richard K. Omole, Remilekun M. Johnson, Qudus O. Uthman, and Olubukola O. Babalola. 2023. "Biopesticides as a Promising Alternative to Synthetic Pesticides: A Case for Microbial Pesticides, Phytopesticides, and Nanobiopesticides." *Frontiers in Microbiology* 14 (February). https://doi.org/10.3389/fmicb.2023.1040901.

Baum, James A., and James K. Roberts. 2014. "Progress towards RNAi-Mediated Insect Pest Management." In *Advances in Insect Physiology*, 249–95. https://doi.org/10.1016/B978-0-12-800197-4.00005-1.

Bawa, A. S., and K. R. Anilakumar. 2013. "Genetically Modified Foods: Safety, Risks and Public Concerns – a Review." *Journal of Food Science and Technology* 50 (6): 1035–46. https://doi.org/10.1007/s13197-012-0899-1.

Berg, Henk van den, Raman Velayudhan, and Rajpal S. Yadav. 2021. "Management of Insecticides for Use in Disease Vector Control: Lessons from Six Countries in Asia and the Middle East." *PLOS Neglected Tropical Diseases* 15 (4): e0009358. https://doi.org/10.1371/journal.pntd.0009358.

Bett, Bosibori, Stephanie Gollasch, Andy Moore, William James, Joel Armstrong, Tom Walsh, Robert Harding, and Thomas J. v. Higgins. 2017. "Transgenic Cowpeas (*Vigna unguiculata* L. Walp) Expressing *Bacillus thuringiensis* Vip3Ba Protein Are Protected against the Maruca Pod Borer (*Maruca vitrata*)." *Plant Cell, Tissue and Organ Culture (PCTOC)* 131 (2): 335–45. https://doi.org/10.1007/s11240-017-1287-3.

Bisht, Deepak Singh, Varnika Bhatia, and Ramcharan Bhattacharya. 2019. "Improving Plant-Resistance to Insect-Pests and Pathogens: The New Opportunities through Targeted Genome Editing." *Seminars in Cell & Developmental Biology* 96 (December): 65–76. https://doi.org/10.1016/j.semcdb.2019.04.008.

Borel, Brooke. 2017. "CRISPR, Microbes and More Are Joining the War against Crop Killers." *Nature* 543 (7645): 302–4. https://doi.org/10.1038/543302a.

Bourtzis, Kostas, and Marc J. B. Vreysen. 2021. "Sterile Insect Technique (SIT) and Its Applications." *Insects* 12 (7): 638. https://doi.org/10.3390/insects12070638.

Cagliari, Deise, Guy Smagghe, Moises Zotti, and Clauvis Nji Tizi Taning. 2020. "RNAi and CRISPR/Cas9 as Functional Genomics Tools in the Neotropical Stink Bug, *Euschistus Heros*." *Insects* 11 (12): 838. https://doi.org/10.3390/insects11120838.

Camargo, Carolina, Ke Wu, Elane Fishilevich, Kenneth E. Narva, and Blair D. Siegfried. 2018. "Knockdown of RNA Interference Pathway Genes in Western Corn Rootworm, *Diabrotica virgifera virgifera*, Identifies No Fitness Costs Associated with Argonaute 2 or Dicer-2." *Pesticide Biochemistry and Physiology* 148 (June): 103–10. https://doi.org/10.1016/j.pestbp.2018.04.004.

Chaitanya, B. N., R. Asokan, T. Sita, K. B. Rebijith, P. Ram Kumar, and N. K. Krishna Kumar. 2017. "Silencing of JHEH and EcR Genes of *Plutella xylostella* (Lepidoptera: Plutellidae) through Double Stranded RNA Oral Delivery." *Journal of Asia-Pacific Entomology* 20 (2): 637–43. https://doi.org/10.1016/j.aspen.2017.03.020.

Chen, Jing-Xiang, Wan-Xue Li, Jun Lyu, Yu-Tao Hu, Gang Huang, and Wen-Qing Zhang. 2021. "CRISPR/Cas9-Mediated Knockout of the NlCSAD Gene Results in Darker Cuticle Pigmentation and a Reduction in Female Fecundity in *Nilaparvata lugens* (Hemiptera: Delphacidae)." *Comparative Biochemistry and Physiology Part A: Molecular & Integrative Physiology* 256 (June): 110921. https://doi.org/10.1016/j.cbpa.2021.110921.

Chen, Qianquan, Jing He, Chuan Ma, Dan Yu, and Le Kang. 2015. "Syntaxin 1A Modulates the Sexual Maturity Rate and Progeny Egg Size Related to Phase Changes in Locusts." *Insect Biochemistry and Molecular Biology* 56 (January): 1–8. https://doi.org/10.1016/j.ibmb.2014.11.001.

Chen, Xu, Guo-Qing Li, Pin-Jun Wan, and Qiang Fu. 2017. "Efficient RNA Interference for Three Neuronally-Expressed Genes in *Nilaparvata lugens* (Stål) (Hemiptera: Delphacidae)." *Journal of Asia-Pacific Entomology* 20 (2): 513–19. https://doi.org/10.1016/j.aspen.2017.03.009.

Chi, Qingping, Fengliang Li, Wei Liu, Qin Xu, Junli Hu, Ying Cheng, Xuejian Jing, and Luogen Cheng. 2016. "Analysis of UB and L40 Resistance Related to Deltamethrin in Drosophila Kc Cells." *Gene* 575 (2): 399–406. https://doi.org/10.1016/j.gene.2015.09.012.

Christiaens, Olivier, Steve Whyard, Ana M. Vélez, and Guy Smagghe. 2020. "Double-Stranded RNA Technology to Control Insect Pests: Current Status and Challenges." *Frontiers in Plant Science* 11 (April). https://doi.org/10.3389/fpls.2020.00451.

Compton, Austin, and Zhijian Tu. 2022. "Natural and Engineered Sex Ratio Distortion in Insects." *Frontiers in Ecology and Evolution* 10 (June). https://doi.org/10.3389/fevo.2022.884159.

Courtier-Orgogozo, Virginie, Baptiste Morizot, and Christophe Boëte. 2017. "Agricultural Pest Control with CRISPR-based Gene Drive: Time for Public Debate." *EMBO Reports* 18 (6): 878–80. https://doi.org/10.15252/embr.201744205.

Devos, Yann, John D. Mumford, Michael B. Bonsall, Ana M. Camargo, Leslie G. Firbank, Debora C. M. Glandorf, Fabien Nogué, Konstantinos Paraskevopoulos, and Ernst A. Wimmer. 2021. "Potential Use of Gene Drive Modified Insects against Disease Vectors, Agricultural Pests and Invasive Species Poses New Challenges for Risk Assessment." *Critical Reviews in Biotechnology* 42 (2): 254–70. https://doi.org/10.1080/07388551.2021.1933891.

Doudna, Jennifer A., and Emmanuelle Charpentier. 2014. "The New Frontier of Genome Engineering with CRISPR-Cas9." *Science* 346 (6213). https://doi.org/10.1126/science.1258096.

Douglas, Angela E. 2018. "Strategies for Enhanced Crop Resistance to Insect Pests." *Annual Review of Plant Biology* 69 (1): 637–60. https://doi.org/10.1146/annurev-arplant-042817-040248.

Fenibo, Emmanuel O., Grace N. Ijoma, and Tonderayi Matambo. 2021. "Biopesticides in Sustainable Agriculture: A Critical Sustainable Development Driver Governed by Green Chemistry Principles." *Frontiers in Sustainable Food Systems* 5 (June). https://doi.org/10.3389/fsufs.2021.619058.

Fu, Xiaozheng, Rui Li, Qiqi Qiu, Mengke Wang, Te Zhao, and Lin Zhou. 2022. "Study on the Function of *Helicoverpa armigera* Wnt1 Gene Using CRISPR/Cas9 System." *Journal of Asia-Pacific Entomology* 25 (1): 101869. https://doi.org/10.1016/j.aspen.2022.101869.

Gantz, Valentino M, and Omar S Akbari. 2018. "Gene Editing Technologies and Applications for Insects." *Current Opinion in Insect Science* 28 (August): 66–72. https://doi.org/10.1016/j.cois.2018.05.006.

Gatehouse, A. M. R., N. Ferry, M. G. Edwards, and H. A. Bell. 2011. "Insect-Resistant Biotech Crops and Their Impacts on Beneficial Arthropods." *Philosophical Transactions of the Royal Society B: Biological Sciences* 366 (1569): 1438–52. https://doi.org/10.1098/rstb.2010.0330.

Gu, Xiaoyong, Lijing Liu, and Huawei Zhang. 2021. "Transgene-Free Genome Editing in Plants." *Frontiers in Genome Editing* 3 (December). https://doi.org/10.3389/fgeed.2021.805317.

Hillary, V. Edwin, Stanislaus Antony Ceasar, and S. Ignacimuthu. 2020. "Genome Engineering in Insects: Focus on the CRISPR/Cas9 System." In *Genome Engineering via CRISPR-Cas9 System*, 219–49. Elsevier. https://doi.org/10.1016/B978-0-12-818140-9.00018-0.

Huang, Yuping, Yazhou Chen, Baosheng Zeng, Yajun Wang, Anthony A. James, Geoff M. Gurr, Guang Yang, Xijian Lin, Yongping Huang, and Minsheng You. 2016. "CRISPR/Cas9 Mediated Knockout of the Abdominal-A Homeotic Gene in the Global Pest, Diamondback Moth (*Plutella xylostella*)." *Insect Biochemistry and Molecular Biology* 75 (August): 98–106. https://doi.org/10.1016/j.ibmb.2016.06.004.

Jain, Ritesh G., Karl E. Robinson, Stephen J. Fletcher, and Neena Mitter. 2020. "RNAi-Based Functional Genomics in Hemiptera." *Insects* 11 (9): 557. https://doi.org/10.3390/insects11090557.

Jiang, Dong, Zhenwu Yu, Yingshi He, Falong Wang, Yucheng Gu, T. G. Emyr Davies, Zhijin Fan, Xingliang Wang, and Yidong Wu. 2024. "Key Role of the Ryanodine Receptor I4790K Mutation in Mediating Diamide Resistance in *Plutella xylostella*." *Insect Biochemistry and Molecular Biology* 168 (May): 104107. https://doi.org/10.1016/j.ibmb.2024.104107.

Kamal Meena, Vijay, Harsh Vardhan Singh Shekhawat, Subhash Chand, Kapil Choudhary, Jitendra Kumar Sharma, and Lekha Lekha. 2023. "Advances in Molecular Marker Technology and Their Significance in Plant Improvement Strategies." In *Recent Trends in Plant Breeding and Genetic Improvement*. IntechOpen. https://doi.org/10.5772/intechopen.1002773.

Kantor, Ariel, Michelle McClements, and Robert MacLaren. 2020. "CRISPR-Cas9 DNA Base-Editing and Prime-Editing." *International Journal of Molecular Sciences* 21 (17): 6240. https://doi.org/10.3390/ijms21176240.

Khabbazi, Saber Delpasand, Afsaneh Delpasand Khabbazi, and Cennet Yaman. 2024. "Different Generations of Genetically Modified Crops for Insect Resistance." *In Plant Resistance to Insects in Major Field Crops*, 279–99. Singapore: Springer Nature Singapore. https://doi.org/10.1007/978-981-99-7520-4_11.

Khalil, Ahmad M. 2020. "The Genome Editing Revolution: Review." *Journal of Genetic Engineering and Biotechnology* 18 (1): 68. https://doi.org/10.1186/s43141-020-00078-y.

Komal, J., H. R. Desai, Ipsita Samal, Andrea Mastinu, R. D. Patel, P. V. Dinesh Kumar, Prasanta Kumar Majhi, Deepak Kumar Mahanta, and Tanmaya Kumar Bhoi. 2023. "Unveiling the Genetic Symphony: Harnessing CRISPR-Cas Genome Editing for Effective Insect Pest Management." *Plants* 12 (23): 3961. https://doi.org/10.3390/plants12233961.

Koutroumpa, Fotini A., Christelle Monsempes, Marie-Christine François, Anne de Cian, Corinne Royer, Jean-Paul Concordet, and Emmanuelle Jacquin-Joly. 2016. "Heritable Genome Editing with CRISPR/Cas9 Induces Anosmia in a Crop Pest Moth." *Scientific Reports* 6 (1): 29620. https://doi.org/10.1038/srep29620.

Kushanov, Fakhriddin N., Ozod S. Turaev, Dilrabo K. Ernazarova, Bunyod M. Gapparov, Barno B. Oripova, Mukhlisa K. Kudratova, Feruza U. Rafieva, et al. 2021a. "Genetic Diversity, QTL Mapping, and Marker-Assisted Selection Technology in Cotton (*Gossypium Spp.*)." *Frontiers in Plant Science* 12 (December). https://doi.org/10.3389/fpls.2021.779386.

Lassoued, Rim, Diego Maximiliano Macall, Stuart J. Smyth, Peter W. B. Phillips, and Hayley Hesseln. 2019. "Risk and Safety Considerations of Genome Edited Crops: Expert Opinion." *Current Research in Biotechnology* 1 (November): 11–21. https://doi.org/10.1016/j.crbiot.2019.08.001.

Lester, Philip J., Mariana Bulgarella, James W. Baty, Peter K. Dearden, Joseph Guhlin, and John M. Kean. 2020. "The Potential for a CRISPR Gene Drive to Eradicate or Suppress Globally Invasive Social Wasps." *Scientific Reports* 10 (1): 12398. https://doi.org/10.1038/s41598-020-69259-6.

Li, F., X. Zhao, M. Li, K. He, C. Huang, Y. Zhou, Z. Li, and J. R. Walters. 2019. "Insect Genomes: Progress and Challenges." *Insect Molecular Biology* 28 (6): 739–58. https://doi.org/10.1111/imb.12599.

Li, Jia-Li, Xiang-Lin Zhuang, Ting-Ting Yuan, Xiao-Ming Cai, Zong-Xiu Luo, Lei Bian, Zong-Mao Chen, Zhao-Qun Li, and Nai-Yong Liu. 2022. "Knockout of the EgriBLOS2 Gene Results in the Transparent Integuments of *Ectropis Grisescens* Larvae." *Journal of Asia-Pacific Entomology* 25 (1): 101876. https://doi.org/10.1016/j.aspen.2022.101876.

Li, Yan, Jie Zhang, Dafeng Chen, Pengcheng Yang, Feng Jiang, Xianhui Wang, and Le Kang. 2016. "CRISPR/Cas9 in Locusts: Successful Establishment of an Olfactory Deficiency Line by Targeting the Mutagenesis of an Odorant Receptor Co-Receptor (Orco)." *Insect Biochemistry and Molecular Biology* 79 (December): 27–35. https://doi.org/10.1016/j.ibmb.2016.10.003.

Liu, Ming, Rongchumu Ge, Lihong Song, Yan Chen, Shuo Yan, and Chunya Bu. 2024. "The Chitinase Genes TuCht4 and TuCht10 Are Indispensable for Molting and Survival of *Tetranychus urticae*." *Insect Biochemistry and Molecular Biology* 171 (August): 104150. https://doi.org/10.1016/j.ibmb.2024.104150.

Liu, Shaoshuai, Maelle Jaouannet, D'Maris Amick Dempsey, Jafargholi Imani, Christine Coustau, and Karl-Heinz Kogel. 2020. "RNA-Based Technologies for Insect Control in Plant Production." *Biotechnology Advances* 39 (March): 107463. https://doi.org/10.1016/j.biotechadv.2019.107463.

Liu, Xiao-Ming, Bang-Xian Zhang, Shi-Guang Li, Xiang-Jun Rao, Dong-Ming Wang, Xiao-Xuan Hu, and Su Liu. 2016. "Knockdown of the Olfactory Co-Receptor Orco Impairs Mate Recognition in *Tenebrio molitor* (Coleoptera: Tenebrionidae)." *Journal of Asia-Pacific Entomology* 19 (2): 503–8. https://doi.org/10.1016/j.aspen.2016.05.005.

Manzoor, Subaya, Sajad Un Nabi, Tariq Rasool Rather, Gousia Gani, Zahoor Ahmad Mir, Ab Waheed Wani, Sajad Ali,

Anshika Tyagi, and Nazia Manzar. 2024. “Advancing Crop Disease Resistance through Genome Editing: A Promising Approach for Enhancing Agricultural Production.” *Frontiers in Genome Editing* 6 (June). https://doi.org/10.3389/fgeed.2024.1399051.

Marec, František, and Marc J. B. Vreysen. 2019. “Advances and Challenges of Using the Sterile Insect Technique for the Management of Pest Lepidoptera.” *Insects* 10 (11): 371. https://doi.org/10.3390/insects10110371.

Marone, Daniela, Anna Maria Mastrangelo, and Grazia Maria Borrelli. 2023. “From Transgenesis to Genome Editing in Crop Improvement: Applications, Marketing, and Legal Issues.” *International Journal of Molecular Sciences* 24 (8): 7122. https://doi.org/10.3390/ijms24087122.

Medina, Raul F. 2018. “Gene Drives and the Management of Agricultural Pests.” *Journal of Responsible Innovation* 5 (sup1): S255–62. https://doi.org/10.1080/23299460.2017.1407913.

Meng, Xiangkun, Chunrui Li, Haibo Bao, Jichao Fang, Zewen Liu, and Yixi Zhang. 2015. “Validating the Importance of Two Acetylcholinesterases in Insecticide Sensitivities by RNAi in *Pardosa pseudoannulata*, an Important Predatory Enemy against Several Insect Pests.” *Pesticide Biochemistry and Physiology* 125 (November): 26–30. https://doi.org/10.1016/j.pestbp.2015.06.006.

Miglani, Gurbachan S. 2017. “Genome Editing in Crop Improvement: Present Scenario and Future Prospects.” *Journal of Crop Improvement* 31 (4): 453–559. https://doi.org/10.1080/15427528.2017.1333192.

Mishra, Jitendra, Venkatesh Dutta, and Naveen Kumar Arora. 2020. “Biopesticides in India: Technology and Sustainability Linkages.” *3 Biotech* 10 (5): 210. https://doi.org/10.1007/s13205-020-02192-7.

Mishra, Riya, Makhan Singh Karada, Dheer Agnihotri, and Niraj Kumar Yadav. 2024c. “Advances in Molecular Genetics: Techniques and Applications.” In *Advances in Genetics and Plant Breeding*, edited by A Tigga, M Shahani, K Modunshim, Longkho K, and Kumari S, 44–65. Stella International Publication.

Mishra, Riya, M. K. Tripathi, M. K. Shrivastava, and Pawan K. Amrate. 2024b. “Genetic Diversity in Crop Improvement: A Cornerstone for Sustainable Agriculture and Global Food Security.” In *Advances in Plant Biotechnology*, edited by Manoj Kumar Tripathi and Niraj Tripathi, 1:1–21. Cornous Publications LLP, Puducherry, India. https://doi.org/https://doi.org/10.37446/volbook032024/1-21.

Mishra, Riya, M. K. Tripathi, Niraj Tripathi, Jagendra Singh, Pramod Kumar Yadav, R. S. Sikarwar, Yogendra Singh, and Sharad Tiwari. 2024a. “Breeding for Major Genes against Drought Stress in Soybean.” In *Advances in Plant Biotechnology*, edited by Manoj Kumar Tripathi and Niraj Tripathi, 1:22–68. Cornous Publications LLP, Puducherry, India. https://doi.org/https://doi.org/10.37446/volbook032024/22-68.

Moon, Tasfia Tasnim, Ishrat Jahan Maliha, Abdullah Al Moin Khan, Moutoshi Chakraborty, Md Sharaf Uddin, Md Ruhul Amin, and Tofazzal Islam. 2022. “CRISPR-Cas Genome Editing for Insect Pest Stress Management in Crop Plants.” *Stresses* 2 (4): 493–514. https://doi.org/10.3390/stresses2040034.

Movahedi, Ali, Soheila Aghaei-Dargiri, Hongyan Li, Qiang Zhuge, and Weibo Sun. 2023. “CRISPR Variants for Gene Editing in Plants: Biosafety Risks and Future Directions.” *International Journal of Molecular Sciences* 24 (22): 16241. https://doi.org/10.3390/ijms242216241.

Naqvi, Rubab Zahra, Hamid Anees Siddiqui, Muhammad Arslan Mahmood, Syed Najeebullah, Aiman Ehsan, Maryam Azhar, Muhammad Farooq, et al. 2022. “Smart Breeding Approaches in Post-Genomics Era for Developing Climate-Resilient Food Crops.” *Frontiers in Plant Science* 13 (September). https://doi.org/10.3389/fpls.2022.972164.

Nolan, Tony. 2021. “Control of Malaria-Transmitting Mosquitoes Using Gene Drives.” *Philosophical Transactions of the Royal Society B: Biological Sciences* 376 (1818): 20190803. https://doi.org/10.1098/rstb.2019.0803.

Palli, Subba Reddy. 2014. “RNA Interference in Colorado Potato Beetle: Steps toward Development of DsRNA as a Commercial Insecticide.” *Current Opinion in Insect Science* 6 (December): 1–8. https://doi.org/10.1016/j.cois.2014.09.011.

Panfilio, Kristen A., Iris M. Vargas Jentzsch, Joshua B. Benoit, Deniz Erezyilmaz, Yuichiro Suzuki, Stefano Colella, Hugh M. Robertson, et al. 2019. “Molecular Evolutionary Trends and Feeding Ecology Diversification in the Hemiptera, Anchored by the Milkweed Bug Genome.” *Genome Biology* 20 (1): 64. https://doi.org/10.1186/s13059-019-1660-0.

Peng, Tianfei, Yiou Pan, Chen Yang, Xiwu Gao, Jinghui Xi, Yongqiang Wu, Xiao Huang, et al. 2016. “Over-Expression of CYP6A2 Is Associated with Spirotetramat Resistance and Cross-Resistance in the Resistant Strain of *Aphis gossypii* Glover.” *Pesticide Biochemistry and Physiology* 126 (January): 64–69. https://doi.org/10.1016/j.pestbp.2015.07.008.

Perez-Pinera, Pablo, David G Ousterout, and Charles A Gersbach. 2012. “Advances in Targeted Genome Editing.” *Current Opinion in Chemical Biology* 16 (3–4): 268–77. https://doi.org/10.1016/j.cbpa.2012.06.007.

Riaz, Saman, Idrees Ahmad Nasir, Muhammad Umar Bhatti, Olawale Samuel Adeyinka, Nida Toufiq, Iqra Yousaf, and Bushra Tabassum. 2020. “Resistance to *Chilo infuscatellus* (Lepidoptera: Pyraloidea) in Transgenic Lines of Sugarcane Expressing Bacillus Thuringiensis Derived Vip3A Protein.” *Molecular Biology Reports* 47 (4): 2649–58. https://doi.org/10.1007/s11033-020-05355-0.

Riga, Maria, Shane Denecke, Ioannis Livadaras, Sven Geibel, Ralf Nauen, and John Vontas. 2020. “Development of Efficient RNAi in *Nezara viridula* for Use in Insecticide Target Discovery.” *Archives of Insect Biochemistry and Physiology* 103 (3). https://doi.org/10.1002/arch.21650.

Rosenberg, Eugene. 2017. “Genetic Engineering.” In *It's in Your DNA*, 81–93. Elsevier. https://doi.org/10.1016/B978-0-12-812502-1.00010-X.

Rouck, Sander de, Antonio Mocchetti, Wannes Dermauw, and Thomas van Leeuwen. 2024. “SYNCAS: Efficient CRISPR/Cas9 Gene-Editing in Difficult to Transform Arthropods.” *Insect Biochemistry and Molecular Biology* 165 (February): 104068. https://doi.org/10.1016/j.ibmb.2023.104068.

Salim, Muhammad, Allah Bakhsh, and Ayhan Gökçe. 2021. “Stacked Insecticidal Genes in Potatoes Exhibit Enhanced Toxicity against Colorado Potato Beetle, *Leptinotarsa decemlineata* (Coleoptera: Chrysomelidae).” *Plant Biotechnology Reports* 15 (2): 197–215. https://doi.org/10.1007/s11816-021-00668-3.

Santos, Dulce, Jozef vanden Broeck, and Niels Wynant. 2014. “Systemic RNA Interference in Locusts: Reverse Genetics and Possibilities for Locust Pest Control.” *Current Opinion in Insect Science* 6 (December): 9–14. https://doi.org/10.1016/j.cois.2014.09.013.

Saurabh, Satyajit. 2021. “Genome Editing: Revolutionizing the Crop Improvement.” *Plant Molecular Biology Reporter* 39 (4): 752–72. https://doi.org/10.1007/s11105-021-01286-7.

Scott, Maxwell J., Fred Gould, Marcé Lorenzen, Nathaniel Grubbs, Owain Edwards, and David O'Brochta. 2017. "Agricultural Production: Assessment of the Potential Use of Cas9-Mediated Gene Drive Systems for Agricultural Pest Control." *Journal of Responsible Innovation* 5 (sup1): S98–120. https://doi.org/10.1080/23299460.2017.1410343.

Singh, Roshan Kumar, Ashish Prasad, Mehanathan Muthamilarasan, Swarup K. Parida, and Manoj Prasad. 2020. "Breeding and Biotechnological Interventions for Trait Improvement: Status and Prospects." *Planta* 252 (4): 54. https://doi.org/10.1007/s00425-020-03465-4.

Singh, Sanchita, Somnath Rahangdale, Shivali Pandita, Gauri Saxena, Santosh Kumar Upadhyay, Geetanjali Mishra, and Praveen C. Verma. 2022. "CRISPR/Cas9 for Insect Pests Management: A Comprehensive Review of Advances and Applications." *Agriculture* 12 (11): 1896. https://doi.org/10.3390/agriculture12111896.

Son, Seungmin, and Sang Ryeol Park. 2022. "Challenges Facing CRISPR/Cas9-Based Genome Editing in Plants." *Frontiers in Plant Science* 13 (May). https://doi.org/10.3389/fpls.2022.902413.

Souza, Dariane, Shawn A. Christensen, Ke Wu, Lyle Buss, Kaylin Kleckner, Constance Darrisaw, Paul D. Shirk, and Blair D. Siegfried. 2022. "RNAi-Induced Knockdown of White Gene in the Southern Green Stink Bug (*Nezara viridula* L.)." *Scientific Reports* 12 (1): 10396. https://doi.org/10.1038/s41598-022-14620-0.

Souza Pacheco, Inaiara de, Anna-Louise A. Doss, Beatriz G. Vindiola, Dylan J. Brown, Cassandra L. Ettinger, Jason E. Stajich, Richard A. Redak, Linda L. Walling, and Peter W. Atkinson. 2022. "Efficient CRISPR/Cas9-Mediated Genome Modification of the Glassy-Winged Sharpshooter *Homalodisca vitripennis* (Germar)." *Scientific Reports* 12 (1): 6428. https://doi.org/10.1038/s41598-022-09990-4.

Tajudeen, Yusuf Amuda, Habeebullah Jayeola Oladipo, Iyiola Olatunji Oladunjoye, Muhammad Kamaldeen Oladipo, Hameedat Damilola Shittu, Imam-Fulani Abdulmumeen, Abdullateef Opeyemi Afolabi, and Mona Said El-Sherbini. 2023. "Transforming Malaria Prevention and Control: The Prospects and Challenges of Gene Drive Technology for Mosquito Management." *Annals of Medicine* 55 (2). https://doi.org/10.1080/07853890.2024.2302504.

Talakayala, Ashwini, Sumalatha Katta, and Mallikarjuna Garladinne. 2020. "Genetic Engineering of Crops for Insect Resistance: An Overview." *Journal of Biosciences* 45. https://doi.org/https://doi.org/10.1007/s12038-020-00081-y.

Tang, Rui, SaLi Li, Junhao Liang, Hao Yi, Xiangfeng Jing, and Tong-Xian Liu. 2022. "Optimization of the Application of the CRISPR/Cas9 System in *Mythimna separata*." *Entomologia Experimentalis et Applicata* 170 (7): 593–602. https://doi.org/10.1111/eea.13184.

Tyagi, Anshika, Zahoor Ahmad Mir, Mohammed A. Almalki, Rupesh Deshmukh, and Sajad Ali. 2024. "Genomics-Assisted Breeding: A Powerful Breeding Approach for Improving Plant Growth and Stress Resilience." *Agronomy* 14 (6): 1128. https://doi.org/10.3390/agronomy14061128.

Tyagi, Shaily, Karthik Kesiraju, Manjesh Saakre, Maniraj Rathinam, Venkat Raman, Debasis Pattanayak, and Rohini Sreevathsa. 2020. "Genome Editing for Resistance to Insect Pests: An Emerging Tool for Crop Improvement." *ACS Omega* 5 (33): 20674–83. https://doi.org/10.1021/acsomega.0c01435.

Wang, Jing, Haonan Zhang, Huidong Wang, Shan Zhao, Yayun Zuo, Yihua Yang, and Yidong Wu. 2016. "Functional Validation of Cadherin as a Receptor of Bt Toxin Cry1Ac in *Helicoverpa armigera* Utilizing the CRISPR/Cas9 System." *Insect Biochemistry and Molecular Biology* 76 (September): 11–17. https://doi.org/10.1016/j.ibmb.2016.06.008.

Wang, Yueqiang, Zhiqian Li, Jun Xu, Baosheng Zeng, Lin Ling, Lang You, Yazhou Chen, Yongping Huang, and Anjiang Tan. 2013. "The CRISPR/Cas System Mediates Efficient Genome Engineering in *Bombyx mori*." *Cell Research* 23 (12): 1414–16. https://doi.org/10.1038/cr.2013.146.

Wani, Shabir H., Mukesh Choudhary, Rutwik Barmukh, Pravin K. Bagaria, Kajal Samantara, Ali Razzaq, Jagdish Jaba, Malick Niango Ba, and Rajeev K. Varshney. 2022. "Molecular Mechanisms, Genetic Mapping, and Genome Editing for Insect Pest Resistance in Field Crops." *Theoretical and Applied Genetics* 135 (11): 3875–95. https://doi.org/10.1007/s00122-022-04060-9.

Weng, Shih-Che, Reem A. Masri, and Omar S. Akbari. 2024. "Advances and Challenges in Synthetic Biology for Mosquito Control." *Trends in Parasitology* 40 (1): 75–88. https://doi.org/10.1016/j.pt.2023.11.001.

Wu, Jian-Jian, Li-Li Mu, Wei-Nan Kang, Long-Ji Ze, Chen-Hui Shen, Lin Jin, Ahmad Ali Anjum, and Guo-Qing Li. 2021. "RNA Interference Targeting Ecdysone Receptor Blocks the Larval – Pupal Transition in *Henos epilachna Vigintiocto punctata*." *Insect Science* 28 (2): 419–29. https://doi.org/10.1111/1744-7917.12777.

Xu, Jun, Xia Xu, Shuai Zhan, and Yongping Huang. 2019. "Genome Editing in Insects: Current Status and Challenges." *National Science Review* 6 (3): 399–401. https://doi.org/10.1093/nsr/nwz008.

Xu, Xia, Yaohui Wang, Honglun Bi, Jun Xu, Zulian Liu, Changying Niu, Lin He, Anthony A. James, Kai Li, and Yongping Huang. 2020. "Mutation of the Seminal Protease Gene, Serine Protease 2, Results in Male Sterility in Diverse Lepidopterans." *Insect Biochemistry and Molecular Biology* 116 (January): 103243. https://doi.org/10.1016/j.ibmb.2019.103243.

Yadav, Priya, Rakesh K. Seth, and Stuart E. Reynolds. 2024. "A Sperm-Activating Trypsin-like Protease from the Male Reproductive Tract of Spodoptera Litura: Proteomic Identification, Sequence Characterization, Gene Expression Profile, RNAi and the Effects of Ionizing Radiation." *Journal of Insect Physiology* 156 (July): 104664. https://doi.org/10.1016/j.jinsphys.2024.104664.

Yadav, Rakesh Kumar, Tripathi Manoj Kumar, Tiwari Susma, Tripathi Niraj, Asati Ruchi, Chauhan Shailja, Tiwari Prakash Narayan, and K. Payasi Devendra. 2023. "Genome Editing and Improvement of Abiotic Stress Tolerance in Crop Plants." *Life* 13 (7): 1456. https://doi.org/10.3390/life13071456

Yan, Ying, Roswitha A. Aumann, Irina Hacker, and Marc F. Schetelig. 2023. "CRISPR-Based Genetic Control Strategies for Insect Pests." *Journal of Integrative Agriculture* 22 (3): 651–68. https://doi.org/10.1016/j.jia.2022.11.003.

Yıldırım, Kubilay, Dragana Miladinović, Jeremy Sweet, Meleksen Akin, Vladislava Galović, Musa Kavas, Milica Zlatković, and Eugenia de Andrade. 2023. "Genome Editing for Healthy Crops: Traits, Tools and Impacts." *Frontiers in Plant Science* 14 (October). https://doi.org/10.3389/fpls.2023.1231013.

Zhang, Hui, Jinshan Zhang, Zhaobo Lang, José Ramón Botella, and Jian-Kang Zhu. 2017. "Genome Editing – Principles and Applications for Functional Genomics Research and Crop Improvement." *Critical Reviews in Plant Sciences* 36 (4): 291–309. https://doi.org/10.1080/07352689.2017.1402989.

Zhu, Tao-Tao, Qing-Wei Meng, Wen-Chao Guo, and Guo-Qing Li. 2015. "RNA Interference Suppression of the Receptor

Tyrosine Kinase Torso Gene Impaired Pupation and Adult Emergence in *Leptinotarsa decemlineata*." *Journal of Insect Physiology* 83 (December): 53–64. https://doi.org/10.1016/j.jinsphys.2015.10.005.

Zhuo, Ji-Chong, Jian Xue, Jia-Bao Lu, Hai-Jian Huang, Hai-Jun Xu, and Chuan-Xi Zhang. 2017. "Effect of RNAi-Mediated Knockdown of NlTOR Gene on Fertility of Male *Nilaparvata lugens*." *Journal of Insect Physiology* 98 (April): 149–59. https://doi.org/10.1016/j.jinsphys.2017.01.002.

Zuma, Mnqobi, Georgina Arthur, Roger Coopoosamy, and Kuben Naidoo. 2023. "Incorporating Cropping Systems with Eco-Friendly Strategies and Solutions to Mitigate the Effects of Climate Change on Crop Production." *Journal of Agriculture and Food Research* 14 (December): 100722. https://doi.org/10.1016/j.jafr.2023.100722.

15 Climate Change Resilience and Entomology

Navigating the Challenges Ahead

Bindu Gudi Ramakrishna[1], *Balaji B. N.*[2], *Gopu Sushma*[3], *Jasti Sri Vishnu Murthy*[1*], *Nikhil Reddy K. S.*[4], *Ranjith M. T.*[1], *and Mani Chellappan*[1*]

[1]Department of Entomology, College of Agriculture, Kerala Agricultural University, Vellanikkara, Thrissur 680656, India

[2]Department of Entomology, College of Agriculture, UAS, GKVK, Bangalore, Karnataka 560065, India

[3]Department of Entomology, College of Agriculture, PJTSAU, Rajendra Nagar, Hyderabad 500030

[4]Department of Entomology, Keladi Shivappa Nayaka University of Agricultural and Horticultural Sciences, Shivamogga 577204, India

*Corresponding author

15.1 INTRODUCTION

Climate change is one of the ongoing concerns for humankind in the 21st century. Global temperatures are predicted to rise by 1.5°C by 2050, and 2 to 4°C by 2100. Climate change impacts insect communities, which have important roles in agriculture, forestry, and ecosystem services (Bonelli et al., 2022; Skendži et al., 2021a; Srivastava et al., 2020). Insects are the foundational members of various ecosystems and are highly susceptible to environmental changes. Insects are pollinators, decomposers, and a staple diet for many species; thus, their health is central to maintaining stable ecosystems and agriculture. The understanding of insects can help understand the response of ecosystems to altered conditions, how they adapt, and how they adjust to new conditions that have been present through early warning signs and changes in the impacts of climate change. Changes in insect populations affect ecosystem services, such as pollination, nutrient cycling, and pest control, which are all significant determinants of food security and biodiversity. Climatic changes can also alter the distributions of insects; such influences promote the dispersal of pests and invasive species that increasingly threaten agriculture and human health (Kingsolver et al., 2011). The dynamic characteristics allow for more distinct disruptions in food webs, along with follow-up on the relative dynamics of ecological changes and the development of strategies for managing biodiversity in a warming world (Ali et al., 2018). Understanding the impact of anthropogenic climate change on natural systems is arguably one of the most significant challenges for ecological sciences in the 21st century (Garcia et al., 2014). One point of interest is insect response to contemporary climate change, since insects represent the most diverse group of multicellular organisms on Earth and are integral to freshwater and terrestrial ecosystem functioning. The issue has become increasingly important with recent reports of worldwide insect population declines (Ali et al., 2023). This chapter covers the impacts of climate change on insect ecosystems, its effects on agricultural systems, the role of invasive species, and technological and ecological strategies for building climate resilience in entomology.

15.2 EFFECTS OF VARIOUS CLIMATE CHANGE FACTORS ON INSECTS

Climate change impacts insect ecosystems through rising temperatures, altered precipitation, and extreme weather events that affect insect development, reproduction, and geographical ranges (Kocmánková et al., 2009). (Figure 15.1). Temperature changes also affect interactions with predators, prey, and host plants, threatening biodiversity and ecosystem services (Shabir et al., 2023). Pollinators, such as bees, face stress from shifting bloom times, which affects crop yields (Goulson et al., 2015). Shifts in precipitation impact water-dependent insects, with droughts reducing habitats for species like dragonflies, while flooding disrupts breeding (Hickling et al., 2006; Deutsch et al., 2008). Increased CO_2 levels alter insect physiology and behavior and impact the plants insects feed, disrupting key ecosystem functions, like soil health, nutrient cycling, and food webs. These changes threaten biodiversity and the stability of ecosystems (Hallmann et al., 2017).

 DOI: 10.1201/9781003516538-15

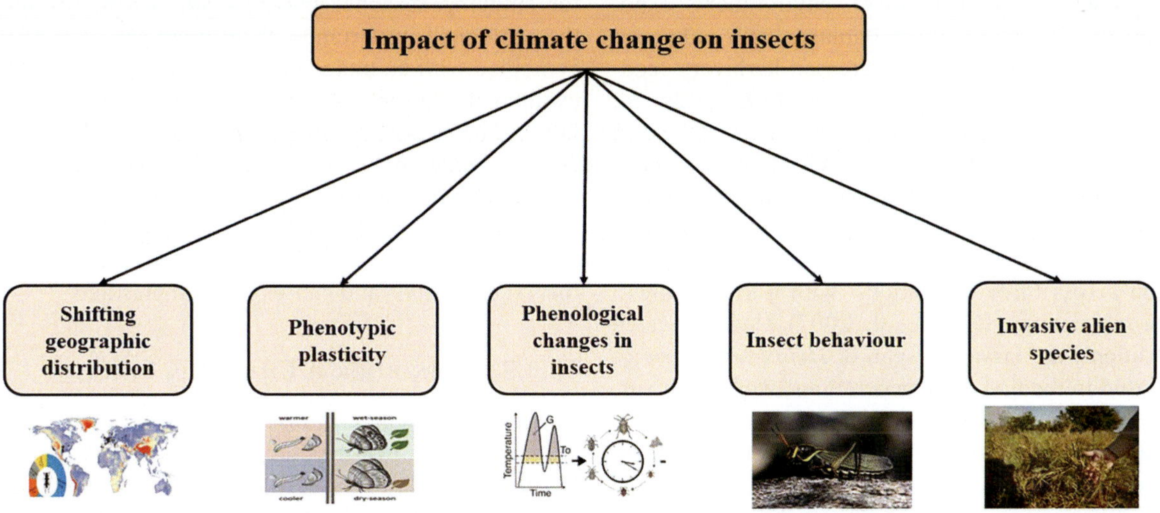

FIGURE 15.1 How climate change influences insect populations through shifting geographic distribution, phenotypic plasticity, phenological changes, altered behavior, and the spread of invasive alien species.

15.2.1 Temperature

Temperature profoundly affects insect life cycles, as most insects are ectothermic. Rising temperatures typically accelerate developmental rates, allowing insects to complete their life cycles more rapidly (Kocmánková et al., 2009) (Figure 15.1). For instance, the European corn borer, *Ostrinia nubilalis* Hübner, has experienced additional generations per year in warmer regions, posing a greater threat to crops by increasing the frequency of infestations (Hutchison et al., 2010). Similarly, the diamondback moth, *Plutella xylostella* Linnaeus, a pest of cruciferous crops, experiences faster development and increased population sizes in warmer conditions, which can lead to enhanced resistance to insecticides (Furlong et al., 2013). In colder regions, rising temperatures can allow invasive species to expand their range and overwinter successfully, affecting native ecosystems. The spotted-wing drosophila, *Drosophila suzukii* (Matsumura), restricted to warmer climates, now survives in temperate regions, causing year-round crop damage (Asplen et al., 2015). The temperature change also influences the synchronization between insects and their food sources, leading to mismatches. For example, the oak processionary moth, *Thaumetopoea processionea* Linnaeus, which relies on oak leaves, may emerge too early or too late to benefit from peak leaf availability, disrupting its life cycle and impacting forest health (Marzano et al., 2020). These examples highlight temperature's critical role in insect life cycles, with significant implications for agriculture, pest management, and ecosystem dynamics.

15.2.2 Altered Precipitation Patterns

Altered precipitation patterns due to climate change significantly impact insect survival, affecting the availability of suitable habitats, breeding success, and overall population dynamics (Robinson et al., 2017; Stange & Ayres, 2020). Increased rainfall and flooding can disrupt terrestrial insect habitats. For instance, the monarch butterfly, *Danaus plexippus* Linnaeus, faces habitat loss and lower survival rates in areas with increased rainfall, as excess moisture can reduce the availability of milkweed, the butterfly's critical host plant (Oberhauser et al., 2017). Alternatively, drought conditions can decrease water-dependent habitats for aquatic insects, like mayflies, which rely on stable, oxygen-rich water bodies to complete their life cycles. The prolonged droughts also affect the mayfly populations and have shown a decline in their population due to reduced stream flow and lower oxygen levels (Bogan & Lytle, 2011). Drought also affects soil-dwelling insects by reducing soil moisture, which can influence the availability of food sources and the effectiveness of microbial decomposers, essential for nutrient cycling. For example, the survival and reproduction of the European earthworm, *Lumbricus terrestris* Linnaeus, an important soil aerator and nutrient recycler, are compromised in drier soils, affecting the ecosystem by altering nutrient availability for plants and other soil organisms (Lubbers et al., 2013). The altered precipitation patterns can influence pest dynamics in the fall armyworm, *Spodoptera frugiperda* Smith, a major agricultural pest, which may thrive in moist climates, as increased humidity supports its larval development, exacerbating crop damage (Nagoshi et al., 2019). These examples highlight that changing precipitation patterns profoundly affect insect survival and ecological balance.

15.2.3 Carbon Dioxide (CO_2)

Elevated CO_2 levels are linked to indirect yet substantial impacts on insect herbivory in crop systems, primarily through changes in host plant chemistry and productivity.

Soybean plants grown in high-CO_2 environments experienced up to 57% more insect damage, with enhanced levels of simple sugars, driving increased herbivory feeding (Hamilton, et al., 2005). Elevated CO_2 levels are linked to increased populations of western corn rootworm (*Diabrotica virgifera*) and soybean aphids (*Aphis glycines*), enhancing herbivory in soybean agroecosystems (Dermody et al., 2008). Bollworms fed on cotton bolls under elevated CO_2 showed a significant increase in consumption and frass production across three generations, with a lower growth rate in each generation (Gang et al., 2007). Under elevated CO_2 conditions, the larval lifespan of *Helicoverpa armigera* increased in the first three generations, while their survival rates decreased, also showing delayed development, higher frass output, and increased consumption of artificial diet (Wu et al., 2006).

15.3 IMPACT OF CLIMATE CHANGE ON INSECTS

15.3.1 Shifting Geographic Distribution

Shifting geographic distribution patterns in insects is a significant response to climate change, as rising temperatures and altered precipitation patterns cause many species to expand their ranges (Hill & Dymock, 1989). These shifts are particularly among pest species and vector-borne disease carriers, with considerable ecological and economic implications. For example, the pine processionary moth, *Thaumetopea pityocampa* Schiff, confined to warmer Mediterranean regions, has expanded northward and to higher elevations in Europe, affecting pine forests. This expansion has increased tree mortality and heightened risks to forest biodiversity (Battisti et al., 2005; Roques et al., 2015). Similarly, the Asian tiger mosquito, *Aedes albopictus* (Skuse), has extended into temperate regions in Europe and North America due to milder winters and warmer summers, raising public health concerns (Bonizzoni et al., 2013). In agricultural fields, shifting distributions of pests like the western corn rootworm, *Diabrotica virgifera virgifera* LeConte, threaten crop productivity. Warmer temperatures allow this pest to establish in areas with longer growing seasons, posing challenges for pest management in European countries (Gray et al., 2009). The range expansion and increased overwintering survival of corn earworms (*Heliothis zea* and *Helicoverpa armigera*) pose a significant yield loss risk and challenge pest management in maize. With global warming, *H. armigera* range extension is likely to continue, adding to pest management complexities.

15.3.2 Exhibit Phenotypic Plasticity

Phenotypic plasticity plays a key role in the organism's response to environmental fluctuations, potentially aiding in its adaptation to rapid changes, including global climate change (Foden et al., 2019). For example, the Afro-tropical butterfly, *Bicyclus anynana*, serves as an important model for studying seasonal polyphenism, where insights into the ecological importance of different seasonal forms are combined with an understanding of the developmental and evolutionary aspects of thermal plasticity (Brakefield et al., 2009). The seasonal plasticity in *B. anynana* likely evolved as an adaptation to the African savannah environment's sharply contrasting wet and dry seasons. The developmental temperature, which serves as a cue for the forthcoming season, produces adult phenotypes suited to the environmental conditions and causes thermal plasticity.

15.3.3 Phenological Changes in Insects

Phenological changes in insects are one of the responses to climate change. Warmer temperatures can prompt insects to emerge earlier in the spring, affecting relationships with their host plants, predators, or prey. For example, the green spruce aphid, *Elatobium abietinum* Walker, emerges earlier in response to warmer winters, leading to increased herbivory on spruce trees before the trees are fully prepared to defend themselves. This early aphid activity stresses the trees and can lead to growth reduction and decreased resilience (Straw et al., 2005). Similarly, the oak leafroller moth, *Tortrix viridana* Linnaeus, is an emerging pest in parts of Europe, leading to asynchrony with the emergence of oak leaves, which impacts larval survival and alters forest ecosystem dynamics (Visser & Holleman, 2001). The Carabid beetle, *Carabus auratus* Linnaeus, a predator of crop pests, may no longer be in sync with the emergence of its prey due to altered phenologies, impacting its effectiveness as a biological control agent in agriculture (Asch & Visser, 2007). These phenological shifts and mismatches signify the complex impacts of climate change on ecological interactions, with cascading effects on ecosystem health and agricultural productivity.

15.3.4 Changes in Insect Behavior

Climate change influences insect behavior through altered temperatures, precipitation, and CO_2 levels. For example, Western horse lubber grasshopper (*Taeniopoda eques*) thermoregulates by shifting between microclimates to avoid extreme temperatures. It alternates between vegetation and soil to maintain an ideal temperature of 35.2°C, roosting plants at night, warming the ground in the morning, cooling vegetation at midday, and returning to plants at dusk (Whitman, 1987). Temperature changes can significantly impact mating behaviors in insects. In the case of the red mason bee (*Osmia bicornis* Linnaeus), female bees sensitive to temperature variations are likely to reject male bees, except those capable of producing steady and appealing signals unaffected by the temperature shifts. Although female bees prefer specific male vibrational signals across different temperatures, they only accept males that consistently produce stable vibrations regardless of changes in the ambient temperature (Conrad et al., 2017). The effects of climate change on locust populations have become increasingly apparent, with notable infestations recently reported in northern India (Joshi et al., 2020). The higher temperatures accelerate the hatching and development of eggs in three grasshopper species by extending the active periods of locusts (Guo et al., 2009). For example, a temperature rise of 1.5°C advances the mid-season activity of *Oedaleus asiaticus* by 4.96 days. Similarly, a combination of warming and increased rainfall shifts the late-season activity of *Camponotus fallax* forward by 5.53 days. This convergence threatens grassland ecosystems and poses risks to agriculture and livelihoods dependent on these environments.

15.3.5 Risk of Invasive Alien Insects

The spread of invasive insect species, driven by climate change, poses significant ecological and economic threats by disrupting agricultural ecosystems (Dukes & Mooney, 1999). These species establish themselves in non-native environments, disrupt local ecosystems, outcompete native species, and affect agricultural productivity (Walther et al., 2009; Lach & Hooper-Bui, 2010). Biogeographic studies revealed that climate variables influence species distributions at large spatial scales (Pearman et al., 2007). Climate change may enhance conditions for new invasive, while altering existing invasive, distributions (Walther et al., 2009). It can also modify competitive interactions, potentially reducing the impacts of primary invaders or facilitating secondary invasions (Pearson et al., 2016). The invasion of insects involves a series of steps, including transport, introduction, establishment, and dispersal of invasive alien species (Skendžić et al., 2021a). Climate change directly impacts the transport and introduction phases, as extreme weather events can push pests into new geographic areas with favorable conditions for establishment (FAO, 2020). For example, the Cactus moth (*Cactoblastis cactorum* Berg) was carried from the Caribbean islands to Mexico during the 2005 hurricane season, threatening over 104 species of prickly pear, 38 of which are endemic, posing ecological and economic risks (Mafokoane et al., 2007). Certain insect species are more likely to spread to new regions, often facilitated by pathways that favor their introduction. This can result in significant economic losses.

15.4 EFFECT OF CLIMATE CHANGE ON BENEFICIAL INSECTS AND POLLINATORS

Insects such as honeybees, bumblebees, wasps, butterflies, and flies are crucial pollinators. Among them, bees pollinate about 73% of the world's cultivated crops, followed by flies (19%), wasps (5%), beetles (5%), and butterflies/moths (4%) (Iqbal et al., 2024). Insect pollination supports one-third of the global food production. However, climate change negatively affects pollinators. Rising temperatures are considered the major factor in climate change on plant–pollinator interactions. The temperature changes can affect honeybees directly by influencing their behavior and physiology. It also alters the quality of the floral resources, which can increase or decrease the foraging capacity and development of the colony (Le Conte and Navajas, 2008). The foraging activity of bees is linked with temperature, as it affects their ability to collect nectar and pollen (Reddy et al., 2013). For instance, the foraging activity of little bees, *Apis florae*, is affected in mango orchards when temperatures exceed 32°C (Reddy et al., 2015). The direct effects of elevated CO_2 on honeybees are more difficult to predict, but indirect effects can be measured. Increased CO_2 concentration may alter the balance of carbon and nitrogen in plant tissues, changing nectar composition and its attractiveness to bees (Rusterholz & Erhardt, 1998). Climate change intensifies the spread and severity of diseases, such as *Varroa destructor*, which quickly infiltrated *Apis mellifera* populations due to climate-driven pathogen dynamics (Solignac et al., 2005).

The other group of beneficial insects includes natural enemies. Climate change severely impacts the abundance, distribution, and seasonal patterns of pests and their natural enemies and alters the effectiveness of biocontrol programs (Thomson et al., 2010). Natural enemies, present in the third trophic level are significantly affected by climate change (Furlong & Zalucki, 2017). For instance, aphids are preyed upon by natural enemies, like parasitic wasps and ladybird beetles (Harrington et al., 1999; Hullé et al., 2010). The global temperature increase led to the aphid population's doubling by reducing their life cycle. Early emergence of aphids can lead to prey shortages for natural enemies and cause local extinctions (Hance et al., 2007). Similarly, higher temperatures disrupt the biological control program of the cereal leaf beetle, *Oulema melanopus* (Evans et al., 2013). CO_2 levels, precipitation, and temperature changes affect plant growth and phenology, indirectly influencing prey availability for natural enemies (Lincoln, 1993; Thomson et al., 2010). For example, *Poa annua* grown under elevated CO_2 and extreme temperatures provides diverse nutritional resources for aphid and parasitoid wasps and affects the fitness of parasitoids (Chen et al., 2007).

15.5 EFFECTS OF CLIMATE CHANGE ON INSECT–PLANT INTERACTIONS

Climate change significantly influences insect–plant interactions, often altering the dynamics between herbivorous insects and their host plants in ways that can impact both agricultural systems and natural ecosystems. Rising temperatures increase CO_2 levels, and shifting precipitation patterns can change plant nutritional quality, defensive compounds, and growth rates, affecting insect feeding and fitness (Chandana et al., 2024). For instance, elevated CO_2 can reduce nitrogen concentration in plants, which can diminish the nutritional value for leaf-feeding insects, such as the cabbage looper, *Trichoplusia ni* (Hübner). This shift often causes herbivores to consume more plant material to meet their nutritional needs, leading to greater plant damage (Docherty et al., 1997). Additionally, some plants may alter their defensive chemical composition in response to climate stressors. Studies on the common milkweed, *Asclepias syriaca* L., show that increased temperatures can enhance concentrations of cardenolides, a group of toxic compounds that deter herbivores, like *D. plexippus* caterpillar, potentially impacting their survival and growth (Züst et al., 2015). Changes in the timing of life cycle events, or phenology, also disrupt insect–plant synchrony. For example, earlier leafing in birch trees, *Betula pendula* Roth, due to warming temperatures can lead to a mismatch with the emergence of specialist herbivores, like the winter moth, *Operophtera brumata* (Linnaeus), which rely on synchrony with budburst for optimal larval development (Asch & Visser, 2007). In agriculture, drought conditions affect plant water content and turgor, making plants more susceptible to certain insect pests, such as aphids, which thrive on stressed crops. As climate change continues to modify the ecological interactions between plants and insects, understanding these effects is crucial for predicting potential impacts on biodiversity and developing adaptive strategies for crop protection.

15.6 ECONOMIC IMPLICATIONS OF CLIMATE CHANGE ON AGRICULTURE

Agriculture is highly vulnerable to climate change due to its scale and sensitivity to weather, resulting in substantial economic impacts (Mendelsohn, 2009). Rising temperatures are likely to reduce crop yields by shortening the growth periods (Mahato, 2014), while the effects of temperature increases, rainfall variability, and CO_2 fertilization vary depending on the crop, location, and magnitude of these changes. The impact of climate change on crop yields also depends on region-specific factors. For instance, the production of staple crops like wheat, rice, and maize is expected to decline in temperate and tropical regions if they experience a 2°C warming (Challinor et al., 2014). Tropical regions are more vulnerable to climate change due to their higher temperatures, which increase crop stress and humidity, which increase pests and disease incidence. Other climatic factors, such as humidity, wind speed, and frequent extreme weather events, also affect crop yields. In wheat, extreme weather events can significantly determine the extent of yield reduction (Powell & Reinhard, 2016). The yield reduction due to climate change may lead to increased food prices, adversely affect global agricultural welfare, and cause a 0.3% annual decline in global GDP by 2100 (Stevanović et al., 2016). In India, temperatures are projected to rise between 2.33 and 4.78°C, along with the increase in CO_2 concentration and prolonged heat waves, which could have substantial effects on agricultural productivity (Kumar & Gautam, 2014). The 1°C global mean surface temperature increase will cause yield losses for rice, maize, and wheat, which are expected to worsen by 10–25% (Deutsch et al., 2018).

Plant responses to climate change vary according to species and developmental stages. The specific threshold level of the plants influences root elongation, root growth angle, and yield (Gray & Brady, 2016). Raised CO_2 levels also have the potential to offset yield losses partially. The increased temperature and the reduced soil moisture cause global yield losses by 4–17%. The variations in crop yields across regions are often due to differences in growing environments (Deryng et al., 2016). Elevated CO_2 also affects crop nutrient concentrations, reducing essential nutrients, like nitrogen, iron, zinc, and sulfur, in non-leguminous C3 crops that impact human health (Uddling et al., 2018). The elevated CO_2 levels lead to increased growth and yield at an optimum temperature of 29°C, while higher temperatures reduce seed set in rice (Madan et al., 2012). Higher CO_2 concentrations also lower zinc and iron content in C3 grain crops and legumes, decreasing protein levels. While C4 plants remain unaffected, these nutrient reductions in crops under higher CO_2 concentrations raise concerns about nutritional quality, particularly for human diets in regions heavily reliant on C3 plants (Myers et al., 2014).

Climate change, such as elevated CO_2 levels, reduces plant transpiration, indirectly increasing air temperature and potentially contributing to surface warming. Sometimes, climate change has positive effects, particularly on C3 plants. C3 crops may experience higher productivity under elevated CO_2; water requirements for both C3 and C4 crops are expected to decrease under non-stressful conditions. The study was conducted using CERES (crop estimation through resources and environmental synthesis) and revealed that under doubled CO_2 levels, wheat and rice yields in northwest India could increase by 28% and 15%, respectively. Similarly, current irrigation practices could maintain rice and wheat yields despite increased CO_2 and thermal stress. However, the combination of temperature stress and water scarcity could still lead to declining crop yields (Lal et al., 1998).

15.7 RESILIENT STRATEGIES TO ADDRESS CLIMATE CHANGE IN ENTOMOLOGY

Climate change significantly affects various ecosystems, with insects being among the most affected organisms, which play an essential role in maintaining ecological balance through pollination, decomposition, and interconnecting food webs. However, rising temperatures have an impact on insects and the agricultural systems. To overcome these challenges, resilient strategies are essential, mainly focused on research monitoring, conservation approaches, and adaptive management (Figure 15.2.).

15.7.1 Research Monitoring

Research monitoring is a vital tool in climate-resilient entomology. It enables tracking insect population trends and detects early warning signs of pest outbreaks through monitoring and high-throughput phenotyping.

15.7.1.1 Detection and Modeling

Detection techniques are crucial in preventing the spread of invasive insects. In New Zealand, biosecurity measures

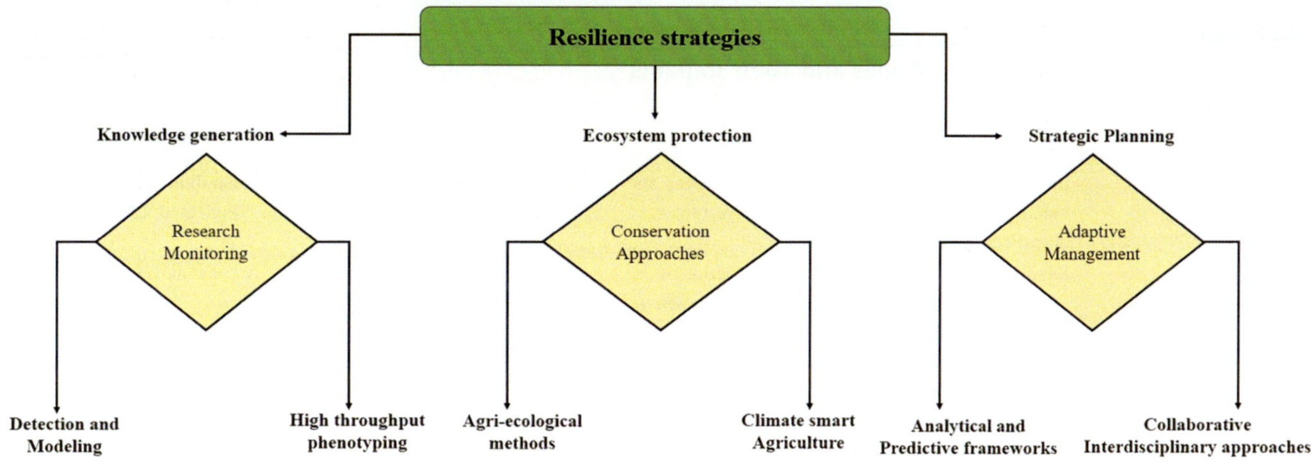

FIGURE 15.2 Framework for enhancing resilience through knowledge generation (research and monitoring), ecosystem protection (conservation approaches), and strategic planning (adaptive management) using various methodologies, including modeling, climate-smart agriculture, and interdisciplinary approaches.

track potential threats and provide reporting guidelines. In China, researchers tested Mediterranean and temperate trees in sentinel areas to detect Asian insect pests by monitoring seedlings over four years. Although they identified potential pests, the process was slow, labor-intensive, and costly. Hence, advanced technology, such as molecular identification, can help overcome this problem. Advanced pest management tools, such as drones equipped with sensors, satellite imagery, and remote sensing, monitor pest populations, crop health, and environmental conditions across large areas. This technology detects early pest infestations and tracks changes in pest dynamics. Internet of Things (IoT) devices power smart traps that capture real-time data on pest behavior and movement, while machine learning helps track the spread and intensity of infestations. AI-based predictive models, using big data, can forecast pest outbreaks with increasing accuracy, analyzing weather and crop growth to predict pests like fall armyworms or bollworms. This allows farmers to apply pest control methods at the optimal time and minimize damage (Heeb et al., 2019). Machine learning and AI algorithms are also being developed to identify pests from images captured by drones, cameras, or smartphones, helping differentiate pest species and assess infestation severity.

15.7.1.2 High-Throughput Phenotyping

In recent years, advanced phenotyping has become a collaborative effort among multidisciplinary research groups to launch high-throughput phenotyping, aimed at accelerating next-generation breeding programs to ensure food security (Fasoula et al., 2020). This technology enables high-resolution imaging of thousands of plants, providing valuable insights into plant phenomics and genetics (Roitsch et al., 2019). It combines plant science, engineering, mathematics, information science, computational tools, and AI and machine learning to explore complex phenotypes across significant genotypes, crucial for developing superior crop cultivars (Furbank et al., 2019). Several cutting-edge phenotyping centers have been set up in Europe, Australia, and the USA. Next-generation breeding technologies offer a robust platform to develop high-yielding and climate-resilient crops. Developing stress-tolerant varieties as a mitigation strategy, such as the SUB1A gene in rice, helps plants withstand the changing environmental stresses (Atlin et al., 2017; Chhogyell et al., 2016).

15.7.2 Ecosystem Conservation

Global temperatures have risen by 0.8°C over the past century, with an expected increase of 1.1 to 5.4°C in the next hundred years. CO_2 levels have also increased significantly and are projected to keep rising. Although elevated CO_2 can stimulate plant growth, higher temperatures and limited water often reduce the benefits, leading to lower yields in essential crops like grains. Additionally, climate change impacts crop productivity by influencing pest and disease patterns, potentially increasing pest-related damage as temperatures and CO_2 levels continue to rise. This underscores the importance of agroecological practices in addressing the growing challenges of climate change. Adapting to climate change is a continuous process involving risk management strategies to mitigate impacts (Howden et al., 2007). Pest management adaptations can be short- or long-term (Juroszek & von Tiedemann, 2011; Olesen & Bindi, 2002).

15.7.2.1 Habitat Conservation and Landscape Connectivity

Traditional and agroecological methods, such as biodiversity practices, soil management, and water harvesting, help build climate-resilient agriculture (Nawaz et al., 2021), by enhancing soil health, reducing erosion, and improving carbon sequestration. Conservation agricultural practices, like zero tillage, no-till farming, cover cropping, nutrient management, and drip irrigation, minimize resource use, lower

TABLE 15.1
Climate-Smart Agricultural Technologies and Their Impacts

Climate-Smart Technology	Impact	Economic Benefit	References
Site-specific nutrient management	Increased partial factor productivity of nitrogen and increased yield by 29% over farmers' fertilizer practices (FFP)	US$34/ha, US$106/ha, US$168/ha, and INR 68,980/ha over FFP	Pampolino et al. (2007); Singh et al. (2015)
Laser land leveling	Saving of 21% irrigation water and reduction of irrigation time, increase in yield, and reduction in electricity charges	INR 23,250/acre, US$143.5/ha/year	Aryal et al. (2015b).
Zero tillage	Higher water productivity, saving of irrigation water, and higher fertilizer use efficiency, enhancing production by 1.88% and lowering cultivation cost, reducing GHGs emissions, and increasing yield	US$97.5/ha	Latif et al. (2013); Tripathi et al. (2013); Aryal et al. (2015a).
Stress-tolerant crop varieties	Increased household income leading to household asset accumulation and investment	Increased income by 83%	Ogada et al. (2020)
Drought-tolerant varieties	Increase in yield by 23%, lower variability, increased share of risk benefits in total benefits	17% reduction in variable cost	Birthal et al. (2012).
Improved crop varieties	Increased net returns	INR 15,712/ha/yr	Khatri-Chhetri et al. (2016).
Drip irrigation	Reduced irrigation and preparation costs, saving of irrigation water and electricity charges, reduced cultivation cost	INR 72,711/acre and 54% higher net returns	Narayanamoorthy and Devika (2018); Narayanamoorthy et al. (2018)
Direct-seeded rice	Saving of irrigation, lesser labor requirement	INR 5,050–INR 8,100/ha over puddled transplanted rice (PTR) – wheat	Bhullar et al. (2018)

greenhouse gas emissions, and reduce cultivation costs (Zou et al., 2013). In South Asia, farmers are adopting zero tillage for wheat cultivation due to the reduction in cultivation cost by 15–16%. Leaf color chart (LCC) has effectively fine-tuned the timing and amount of fertilizer application, resulting in rice yields comparable to the standard recommended fertilizer dose (Ladha et al., 2007). Technologies like laser land leveling, which increases crop yields and reduces input costs, help mitigate losses from climate variability. For example, in Karnataka's Raichur district, LLL increased paddy yields by 0.5 t per hectare, raising net farm income by INR 5,000 annually. Sequential cropping systems and adjusting planting dates in response to climate patterns help reduce crop losses (Verchot et al., 2007).

15.7.2.2 Climate-Smart Agriculture

Climate-smart agriculture (CSA) practices, such as water-smart and nutrient-smart methods, help build resilience in agriculture, improve productivity, and enhance resource efficiency. Climate-smart agriculture technologies aid in the higher productivity of cotton with higher returns and resource-use efficiency (Imran et al., 2018). Agroforestry practices, such as alternate drying in rice, mid-season drainage, improved livestock diet, increasing N-use efficiency, and soil carbon, play a role in reducing atmospheric greenhouse gases and helping farmers cope with climate variability. Practices like aerobic rice cultivation save water by 73% during land preparation and 56% during the crop period. It also reduces methane emissions (Parthasarathi et al., 2012).

Climate-resilient biopesticides, developed from bacteria, fungi, or viruses, can withstand temperature extremes and remain effective. These biopesticides provide an eco-friendly alternative to chemical pesticides, preserving natural pest control agents in the ecosystem (Tan et al., 2022). Scientists are exploring combinations of beneficial microbes that provide broader pest resistance and can tolerate diverse environmental conditions and protect crops without disrupting ecological balance (Fite et al., 2023). In tomatoes, *Beauveria bassiana* successfully reduced damage from *Helicoverpa armigera* (Toffa et al., 2021). Similarly, in common beans (*Phaseolus vulgaris*), *Metarhizium anisopliae* ICIPE 78 effectively lowered damage caused by the bean stem maggot (*Ophiomyia phaseoli*) (Mutune et al., 2016). The successfully implemented climate-smart agricultural technologies are presented in Table 15.1.

15.7.3 Adaptive Management

15.7.3.1 Analytical Frameworks and Policy Integration

Effective climate adaptation in agriculture relies on disseminating technology, market integration, public research support, and capacity building. Policies promoting agricultural research, market access, and extension services help shape farmers' responses to climate challenges. Public policies tailored to address climate variability and socioeconomic, political, and economic support are critical in aiding farmers, mainly smallholders, to adapt to climate extremes. Capacity-building initiatives and agricultural financing linked to climate adaptation are essential to ensure that small and marginal farmers can access climate-resilient strategies, thus improving their resilience and reducing susceptibility to climate-related losses. Technology diffusion is critical in shaping farmers' response to climate change. The main priority areas are market integration and support of public research and capacity-building (Lybbert & Sumner, 2012). The means to adapt to climate change are mainly modified farm practices. They are influenced greatly by the policy decisions suiting climatic variability, climate extremes, and social, political, and economic conditions (Smit & Skinner, 2002).

15.7.3.2 Collaborative Interdisciplinary Approach

Traditional methods have challenges, like long selection times, hybridization barriers, and genetic erosion. Advanced techniques, such as marker-assisted breeding, transcriptomics, metabolomics, phenomics, speed breeding, and genome editing, provide more accurate insights into gene function and regulation. Tools like CRISPR are particularly transformative, enabling precise genetic engineering to enhance crop yield, quality, and stress tolerance (Tejasree et al., 2024). Shuttle breeding, pioneered by Dr. Norman E. Borlaug at CIMMYT, speeds up breeding cycles by growing crops in multiple locations, allowing an extra generation each year. This approach also tests crops for resistance to biotic stresses (Lenaerts et al., 2019). The world is attracted by the most fascinating technology of speed breeding. Watson et al. (2018) from the University of Queensland, inspired by NASA to grow wheat plants in space under artificial lights, successfully developed the protocols for different plant species under a speed breeding system. *Speed breeding* is a powerful strategy to shorten the crop generation time and expedite the breeding programs for crop improvement (Watson et al., 2018). Speed breeding extends the photoperiod to about 22 hours using various light sources, mimicking daily dawn and dusk. This, along with controlled temperature and optimal light intensity, boosts photosynthesis, leading to faster flowering and early seed development, thus reducing generation time (Ghosh et al., 2018). In conventional breeding, only one to two generations per season of any crop can be achieved, but in speed breeding, up to four generations of *Brassica napus* and six generations of *Hordeum vulgare*, *Triticum aestivum*, *Pisum sativum*, *Cicer arietinum*, and *B. distachyon* (Watson et al., 2018); five in winter wheat (Cha et al., 2021), pea (Mobini and Warkentin, 2016), and soybean; four in canola (Watson et al., 2018) and pigeon pea (Saxena et al., 2019); and three to four in groundnut (O'Connor et al., 2013).

Genomic studies of insects reveal the behavior of insects to climate change and can aid in breeding resilient plants or designing targeted control agents (Tejasree et al., 2024). Advances in gene editing have opened doors for developing pest-resistant crop varieties. By targeting genes responsible for pest susceptibility, scientists can create crops that are less affected by pests and are likely to thrive in a warmer climate. Unlike traditional GM methods, genome editing enables precise genetic modifications without adding foreign DNA, such as antibiotic-resistant genes. Techniques like TALENs and CRISPR-Cas allow targeted edits, reducing concerns about external DNA in the final crop (Liang et al., 2018). The genome-editing approaches, such as CRISPR with ribonucleoprotein (RNP) complexes or virus-based DNA replicons, can make precise edits without T-DNA insertion. One such exploration was the use of CRISPR/Cas9 to knock out the *OsRR22* gene, which is linked to salt sensitivity in rice (Zhang et al., 2019). This modification enhanced the rice plant's performance in high-salinity conditions (0.75% NaCl) without reducing grain yield, biomass, or quality. Wang et al. (2016) applied CRISPR/Cas9 in *H. armigera*, demonstrating that *Cry1Ac* resistance depends on the *HaCad* gene, a key *Cry1Ac* receptor. By injecting eggs with Cas9 mRNA and sgRNA, they achieved efficient *HaCad* editing, advancing Bt toxin use. The phytophagous pest *Mythimna separata* was recently targeted with gene editing for the first time. Tang et al. (2022) used CRISPR/Cas9 to edit the *NPC1b* gene, essential for intestinal absorption and sterol transport, by microinjecting pest eggs with a Cas9 and sgRNA complex. Another genomic method used in pest management was RNA interference (RNAi). It is an innate defense and gene regulation mechanism that can disrupt specific gene functions, selectively targeting pest species. RNAi has shown high sequence specificity, enabling targeted degradation of transcripts through precise dsRNA guidance. Targeting a cytochrome P450 gene responsible for producing a detoxifying enzyme in the cotton bollworm (*Helicoverpa armigera*), which can tolerate gossypol and similar toxic sesquiterpene aldehydes, GM plants were developed that express hairpin RNA (hpRNA) to suppress *CYP6AE14* in the cotton bollworm, effectively reducing the larvae's tolerance to gossypol (Mao et al., 2007).

15.8 FUTURE DIRECTIONS IN CLIMATE RESILIENCE FOR ENTOMOLOGY

Climate change continues to accelerate, and establishing a climate resilience area within entomology is becoming more imperative. Here are some promising future directions that would potentially enhance resilience in managing insect populations under changing environmental conditions:

1. *Predictive modeling and early warning systems.* Developing sophisticated predictive models will outline shifts in insect populations, pest outbreaks, and emerging threats to agriculture and biodiversity. Thus, through the incorporation of climate models along with data on insect population and distribution, there is every indication

that practitioners will be equipped for proactive responses to pest management and conservation needs.

2. *Climate-resilient pest management.* Climate-resistant pest management strategies, such as climate-resilient biological control methods, are, therefore, wanted to consider the changing interactions between the pests and the natural enemies and crops. This includes optimizing biological control agents and developing resistant crop varieties.
3. *Landscapes and habitat management.* Preserving and managing diverse habitats may promote natural pest control and support pollinators and other beneficial insect populations. Restoring ecosystems and habitat connectivity ensures beneficial insects can thrive and promote ecological balance while climate conditions change.
4. *Technology-based monitoring and data collection.* Technology such as remote sensing, drones, and IoT sensors allow for real-time monitoring of insects at population levels over large areas. We can better understand the impacts of climate on insects and inform adaptive management by using data analytics and machine learning.
5. *Resilience of pollinators.* With the significant risks to pollinators by climate change, research that increases resilience in key pollinator species is very important. This includes studying the interactions of plants with pollinators under climate stress, testing for robust pollinator species, and promoting pollinator-friendly agricultural practices.
6. *International collaboration and policy.* Climate change is a worldwide issue, so its management will require international cooperation. Cross-border sharing of research, data, and innovations can contribute to the development of a single approach to insect management. In addition, policies that support climate resilience in agriculture and forestry improve research and application for sustainable practice.

This will go a long way in helping entomology overcome the challenges posed by climate change more effectively while at the same time building resilient agricultural, natural, and urban ecosystems. It is a forward approach to ensuring food security, biodiversity, and ecosystem stability in a changing climate.

15.9 CONCLUSION

The impacts of climate change on agriculture and its insects are well established. Still, many uncertainties remain for increases in localized temperatures, CO_2 levels in the atmosphere, changed rainfall regimes, moisture changes, and other environmental factors. The dispersity of responses of insects to global warming is disparate around the globe and, in turn, reflects the great diversity of species of insects as well as their host plants and climate differences at the regional level. The relationship created by climate change is complex. It benefits one insect species while harming another, affecting their geographical distribution, the diversity and size of their populations, the development cycles, the growth patterns, and the seasonal timing of life events. Experts foresee several alarming trends: increases in the incidence and geographical range of outbreaks, more remarkable survival during the winter, increased numbers of generations per season, increased invasion success of immigrant species invading new habitats, increased vectors of plant pathogens by insects, and decreased efficacy of biological control by natural enemies. These developments seriously affect agricultural productivity and food security, with considerable economic implications. We should focus only on such a forward-looking scientific approach – certainly transformed IPM, improved climate, and pest monitoring systems, and advanced techniques for predictions and management – from the stage of understanding to formulating practical adaptation and mitigation strategies.

REFERENCES

Ali, H., Muhammad, A., Islam, S.U., Islam, W., & Hou, Y., 2018. A novel bacterial symbiont association in the hispid beetle, Octodonta nipae (Coleoptera: Chrysomelidae), their dynamics and phylogeny. *Microbial Pathogenesis, 118*, 378–386.

Ali, H., Hou, Y., & Tahir, M.B. (Eds.). (2023). Climate Change and Insect Biodiversity: *Challenges and Implications* (1st ed.). CRC Press. https://doi.org/10.1201/9781003382089

Aryal, J. P., Mehrotra, M. B., Jat, M. L., & Sidhu, H. S. (2015a). Impacts of laser land leveling in rice – wheat systems of the north – western Indo-Gangetic plains of India. *Food Security, 7*(3), 725–738. https://doi.org/10.1007/s12571-015-0460-y

Aryal, J. P., Sapkota, T. B., Jat, M. L., & Bishnoi, D. K. (2015b). On-farm economic and environmental impact of zero-tillage wheat: A case of North-West India. *Experimental Agriculture, 51*(1), 1–16. https://doi.org/10.1017/S001447971400012X

Asch, M. V., & Visser, M. E. (2007). Phenology of forest caterpillars and their host trees: The importance of synchrony. *The Annual Review of Entomology, 52*(1), 37–55. https://doi.org/10.1146/annurev.ento.52.110405.091418

Asplen, M. K., et al. (2015). Invasion biology of spotted wing Drosophila (*Drosophila suzukii*): A global perspective and future priorities. *Journal of Pest Science, 88*, 469–494. https://doi.org/10.1007/s10340-015-0681-z

Atlin, G. N., Cairns, J. E., & Das, B. (2017). Rapid breeding and varietal replacement are critical to adaptation of cropping systems in the developing world to climate change. *Global Food Security, 12*, 31–37. https://doi.org/10.1016/j.gfs.2017.01.008

Battisti, A., Stastny, M., Netherer, S., Robinet, C., Schopf, A., Roques, A., & Larsson, S. (2005). Expansion of geographic range in the pine processionary moth caused by increased winter temperatures. *Ecological Applications, 15*(6), 2084–2096. https://doi.org/10.1890/04-1903

Bhullar, M. S., Singh, S., Kumar, S., & Gill, G. (2018). Agronomic and economic impacts of direct seeded rice in Punjab. *Agricultural Research Journal*, *55*(2). https://doi.org/10.5958/2395-146X.2018.00038.8

Birthal, P. S., & Negi, D. S. (2012). Livestock for higher, sustainable and inclusive agricultural growth. *Economic and Political Weekly, 30*, 89–99.

Bogan, M. T., & Lytle, D. A. (2011). Severe drought drives novel community trajectories in desert stream pools. *Freshwater Biology*, *56*(10), 2070–2081. https://doi.org/10.1111/j.1365-2427.2011.02638.x

Bonelli, M., Eustacchio, E., Avesani, D., Michelsen, V., Falaschi, M., Caccianiga, M., Gobbi, M., & Casartelli, M. (2022). The early season community of flower-visiting arthropods in a high-altitude alpine environment. *Insects*, *14*, 393. https://doi.org/10.3390/insects13040393

Bonizzoni, M., Gasperi, G., Chen, X., & James, A. A. (2013). The invasive mosquito species Aedes albopictus: Current knowledge and future perspectives. *Trends in Parasitology*, *29*(9), 460–468. 10.1016/j.pt.2013.07.003

Brakefield, P. M., Beldade, P., & Peralta, C. M. (2009). Seasonal polyphenism in Bicyclus anynana. *Ecology and Evolutionary Biology*, *15*, 23–33.

Cha, J. K., O'Connor, K., Alahmad, S., Lee, J. H., Dinglasan, E., Park, H., Lee, S. M., Hirsz, D., Kwon, S. W., Kwon, Y., & Kim, K. M. (2021). A new protocol for speed vernalisation of winter cereals. *bioRxiv*, 2021–2012. https://doi.org/10.1101/2021.12.01.470717

Challinor, A. J., Watson, J., Lobell, D. B., Howden, S. M., Smith, D. R., & Chhetri, N. (2014). A meta-analysis of crop yield under climate change and adaptation. *Nature Climate Change*, *4*(4), 287–291. https://doi.org/10.1038/nclimate2153

Chandana, C. R., Nadagouda, S., Sreenivas, A. G., Chandana, T. P., & Hallikeri, V. F. (2024). Climate-smart pest management strategies: Under changing climatic scenarios. *Journal of Advances in Biology & Biotechnology*, *27*(6), 364–377. 10.9734/jabb/2024/v27i6896

Chen, F., Wu, G., Parajulee, M. N., & Ge, F. (2007). Impact of elevated CO2 on the third trophic level: A predator Harmonia axyridis and a parasitoid Aphidius picipes. *Biocontrol Science and Technology*, *17*(3), 313–324. https://doi.org/10.1080/09583150701211814

Chhogyell, N., Pradhan, N., Ghimiray, M., & Bajgai, Y. (2016). Evaluation of short duration rice (*Oryza sativa*) varieties as a strategy to cope with climate change. *Proc Bhutan Ecol Soc*, *1*, 91–103.

Conrad, T., Stöcker, C., & Ayasse, M. (2017). The effect of temperature on male mating signals and female choice in the red mason bee, Osmia bicornis (L.). *Ecology and Evolution*, *7*(21), 8966–8975. https://doi.org/10.1002/ece3.3331

Dermody, O., O'Neill, B. F., Zangerl, A. R., Berenbaum, M. R., & DeLucia, E. H. (2008). Effects of elevated CO_2 and O_3 on leaf damage and insect abundance. *Journal of Experimental Botany*, *59*(2), 279–287. https://doi.org/10.1007/s11829-008-9045-4

Deryng, D., Elliott, J., Folberth, C., Müller, C., Pugh, T. A., Boote, K. J., Conway, D., Ruane, A. C., Gerten, D., Jones, J. W., & Rosenzweig, C. (2016). Regional disparities in the beneficial effects of rising CO2 concentrations on crop water productivity. *Nature Climate Change*, *6*(8), 786–790. https://doi.org/10.1038/nclimate2995

Deutsch, C. A., Tewksbury, J. J., Huey, R. B., Sheldon, K. S., Ghalambor, C. K., Haak, D. C., & Martin, P. R. (2008). Impacts of climate warming on terrestrial ectotherms across latitude. *Proceedings of the National Academy of Sciences*, *105*(18), 6668–6672. https://doi.org/10.1073/pnas.0709472105

Deutsch, C. A., Tewksbury, J. J., Tigchelaar, M., Battisti, D. S., Merrill, S. C., Huey, R. B., & Naylor, R. L. (2018). Increase in crop losses to insect pests in a warming climate. *Science*, *361*(6405), 916–919. https://doi.org/10.1126/science.aat3466

Docherty, M., Wade, F., Hurst, D., Whittaker, J., & Lea, P. (1997). Responses of tree sap-feeding herbivores to elevated CO2. *Global Change Biology*, *3*(1), 51–59. https://doi.org/10.1046/j.1365-2486.1997.00096.x

Dukes, J. S., & Mooney, H. A. (1999). Does global change increase the success of biological invaders? *Trends in Ecology & Evolution*, *14*(4), 135–139. https://doi.org/10.1016/S0169-5347(98)01554-7

Evans, E. W., Carlile, N. R., Innes, M. B., & Pitigala, N. (2013). Warm springs reduce parasitism of the cereal leaf beetle through phenological mismatch. *Journal of Applied Entomology*, *137*(5), 383–391. https://doi.org/10.1111/jen.12028

FAO. *Food and Agriculture Organization Plant Pests and Diseases in the Context of Climate Change and Climate Variability, Food Security and Biodiversity Risks*. Available online: www.fao.org/3/nb088/nb088.pdf (accessed on 12 January 2020).

Fasoula, D. A., Ioannides, I. M., & Omirou, M. (2020). Phenotyping and plant breeding: Overcoming the barriers. *Frontiers in Plant Science*, *10*, 1713. https://doi.org/10.3389/fpls.2019.01713

Fite, T., Kebede, E., Tefera, T., & Bekeko, Z. (2023). Endophytic fungi: Versatile partners for pest biocontrol, growth promotion, and climate change resilience in plants. *Frontiers in Sustainable Food Systems*, *7*, 1322861. https://doi.org/10.3389/fsufs.2023.1322861

Foden, W., et al. (2019). Phenotypic plasticity and adaptation to climate change. *Nature Climate Change*, *9*, 595–601. https://doi.org/10.1016/j.cois.2019.06.007

Furbank, R. T., Jimenez-Berni, J. A., George-Jaeggli, B., Potgieter, A. B., & Deery, D. M. (2019). Field crop phenomics: Enabling breeding for radiation use efficiency and biomass in cereal crops. *New Phytologist*, *223*, 1714–1727. https://doi.org/10.1111/nph.15817

Furlong, M. J., Wright, D. J., & Dosdall, L. M. (2013). Diamondback moth ecology and management: Problems, progress, and prospects. *Annual Review of Entomology*, *58*(1), 517–541. https://doi.org/10.1146/annurev-ento-120811-153605

Furlong, M. J., & Zalucki, M. P. (2017). Climate change and biological control: The consequences of increasing temperatures on host – parasitoid interactions. *Current Opinion in Insect Science*, *20*, 39–44. https://doi.org/10.1016/j.cois.2017.03.006

Gang, L., et al. (2007). Bollworm responses to elevated CO_2. *Entomological Research*, *37*(2), 159–165.

Garcia, R. A., Cabeza, M., Rahbek, C., & Araújo, M. B. (2014). Multiple dimensions of climate change and their implications for biodiversity. *Science*, *344*(6183), 1247579. https://doi.org/10.1126/science.1247579

Ghosh, S., Watson, A., Gonzalez-Navarro, O. E., Ramirez-Gonzalez, R. H., Yanes, L., Mendoza-Suárez, M., Simmonds, J., Wells, R., Rayner, T., Green, P., & Hafeez, A. (2018). Speed breeding in growth chambers and glasshouses for crop breeding and model plant research. *Nature Protocols*, *13*(12), 2944–2963. https://doi.org/10.1038/s41596-018-0072-z

Goulson, D., Nicholls, E., Botías, C., & Rotheray, E. L. (2015). Bee declines driven by combined stress from parasites, pesticides, and lack of flowers. *Science*, *347*(6229), 1255957. https://doi.org/10.1126/science.1255957

Gray, M. E., Sappington, T. W., Miller, N. J., Moeser, J., & Bohn, M. O. (2009). Adaptation and invasiveness of western corn rootworm: Intensifying research on a worsening pest. *Annual Review of Entomology*, *54*(1), 303–321. https://doi.org/10.1146/annurev.ento.54.110807.090434

Gray, S. B., & Brady, S. M. (2016). Plant developmental responses to climate change. *Developmental Biology*, *419*(1), 64–77. https://doi.org/10.1016/j.ydbio.2016.07.023

Guo, K. U. N., HAO, S. G., Sun, O. J., & Kang, L. E. (2009). Differential responses to warming and increased precipitation among three contrasting grasshopper species. *Global Change Biology*, *15*(10), 2539–2548. https://doi.org/10.1111/j.1365-2486.2009.01861.x

Hallmann, C. A., et al. (2017). More than 75 percent decline over 27 years in total flying insect biomass in protected areas. *PloS One*, *12*(10), e0185809. https://doi.org/10.1371/journal.pone.0185809

Hamilton, J. G., Dermody, O., Aldea, M., Zangerl, A. R., Rogers, A., & DeLucia, E. H. (2005). Elevated CO_2 increases the activity of leaf-chewing insects in a soybean agroecosystem. *Ecological Applications*, *15*(5), 1675–1683.

Hance, T., van Baaren, J., Vernon, P., & Boivin, G. (2007). Impact of extreme temperatures on parasitoids in a climate change perspective. *The Annual Review of Entomology*, *52*(1), 107–126. https://doi.org/10.1146/annurev.ento.52.110405.091333

Harrington, R., Woiwod, I., & Sparks, T. (1999). Climate change and trophic interactions. *Trends in Ecology & Evolution*, *14*(4), 146–150. https://doi.org/10.1016/S0169-5347(99)01604-3

Heeb, L., Jenner, E., & Cock, M. J. (2019). Climate-smart pest management: Building resilience of farms and landscapes to changing pest threats. *Journal of Pest Science*, *92*(3), 951–969. https://doi.org/10.1007/s10340-019-01083-y

Hickling, R., Roy, D. B., Hill, J. K., Fox, R., & Thomas, C. D. (2006). The distributions of a wide range of taxonomic groups are expanding polewards. *Global Change Biology*, *12*(3), 450–455. https://doi.org/10.1111/j.1365-2486.2006.01116.x

Hill, J. O., & Dymock, J. J. (1989). Shifts in the manifestations of insect pests under predicted climatic changes. *Journal of Applied Ecology*, *26*(3), 685–692. https://doi.org/10.1002/9781118945568.ch21

Howden, S. M., Soussana, J. F., Tubiello, F. N., Chhetri, N., Dunlop, M., & Meinke, H. (2007). Adapting agriculture to climate change. *Proceedings of the National Academy of Sciences*, *104*(50), 19691–19696.https://doi.org/10.1073/pnas.0701890104

Hullé, M., d'Acier, A. C., Bankhead-Dronnet, S., & Harrington, R. (2010). Aphids in the face of global changes. *Comptes Rendus. Biologies*, *333*(6–7), 497–503. 10.1016/j.crvi.2010.03.005

Hutchison, W. D., et al. (2010). Areawide suppression of European corn borer with Bt maize reaps savings to non-Bt maize growers. *Science*, *330*(6001), 222–225. https://doi.org/10.1126/science.1190242

Imran, M. A., Ali, A., Ashfaq, M., Hassan, S., Culas, R., & Ma, C. (2018). Impact of Climate Smart Agriculture (CSA) practices on cotton production and livelihood of farmers in Punjab, Pakistan. *Sustainability*, *10*(6), 2101. https://doi.org/10.3390/su10062101

Iqbal, S., Zhuo, Z., Ali, H., Xu, D., Niaz, Y., Shah, A.N., Almoallim, H.S., Ansari, M.J., Bodlah, M.A., Ali, B., & Nawaz, M. (2024). Exploring the synergistic toxicity of synthetic pesticides and their impact on development and behavior of Honeybee (Apis mellifera L.). *Emirates Journal of Food and Agriculture, 36*, 1–7.

Joshi, M., Varadharasu, P., Solanki, C., & Birari, V. (2020). Desert locust (*Schistocera gregaria* F.) outbreak in Gujarat (India). *Agriculture and Food: E-Newsletter*, *2*(6), 691–693.

Juroszek, P., & Von Tiedemann, A. (2011). Potential strategies and future requirements for plant disease management under a changing climate. *Plant Pathology*, *60*(1), 100–112. https://doi.org/10.1111/j.1365-3059.2010.02410.x

Khatri-Chhetri, A., Aryal, J. P., Sapkota, T. B., & Khurana, R. (2016). Economic benefits of climate-smart agricultural practices to smallholder farmers in the Indo-Gangetic Plains of India. *Current Science*, *10*, 251–256.

Kingsolver, J. G., Arthur Woods, H., Buckley, L. B., Potter, K. A., MacLean, H. J., & Higgins, J. K. (2011). Complex life cycles and the responses of insects to climate change. *Integrative and Comparative Biology*, *51*(5), 719–732. https://doi.org/10.1093/icb/icr015

Kocmánková, E., Trnka, M., Juroch, J., Dubrovský, M., Semerádová, D., Možný, M., & Žalud, Z. (2009). Impact of climate change on the occurrence and activity of harmful organisms. *Plant Protection Science*, *45*(Special Issue), S48–52. https://doi.org/10.17221/2835-PPS

Kumar, R., & Gautam, H. R. (2014). Climate change and its impact on agricultural productivity in India. *Journal of Climatology & Weather Forecasting*, *2*(1), 1–3. https://doi.org/10.4172/2332-2594.1000109

Lach, L., & Hooper-Bui, L. M. (2010). Consequences of ant invasions. *Ant Ecology*, *261*, 286.

Ladha, J. K., Bains, J. S., Gupta, R. K., & Balasubramanian, V. (2007). On-farm evaluation of leaf color chart for need-based nitrogen management in irrigated transplanted rice in northwestern India. *Nutrient Cycling in Agroecosystems*, *78*, 167–176. https://doi.org/10.1007/s10705-006-9082-2

Lal, M., Singh, K. K., Rathore, L. S., Srinivasan, G., & Saseendran, S. A. (1998). Vulnerability of rice and wheat yields in NW India to future changes in climate. *Agricultural and Forest Meteorology*, *89*(2), 101–114. https://doi.org/10.1016/S0168-1923(97)00064-6

Latif, A., Shakir, A. S., & Rashid, M. U. (2013). Appraisal of economic impact of zero tillage, laser land levelling and bed-furrow interventions in Punjab, Pakistan. *Pakistan Journal of Engineering and Applied Sciences*.

Le Conte, Y., & Navajas, M. (2008). Climate change: Impact on honey bee populations and diseases. *Revue Scientifique et Technique-Office International des Epizooties*, *27*(2), 499–510.

Lenaerts, B., Collard, B. C., & Demont, M. (2019). Improving global food security through accelerated plant breeding. *Plant Science*, *287*, 110207. https://doi.org/10.1016/j.plantsci.2019.110207

Liang, Z., Chen, K., Zhang, Y., Liu, J., Yin, K., Qiu, J. L., & Gao, C. (2018). Genome editing of bread wheat using biolistic delivery of CRISPR/Cas9 in vitro transcripts or ribonucleoproteins. *Nature Protocols*, *13*(3), 413–430. https://doi.org/10.1038/nprot.2017.145

Lincoln, D. E. (1993). The influence of plant carbon dioxide and nutrient supply on susceptibility to insect herbivores. *Vegetation*, *104*, 273–280. https://doi.org/10.1007/BF00048158

Lubbers, I. M., Van Groenigen, K. J., Fonte, S. J., Six, J., Brussaard, L., & Van Groenigen, J. W. (2013). Greenhouse gas emissions from soils increased by earthworms. *Nature Climate Change*, *3*(3), 187–194. https://doi.org/10.1038/nclimate1692

Lybbert, T. J., & Sumner, D. A. (2012). Agricultural technologies for climate change in developing countries: Policy options for innovation and technology diffusion. *Food Policy*, *37*(1), 114–123. https://doi.org/10.1016/j.foodpol.2011.11.001

Madan, P., Jagadish, S. V. K., Craufurd, P. Q., Fitzgerald, M., Lafarge, T., & Wheeler, T. R. (2012). Effect of elevated CO2 and high temperature on seed-set and grain quality of rice. *Journal of Experimental Botany*, *63*(10), 3843–3852. https://doi.org/10.1093/jxb/ers077

Mafokoane, L. D., Zimmermann, H. G., & Hill, M. P. (2007). Development of Cactoblastis cactorum (Berg) (Lepidoptera: pyralidae) on six north American Opuntia species. *African Entomology*, *15*(2), 295–299. https://hdl.handle.net/10520/EJC32740

Mahato, A. (2014). Climate change and its impact on agriculture. *International Journal of Scientific and Research Publications*, *4*(4), 1–6.

Mao, Y. B., Cai, W. J., Wang, J. W., Hong, G. J., Tao, X. Y., Wang, L. J., Huang, Y. P., & Chen, X. Y. (2007). Silencing a cotton bollworm P450 monooxygenase gene by plant-mediated RNAi impairs larval tolerance of gossypol. *Nature Biotechnology*, *25*(11), 1307–1313. https://doi.org/10.1038/nbt1352

Marzano, M., Ambrose-Oji, B., Hall, C., & Moseley, D. (2020). Pests in the city: Managing public health risks and social values in response to oak processionary moth (*Thaumetopoea processionea*) in the United Kingdom. *Forests*, *11*(2), 199. https://doi.org/10.3390/f11020199

Mendelsohn, R. (2009). The impact of climate change on agriculture in developing countries. *Journal of Natural Resources Policy Research*, *1*(1), 5–19. https://doi.org/10.1080/19390450802495882

Mobini, S. H., & Warkentin, T. D. (2016). A simple and efficient method of in vivo rapid generation technology in pea (Pisum sativum L.). *In Vitro Cellular & Developmental Biology-Plant*, *52*, 530–536. https://doi.org/10.1007/s11627-016-9772-7

Mutune, B., Ekesi, S., Niassy, S., Matiru, V., Bii, C., & Maniania, N. K. (2016). Fungal endophytes as promising tools for the management of bean stem maggot *Ophiomyia phaseoli* on beans *Phaseolus vulgaris*. *Journal of Pesticide Science*, *89*, 993–1001. https://doi.org/10.1007/s10340-015-0725-4

Myers, S. S., Zanobetti, A., Kloog, I., Huybers, P., Leakey, A. D., Bloom, A. J., Carlisle, E., Dietterich, L. H., Fitzgerald, G., Hasegawa, T., & Usui, Y. (2014). Increasing CO2 threatens human nutrition. *Nature*, *510*(7503), 139–142. https://doi.org/10.1038/nature13179

Nagoshi, R. N., Goergen, G., Plessis, H. D., van den Berg, J., & Meagher Jr, R. (2019). Genetic comparisons of fall armyworm populations from 11 countries spanning sub-Saharan Africa provide insights into strain composition and migratory behaviors. *Scientific Reports*, *9*(1), 8311. https://doi.org/10.1038/s41598-019-44744-9

Narayanamoorthy, A., Bhattarai, M., & Jothi, P. (2018). An assessment of the economic impact of drip irrigation in vegetable production in India. *Agricultural Economics Research Review*, *31*(1), 105–112. https://doi.org/10.5958/0974-0279.2018.00010.1

Narayanamoorthy, A., & Devika, N. (2018). Economic and resource impacts of drip method of irrigation on okra cultivation: An analysis of field survey data. *Journal of Land and Rural Studies*, *6*(1), 15–33. https://doi.org/10.1177/2321024917731840

Nawaz, A., Ali, H., Fiaz, S., Sufyan, M., Qayyum, M. A., Azad, R., Niaz, Y., Bodlah, M. A., & Khan, K. A. (2021). Analysis of food resources, host availability and insecticidal impacts on the fecundity, longevity and parasitism efficiency of Diaertiella rapae (M'intosh). *International Journal of Tropical Insect Science*, *41*(4), 2883–2896.

Oberhauser, K., et al. (2017). Impacts of climate change on monarch butterflies. *Journal of Insect Conservation*, *21*, 123–136. https://doi.org/10.1007/s10841-024-00644-6

O'Connor, D. J., Wright, G. C., Dieters, M. J., George, D. L., Hunter, M. N., Tatnell, J. R., & Fleischfresser, D. B. (2013). Development and application of speed breeding technologies in a commercial peanut breeding program. *Peanut science*, *40*(2), 107–114.

Ogada, M. J., Rao, E. J., Radeny, M., Recha, J. W., & Solomon, D. (2020). Climate-smart agriculture, household income and asset accumulation among smallholder farmers in the Nyando basin of Kenya. *World Development Perspectives*, *18*, 100203 https://doi.org/10.1016/j.wdp.2020.100203

Olesen, J. E., & Bindi, M. (2002). Consequences of climate change for European agricultural productivity, land use and policy. *European Journal of Agronomy*, *16*(4), 239–262. https://doi.org/10.1016/S1161-0301(02)00004-7

Pampolino, M. F., Manguiat, I. J., Ramanathan, S., Gines, H. C., Tan, P. S., Chi, T. T. N., Rajendran, R., & Buresh, R. J. (2007). Environmental impact and economic benefits of site-specific nutrient management (SSNM) in irrigated rice systems. Agricultural Systems, 93(1–3), 1–24. https://doi.org/10.1016/j.agsy.2006.04.002

Parthasarathi, K., et al. (2012). Aerobic rice cultivation: Benefits and impact on water and methane emissions. *Journal of Agricultural Research*, *43*, 153–160.

Pearman, P. B., Guisan, A., Broennimann, O., & Randin, C. F. (2008). Niche dynamics in space and time. *Trends in Ecology & Evolution*, *23*(3), 149–158. https://doi.org/10.1016/j.tree.2007.11.005

Pearson, D. E., Ortega, Y. K., Runyon, J. B., & Butler, J. L. (2016). Secondary invasion: The bane of weed management. *Biological Conservation*, *197*, 8–17. https://doi.org/10.1016/j.biocon.2016.02.029

Powell, T., & Reinhard, S. (2016). Climate impacts on wheat yields. *Agricultural Meteorology*, *22*, 101–112.

Reddy, P. R., Rashmi, T., & Verghese, A. (2015). Foraging activity of Indian honey bee, Apis cerana in relation to ambient climate variables under tropical conditions. *Journal of Environmental Biology*, *36*(3), 577.

Reddy, P. V. R., Verghese, A., Sridhar, V., & Rajan, V. V. (2013). Plant-pollinator interactions: A highly evolved synchrony at risk due to climate change. *Climate-resilient Horticulture: Adaptation and Mitigation Strategies*, 295–302. https://doi.org/10.1007/978-81-322-0974-4_26

Robinson, W. S., Stange, E. E., & Ayres, M. P. (2017). Impact of climate change on the survival of insect species. *Journal of Insect Conservation*, *21*(5), 439–455.

Roitsch, T., et al. (2019). New sensors and data-driven approaches – a path to next generation phenomics. *Plant Science*, *282*, 2–10. https://doi.org/10.1016/j.plantsci.2019.01.011

Roques, A., Fan, J. T., Courtial, B., Zhang, Y. Z., Yart, A., Auger-Rozenberg, M. A., Denux, O., Kenis, M., Baker, R., & Sun, J. H. (2015). Planting sentinel European trees in Eastern Asia as a novel method to identify potential insect pest invaders. *PloS One*, *10*(5), e0120864. https://doi.org/10.1371/journal.pone.0120864

Rusterholz, H. P., & Erhardt, A. (1998). Effects of elevated CO 2 on flowering phenology and nectar production of nectar plants important for butterflies of calcareous grasslands. *Oecologia*, *113*, 341–349. https://doi.org/10.1007/s004420050385

Saxena, K. B., Saxena, R. K., Hickey, L. T., & Varshney, R. K. (2019). Can a speed breeding approach accelerate genetic gain in pigeonpea? *Euphytica*, *215*(12), 202.

Shabir, A., Sarwar, Z. M., & Ali, H. (2023). Eco-friendly approaches of zinc oxide and silver nitrate nanoparticles along with plant extracts against Spodoptera litura (Fabricius) under laboratory conditions. *Science progress, 106*(4), 00368504231219171.

Singh, V. K., Shukla, A. K., Singh, M. P., Majumdar, K., Mishra, R. P., Rani, M., & Singh, S. K. (2015). Effect of site-specific nutrient management on yield, profit and apparent nutrient balance under pre-dominant cropping systems of Upper Gangetic Plains. *The Indian Journal of Agricultural Sciences, 85*(3), 335–343.

Skendžić, S., Zovko, M., Živković, I. P., Lešić, V., & Lemić, D. (2021a). The impact of climate change on agricultural insect pests. *Insects, 12,* 440. https://doi.org/10.3390/insects12050440

Skendžić, S., Zovko, M., Živković, I. P., Lešić, V., & Lemić, D. (2021b). Effect of climate change on introduced and native agricultural invasive insect pests in Europe. *Insects, 12*(11), 985. https://doi.org/10.3390/insects12110985

Smit, B., & Skinner, M. W. (2002). Adaptation options in agriculture to climate change: A typology. *Mitigation and Adaptation Strategies for Global Change, 7*(1), 85–114. https://doi.org/10.1023/A:1015862228270

Solignac, M., Cornuet, J. M., Vautrin, D., Le Conte, Y., Anderson, D., Evans, J., Cros-Arteil, S., & Navajas, M. (2005). The invasive Korea and Japan types of Varroa destructor, ectoparasitic mites of the Western honeybee (Apis mellifera), are two partly isolated clones. *Proceedings of the Royal Society of London: Biological Sciences, 272*(1561), 411–419. https://doi.org/10.1098/rspb.2004.2853

Srivastava, V., Liang, W., Keena, M. A., Roe, A. D., Hamelin, R. C., & Griess, V. C. (2020). Assessing niche shifts and conservatism by comparing the native and post-invasion niches of major forest invasive species. *Insects, 11*, 479. https://doi.org/10.3390/insects11080479

Stange, E. E., & Ayres, M. P. (2020). The effects of changing precipitation patterns on insect populations. *Environmental Entomology, 49*(2), 412–426.

Stevanović, M., Popp, A., Lotze-Campen, H., Dietrich, J. P., Müller, C., Bonsch, M., Schmitz, C., Bodirsky, B. L., Humpenöder, F., & Weindl, I. (2016). The impact of high-end climate change on agricultural welfare. *Science Advances, 2*(8), e1501452. https://doi.org/10.1126/sciadv.1501452

Straw, N. A., Fielding, N. J., Green, G., & Price, J. (2005). Defoliation and growth loss in young Sitka spruce following repeated attack by the green spruce aphid, *Elatobium abietinum* (Walker). *Forest Ecology and Management, 213*(1–3), 349–368. https://doi.org/10.1016/j.foreco.2005.04.002

Tan, C., Kalhoro, M. T., Faqir, Y., Ma, J., Osei, M. D., & Khaliq, G. (2022). Climate-resilient microbial biotechnology: A perspective on sustainable agriculture. *Sustainability, 14*(9), 5574. https://doi.org/10.3390/su14095574

Tang, R., Li, S., Liang, J., Yi, H., Jing, X., & Liu, T. X. (2022). Optimization of the application of the CRISPR/Cas9 system in *Mythimna separata. Entomologia Experimentalis et Applicata, 170*(7), 593–602. https://doi.org/10.1111/eea.13184

Tejasree, P., Ramadevi, S., Eshwar Sai Prasad, Ch., Praisy, T., Keerthana, D., Sindhura, N. R. H., Koushik, P. V., & Ramya Harika, K. (2024). Breeding for resilience – adapting to climate change. *Frontiers in Plant Protection Current Status and Management; Ninetales Publishings: Ghaziabad, India, 3*, 191.

Thomson, L. J., Macfadyen, S., & Hoffmann, A. A. (2010). Predicting the effects of climate change on natural enemies of agricultural pests. *Biological Control, 52*(3), 296–306. https://doi.org/10.1016/j.biocontrol.2009.01.022

Toffa, J., et al. (2021). Endophytic colonization of tomato plants by *Beauveria bassiana* Vuillemin (Ascomycota: Hypocreales) and leaf damage in *Helicoverpa armigera* (Hübner) (Lepidoptera: Noctuidae) larvae. *Egyptian Journal of Biological Pest Control, 31*, 82. https://doi.org/10.1186/s41938-021-00431-4

Tripathi, R. S., Raju, R., & Thimmappa, K. (2013). Impact of zero tillage on economics of wheat production in Haryana. *Agricultural Economics Research Review, 26*(1), 101–108.

Uddling, J., Broberg, M. C., Feng, Z., & Pleijel, H. (2018). Crop quality under rising atmospheric CO2. *Current Opinion in Plant Biology, 45*, 262–267. https://doi.org/10.1016/j.pbi.2018.06.001

Verchot, L. V., Van Noordwijk, M., Kandji, S., Tomich, T., Ong, C., Albrecht, A., Mackensen, J., Bantilan, C., Anupama, K. V., & Palm, C. (2007). Climate change: Linking adaptation and mitigation through agroforestry. *Mitigation and Adaptation Strategies for Global Change, 12*, 901–918. https://doi.org/10.1007/s11027-007-9105-6

Visser, M. E., & Holleman, L. J. (2001). Warmer springs disrupt the synchrony of oak and winter moth phenology. *Proceedings of the Royal Society of London. Series B: Biological Sciences, 268*(1464), 289–294. https://doi.org/10.1098/rspb.2000.1363

Walther, G. R., et al. (2009). Alien species in a warmer world: Risks and opportunities. *Trends in Ecology & Evolution, 24*(12), 686–693. https://doi.org/10.1016/J.TREE.2009.06.008

Wang, J., Zhang, H., Wang, H., Zhao, S., Zuo, Y., Yang, Y., & Wu, Y. (2016). Functional validation of cadherin as a receptor of Bt toxin Cry1Ac in *Helicoverpa armigera* utilizing the CRISPR/Cas9 system. *Insect Biochemistry and Molecular Biology, 76*, 11–17. https://doi.org/10.1016/j.ibmb.2016.06.008

Watson, A., Ghosh, S., Williams, M. J., Cuddy, W. S., Simmonds, J., Rey, M. D., Asyraf Md Hatta, M., Hinchliffe, A., Steed, A., Reynolds, D., & Adamski, N. M. (2018). Speed breeding is a powerful tool to accelerate crop research and breeding. *Nature Plants, 4*(1), 23–29. https://doi.org/10.1038/s41477-017-0083-8

Whitman, D. W. (1987). Thermoregulation and daily activity patterns in a black desert grasshopper, *Taeniopoda eques. Animal Behaviour, 35*(6), 1814–1826. https://doi.org/10.1016/S0003-(87)80074-X

Wu, G., et al. (2006). Effects of elevated CO_2 on Helicoverpa armigera. *Entomological Science, 9*, 115–121.

Zhang, A., et al. (2019). Enhanced rice salinity tolerance via CRISPR/Cas9-targeted mutagenesis of the OsRR22 Gene. *Molecular Breeding, 39*, 47. https://doi.org/10.1007/s11032-019-0954-y

Zou, X., Cremades, R., Gao, Q., Wan, Y., & Qin, X. (2013). Cost-effectiveness analysis of water-saving irrigation technologies based on climate change response: A case study of China. *Agricultural Water Management, 129*, 9–20. https://doi.org/10.1016/j.agwat.2013.07.004

Züst, T., Rasmann, S., & Agrawal, A. A. (2015). Growth – defense tradeoffs for two major anti-herbivore traits of the common milkweed *Asclepias syriaca. Oikos, 124*(10), 1404–1415. https://doi.org/10.1111/oik.02075

16 Nanotechnology in Entomology
Tiny Solutions for Big Challenges

F. Yarahmadi[1*] *and A. Rajabpour*[1]
[1]Department of Plant Protection, Faculty of Agriculture, Agricultural Sciences and Natural Resources University of Khuzestan, Bavi, Iran
[*]Corresponding author

16.1 INTRODUCTION

Nanotechnology refers to the manipulation of matter on an atomic, molecular, and supramolecular scale, typically within the range of 1 to 100 nm. At this scale, materials can exhibit unique physical and chemical properties different from their bulk counterparts. These properties can include increased strength, lighter weight, greater chemical reactivity, and increased conductivity (Kannan et al., 2022). Nanotechnology encompasses a broad range of practices and applications, from developing new materials and pharmaceuticals to enhancing energy production and storage. It represents a cutting-edge field of research that has the potential to revolutionize numerous sectors, including medicine, materials science, and environmental science (Shahid et al., 2021; Zannat et al., 2021).

In the context of entomology, nanotechnology has emerged as a novel and significant tool. Insect pests pose major challenges in agricultural systems, contributing to reduced crop yields and increased costs for farmers through the need for pest control measures (Bhan et al., 2018). Furthermore, many insect species are vectors for serious human diseases, making them a critical public health concern. By leveraging the unique properties of nanomaterials, researchers in entomology aim to develop innovative strategies for pest control that are more efficient, targeted, and environmentally friendly than traditional chemical pesticides (Dwivedi & Nameirakpam, 2022; Nimisha & Deepthy, 2020).

Nanotechnology opens new avenues for interventions that can be less harmful to non-target species, minimize environmental impacts, and address the issue of pesticide resistance that has become prevalent in many pest populations (El-Helaly & Sayed, 2015; Basu & Bhattacharya, 2021). The relevance of nanotechnology in entomology extends beyond pest control; it also encompasses the understanding of insect behavior and physiology at the nanoscale, potentially leading to advancements in crop protection and pest management. For instance, nanoparticles can be designed to target specific physiological pathways in insects, allowing for the development of more effective and less harmful insecticides. Additionally, nanoparticles can facilitate the delivery of biologically active compounds that enhance plant defenses against pests or that disrupt the life cycles of harmful insect species (Álvarez et al., 2019; Margulis-Goshen & Magdassi, 2013).

Addressing pest challenges in agriculture is critical for a variety of reasons. The global population is projected to reach nearly 10 billion by 2050, requiring a significant increase in food production to meet this growing demand. Simultaneously, the impact of climate change on agricultural systems presents additional challenges, including shifts in pest populations and distributions. Insects can affect up to 40% of crop yields globally, leading to considerable economic losses and threatening food security (Yousef et al., 2023). Furthermore, the economic implications extend beyond direct crop losses; they influence food prices, farmers' livelihoods, and community stability (Kumar et al., 2017).

From a public health perspective, the situation is equally urgent. Many insect species, such as mosquitoes, ticks, and fleas, are vectors for diseases that affect millions of people worldwide (Shukla et al., 2020). Diseases such as malaria, dengue fever, and Zika virus are transmitted by insects, leading to significant morbidity and mortality. According to the World Health Organization (WHO), vector-borne diseases account for more than 700,000 deaths annually. Effective control of insect vector populations is crucial in reducing the burden of these diseases and enhancing overall public health (Elmer & White, 2018).

The economic impact of pest-related challenges in agriculture and public health is staggering, with the FAO estimating that pests cause global crop yield losses of between 20 and 40% each year (Sarmah et al., 2023). In monetary terms, this translates into hundreds of billions of dollars lost annually due to pest-related crop damage and disease transmission. The need for innovative solutions to pest management is evident, particularly in light of the limitations and environmental impacts associated with traditional chemical pesticides (Ali et al., 2023). The landscape of pest management has historically relied heavily on chemical interventions, primarily conventional pesticides. While these products can be effective in controlling pest populations, they often come with significant drawbacks, including the development of pesticide-resistant pest strains, adverse

DOI: 10.1201/9781003516538-16

effects on non-target species, and environmental contamination. Over time, the repeated application of chemical pesticides has triggered adaptations in insect populations, leading to the phenomenon of resistance. As pests develop resistance, the efficacy of current chemical controls diminishes, necessitating higher doses or the introduction of new chemicals to maintain control, which in turn exacerbates environmental pollution.

Moreover, the extensive use of chemicals in agriculture can lead to negative repercussions for ecosystems. Pesticides often contaminate soil and water resources, harming beneficial organisms and disrupting natural ecological balances. Non-target insects, including pollinators, such as bees, are particularly vulnerable to pesticide exposure, leading to declines in populations crucial for agriculture. The loss of biodiversity associated with pesticides can undermine the resilience of ecosystems, impairing their ability to provide essential services, including nutrient cycling and pest regulation (Kumar et al., 2017; Nimisha & Deepthy, 2020).

Human health risks associated with pesticide exposure are also a fundamental concern. Farmers and agricultural workers are often at heightened risk of experiencing acute poisoning from direct exposure to pesticides, which can result in immediate health issues. Long-term exposure is associated with a range of chronic health effects, including respiratory problems, neurological disorders, and even an increased risk of certain cancers. Pesticide residues remaining on food products pose risks to consumers, raising public concerns about food safety (Bhattacharya et al., 2020; Shukla et al., 2020).

In light of these challenges, the integration of nanotechnology into pest management practices offers an innovative approach. Nanotechnology has the potential to revolutionize the way pests are managed by providing more precise, effective, and sustainable solutions. The application of nanotechnology in pest control can involve various strategies, including the development of nanoparticle-based insecticides, nanocarriers for targeted delivery of biopesticides, and materials designed to repel pests or disrupt their life cycles (Bhan et al., 2018; Dwivedi & Nameirakpam, 2022; Elmer & White, 2018).

One of the key advantages of using nanotechnology in pest management is the ability to create nanoparticles that can function as insecticides with specific modes of action. For instance, nanoparticles can be engineered to disrupt cellular processes in target insects while having minimal toxicity to non-target species and beneficial insects (Álvarez et al., 2019). By targeting biological pathways unique to pests, these nanoparticle formulations can reduce the environmental footprint and enhance the specificity of pest control efforts.

Another promising application of nanotechnology in entomology is the use of nanoscale carriers for the delivery of biologically active compounds, such as plant extracts or biopesticides. These carriers can improve the efficacy of these compounds by enhancing their stability and controlling their release, allowing for targeted application and reduced overall quantities needed for effective pest management (Margulis-Goshen & Magdassi, 2013; Kannan et al., 2022). Such innovations could mitigate the environmental impact of pest control while ensuring food security and protection of crops from harmful pests.

Furthermore, the development of nanoscale surfaces that exhibit repellent properties can prevent pests from landing on plants or crops, thereby reducing pest pressure without relying on chemicals. These surfaces can mimic the physical structures found in nature, such as those seen in insect-resistant plants or certain insect species. This biomimetic approach taps into the inherent properties of materials at the nanoscale to provide a passive means of pest management that complements active chemical measures (Álvarez et al., 2019; Kannan et al., 2022).

Nanotechnology presents innovative applications for enhancing the use of pheromones in integrated pest management (IPM). This technology allows for the creation of controlled-release systems that encapsulate pheromones, ensuring a sustained release over time and reducing the need for frequent replacements. Moreover, nanoparticles enable the targeted delivery of pheromones in response to specific stimuli, thereby increasing effectiveness while minimizing impact on non-target species (Zannat et al., 2021). Enhanced stability is achievable through protective coatings that preserve pheromones in challenging environmental conditions. Furthermore, nanotechnology can improve detection systems for pests by integrating pheromones with nanosensors for real-time monitoring, facilitating timely interventions. The combination of pheromones with biopesticides via nanoparticles can result in a synergistic effect, while advances in synthetic pheromone production can make these solutions more cost-effective. Additionally, nanoparticles can manipulate pest behavior through pheromone signaling, and integrating this technology with smart devices allows for data-driven optimizations in pheromone usage (Martinazzo et al., 2022) (Figure 16.1).

The importance of interdisciplinary collaboration in advancing nanotechnology for pest management cannot be overstated. The effective integration of nanotechnology into entomology will require cooperation among specialists in various fields, including material scientists, entomologists, agronomists, and policy experts (Dwivedi & Nameirakpam, 2022). Collaborative research efforts can help bridge the gap between technological innovation and practical applications, addressing challenges and identifying opportunities for effective pest management (El-Helaly & Sayed, 2015; Nimisha & Deepthy, 2020).

While the potential of nanotechnology is vast, it is essential to approach its application responsibly. The regulatory landscape for nanotechnology is evolving, and it is crucial that researchers adhere to evolving guidelines to ensure the safety and efficacy of nanomaterials used in pest management (Bhan et al., 2018; Yousef et al., 2023). As the field progresses, ongoing risk assessment, monitoring, and evaluation of the environmental and health impacts of

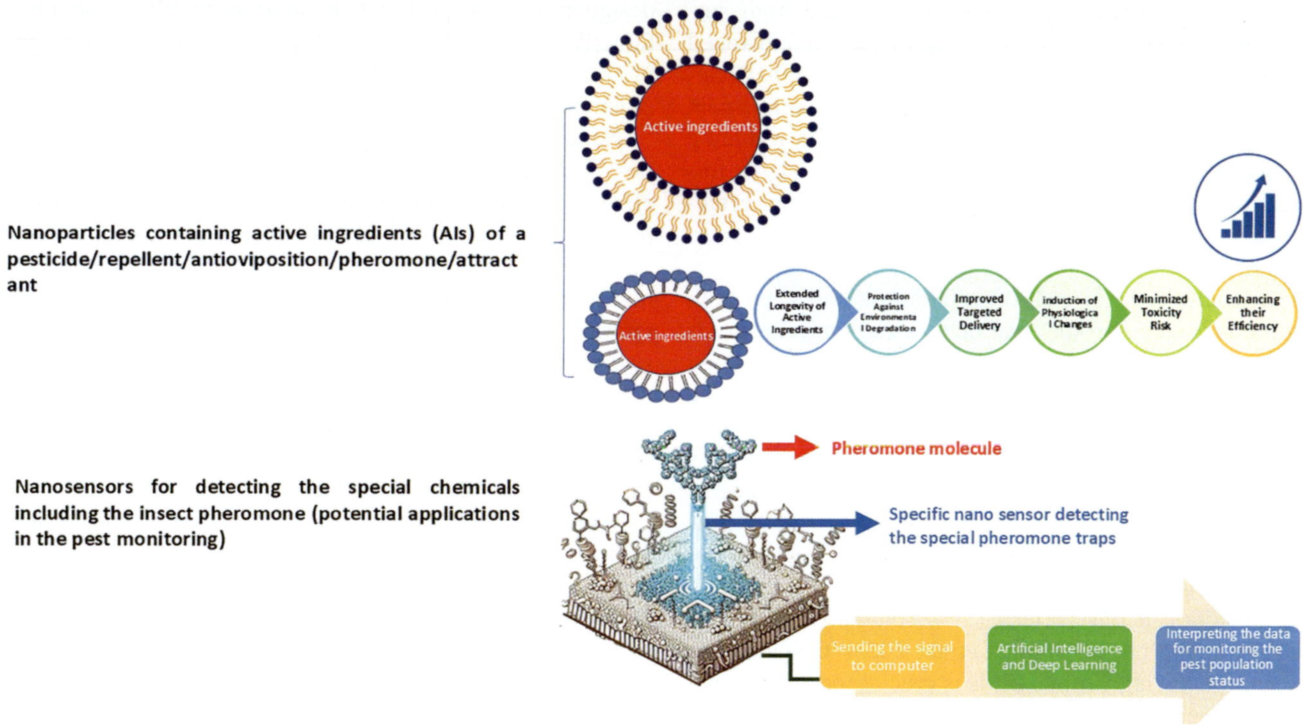

FIGURE 16.1 Potential applications of nanotechnologies in pest management programs, including nano-encapsulated pesticides for targeted delivery, nanosensors for early pest detection, nanocarriers for gene delivery in pest control, nanoformulations to enhance pesticide stability and efficiency, and nanomaterials for pest-repellent coatings.

nanoscale products will be essential to maintain public trust and ensure safe use.

Public acceptance is another key factor influencing the successful implementation of nanotechnology in pest management. Engaging stakeholders, including farmers, consumers, and policymakers, is critical in fostering understanding and support for nanotechnology innovations (Shahid et al., 2021; Zannat et al., 2021). Education and transparent communication about the benefits, risks, and safety of nanotechnology solutions can help generate enthusiasm and trust in these new methods, leading to more widespread adoption.

16.2 SUSTAINABLE PEST CONTROL: THE ROLE OF INTEGRATED PEST MANAGEMENT AND NANOTECHNOLOGY

IPM is an ecologically based strategy that combines various practices and technologies to address pest control effectively and sustainably. It focuses on long-term prevention of pests and their damage while minimizing detrimental environmental and economic effects (Yousef et al., 2023). Central to IPM is the understanding that no single method is sufficient for effective pest control. Instead, it relies on an assortment of biological, cultural, physical, chemical, and regulatory approaches to manage pest populations efficiently (Dwivedi & Tomer, 2020). This comprehensive nature allows for restraint in the use of chemical pesticides, promoting more sustainable agricultural practices (Bhan et al., 2018).

IPM principles revolve around understanding pest biology, ecology, and environments to effectively manage pests (Ndlovu et al., 2020). Cultural practices like crop rotation and selecting pest-resistant varieties create unfavorable conditions for pests, enhancing ecosystem health (Ghouri et al., 2020; Nimisha & Deepthy, 2020). Biological control entails using natural enemies to reduce reliance on chemical pesticides, such as using lady beetles to control aphids (Bhattacharya et al., 2020; Basu & Bhattacharya, 2021). Physical control methods, like traps and barriers, serve as non-toxic alternatives that complement IPM strategies (Chandrashekharaiah et al., 2015). Overall, IPM aims to balance pest management with ecosystem health (Humbal & Pathak, 2023). When other strategies are insufficient, IPM advocates for the careful use of targeted, less harmful pesticides, including biopesticides and reduced-risk chemicals (Bhan et al., 2018). Timing and application methods are vital to minimize non-target impacts and prevent resistance (Kumar & Gong, 2018). Emerging technologies, such as remote sensing, drones, and soil sensors, enhance IPM efficiency by enabling early detection of pest infestations and real-time data collection (Rajabpour & Yarahmadi, 2024).

Nanotechnology represents a groundbreaking frontier within the IPM framework, offering innovative solutions that enhance pest management effectiveness while minimizing environmental impact (Álvarez et al., 2019).

Its unique properties, such as small size and high surface area, make nanomaterials invaluable for advanced pest control methods (Singh et al., 2020). Nanoparticle-based insecticides are revolutionizing pest management by improving the efficacy of pesticides while reducing environmental impact (Ndlovu et al., 2020). These formulations release pesticides slowly, enhancing effectiveness and minimizing harm to non-target species (Bhattacharya et al., 2020). Furthermore, nanoscale carriers enhance the stability of biopesticides, protecting them from environmental factors and targeting pests more effectively (Kannan et al., 2022). Nanotechnology also offers the creation of pest-repellent surfaces without harmful chemicals, reducing pest populations sustainably (Elmer & White, 2018). Nanotechnology can play an important role in the enhancement of the insect pheromone efficiencies and their applications in early time detection of pests (Martinazzo et al., 2022).

The integration of nanotechnology aligns with sustainable practices in IPM, offering new solutions for agriculture's evolving needs (Sarmah et al., 2023). Collaborative research involving various fields is crucial for the successful implementation of nanotechnology in IPM (Dwivedi & Nameirakpam, 2022). Engaging agricultural practitioners and addressing safety concerns are key to advancing practical applications of IPM to evolve with new technologies, like nanotechnology, providing innovative solutions for sustainable agriculture (Yousef et al., 2023). The integration of these approaches within the IPM framework is essential for a resilient and sustainable future in global food systems.

16.3 NANOMATERIALS AND THEIR UNIQUE PROPERTIES

Nanotechnology is a rapidly advancing field that manipulates materials at the nanoscale, from 1 to 100 nm, where nanomaterials exhibit unique properties different from those of their bulk counterparts, making them valuable in agriculture and IPM. Nanomaterials are defined by their nanoscale structures, including nanoparticles, nanosensors, and nanocarriers, each serving specific functions with distinct advantages (Bhan et al., 2018; Kumar et al., 2017). Nanoparticles, commonly composed of metals or polymers, display unique physical and chemical properties compared to larger particles. For instance, silver nanoparticles are known for their antibacterial properties, suitable for agricultural use to control microbial growth (Álvarez et al., 2019). They can inhibit pathogens while being less harmful to beneficial microorganisms, revealing their potential in pest management strategies (Humbal & Pathak, 2023; Ghouri et al., 2020; Das et al., 2021).

Nanosensors are nanomaterial devices that can detect and measure physical or chemical changes at the nanoscale, useful for monitoring crop health and pest presence in agriculture. They provide real-time data for informed decision-making on pest management (Gamage et al., 2024). Nanocarriers are another important nanomaterial category designed to transport active substances like pesticides directly to targeted sites within plants or pests, enhancing product effectiveness and reducing environmental impact (Kannan et al., 2022).

Nanomaterials' unique size gives them distinct properties, with materials at the nanoscale exhibiting quantum effects that alter their optical, electrical, and magnetic properties. For example, gold nanoparticles appear red or purple due to their unique behavior (Singh et al., 2020). The high surface area of nanomaterials increases reactivity and efficacy, allowing for enhanced interactions with biological systems. This characteristic is beneficial in pest management, as nanoparticles can have greater contact with target pests or plant surfaces, leading to successful pest control with lower active ingredient dosages and reduced environmental impact (Bhan et al., 2018).

Overall, nanomaterials offer innovative solutions for agricultural applications, improving efficiency and sustainability in pest management practices (Nimisha & Deepthy, 2020). Reactivity is another defining property of nanomaterials that contributes to their effectiveness in various applications. At the nanoscale, the chemical reactivity of materials can differ significantly from that at larger scales. This heightened reactivity can be attributed to the increased number of active sites available on the surface of nanoparticles. For instance, titanium dioxide nanoparticles, commonly used as a photocatalyst, can enhance the degradation of organic pollutants when exposed to light (Shukla et al., 2020). In pest management, the reactive nature of certain nanoparticles can facilitate targeted interactions with pests, potentially leading to the development of environmentally friendly pest control agents (Zannat et al., 2021).

Nanomaterials have the potential to revolutionize agriculture and pest management due to their unique size, surface area, and reactivity (Basu & Bhattacharya, 2021). They can be used for more than just direct pest control, with nanosensors able to detect plant stress indicators and provide crucial information for crop health monitoring (Gamage et al., 2024). Nanocarriers can protect and deliver biopesticides efficiently, reducing the need for synthetic chemicals and promoting sustainable agricultural practices (Dwivedi & Nameirakpam, 2022; Dwivedi & Tomer, 2020). The development of multifunctional pest control solutions combining different functionalities, such as pest detection, biocontrol agent release, and insect repelling, is being explored (Shahid et al., 2021). These comprehensive approaches enhance pest control measures and support integrated pest management strategies that prioritize ecological balance and sustainability (Bhan et al., 2018). However, as nanomaterials are further researched for agricultural use, safety and environmental impact considerations are vital (Yousef et al., 2023). Thorough evaluation of their effects on human health and ecosystems is necessary to ensure they do not harm beneficial organisms or create new environmental challenges unintentionally (Ndlovu et al., 2020; Nimisha & Deepthy, 2020).

16.4 ADVANCEMENTS IN NANOTECHNOLOGY FOR INSECT PEST MANAGEMENT

Nanotechnology is transforming entomology with cutting-edge solutions for pest control. Advancements include nanosensors for monitoring pests in real time, targeted pesticide delivery using nanocarriers, and biocompatible nanoparticles to boost effectiveness of biological control agents for sustainable agriculture.

16.4.1 Control of Insect Pests

The control of insect pests is a critical aspect of agricultural practices and ecosystem management. Traditional methods often involve the widespread application of chemical pesticides, which can lead to several drawbacks, including environmental contamination, the development of pest resistance, and negative impacts on non-target organisms (Bhan et al., 2018; Ghouri et al., 2020). In this context, nanotechnology presents innovative approaches that enhance the efficacy and sustainability of pest control measures (Kannan et al., 2022; Kumar et al., 2017).

One of the most promising applications of nanotechnology in this realm is the development of nanopesticides. These are formulations of pesticides that incorporate nanomaterials to improve their performance (Shahid et al., 2021). The incorporation of nanoparticles can enhance the solubility, stability, and bioavailability of active ingredients, allowing for reduced dosages and minimized environmental impact (Álvarez et al., 2019; Das et al., 2021). Nanopesticides can also offer controlled release properties, ensuring that active substances remain effective over a longer period and reduce the frequency of application. For instance, encapsulating pesticides in biodegradable nanoparticles can provide a slow-release mechanism that maintains effective concentrations of the active ingredient at the target site, thus improving pest management efficacy while decreasing the likelihood of runoff into surrounding soil and water resources (Dwivedi & Nameirakpam, 2022; Nimisha & Deepthy, 2020).

Moreover, nanopesticides can be tailored to target specific pests or physiological systems, further enhancing pest control. This specificity can be achieved through surface modification of nanoparticles, which allows for the attachment of ligands or molecules that preferentially interact with target insects (Álvarez et al., 2019; Bhattacharya et al., 2020). Such design strategies enable the selective targeting of pests, minimizing the impact on beneficial organisms such as pollinators and natural enemies of pests. Experimental and field studies have demonstrated that nanopesticides can show enhanced insecticidal activity and reduced toxicity to non-target species compared to their conventional counterparts, underscoring the potential of this technology in IPM (Bhan et al., 2018; Nimisha & Deepthy, 2020; Ya-Ali et al., 2020).

Alongside nanopesticides, targeted delivery systems for active ingredients represent another significant advancement in pest control. These systems utilize nanoscale carriers to transport and deliver active substances precisely to the sites where they are needed (Dwivedi & Nameirakpam, 2022). The ability to create highly specific delivery systems enhances the overall efficacy of pest control strategies while minimizing off-target effects (Ghouri et al., 2020). For example, nanoparticles can be engineered to recognize certain chemical signals emitted by pests, ensuring that the release of the pesticide occurs exactly at the moment when the pest is most vulnerable. This precise timing greatly increases the likelihood of successful control while reducing the amount of active ingredient used, thus lowering costs and environmental risks (Kannan et al., 2022; Sarmah et al., 2023).

Targeted delivery systems can enhance the effectiveness of pesticides by prolonging their activity against insect pests. Researchers can regulate the degradation of nanocarriers to optimize the duration of pesticide efficacy, protecting crops during critical phases of pest life cycles (Bhan et al., 2018; Ghouri et al., 2020). Integrating nanotechnology in pest management not only improves efficiency and reduces environmental harm but also combats the development of resistance in pest populations (Kumar, M. et al., 2017; Yousef et al., 2023). Unlike traditional pesticides, nanopesticides require lower application rates due to their precise delivery, potentially delaying resistance by exposing pests to sublethal doses intermittently (Shahid et al., 2021; Zannat et al., 2021). This innovative approach combines enhanced effectiveness with environmental consciousness, paving the way for sustainable agricultural practices that prioritize ecological balance. Embracing nanotechnology can lead to more resilient agricultural systems that effectively manage pest challenges while maintaining productivity and environmental stewardship (Dwivedi & Nameirakpam, 2022; Elmer & White, 2018). The ongoing research and advancement in nanopesticides and targeted delivery systems offer a promising solution for the future of insect pest control in agriculture (Kannan et al., 2022; Nimisha & Deepthy, 2020).

16.4.2 Insect Monitoring and Management

Insect monitoring and management are essential components of effective pest control strategies in agriculture and environmental conservation. Accurate and timely detection of pest populations allows for informed decision-making regarding pest management interventions, significantly reducing the reliance on broad-spectrum pesticides. One of the most exciting advancements in this arena is the application of nanotechnology, particularly the development of nanosensors and remote sensing technologies (Kannan et al., 2022; Ghouri et al., 2020).

Nanosensors, small devices using nanomaterials, detect specific biological or chemical signals, aiding in pest population monitoring (Gamage et al., 2024). They offer high sensitivity, identifying pests at lower concentrations than traditional methods. Nanosensors can detect pheromones, chemicals insects use for communication (Bhan et al., 2018). Recognizing these pheromones confirms pest presence, enabling quicker response to potential infestations (Nimisha & Deepthy, 2020). Nanosensors are versatile, incorporating multiple detection methods in one device. Some can detect various insect species simultaneously by integrating different receptors for distinct chemical signatures (Chandrashekharaiah et al., 2015). This multiplexing capability improves pest monitoring efficiency, providing a comprehensive understanding of pest dynamics in agriculture (Singh et al., 2020). Overall, nanosensors enhance pest control measures, reducing crop damage and ensuring a timely response to pest threats (Singh et al., 2023).

Remote sensing technologies are essential for monitoring and managing insects in agriculture (Rajabpour & Yarahmadi, 2024). By utilizing aerial imagery and satellite data, stakeholders can gather information on crop conditions, pest infestations, and environmental factors affecting pest populations (Rajabpour & Yarahmadi, 2024). Integrating remote sensing with nanosensor data provides a comprehensive understanding of pest dynamics, enabling timely interventions (Elmer & White, 2018). Drones with multispectral imaging capabilities can evaluate plant health, detect stress levels, and identify early signs of pest damage (Humbal & Pathak, 2023). This aerial surveillance can pinpoint areas of unhealthy foliage caused by pest feeding, allowing targeted interventions (Shukla et al., 2020). The combination of nanosensors for localized data and remote sensing for broader coverage improves the effectiveness and sustainability of pest management practices in agriculture (Dwivedi & Nameirakpam, 2022; Yousef et al., 2023).

The integration of nanosensors with remote sensing technologies enables precision agriculture practices that optimize resource use by providing real-time data on pest presence and crop health (Kannan et al., 2022; Ghouri et al., 2020). This allows farmers to apply interventions like pesticides only when necessary, reducing chemical inputs and environmental risks while preserving beneficial insect populations (Basu & Bhattacharya, 2021; Nimisha & Deepthy, 2020). Advances in data analytics and artificial intelligence further enhance insect monitoring systems by analyzing vast amounts of data for predictive modeling and actionable insights (Roy, 2020; Rajabpour & Yarahmadi, 2024). These technologies facilitate proactive management strategies, enabling early detection of pests and adapting strategies accordingly (Shahid et al., 2021; Elmer & White, 2018). Overall, the use of nanosensors and remote sensing technologies revolutionizes pest identification and control, improving pest management outcomes and promoting environmentally friendly practices (Ali et al., 2025; Singh et al., 2023). As these technologies evolve, they have the potential to transform pest management strategies in the face of agricultural intensification and climate change (Das et al., 2021; Zannat et al., 2021). By focusing on sustainability, integrating nanotechnology in pest monitoring can enhance food security while safeguarding crops and ecosystems (Dwivedi & Tomer, 2020; Gamage et al., 2024).

16.4.3 Pest Resistance Management

Pest resistance management is a critical aspect of agricultural sustainability, addressing the growing challenge posed by pests that increasingly adapt to conventional control methods (Ya-Ali et al., 2020). As pests develop resistance to pesticides and other management strategies, the efficacy of these tools diminishes, often leading to increased usage of chemicals, economic losses, and greater environmental impacts (Ndlovu et al., 2020; Yousef et al., 2023). In this context, nanotechnology has emerged as a promising approach to help mitigate resistance development and enhance the durability of pest management strategies (Kumar et al., 2017; Zannat et al., 2021).

Nanotechnology is essential for managing pest resistance by improving the effectiveness of pest control methods and reducing pesticide use (Ghouri et al., 2020). This approach is crucial for promoting ecological balance and sustainability in agriculture, especially as integrated pest management becomes more important (Shahid et al., 2021). By incorporating nanomaterials into pesticides, the bioavailability and efficacy of active ingredients can be increased (Bhattacharya et al., 2020). Nanoparticles allow for the precise delivery of these ingredients to target organisms, resulting in lower doses and lesser environmental impact (Bhan et al., 2018; Dwivedi & Tomer, 2020). Additionally, engineered nanoparticles can enhance the stability and persistence of active ingredients, ensuring prolonged effectiveness and reducing the emergence of resistant pest populations (Kannan et al., 2022). By using nanotechnology to optimize insecticide applications, the development of resistance can be disrupted by extending treatment intervals, ultimately improving pest control outcomes (Margulis-Goshen & Magdassi, 2013; Nimisha & Deepthy, 2020).

Another aspect in which nanotechnology contributes to resistance management is through the development of nanopesticides with novel modes of action. Many traditional pesticides act on specific biological pathways in pests, and over time, pests can adapt and evolve to circumvent these mechanisms (Soylu et al., 2025; Yousef et al., 2023). Nanotechnology enables the formulation of pesticide compounds that may affect multiple targets or exhibit novel modes of action not previously utilized in conventional products (Álvarez et al., 2019; Shukla et al., 2020). This diversity in mechanisms can hinder the ability of pests to adapt and develop resistance, as they would need to evolve multiple resistance pathways simultaneously (Ndlovu et al., 2020; Singh et al., 2023).

The integration of nanosensors in agroecosystems allows for precise monitoring of pest populations, providing real-time data on densities and susceptibility to management tactics (Gamage et al., 2024; Kannan et al., 2022). This data-driven approach enables growers to make informed decisions on the timing of pest control measures, reducing reliance on calendar-based pesticide applications and promoting targeted interventions based on actual pest presence and life cycle stages (Dwivedi & Nameirakpam, 2022; Rajabpour & Yarahmadi, 2024).

Nanotechnology also enhances resistance management strategies by incorporating cultural practices and biological control agents, offering innovative solutions alongside traditional methods (Das et al., 2021; Kumar et al., 2017). By developing nanoformulations of biopesticides and beneficial organisms, the efficacy of alternative control strategies can be improved, further reducing the need for synthetic chemicals (Basu & Bhattacharya, 2021; Shahid et al., 2021). Overall, nanotechnology transforms pest resistance management in agriculture, improving control measures, reducing application frequencies, and creating more resilient agricultural systems. By combining targeted interventions, real-time monitoring, and innovative formulations, farmers can effectively manage pest populations while fostering sustainability and mitigating resistance development (Kumar & Gong, 2018; Nimisha & Deepthy, 2020). As research advances, nanotechnology is poised to play a significant role in shaping the future of pest management, aligning with the goals of resilience and sustainability in agriculture (El-Helaly & Sayed, 2015; Ghouri et al., 2020).

16.4.4 Vector Control for Public Health

A vector control is a fundamental strategy in public health aimed at reducing the transmission of diseases by managing the populations of organisms that transmit pathogens, such as mosquitoes. Mosquitoes are well-known as vectors for a range of diseases, including malaria, dengue fever, Zika virus, and West Nile virus, which collectively pose significant challenges to global health (Yousef et al., 2023). Effective vector control is crucial for preventing outbreaks and minimizing the health burdens associated with these diseases. Advances in technology, including targeted insecticides, biological controls, and genetically modified organisms, have transformed how public health officials approach this critical issue (Kumar & Gong, 2018; Elmer & White, 2018).

The application of vector control strategies has primarily focused on two main areas: reducing mosquito populations and interrupting the transmission of pathogens. One of the most common methods for mosquito control involves the use of chemical insecticides, which can be deployed in various forms, such as sprays, fogs, or residual treatments (Dwivedi & Tomer, 2020). These insecticides target adult mosquitoes as well as their larvae, significantly impacting mosquito populations in both urban and rural settings. Indoor residual spraying (IRS) and space spraying during outbreaks are widely implemented to quickly reduce adult mosquito numbers, thereby decreasing their capacity to transmit viruses and parasites to humans (Ndlovu et al., 2020).

However, reliance on chemical insecticides has led to increased concerns about the development of resistance among mosquito populations (Bhan et al., 2018). As a result, integrated vector management (IVM) strategies have gained prominence, combining multiple approaches to enhance efficacy while minimizing environmental and health impacts (Shahid et al., 2021). IVM incorporates environmental management, biological control agents, community participation, and the responsible use of chemical interventions. For instance, managing standing water to reduce mosquito breeding sites, promoting the use of larvivorous fish in ponds, and employing natural predators can effectively decrease mosquito populations while maintaining ecological balance (Chandrashekharaiah et al., 2015; Zannat et al., 2021).

In addition to traditional vector control methods, biotechnological approaches like genetically modified mosquitoes are being used to combat vector-borne diseases. For instance, the sterile insect technique involves releasing sterilized male mosquitoes to reduce viable offspring (Basu & Bhattacharya, 2021). Another approach utilizes Wolbachia-infected mosquitoes to disrupt virus transmission. These innovative methods have shown promise in field trials, indicating the potential of genetic strategies in public health vector control (Kumar & Gong, 2018). Studies have confirmed the effectiveness of various control strategies, such as IRS and larviciding, in reducing malaria transmission rates (Ndlovu et al., 2020). IVM, which combines multiple strategies, has been shown to further improve control efforts by addressing environmental and social factors (Shahid et al., 2021). IVM has proven effective in achieving long-term control and prevention of vector-borne diseases.

Education and community engagement are also critical components of successful vector control programs. Raising awareness about vector-borne diseases and encouraging participation in control measures are integral to the effectiveness of these programs (Ndlovu et al., 2020; Nimisha & Deepthy, 2020). Promoting personal protective actions, such as using mosquito nets and repellents, empowers communities to take an active role in safeguarding public health (Ghouri et al., 2020). Furthermore, fostering collaboration between local health agencies, government bodies, and the community can enhance the implementation of vector control strategies (Kumar & Gong, 2018). This collaborative approach can foster a culture of vigilance and proactive measures against vector-related risks (Margulis-Goshen & Magdassi, 2013).

Vector control remains an essential pillar of public health strategies aimed at mitigating the impact of mosquito-borne diseases. By employing a combination of chemical, biological, and ecological approaches, along with integrating community participation, health authorities can effectively target disease vectors and interrupt pathogen transmission (Bhan et al., 2018; Nimisha & Deepthy, 2020). The ongoing development of innovative technologies presents exciting prospects for enhancing vector control outcomes. However, the sustainability of these strategies relies on their adaptability to local contexts, the management of resistance, and continuous engagement with communities (Elmer & White, 2018; Yousef et al., 2023). As public health agencies work to combat vector-borne diseases, robust vector control measures play a critical role in promoting health and well-being in populations worldwide (Bhattacharya et al., 2020; Kannan et al., 2022). Effective vector control strategies not only reduce the incidence of diseases like malaria and dengue but also contribute to overall public health resilience (Ndlovu et al., 2020). Implementing community-centric approaches that involve education and awareness can enhance the effectiveness of these measures (Ghouri et al., 2020; Nimisha & Deepthy, 2020).

16.5 NANOTECHNOLOGY IN IPM: ENHANCING EFFICACY, REDUCING ENVIRONMENTAL IMPACT

Nanotechnology is revolutionizing agriculture, particularly in IPM. By using nanoparticles, pest control strategies become more effective and efficient. Nanoparticles can improve the delivery and effectiveness of insecticides and pesticides, leading to better results with lower doses (Dwivedi & Nameirakpam, 2022; Yousef et al., 2023). This targeted approach reduces environmental impacts and improves crop productivity (Das et al., 2021; Bhan et al., 2018).

Nanotechnology enhances the stability and efficacy of active ingredients, reducing waste and allowing for more precise application methods (Basu and Bhattacharya, 2021; El-Helaly and Sayed, 2015). Compared to traditional chemical applications, nanotechnology minimizes environmental risks by targeting pests more effectively (Kannan et al., 2022; Ghouri et al., 2020). The reduced chemical usage and targeted approach of nanopesticides lower the environmental footprint, benefiting sensitive ecosystems and promoting biodiversity (Sarmah et al., 2023; Nimisha & Deepthy, 2020).

One of the key advantages of nanotechnology in IPM is target specificity, which minimizes unintended consequences on non-target organisms (Elmer & White, 2018; Shahid et al., 2021). Nanocapsules can be engineered to release ingredients only when triggered by specific pests, ensuring that beneficial insects are not harmed (Bhattacharya et al., 2020). This specificity also extends to pathogen control, where targeted antimicrobial agents combat plant diseases without affecting nearby flora (Garde-Cerdán et al., 2021; Humbal & Pathak, 2023).

Moreover, nanotechnology promotes safer agricultural practices by reducing chemical loads in farming systems (Dwivedi & Nameirakpam, 2022; Yousef et al., 2023). As concerns about pesticide residues and consumer health grow, nanotechnology offers a sustainable alternative that aligns with the demand for eco-friendly practices (Das et al., 2021; Bhan et al., 2018). By minimizing chemical applications while maintaining efficacy, nanotechnology supports food security and human health (Basu & Bhattacharya, 2021; El-Helaly & Sayed, 2015).

Overall, the integration of nanotechnology into IPM offers numerous benefits, including enhanced efficacy and efficiency, reduced environmental impact, and improved target specificity (Kannan et al., 2022; Ghouri et al., 2020). By harnessing the unique properties of nanoparticles, farmers can optimize productivity and promote ecological sustainability (Sarmah et al., 2023; Nimisha & Deepthy, 2020). As research in nanotechnology progresses, the potential for transforming pest management practices in agriculture grows, paving the way for more resilient and sustainable farming systems (Elmer & White, 2018; Shahid et al., 2021).

16.6 NAVIGATING CHALLENGES AND BALANCING INNOVATION IN ADDRESSING REGULATORY AND SAFETY LIMITATIONS OF NANOTECHNOLOGY IN IPM

While the application of nanotechnology in integrated pest management (IPM) presents numerous advantages, it is crucial to acknowledge the challenges and limitations that accompany its implementation. As with any emerging technology, especially one involving chemical substances and biological interactions, there are regulatory and safety concerns surrounding nanomaterials that must be addressed (Kannan et al., 2022; Ghouri et al., 2020). These concerns reflect the complexities of ensuring that nanotechnology is both effective and safe for agricultural use.

One of the primary challenges involves navigating the regulatory landscape that governs the use of new materials in agriculture. Nanomaterials, due to their unique properties and behaviors at the nanoscale, may not fit neatly within existing regulatory frameworks that were largely developed for conventional chemical substances (Elmer & White, 2018). This poses a significant challenge for manufacturers and researchers seeking to bring nanotechnology-based solutions to market. Regulatory agencies need to evaluate the safety and efficacy of these new materials while also considering their environmental impact (Shahid et al., 2021). However, the current regulatory processes may not fully capture the potential risks associated with nanomaterials, leaving gaps in oversight. Consequently, this uncertainty can hinder innovation and reduce the willingness of stakeholders to adopt nanotechnology in pest management (Yousef et al., 2023).

Moreover, there are potential environmental and human health risks associated with the use of nanotechnology in agriculture. Although nanomaterials can offer enhanced targeting and reduced chemical usage, the introduction of nanomaterials into ecosystems raises questions about their behavior, fate, and effects (Bhan et al., 2018; Basu & Bhattacharya, 2021). For example, once released into the environment, nanoparticles could interact with soil, water, and biological systems in unforeseen ways. Their small size and unique characteristics might lead to bioaccumulation and unintended toxic effects, raising concerns about the long-term implications for non-target organisms, including beneficial insects and other wildlife (Dwivedi & Nameirakpam, 2022). Additionally, there are potential risks for human health, especially for farmworkers who may be exposed to nanotechnology applications during their handling and application (Das et al., 2021). Ensuring safety for all stakeholders is paramount, and without comprehensive studies assessing these risks, the integration of nanotechnology into pest management could face resistance from both regulatory bodies and the public (El-Helaly & Sayed, 2015; Garde-Cerdán et al., 2021).

Another critical challenge lies in the knowledge gaps that currently exist regarding nanotechnology's application in IPM. While research is progressing, there is still much to learn about the interactions between nanomaterials and biological systems (Sarmah et al., 2023). Understanding how nanoparticles behave in various agricultural settings requires in-depth investigations into their physical and chemical properties, their interactions with organisms, and their environmental processes. Additionally, there is a strong need for standardized protocols to assess the efficacy and safety of nanotechnology-based products (Nimisha & Deepthy, 2020). As the field evolves, researchers must conduct comprehensive studies that address these knowledge gaps to provide the evidence needed for regulatory approval and consumer acceptance. Furthermore, interdisciplinary collaboration among scientists, regulatory agencies, and agricultural stakeholders is essential to foster a more profound understanding of these technologies and to develop best practices (El-Helaly & Sayed, 2015).

The need for further research extends beyond safety and efficacy; it includes exploring the economic viability of nanotechnology in pest management. Understanding the cost–benefit dynamics of implementing nanotechnology solutions is critical for widespread adoption by farmers, especially smallholders who may lack the resources to transition to new technologies (Ghouri et al., 2020). Validating the economic advantages of nanotechnology through field trials and real-world applications will play a significant role in advancing its acceptance and integration into existing agricultural practices (Kannan et al., 2022).

16.7 SMART SOLUTIONS FOR SUSTAINABLE FARMING: ADVANCEMENTS IN NANOTECHNOLOGY FOR PEST CONTROL

As the future of agriculture and pest management is considered, exciting opportunities presented by the role of nanotechnology in entomology could revolutionize traditional practices (Liu et al., 2025). A rapidly evolving landscape is indicated by emerging trends, where nanotechnology applications are being increasingly integrated into pest management strategies, providing innovative solutions to long-standing challenges in the field (Ali et al., 2023). Enhanced pest control is promised by these developments, which also align with the broader goals of sustainability and efficiency in agricultural practices (Fu et al., 2024).

One notable trend in nanotechnology is the advancement of smart delivery systems tailored for pest management. Innovations in this area aim to optimize the timing and location of pesticide application, thereby enhancing efficacy while minimizing environmental impact. For instance, researchers are developing nanoparticles and nanoencapsulated formulations that respond to specific stimuli, such as changes in pH and temperature or the presence of target pests (Qian et al., 2024). These smart delivery systems can ensure that pesticides are released only when needed, significantly reducing the volume of chemicals applied and decreasing the likelihood of non-target organism exposure. This approach not only promotes sustainable agricultural practices but may also enhance the protective effects of biological controls, improving overall pest resistance management (Iqbal et al., 2024).

Furthermore, advancements in nanotechnology are not limited to crop protection. The field of biosensing is experiencing innovations that leverage nanoscale materials for early pest detection and monitoring (Ali et al., 2018a). Nanosensors capable of detecting the chemical signatures of pest infestations from a distance can provide farmers with real-time information about pest activity. This data enables timely interventions, reducing reliance on preventative pesticide applications and thereby minimizing chemical use. As these technologies mature, the integration of nanosensing with precision agriculture tools promises to transform pest management from reactive to proactive (Ali et al., 2018b).

Collaboration between researchers, industry, and policymakers will be pivotal in shaping the future of nanotechnology applications in entomology. The interdisciplinary nature of nanotechnology necessitates a cooperative approach, bringing together expertise from fields such as material science, entomology, environmental science, and agricultural economics (Yousaf et al., 2024). By fostering collaborations, stakeholders can ensure that research aligns with real-world applications, leading to solutions that are both scientifically sound and practically viable for farmers (Nimisha & Deepthy, 2020; Kannan et al., 2022). Policymakers are essential in fostering collaboration by creating a regulatory framework that encourages innovation while protecting public health and the environment. As nanotechnology advances pest management, regulatory agencies must adapt their methods to assess the safety and effectiveness of new nanomaterials, necessitating ongoing communication between researchers and regulators to align safety assessments with technological progress without hindering innovation. Additionally, the successful integration of nanotechnology into pest management requires investments in education and outreach, especially for smallholder farmers who may be hesitant to adopt new technologies. Providing access to information and training programs can empower these farmers to utilize nanotechnology effectively.

16.8 CONCLUSION

Nanotechnology is recognized as a transformative force in addressing the significant challenges faced in entomology and IPM. Promising solutions are offered by this innovative field across various dimensions, including pest control, insect monitoring, and vector management, ultimately aligning with the need for sustainable agricultural practices and improvements in public health. The potential of nanotechnology to revolutionize pest control is primarily attributed to its ability to enhance the efficacy and specificity of interventions. The development of nanopesticides is exemplified by advancements that improve the solubility and stability of active ingredients, allowing for lower application rates and reduced environmental impact. Harm to non-target species is minimized, fostering a more balanced ecosystem.

Furthermore, the integration of nanosensors and remote sensing technologies amplifies the capabilities of pest monitoring and management. Real-time data collection enables earlier detection of pest populations and facilitates swifter responses, thereby reducing reliance on broad-spectrum pesticides and maintaining the ecological integrity of agricultural systems. As pest resistance management is considered, significant promise is held by nanotechnology. Varied dosages can be allowed, and the stability and persistence of active ingredients can be enhanced, extending the life of conventional pesticides while mitigating the acceleration of resistance development in pests. The incorporation of innovative modes of action through nanoformulations could also thwart pest adaptation, further improving pest management strategies. The application of nanotechnology is not limited to agricultural settings; it is also extended into public health through vector control for disease prevention. As challenges such as malaria and dengue fever persist, the role of nanotechnology in developing targeted insecticides and biocontrol agents is expected to enhance vector management while reducing the burden of these diseases on global health systems. As the future unfolds, collaboration among researchers, industry leaders, and policymakers will be deemed critical to fully realize the potential of nanotechnology in entomology and IPM. Continued investment in research and development, coupled with supportive regulatory frameworks, will be encouraged to foster innovation while ensuring safety and effectiveness. In summary, tiny solutions are offered by nanotechnology to address big challenges within entomology, enabling more sustainable practices and improved outcomes for agriculture and public health.

REFERENCES

Ali, H., Muhammad, A., Islam, S.U., Islam, W. and Hou, Y., 2018a. A novel bacterial symbiont association in the hispid beetle, Octodonta nipae (Coleoptera: Chrysomelidae), their dynamics and phylogeny. *Microbial Pathogenesis, 118*, 378–386.

Ali, H., Muhammad, A., Bala, N.S., Wang, G., Chen, Z., Peng, Z. and Hou, Y., 2018b. Genomic evaluations of Wolbachia and mtDNA in the population of coconut hispine beetle, Brontispa longissima (Coleoptera: Chrysomelidae). *Molecular Phylogenetics and Evolution*, 127, 1000–1009.

Ali, H., Noor Shah, A., Bilal Tahir, M., Fiaz, S., Ali, B., editors. 2023. Insecticides – Advances in Insect Control and Sustainable Pest Management. *IntechOpen*; Available from: http://dx.doi.org/10.5772/intechopen.104285

Ali, H., Xiao, Y. (Eds.). 2025. *Genome Editing for Pest Management* (1st ed.). CRC Press. https://doi.org/10.1201/9781003478706

Ali, H., Youming, H., and Tahir, M. B. (2023). Climate change and insect biodiversity: Challenges and implications. Taylor & Francis. https://doi.org/10.1201/9781003382089

Álvarez, S. P., Tapia, M. A. M., Ardisana, E. F. H. 2019. Nanotechnology and entomopathogenic microorganisms in modern agriculture. In *Nanoscience for Sustainable Agriculture*, eds. R. N. Pudake, N. Chuhan, C. Kole (pp. 171–187). Springer, Germany.

Basu, P., Bhattacharya, S. 2021. Nanotechnology and polymer science: A novel approach in vector control. In *Genetically Modified and other Innovative Vector Control Technologies: Eco-bio-social Considerations for Safe Application*, ed. B. K. Tyagi (pp. 347–358). Springer, Singapore.

Bhan, S., Mohan, L., Srivastava, C. N. 2018. Nanopesticides: A recent novel ecofriendly approach in insect pest management. *Journal of Entomological Research 42*(2): 263–270.

Bhattacharya, A., Epidi, T. T., Kannan, M. 2020. Nano-technology applications in pest management. In *Innovative Pest Management Approaches for the 21st Century: Harnessing Automated Unmanned Technologies*, ed. A. K. Chakravarthy (pp. 391–401). Springer Singapore.

Chandrashekharaiah, M., Kandakoor, S. B., Basana Gowda, G., Kammar, V., Chakravarthy, A. K. 2015. Nanomaterials: A review of their action and application in pest management and evaluation of DNA-tagged particles. In *New Horizons in Insect Science: Towards Sustainable Pest Management*, ed. A. K. Chakravarthy (pp. 113–126). Springer, New Delhi.

Das, K., Jhan, P. K., Das, S. C., Aminuzzaman, F. M., Ayim, B. Y. 2021. Nanotechnology: Past, present and future prospects in crop protection. In *Technology in Agriculture*, eds. F. Ahmad, M. Sultan (pp. 1–22). IntechOpen.

Dwivedi, S. A., Nameirakpam, L. 2022. The role of nanotechnology in insect pest management. In *Environmental Management Technologies*, eds. P. Chawdhary, V. Kumar, S. Kumar, V. Hare (pp. 273–294). CRC Press.

Dwivedi, S. K., Tomer, A. 2020. Applications for nanotechnology in the polyphagous destructive insect pest management of agricultural crops. In *Nanotechnology for Food, Agriculture, and Environment*, eds. D. Thangadurai, J. Sangeetha, R. Prasad (pp. 181–188). Springer, Cham.

El-Helaly, A., Sayed, W. A. A. 2015. New approach based on nanotechnology in baculovirus protection. *Egyptian Academic Journal of Biological Sciences, F. Toxicology & Pest Control 7*(1): 163–175.

Elmer, W., White, J. C. 2018. The future of nanotechnology in plant pathology. *Annual Review of Phytopathology 56*(1): 111–133.

Fu, C., Wan, S., Yang, P., Zhao, X., Yan, Y., Jiang, S. and Ali, H., 2024. Identification of the Ilex macrocarpa anthracnose pathogen and the antifungal potential of the cell-free supernatant of Bacillus velezensis against Colletotrichum fioriniae. *Frontiers in Microbiology*, 15: 1419436.

Gamage, V., Nishshanka, U., Thiripuranathar, G., Priyantha, N., Gunewardene, S., Jayanetti, S. 2024. Pesticides detection by nanotechnology-based sensors. In *Nanotechnology-Based Sensors for Detection of Environmental Pollution*, eds. F. M. P. Tonelli, M. Ozturk (pp. 215–236). Elsevier.

Garde-Cerdán, T., Souza-da Costa, B., Rubio-Bretón, P., Pérez-Álvarez, E. P. 2021. Nanotechnology: Recent advances in viticulture and enology. *Journal of the Science of Food and Agriculture 101*(15): 6156–6166.

Ghouri, M. Z., Khan, Z., Khan, S. H., Ismail, M., Aftab, S. O., Sultan, Q., Ahmad, A. 2020. Nanotechnology: Transformation of agriculture and food security. *Bioscience 3*: 19.

Humbal, A., Pathak, B. 2023. Application of nanotechnology in plant growth and diseases management: Tool for sustainable agriculture. In *Agricultural and Environmental Nanotechnology: Novel Technologies and Their Ecological Impact*, eds. J. K. Patra, G. Das (pp. 145–168). Springer Singapore.

Iqbal, S., Zhuo, Z., Ali, H., Xu, D., Niaz, Y., Shah, A. N., Almoallim, H. S., Ansari, M. J., Bodlah, M. A., Ali, B., Nawaz, M. 2024. Exploring the synergistic toxicity of synthetic pesticides and their impact on development and behavior of Honeybee (Apis mellifera L.). *Emirates Journal of Food and Agriculture, 36*: 1–7.

Kannan, M., Mohan, M., Elango, K., Govindaraju, K., Mani, M. 2022. Principles and application of nanotechnology in pest management. In *Trends in Horticultural Entomology*, ed. M. Mani (pp.49–79). Springer, Singapore.

Kumar, D., Gong, C. 2018. *Trends in Insect Molecular Biology and Biotechnology*. Springer.

Kumar, M., Shamsi, T. N., Parveen, R., Fatima, S. 2017. Application of nanotechnology in enhancement of crop productivity and integrated pest management. In *Nanotechnology: An Agricultural Paradigm*, eds. R. Prasad, M. Kumar, V. Kumar (pp. 361–371). Springer, Singapore.

Liu, Q., Chen, G., Zhang, J., Xu, D., Ali, H., Wang, X., 2025. Rhizosphere and phyllosphere microbial communities of male and female plants of Morus macroura. *Physiological and Molecular Plant Pathology, 136*, 102517.

Margulis-Goshen, K., Magdassi, S. 2013. Nanotechnology: An advanced approach to the development of potent insecticides. In *Advanced Technologies for Managing Insect Pests*, eds. I. Ishaaya, S. R. Palli, A. R. Horowitz (pp. 295–314). Dordrecht: Springer Netherlands.

Ndlovu, N., Mayaya, T., Muitire, C., Munyengwa, N. 2020. Nanotechnology applications in crop production and food systems. *International Journal of Plant Breeding 7*(1): 624–634.

Nimisha, T., Deepthy, K. B. 2020. Nanotechnology applications in insect pest management-A review. *Journal of Eco-friendly Agriculture 15*(1): 1–6.

Rajabpour, A., Yarahmadi, F. 2024. *Decision System in Agricultural Pest Management*. Springer, Singapore.

Roy, N. 2020. Modern strategies of integrated pest management: A review. *JAST 4*.

Sarmah, K., Borah, R., Boruah, S., Sarmah, D. 2023. Nanopesticides: Revolutionizing pest management with nanotechnology. *Biological Forum – An International Journal* 15(7): 210–218.

Shahid, M., Naeem-Ullah, U., Khan, W., Saeed, D. S., Razzaq, K. 2021. Application of nanotechnology for insect pests management: A review. *Journal of Innovative Science 7*: 28–39.

Shukla, N., Meghvanshi, K., Shukla, J. N. 2020. Role of nanotechnology in the management of agricultural pests. In *Biogenic Nano-particles and Their Use in Agro-ecosystems*, eds. M. Ghorbanpour, P. Bhargava, A. Varma, D. K. Choudhary (pp. 85–98). Springer, Singapore.

Singh, A. K., Chaudhary, B. K., Kumar, V. 2020. Potential use of nanotechnology in agriculture. *International Journal of Engineering Science Advances 6*(1): 27–31.

Singh, A. K., Mishra, K. K., Jeevan, B., Paschapur, A. U., Verma, G., Kant, L. 2023. Nanotechnology in biotic stress management: Future challenges and opportunities. In *Biotic Stress Management of Crop Plants Using Nanomaterials*, eds. K. K. Mishra, S. Kumar (pp. 173–192). CRC Press.

Soylu, S., Bozkurt, A., Ali, H. et al. 2025. Evaluation of resistance of bean genotypes to common bean bacterial blight disease agents and characterisation of plant defence mechanisms involved in resistance. *Eur J Plant Pathol*, 173, 387–411. https://doi.org/10.1007/s10658-025-03066-5

Qian, Q., Deng, X., Mureed, S., Gan, Y., Xu, D., Wang, X., Ali, H. 2024. Integrating transcriptomics and metabolomics to analyze the defense response of Morus notabilis to mulberry ring rot disease. *Frontiers in Microbiology, 15*: 1373827.

Ya-Ali, P., Yarahmadi, F., Mehrnia, M. A. 2020. Efficacies of two nano-formulations of Tasmanian blue gum essential oil to control Callosobruchus maculatus. *Journal of Economic Entomology 113*(3): 1555–1562.

Yousef, H. A., Fahmy, H. M., Arafa, F. N., Abd Allah, M. Y., Tawfik, Y. M., El Halwany, K. K., Bassily, M. E. 2023. Nanotechnology in pest management: Advantages, applications, and challenges. *International Journal of Tropical Insect Science 43*(5): 1387–1399.

Yousuf HM, Yasin M, Ali H, Naveed K, Riaz A, AlGarawi AM, Hatamleh AA, Shan Y. 2024. Evaluation of different Kabuli chickpea genotypes against Helicoverpa armigera (Hübner) (Lepidoptera: Noctuidae) in relation to biotic and abiotic factors. *PeerJ. 12*: e16944. https://doi.org/10.7717/peerj.16944

Zannat, R., Rahman, M. M., Afroz, M. 2021. Application of nanotechnology in insect pest management: A review. *SAARC Journal of Agriculture 19*(2): 1–11.

17 Innovations in Biological Pest Control

Moazam Hyder[1*], *Inzamam Ul Haq*[1], *Muhammad Rehan Akhtar*[1], *Muhammad Adeel Ghafar*[1], *Asalam Bukero*[2], *Habib Ali*[3], *Muhammad Nawaz*[3], *and Adnan Noor Shah*[3]

[1]State Key Laboratory of Ecological Pest Control for Fujian and Taiwan Crops, Key Laboratory of Biopesticides and Chemical Biology, MOE, College of Plant Protection, Fujian Agriculture and Forestry University, Fuzhou 350002, China

[2]Department of Entomology, Sindh Agriculture University, Tandojam 70080, Pakistan

[3]Department of Agricultural Engineering, Khwaja Fareed University of Engineering and Information Technology, Rahim Yar Khan, Pakistan

[*]Corresponding author

17.1 INTRODUCTION

17.1.1 Overview of Biological Pest Control

Biological pest control is a vital component of integrated pest management (IPM) that leverages natural predators, parasitoids, pathogens, and competitors to manage pest populations. This approach is not new; its roots can be traced back to ancient agricultural practices, where farmers relied on nature's balance to control pests. For instance, the introduction of the predatory weaver ant, *Oecophylla smaragdina*, to citrus orchards in China over 1,700 years ago is an early documented example of biological control (Norton et al. 2010).

The historical significance of biological pest control has grown over the centuries, evolving into a more systematic and scientific approach. Today, it is recognized as a critical method within IPM, which aims to minimize the use of synthetic pesticides and reduce the environmental impact of agriculture (Bale, van Lenteren, and Bigler 2008). The adoption of biological pest control is particularly significant in the context of sustainable agriculture, where there is a pressing need for innovative strategies that can address the challenges posed by climate change, pesticide resistance, and the loss of biodiversity (Barratt et al. 2018).

Innovations in biological pest control are increasingly being sought to enhance its efficacy and expand its application across different agroecosystems. These innovations include the discovery of new biocontrol agents, advancements in mass-rearing techniques, and the integration of modern technologies like genetic engineering and precision agriculture tools (Parra 2014). Such developments are critical to ensuring that biological pest control can meet the growing demands of sustainable agricultural practices in a world where environmental concerns are becoming increasingly paramount.

17.1.2 The Role of Biological Pest Control in Sustainable Agriculture

Biological pest control plays a crucial role in promoting sustainable agricultural practices by reducing the dependency on chemical pesticides, which are known to have adverse effects on the environment and human health (van Lenteren 2012). Chemical pesticides, while effective in the short term, often lead to issues such as pesticide resistance, the resurgence of pest populations, and the destruction of non-target species, including beneficial insects (Geiger et al. 2010). In contrast, biological control methods harness the power of natural enemies to maintain pest populations at levels that do not cause significant harm to crops, thereby preserving the ecological balance and enhancing biodiversity (Cock et al. 2010).

One of the primary environmental benefits of biological pest control is its ability to minimize the ecological footprint of agriculture. By reducing the need for synthetic chemicals, biological control methods contribute to the conservation of soil health, water quality, and non-target organisms, including pollinators and natural predators (Garcia et al. 2020). Moreover, the growing global demand for eco-friendly pest control solutions is driving the adoption of biological pest control practices. As consumers and policymakers alike become more aware of the negative impacts of chemical pesticides, there is an increasing push for sustainable alternatives that align with the principles of organic farming and integrated pest management (IPM) (Meyling and Eilenberg 2007).

Furthermore, biological pest control is integral to the long-term sustainability of agricultural systems. It supports the resilience of agroecosystems by enhancing the natural regulatory processes that keep pest populations in check, thereby reducing the likelihood of pest outbreaks and the

DOI: 10.1201/9781003516538-17

need for emergency chemical interventions (Barbosa 1998). The integration of biological control into IPM programs is essential for achieving sustainable agricultural productivity, especially in the face of global challenges, such as climate change, which is expected to alter pest dynamics and increase the vulnerability of crops to pest infestations (Rosenheim, Wilhoit, and Armer 1993).

In conclusion, the role of biological pest control in sustainable agriculture cannot be overstated. It not only provides an effective and environmentally friendly alternative to chemical pesticides but also contributes to the broader goals of sustainability by preserving biodiversity, enhancing ecosystem services, and promoting the long-term health of agricultural landscapes (Inbar and Gerling 2008). As the global demand for sustainable farming practices continues to rise, innovations in biological pest control will play a pivotal role in shaping the future of agriculture.

17.2 INNOVATIONS IN BIOLOGICAL PEST CONTROL METHODS

17.2.1 Advances in Parasitoid-Based Control

Recent innovations in parasitoid-based control have significantly enhanced the efficacy and scope of biological pest management. Parasitoids, which are organisms that spend a significant portion of their life cycle attached to or within a single host organism, leading to its death, are crucial in managing various agricultural pests. Advances in this field have focused on the discovery of new parasitoid species, genetic enhancement of existing ones, and improvements in mass-rearing techniques, all of which contribute to more effective pest control strategies.

One of the key innovations is the discovery of new parasitoid species that target specific pests, providing farmers with more targeted and efficient biological control options. For example, the recent identification of *Trichogramma dendrolimi* as an effective parasitoid against the fall armyworm (*Spodoptera frugiperda*), a major global pest, has shown promising results in reducing pest populations in maize crops (Dong et al. 2021). The introduction of such novel species into IPM programs has expanded the toolkit available for pest management, allowing for more precise control measures tailored to specific pests and regions.

Genetic enhancement of parasitoids has also emerged as a vital innovation, aimed at improving the adaptability, effectiveness, and resilience of these natural enemies. Techniques such as selective breeding and genetic modification are being used to enhance desirable traits in parasitoids, such as increased host specificity, tolerance to environmental stresses, and higher reproductive rates (Leung et al. 2019). For instance, the genetic improvement of *Aphidius colemani*, a widely used parasitoid against aphids, has led to strains that are more resistant to temperature fluctuations, thereby ensuring more consistent pest control across different climatic conditions (Jerbi-Elayed et al. 2015). These genetically enhanced parasitoids offer a more robust solution to pest management, particularly in the face of climate change and the evolving nature of agricultural pests.

Mass-rearing of parasitoids is another area where significant advancements have been made. The ability to produce large numbers of parasitoids in controlled environments is crucial for their widespread use in commercial agriculture. Innovations in rearing techniques, including the use of artificial diets and automated rearing systems, have dramatically increased the scale and efficiency of parasitoid production (Khokhar and Kumar 2024). These advancements have made it possible to release large quantities of parasitoids at critical times during the growing season, leading to more effective suppression of pest populations and reduced crop damage.

The effectiveness of these innovations in controlling specific pest populations and reducing crop damage is well documented. For example, the mass release of *Trichogramma* species has been shown to significantly reduce the infestation levels of corn borers in maize fields, leading to higher crop yields and reduced reliance on chemical pesticides (Moraes 2022). Similarly, the use of genetically enhanced parasitoids has resulted in more sustainable pest management practices, as these organisms can better withstand environmental challenges and maintain their efficacy over time (Force 1967).

17.2.2 Enhanced Use of Predators

Advancements in predator-based pest control have been a crucial element of integrated pest management (IPM) strategies, offering an environmentally friendly alternative to chemical pesticides. The introduction of novel predator species and the refinement of predator release strategies have significantly improved the effectiveness of predator-based control methods in managing agricultural pests. These innovations are particularly important as they provide sustainable solutions to pest management, reduce the ecological footprint of agriculture, and promote biodiversity.

One of the most notable advancements in this field is the introduction of novel predator species that are highly effective against specific pests. For instance, the introduction of the predatory mite *Amblyseius swirskii* has proven to be an effective control agent against thrips, whiteflies, and spider mites in various horticultural crops (Calvo et al. 2015). This species has been widely adopted in greenhouse environments, where it has successfully reduced pest populations without the need for chemical interventions. Similarly, the predatory beetle *Delphastus catalinae* has been introduced as a biocontrol agent against the whitefly, *Bemisia tabaci*, a major pest in many vegetable and ornamental crops (Naranjo and Ellsworth 2009). These novel predator species offer targeted pest control, reducing the likelihood of non-target impacts and enhancing the sustainability of IPM programs.

Improvements in predator release strategies have also played a pivotal role in the success of predator-based pest control. Traditional release methods often faced challenges, such as uneven distribution of predators, poor establishment, and insufficient coverage, which limited their effectiveness. Recent advancements have addressed these issues through the development of more sophisticated release techniques, including the use of drones for precision release, improved packaging, and storage conditions that enhance predator survival and effectiveness. For example, drones equipped with precision delivery systems have been used to release predatory mites in large agricultural fields, ensuring even distribution and effective pest control (Moses-Gonzales and Brewer 2021). These innovations have significantly increased the efficiency of predator release, resulting in better pest control outcomes and reduced crop damage.

However, integrating predators into IPM programs is not without its challenges. One of the primary obstacles is the variability in predator performance under different environmental conditions. Factors such as temperature, humidity, and the availability of alternative prey can influence predator efficacy, making it difficult to achieve consistent results across different settings (Snyder and Ives 2009). Additionally, the potential for intraguild predation, where introduced predators prey on other beneficial species, can complicate the dynamics of biological control (Sunderland 1996). Addressing these challenges requires a nuanced understanding of predator ecology and the development of strategies that mitigate negative interactions while maximizing the benefits of predator-based control.

Despite these challenges, the successes achieved in integrating predators into IPM programs are noteworthy. The use of predators has led to significant reductions in pest populations, decreased reliance on chemical pesticides, and improved crop yields (Gurr et al. 2017). For example, the successful integration of the predatory beetle *Harmonia axyridis* into IPM programs for aphid control in soybean crops has resulted in a substantial decrease in aphid populations and minimized the need for chemical treatments (Plata-Rueda et al. 2022). These successes highlight the potential of predator-based pest control as a key component of sustainable agriculture.

17.2.3 Entomopathogens and Microbial Control

Entomopathogens, including fungi, bacteria, and viruses, have long been recognized as powerful tools in biological pest control. Recent innovations in their application have significantly enhanced their effectiveness, making them a vital component of integrated pest management (IPM) strategies. These advancements have primarily focused on improving the formulation and delivery systems of entomopathogens, thereby increasing their stability, efficacy, and ease of use in various agricultural settings.

One of the key innovations in the use of entomopathogenic fungi is the development of more effective formulations that enhance their viability and persistence in the field. For instance, research has shown that oil-based formulations of *Beauveria bassiana*, a widely used entomopathogenic fungus, improve its stability and efficacy against pests, such as the coffee berry borer (*Hypothenemus hampei*), under tropical conditions. These oil-based formulations protect the fungal spores from desiccation and ultraviolet (UV) radiation, thereby extending their shelf life and field performance. Additionally, the use of nanoparticle-based carriers has been explored to improve the delivery of fungal spores, ensuring a more uniform distribution and better adherence to pest surfaces.

Innovations in the application of entomopathogenic bacteria, such as *Bacillus thuringiensis* (Bt), have also made significant strides. New Bt strains have been engineered to produce multiple toxins, broadening their spectrum of activity against various pests while reducing the likelihood of resistance development (Bravo et al. 2017). Moreover, microencapsulation techniques have been developed to protect Bt spores and toxins from environmental degradation, thereby enhancing their persistence and efficacy in the field (Baharudin, Ahmad, and Hossain 2024). These advancements have enabled more targeted and sustained pest control, particularly in crops like maize and cotton, where Bt-based products are widely used.

Entomopathogenic viruses, particularly baculoviruses, have also seen considerable innovation in their formulation and delivery. The use of adjuvants and synergists in viral formulations has improved their effectiveness by enhancing the infection rates and overcoming the host's immune defenses (Reid et al. 2023). For example, the incorporation of UV protectants into formulations of *Helicoverpa armigera* nucleopolyhedrovirus (HaNPV) has significantly improved its efficacy against this major pest in cotton crops, even under high sunlight conditions (Zhu et al. 2008). Furthermore, advances in genetic engineering have allowed the development of recombinant viruses with enhanced virulence and broader host ranges, providing more versatile tools for pest management (Khattab et al. 2022).

Case studies across different crops have demonstrated the successful use of these innovative entomopathogen-based methods. In Kenya, the application of *Metarhizium anisopliae* in combination with an attractant pheromone has shown remarkable success in controlling locust swarms, significantly reducing the need for chemical insecticides (Haarith, Bushley, and Chen 2020). Similarly, the use of Bt formulations in rice paddies in India has led to a substantial reduction in the population of rice stem borers, contributing to higher yields and lower pesticide residues in the environment. These examples underscore the potential of entomopathogens in providing effective, sustainable, and environmentally friendly pest management solutions.

17.3 EMERGING TECHNOLOGIES IN BIOLOGICAL PEST CONTROL

17.3.1 Genetic Engineering and CRISPR/Cas9 in Pest Control

Genetic engineering, particularly the advent of CRISPR/Cas9 technology, has revolutionized biological pest control by enabling precise modifications to both pest organisms and their natural enemies. This technology offers unprecedented opportunities to enhance the effectiveness of biological control agents and develop pest-resistant crops, significantly improving the sustainability and efficiency of pest management practices.

CRISPR/Cas9 technology allows for targeted gene editing, which can be used to alter specific traits in pests or their natural enemies, making them more susceptible to biological control or enhancing their effectiveness as control agents. One of the most promising applications of CRISPR/Cas9 is in the development of pest-resistant crops. For example, researchers have successfully used CRISPR/Cas9 to introduce mutations in the *OsSWEET* gene family in rice, which confers resistance to bacterial blight, a devastating disease caused by *Xanthomonas oryzae* (Li et al. 2012). This resistance reduces the reliance on chemical pesticides, thereby contributing to more sustainable agricultural practices.

In addition to creating pest-resistant crops, CRISPR/Cas9 has been employed to enhance the effectiveness of natural enemies in pest control. For instance, the technology has been used to genetically modify predatory insects and parasitoids to improve their ability to locate and eliminate pests. A notable example is the modification of *Nasonia vitripennis*, a parasitoid wasp, to increase its ability to parasitize the eggs of agricultural pests, such as the Mediterranean fruit fly (*Ceratitis capitata*) (Komal et al. 2023). By enhancing traits such as host-seeking behavior and reproductive capacity, CRISPR/Cas9 has the potential to make natural enemies more efficient and reliable in controlling pest populations.

Another significant application of CRISPR/Cas9 is in the creation of gene drives, which can spread a genetic modification rapidly through a pest population. This approach has been explored in mosquitoes, where a gene drive was designed to spread a gene that confers resistance to malaria parasites, effectively reducing the mosquito population and the transmission of the disease (Gantz and Bier 2015). Although primarily focused on disease vectors, this technology holds promise for agricultural pest management as well, offering a potential solution to controlling invasive species or reducing the spread of pests that are difficult to manage through traditional methods.

The integration of CRISPR/Cas9 technology into biological pest control strategies has shown significant success in field trials. For example, the deployment of genetically modified *Anopheles gambiae* mosquitoes with a CRISPR/Cas9-based gene drive resulted in the rapid spread of a mutation that sterilizes females, leading to a dramatic decline in mosquito populations in experimental settings (Kyrou et al. 2018). Similar strategies could be applied to agricultural pests, potentially offering a powerful tool for managing pests that are resistant to conventional control methods.

Despite its potential, the application of CRISPR/Cas9 in pest control is not without challenges. Ethical considerations, regulatory hurdles, and the potential for unintended ecological consequences must be carefully addressed (Oye et al. 2014). There is also the risk of resistance developing against gene drives or other genetic modifications, which could undermine the long-term effectiveness of these strategies. Therefore, ongoing research is essential to refine these technologies and ensure their safe and effective implementation in pest management.

17.3.2 RNA Interference (RNAi) Technology

RNA interference (RNAi) technology has emerged as a powerful tool for gene silencing, offering innovative approaches to pest control by targeting and suppressing specific genes essential for the survival or reproduction of pests. RNAi operates by introducing double-stranded RNA (dsRNA) molecules into the pest organism, which triggers the degradation of messenger RNA (mRNA) corresponding to the target gene, effectively silencing the gene and disrupting vital biological processes (Baum et al. 2007). This method has gained considerable attention in biological control strategies due to its specificity, flexibility, and potential to address challenges posed by traditional pest management methods.

One of the primary advantages of RNAi technology is its high specificity. Unlike conventional pesticides that often affect a broad range of organisms, including beneficial species, RNAi can be designed to target only the pest of interest, thereby minimizing non-target effects and reducing environmental impact (Whyard, Singh, and Wong 2009). For instance, RNAi-based approaches have been developed to target genes in the Colorado potato beetle (*Leptinotarsa decemlineata*), a major pest of potato crops, with minimal impact on non-target organisms like pollinators (Zhu et al. 2011). This level of precision makes RNAi an attractive option for integrated pest management (IPM), where preserving beneficial species is crucial for maintaining ecosystem balance.

In addition to specificity, RNAi technology offers the flexibility to target multiple genes simultaneously or sequentially, which can be particularly useful in managing pests that develop resistance to single-gene silencing (Gordon and Waterhouse 2007). This multi-targeting capability allows for the development of more robust pest control strategies that can adapt to evolving pest populations. For example, RNAi has been used to simultaneously silence multiple genes in the western corn rootworm (*Diabrotica virgifera virgifera*), a notorious pest in maize cultivation, leading to significant reductions in pest survival and crop damage (Lundgren and Duan 2013). Such strategies not only enhance the effectiveness of pest control but also help mitigate the risk of resistance development.

Despite its potential, the application of RNAi in pest management faces several challenges. One of the main obstacles is the efficient delivery of dsRNA to the target pests. The ingestion of dsRNA through feeding has shown varying degrees of success depending on the pest species, with some pests exhibiting strong RNAi responses while others show limited or no response (Gu and Knipple 2013). This variability in response necessitates the development

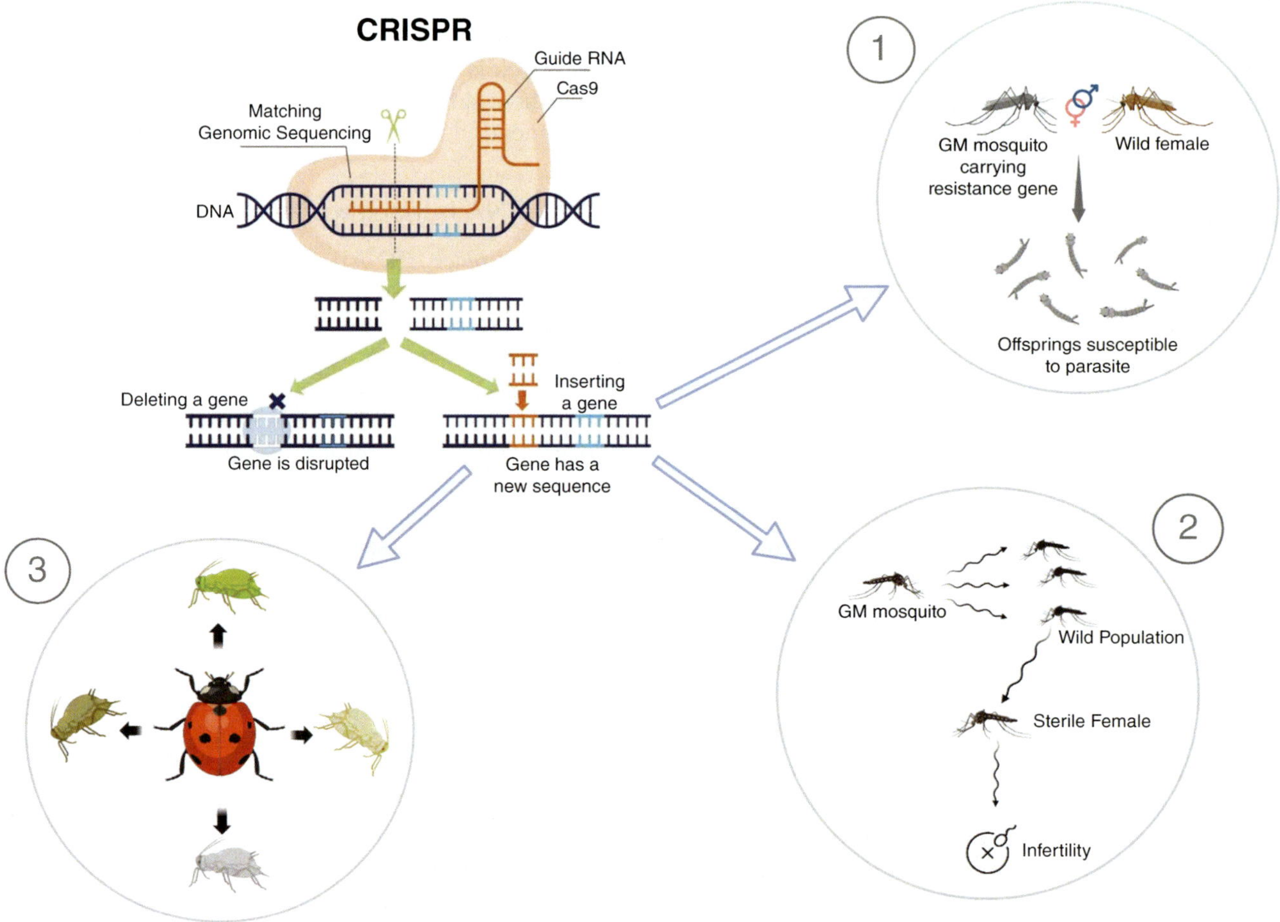

FIGURE 17.1 The deployment of genetically modified *Anopheles gambiae* mosquitoes with a CRISPR/Cas9-based gene drive resulted in the rapid spread of a mutation that sterilizes females, leading to a dramatic decline in mosquito populations in experimental settings. (a) This approach has been explored in mosquitoes, where a gene drive was designed to spread a gene that confers resistance to malaria parasites, effectively reducing the mosquito population and the transmission of the disease. (b) The technology has been used to genetically modify predatory insects and parasitoids to improve their ability to locate and eliminate pests

of optimized delivery systems, such as nanoparticle-based carriers or transgenic plants expressing dsRNA, to ensure consistent and effective gene silencing across different pest species (Zhang, Li, and Miao 2013). Additionally, the cost of producing and delivering RNAi-based products on a commercial scale remains a significant challenge, particularly when compared to traditional chemical pesticides.

Another challenge lies in the potential for off-target effects, where unintended genes in the pest or non-target species could be silenced, leading to unforeseen ecological consequences (Lundgren and Duan 2013). Although RNAi technology is generally considered highly specific, the design of dsRNA sequences must be carefully optimized to avoid such effects. Furthermore, the persistence of dsRNA in the environment and its impact on non-target organisms, including soil microbes and aquatic life, requires thorough assessment to ensure the long-term safety of RNAi-based pest control strategies.

In terms of cost, RNAi technology is currently more expensive than traditional chemical pesticides due to the complexities involved in dsRNA synthesis and delivery. However, as the technology advances and production methods become more efficient, the cost is expected to decrease, making RNAi a more economically viable option for pest management in the future (Head et al. 2017). Moreover, the reduced environmental impact and potential for sustainable pest control offered by RNAi could offset the initial costs, particularly in regions where pesticide resistance and environmental regulations are major concerns.

17.3.3 Use of Drones and Precision Agriculture Tools

The integration of drones and precision agriculture technologies into biological pest control represents a significant advancement in pest management practices. These technologies have revolutionized how biocontrol agents are monitored and applied, enabling more accurate, efficient, and targeted pest control interventions. Drones, in particular, have become invaluable tools in precision agriculture,

TABLE 17.1
Comparison of Traditional Pest Control Methods with RNAi-Based Approaches

Aspect	Traditional Pest Control	RNAi-Based Approaches
Efficiency	High, but with potential for resistance development	High, with potential for multi-targeting
Environmental impact	Often significant, with non-target effects	Minimal, with species-specific targeting
Cost	Generally low, with widespread use	Currently higher, but decreasing with technological advancements
Delivery methods	Spraying of chemical pesticides	Ingestion through dsRNA-treated plants or direct application
Resistance risk	High, due to overuse and single-mode action	Lower, with potential for multi-gene targeting
Regulatory hurdles	Well-established, but tightening due to environmental concerns	Emerging, with evolving regulations

offering a rapid and cost-effective means to survey large agricultural areas, identify pest hotspots, and deliver biocontrol agents precisely where they are needed.

One of the primary applications of drones in biological pest control is in the targeted release of natural enemies, such as predatory insects or parasitoids, directly into pest-infested areas. This method has been shown to improve the effectiveness of biological control by ensuring that the agents are deployed exactly where they are most needed, reducing the waste of resources and enhancing the overall impact on pest populations (Emimi, Khaleel, and Alkrash 2023). For instance, drones have been used to release *Trichogramma* wasps for controlling corn borers in maize fields, resulting in more uniform coverage and improved pest suppression compared to manual release methods (Dutta and Goswami 2020).

Precision agriculture tools, such as GPS-guided equipment and remote sensing technologies, also play a critical role in monitoring pest populations and environmental conditions that influence pest dynamics. These tools allow for the real-time collection and analysis of data, enabling farmers to make informed decisions about when and where to apply biocontrol agents. For example, multispectral imaging from drones can detect early signs of pest infestations or crop stress, allowing for timely interventions that can prevent the spread of pests and minimize crop damage (Singh et al. 2022). This level of precision not only improves the efficacy of biological pest control but also reduces the environmental impact by minimizing the need for blanket applications of biocontrol agents or chemical pesticides.

The use of drones and precision agriculture tools has significantly enhanced the accuracy and efficiency of biological pest control. By enabling targeted applications and real-time monitoring, these technologies are helping optimize pest management practices, contributing to more sustainable and environmentally friendly agricultural systems.

17.3.4 Artificial Intelligence and Machine Learning in Biocontrol

Artificial intelligence (AI) and machine learning (ML) are increasingly being applied to optimize biological pest control, offering new ways to predict pest outbreaks, select the most effective control agents, and automate monitoring processes. These technologies enable the analysis of large datasets, identifying patterns and trends that would be difficult or impossible for humans to discern, thereby improving the precision and effectiveness of pest management strategies. One of the key applications of AI in biological pest control is in the prediction of pest outbreaks. Machine learning algorithms can analyze historical data on weather conditions, crop cycles, and pest behavior to predict when and where pest populations are likely to surge (Domingues, Brandão, and Ferreira 2022). For example, AI-driven models have been used to predict the outbreaks of the fall armyworm (*Spodoptera frugiperda*), allowing for early interventions and reducing the potential for widespread crop damage (Shaurub 2024). This predictive capability is crucial for implementing timely biological control measures, thereby reducing the reliance on chemical pesticides.

AI and ML are also being used to enhance the selection and deployment of biocontrol agents. By analyzing data on pest species, environmental conditions, and the performance of various biocontrol agents, AI systems can recommend the most effective agents for specific situations, optimizing pest control outcomes (Dalla Corte et al. 2020). Additionally, AI-driven robotic systems are being developed to automate the release of biocontrol agents and monitor their effectiveness in the field, further improving the efficiency and scalability of biological pest control (Su 2020).

Current research is exploring the integration of AI with precision agriculture tools, such as drones and remote sensors, to create fully automated pest management systems. These systems have the potential to revolutionize agricultural practices by providing continuous monitoring and real-time responses to pest threats, significantly enhancing the sustainability and effectiveness of biological pest control (Kumar et al. 2022). As AI and ML technologies continue to advance, their role in biological pest control is expected to grow, offering new possibilities for managing pests in a more precise, efficient, and environmentally responsible manner.

AI and machine learning are playing an increasingly important role in optimizing biological pest control. By improving the prediction of pest outbreaks, enhancing the selection of control agents, and automating monitoring processes, these technologies are helping create more effective and sustainable pest management strategies. The future of biological pest control will likely be shaped by these innovations, leading to a new era of precision agriculture.

CASE STUDIES OF SUCCESSFUL BIOLOGICAL PEST CONTROL PROGRAMS

CASE STUDY 1: BIOLOGICAL CONTROL OF FALL ARMYWORM

The fall armyworm (*Spodoptera frugiperda*) is one of the most devastating agricultural pests, particularly affecting maize and other staple crops across the globe. Originally native to the Americas, this pest has rapidly spread to Africa, Asia, and Australia, posing a significant threat to food security (Day et al. 2017). In response to its widespread impact, several biological control programs have been developed and implemented with varying degrees of success. One of the most successful programs has been carried out in Africa, where innovative strategies and technologies have played a crucial role in managing fall armyworm populations.

The biological control program targeting fall armyworm in Kenya is a prime example of a successful initiative that leverages the power of natural enemies. The program primarily employed *Cotesia icipe*, a parasitoid wasp species native to East Africa and effective against various Lepidoptera pests (Idrees et al. 2023). Researchers from the International Centre of Insect Physiology and Ecology (ICIPE) identified and mass-reared this wasp, which was then released in infested maize fields. The wasps lay their eggs inside the fall armyworm larvae, and upon hatching, the larvae are killed, effectively reducing the pest population.

One of the significant challenges faced during the program was the initial scarcity of *Cotesia icipe* in areas heavily infested by fall armyworm. To overcome this, the program included a mass-rearing and release strategy, which involved breeding large numbers of the wasps in controlled environments before releasing them in the field. The

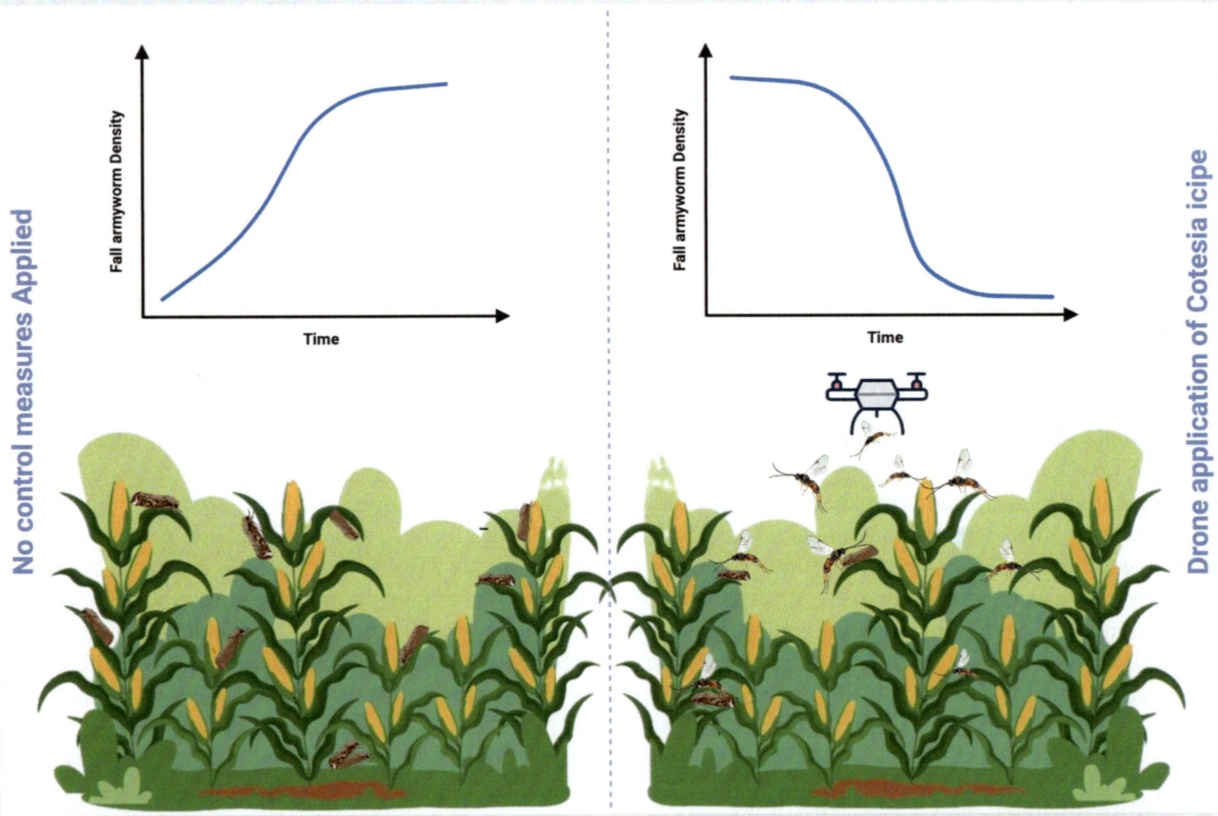

FIGURE 17.2 Graphical representation of the population dynamics of fall armyworm before and after the implementation of biological control measures in Kenya provides a clear visualization of the program's success. The figure shows a sharp decline in the pest population following the introduction of *Cotesia icipe* and the strategic use of drones for parasitoid release.

mass-rearing of natural enemies was complemented by targeted release strategies, facilitated by drone technology. Drones were used to disperse the parasitoids evenly across large maize fields, ensuring that the wasps reached the areas most affected by fall armyworm (Monica et al. 2024). This approach not only enhanced the coverage of the biological control agents but also optimized the use of resources.

The outcomes of this biological control program have been highly promising. Studies conducted before and after the implementation of the program demonstrated a significant reduction in fall armyworm populations, leading to improved maize yields and reduced reliance on chemical pesticides (Midega et al. 2018). The success of the program has been attributed to the innovative use of technology, particularly the integration of drones for the precise release of biocontrol agents, and the selection of an effective natural enemy well-adapted to the local environment.

Furthermore, the program's success has encouraged the expansion of similar biological control strategies to other regions affected by fall armyworm. The adaptability of the approach, including the use of locally sourced natural enemies and cutting-edge technology, has made it a model for future pest management programs in Africa and beyond (Harrison et al. 2019). The program's results are well documented, with studies highlighting the sustained reduction in pest populations and the positive economic impact on smallholder farmers (Day et al. 2017).

CASE STUDY 2: INTEGRATED BIOCONTROL IN GREENHOUSE CROPS

The integration of biological control in greenhouse crops has proven to be an effective and sustainable approach to managing pest populations, particularly in environments where chemical pesticide use is restricted or undesirable. This case study focuses on the successful implementation of integrated biocontrol strategies in greenhouse tomato crops in the Netherlands, where a combination of different biocontrol agents and the adoption of emerging technologies have led to significant economic and environmental benefits.

In this greenhouse biocontrol program, a variety of natural enemies were employed to manage multiple pest species that commonly affect tomato crops. The primary pests targeted included the whitefly (*Trialeurodes vaporariorum*), thrips (*Frankliniella occidentalis*), and two-spotted spider mite (*Tetranychus urticae*) (van Lenteren et al. 2018). The biocontrol agents used in this program were selected based on their effectiveness against specific pests and their ability to coexist without negatively impacting each other. These agents included the predatory mite *Phytoseiulus persimilis* for spider mite control, the parasitic wasp *Encarsia formosa* for whitefly management, and *Amblyseius swirskii* for thrips control (Vangansbeke et al. 2016).

One of the key factors contributing to the success of this integrated biocontrol approach was the careful timing and method of application. Biocontrol agents were introduced at critical points in the growing season when pest populations were beginning to establish but had not yet reached damaging levels (Messelink et al. 2021). This proactive strategy allowed the natural enemies to suppress pest populations early, preventing outbreaks and minimizing crop damage. Additionally, the use of banker plants, which are non-crop plants introduced to support the establishment and reproduction of biocontrol agents, played a crucial role in maintaining a stable population of beneficial insects throughout the growing season (Wäckers, van Rijn, and Bruin 2005).

Emerging technologies, such as climate-controlled environments and automated monitoring systems, further enhanced the effectiveness of this biocontrol program. Climate control systems in greenhouses were used to create optimal conditions for both the crops and the biocontrol agents, ensuring that natural enemies could thrive and effectively suppress pests (Benke and Tomkins 2017). Automated monitoring systems, including sticky traps combined with digital imaging software, allowed for real-time monitoring of pest and biocontrol agent populations, enabling timely interventions and adjustments to the biocontrol strategy as needed (Sanchez et al. 2021). These technologies reduced the labor required for manual monitoring and increased the accuracy of pest population assessments.

The economic and environmental benefits of this integrated biocontrol approach were substantial. Economically, the program led to a significant reduction in the use of chemical pesticides, resulting in lower production costs and higher market value for the crops due to their pesticide-free status. The reduced need for chemical inputs also minimized the risk of pesticide resistance developing in pest populations, ensuring the long-term sustainability of the biocontrol program (Pijnakker et al. 2020). Environmentally, the program contributed to the conservation of biodiversity within the greenhouse ecosystem by fostering a balance between pest and beneficial insect populations. The elimination of chemical pesticides also reduced the risk of contamination to soil and water sources, aligning with the goals of sustainable agriculture and environmental protection (Stenberg 2017).

TABLE 17.2
Biocontrol Agents Used in the Greenhouse Crop Case Study

Biocontrol Agent	Target Pest	Method of Application	Outcomes
Phytoseiulus persimilis	Two-spotted spider mite (*Tetranychus urticae*)	Release at early signs of mite presence	Effective suppression of mite populations
Encarsia formosa	Whitefly (*Trialeurodes vaporariorum*)	Regular releases throughout the season	Significant reduction in whitefly populations
Amblyseius swirskii	Thrips (*Frankliniella occidentalis*)	Banker plants and direct releases	Prevented thrips outbreaks and maintained low population levels

Note: The table highlights the target pests, methods of application, and outcomes, demonstrating the effectiveness of an integrated approach to pest management.

17.4 CHALLENGES AND FUTURE DIRECTIONS

17.4.1 Overcoming Barriers to Adoption

Despite the demonstrated effectiveness and environmental benefits of biological pest control, its widespread adoption faces several significant challenges. These barriers include regulatory hurdles, cost considerations, and a lack of awareness among farmers. Addressing these challenges is crucial to promoting broader adoption of biological control methods and realizing their full potential in sustainable agriculture.

One of the primary challenges to the adoption of biological pest control is the complex and often stringent regulatory environment. In many countries, the registration and approval process for biocontrol agents can be lengthy and expensive, particularly when compared to the approval process for chemical pesticides (Broughton and Herron 2009). Regulatory frameworks often require extensive testing to demonstrate the safety and efficacy of biological control agents, which can be a significant financial and logistical burden for developers, particularly small and medium-sized enterprises (SMEs). Moreover, the lack of harmonization in regulations across different countries and regions can further complicate the process, limiting the international availability and use of biocontrol products. To overcome these regulatory barriers, there is a need for streamlined approval processes that are tailored to the unique characteristics of biological control agents. Governments and regulatory bodies should work toward harmonizing regulations globally and implementing risk-based approaches that prioritize safety while reducing unnecessary delays in the approval process (Kearns et al. 2021).

Cost considerations also pose a significant barrier to the adoption of biological pest control. Biocontrol agents often require more complex production, storage, and application methods than chemical pesticides, leading to higher up-front costs for farmers (Hajek and Eilenberg 2018). Additionally, the need for repeated applications and the requirement for specific environmental conditions can further increase costs, particularly for smallholder farmers in developing countries (Pretty et al. 2018). To address these cost challenges, research and development efforts should focus on improving the cost-effectiveness of biocontrol agents. This could include the development of more efficient mass-rearing techniques; the use of innovative delivery systems, such as encapsulation or slow-release formulations; and the integration of biocontrol with other sustainable farming practices to reduce overall input costs (Daam et al. 2019). Additionally, government subsidies or incentive programs could be introduced to offset the initial costs of adopting biological control methods, making them more accessible to a broader range of farmers.

Farmer awareness and education are also critical challenges in the adoption of biological pest control. Many farmers, particularly in regions with limited access to agricultural extension services, may be unfamiliar with biological control methods or skeptical of their effectiveness compared to conventional pesticides (Gurr et al. 2017). This lack of awareness can be a significant impediment to adoption, as farmers are less likely to invest in practices they do not fully understand or trust. To promote broader adoption, it is essential to implement comprehensive education and outreach programs that provide farmers with the knowledge and skills needed to effectively implement biological control strategies (Heimpel and Mills 2017). Extension services, agricultural cooperatives, and NGOs can play a vital role in delivering these programs, offering training workshops, demonstration projects, and access to resources that highlight the benefits and best practices of biological pest control. Additionally, fostering peer-to-peer learning networks among farmers can help build trust and encourage the sharing of successful experiences with biological control (McDougall).

In conclusion, overcoming the barriers to the adoption of biological pest control requires a multifaceted approach that addresses regulatory, economic, and educational challenges. By streamlining regulations, reducing costs, and enhancing farmer awareness, it is possible to promote the broader use of biological control methods and contribute to more sustainable agricultural systems.

17.4.2 Future Innovations and Research Directions

The future of biological pest control is promising, with numerous emerging research areas and potential breakthroughs that could significantly enhance the effectiveness and scalability of these methods. Continuous innovation is essential to address the evolving challenges in agriculture, such as climate change, pest resistance, and the need for more sustainable farming practices.

One of the most exciting areas of research in biological pest control is the development of next-generation biocontrol agents through genetic engineering and synthetic biology. Advances in CRISPR/Cas9 technology, for example, offer the potential to create more effective and targeted biocontrol agents by modifying the genomes of natural enemies or developing entirely new organisms with enhanced pest control capabilities (Enríquez 2016). These innovations could lead to the creation of biocontrol agents that are more resilient to environmental stresses, have increased reproductive rates, or possess the ability to target specific pest populations with unprecedented precision (Leung et al. 2019). However, the use of genetic engineering in biological control also raises ethical and ecological concerns that must be carefully considered, including the potential impact on non-target species and the environment.

Another promising area of research is the integration of biological pest control with precision agriculture technologies. The use of drones, sensors, and AI-driven platforms to monitor pest populations and optimize the timing and application of biocontrol agents can significantly improve the efficiency and effectiveness of biological control programs (Sishodia et al., 2020). Precision agriculture tools enable real-time data collection and analysis, allowing for more accurate predictions of pest outbreaks and more targeted interventions. This integration not only enhances pest control outcomes but also reduces the environmental impact by minimizing the use of inputs and conserving resources (Finger et al. 2019). As technology continues to advance, the combination of biological control and precision agriculture has the potential to revolutionize pest management practices, making them more sustainable and adaptable to changing environmental conditions.

Global collaboration and knowledge-sharing will also play a crucial role in advancing the field of biological pest control. The challenges faced by agriculture are global in nature, and addressing them requires coordinated efforts across borders. International research partnerships, collaborative field trials, and the sharing of best practices can accelerate the development and dissemination of innovative biocontrol methods (Pretty et al. 2018). Additionally, creating global networks of researchers, policymakers, and practitioners can help harmonize regulatory frameworks, facilitate the exchange of biocontrol agents, and promote the adoption of sustainable pest management practices worldwide. These collaborative efforts will be essential in addressing the complex and interconnected challenges of modern agriculture.

17.5 CONCLUSION

Biological pest control has emerged as a cornerstone of sustainable agriculture, offering an environmentally friendly alternative to chemical pesticides. This chapter has highlighted the significant role that biological control plays in preserving ecological balance, reducing reliance on harmful chemicals, and promoting long-term agricultural productivity. The environmental and economic benefits of these approaches underscore their importance in modern farming practices. The innovations and emerging technologies discussed in this chapter, including genetic engineering, RNA interference (RNAi) technology, the use of drones, and the integration of artificial intelligence (AI) and machine learning, represent significant advancements in the field of biological pest control. These technologies are not only enhancing the precision and effectiveness of pest management strategies but are also making them more scalable and adaptable to various agricultural settings. By leveraging these advancements, farmers can achieve more targeted and efficient pest control, leading to higher yields and reduced environmental impact. Despite the clear advantages, the widespread adoption of biological pest control still faces several challenges. Regulatory hurdles, cost considerations, and a lack of awareness among farmers remain significant barriers. To overcome these obstacles, there is a need for continued innovation, particularly in developing cost-effective and accessible biocontrol solutions. Furthermore, global collaboration and knowledge-sharing are essential to harmonize regulations, improve the availability of biocontrol agents, and promote their adoption worldwide. The future of pest management lies in the continued research, development, and adoption of biological pest control methods. Policymakers, researchers, and farmers must work together to overcome existing challenges and embrace these sustainable practices. By doing so, we can ensure the long-term health and productivity of agricultural systems, ultimately contributing to global food security and environmental sustainability.

REFERENCES

Baharudin, Nurul Syuhada, Harlina Ahmad, and Md Sohrab Hossain. 2024. "Understanding the degradation of carbofuran in agricultural area: A review of fate, metabolites, and toxicity." *Pertanika Journal of Science & Technology* 32 (1).

Bale, JS, JC van Lenteren, and F Bigler. 2008. "Biological control and sustainable food production." *Philosophical Transactions of the Royal Society B: Biological Sciences* 363 (1492):761–776.

Barbosa, Pedro A. 1998. *Conservation biological control*. Elsevier.

Barratt, BIP, VC Moran, Franz Bigler, and JC van Lenteren. 2018. "The status of biological control and recommendations for improving uptake for the future." *BioControl* 63:155–167.

Baum, James A, Thierry Bogaert, William Clinton, Gregory R Heck, Pascale Feldmann, Oliver Ilagan, Scott Johnson, Geert Plaetinck, Tichafa Munyikwa, and Michael Pleau. 2007. "Control of coleopteran insect pests through RNA interference." *Nature Biotechnology* 25 (11):1322–1326.

Benke, Kurt, and Bruce Tomkins. 2017. "Future food-production systems: Vertical farming and controlled-environment agriculture." *Sustainability: Science, Practice and Policy* 13 (1):13–26.

Bravo, Armando, Mathias Brands, Vera Wewer, Peter Dörmann, and Maria J Harrison. 2017. "Arbuscular mycorrhiza-specific enzymes FatM and RAM 2 fine-tune lipid biosynthesis to promote development of arbuscular mycorrhiza." *New Phytologist* 214 (4):1631–1645.

Broughton, S, and GA Herron. 2009. "Potential new insecticides for the control of western flower thrips (Thysanoptera: Thripidae) on sweet pepper, tomato, and lettuce." *Journal of Economic Entomology* 102 (2):646–651.

Calvo, F Javier, Markus Knapp, Yvonne M van Houten, Hans Hoogerbrugge, and José E Belda. 2015. "Amblyseius swirskii: What made this predatory mite such a successful biocontrol agent?" *Experimental and Applied Acarology* 65 (4):419–433.

Cock, Matthew JW, Joop C van Lenteren, Jacques Brodeur, Barbara IP Barratt, Franz Bigler, Karel Bolckmans, Fernando L Cônsoli, Fabian Haas, Peter G Mason, and José Roberto P Parra. 2010. "Do new access and benefit sharing procedures under the convention on biological diversity threaten the future of biological control?" *BioControl* 55:199–218.

Daam, Michiel A, Sónia Chelinho, Júlia C Niemeyer, Olugbenga J Owojori, P Mangala CS De Silva, Jóse Paulo Sousa, Cornelis AM van Gestel, and Jörg Römbke. 2019. "Environmental risk assessment of pesticides in tropical terrestrial ecosystems: Test procedures, current status and future perspectives." *Ecotoxicology and Environmental Safety* 181:534–547.

Dalla Corte, Ana Paula, Deivison Venicio Souza, Franciel Eduardo Rex, Carlos Roberto Sanquetta, Midhun Mohan, Carlos Alberto Silva, Angelica Maria Almeyda Zambrano, Gabriel Prata, Danilo Roberti Alves de Almeida, and Jonathan William Trautenmüller. 2020. "Forest inventory with high-density UAV-Lidar: Machine learning approaches for predicting individual tree attributes." *Computers and Electronics in Agriculture* 179:105815.

Day, Roger, Phil Abrahams, Melanie Bateman, Tim Beale, Victor Clottey, Matthew Cock, Yelitza Colmenarez, Natalia Corniani, Regan Early, and Julien Godwin. 2017. "Fall armyworm: Impacts and implications for Africa." *Outlooks on Pest Management* 28 (5):196–201.

Domingues, Tiago, Tomás Brandão, and João C Ferreira. 2022. "Machine learning for detection and prediction of crop diseases and pests: A comprehensive survey." *Agriculture* 12 (9):1350.

Dong, Hui, Kai-hui Zhu, Qian Zhao, Xue-ping Bai, Jin-cheng Zhou, and Li-sheng Zhang. 2021. "Morphological defense of the egg mass of Spodoptera frugiperda (Lepidoptera: Noctuidae) affects parasitic capacity and alters behaviors of egg parasitoid wasps." *Journal of Asia-Pacific Entomology* 24 (3):671–678.

Dutta, Gopal, and Purba Goswami. 2020. "Application of drone in agriculture: A review." *International Journal of Chemical Studies* 8 (5):181–187.

Emimi, Mohamed, Mohamed Khaleel, and Abobakr Alkrash. 2023. "The current opportunities and challenges in drone technology." *The International Journal of Electrical Engineering and Sustainability*:74–89.

Enríquez, Paul. 2016. "Genome editing and the jurisprudence of scientific empiricism." *Vanderbilt Journal of Entertainment and Technology Law* 19:603.

Finger, Robert, Scott M Swinton, Nadja El Benni, and Achim Walter. 2019. "Precision farming at the nexus of agricultural production and the environment." *Annual Review of Resource Economics* 11 (1):313–335.

Force, Don C. 1967. "Genetics in the colonization of natural enemies for biological control." *Annals of the Entomological Society of America* 60 (4):722–729.

Gantz, Valentino M, and Ethan Bier. 2015. "The mutagenic chain reaction: A method for converting heterozygous to homozygous mutations." *Science* 348 (6233):442–444.

Garcia, Flávio RM, Sérgio M Ovruski, Lorena Suárez, Jorge Cancino, and Oscar E Liburd. 2020. "Biological control of tephritid fruit flies in the Americas and Hawaii: A review of the use of parasitoids and predators." *Insects* 11 (10):662.

Geiger, Flavia, Jan Bengtsson, Frank Berendse, Wolfgang W Weisser, Mark Emmerson, Manuel B Morales, Piotr Ceryngier, Jaan Liira, Teja Tscharntke, and Camilla Winqvist. 2010. "Persistent negative effects of pesticides on biodiversity and biological control potential on European farmland." *Basic and Applied Ecology* 11 (2):97–105.

Gordon, Karl HJ, and Peter M Waterhouse. 2007. "RNAi for insect-proof plants." *Nature Biotechnology* 25 (11):1231–1232.

Gu, Liuqi, and Douglas C Knipple. 2013. "Recent advances in RNA interference research in insects: Implications for future insect pest management strategies." *Crop Protection* 45:36–40.

Gurr, Geoff M, Steve D Wratten, Douglas A Landis, and Minsheng You. 2017. "Habitat management to suppress pest populations: Progress and prospects." *Annual Review of Entomology* 62 (1):91–109.

Haarith, Deepak, Kathryn E Bushley, and Senyu Chen. 2020. "Fungal communities associated with and their potential in biological control: A current update." *Journal of Nematology* 52 (1):1–17.

Hajek, Ann E, and Jørgen Eilenberg. 2018. *Natural enemies: An introduction to biological control*. Cambridge University Press.

Harrison, Rhett D, Christian Thierfelder, Frédéric Baudron, Peter Chinwada, Charles Midega, Urs Schaffner, and Johnnie Van Den Berg. 2019. "Agro-ecological options for fall armyworm (Spodoptera frugiperda JE Smith) management: Providing low-cost, smallholder friendly solutions to an invasive pest." *Journal of Environmental Management* 243:318–330.

Head, Graham P, Matthew W Carroll, Sean P Evans, Dwain M Rule, Alan R Willse, Thomas L Clark, Nicholas P Storer, Ronald D Flannagan, Luke W Samuel, and Lance J Meinke. 2017. "Evaluation of SmartStax and SmartStax PRO maize against western corn rootworm and northern corn rootworm: Efficacy and resistance management." *Pest Management Science* 73 (9):1883–1899.

Heimpel, George E, and Nicholas J Mills. 2017. *Biological control*. Cambridge University Press.

Idrees, Atif, Ayesha Afzal, Ziyad Abdul Qadir, and Jun Li. 2023. "Virulence of entomopathogenic fungi against fall armyworm, *Spodoptera frugiperda* (Lepidoptera: Noctuidae) under laboratory conditions." *Frontiers in Physiology* 14:1107434.

Inbar, Moshe, and Dan Gerling. 2008. "Plant-mediated interactions between whiteflies, herbivores, and natural enemies." *Annual Review of Entomology* 53 (1):431–448.

Jerbi-Elayed, Mey, Kaouthar Lebdi-Grissa, Vincent Foray, Frédéric Muratori, and Thierry Hance. 2015. "Using multiple traits to estimate the effects of heat shock on the fitness of A phidius colemani." *Entomologia Experimentalis et Applicata* 155 (1):18–27.

Kearns, Peter WE, Gijs A Kleter, Hans EN Bergmans, and Harry A Kuiper. 2021. "Biotechnology and biosafety policy at OECD: Future trends." *Trends in Biotechnology* 39 (10):965–969.

Khattab, OM, Mashaly MM AbdelmegeedHK, Hagag N HamdyME, A Hamed, HA Fahmy, E Ibrahim, MA Shahein, and EM Ahmed. 2022. "Equine herpes virus 4 (EHV4) investigation in aborted Egyptian Mares; molecular detection, isolation, and phylogeny for viral glycoprotein B." *Advances in Animal and Veterinary Sciences* 10: 1907–15.

Khokhar, Harsh, and Chitranjan Kumar. 2024. "Safeguarding Tomato Cultivation: Challenges and Integrated Pest Management Strategies in North India." BIO Web of Conferences.

Komal, J, HR Desai, Ipsita Samal, Andrea Mastinu, RD Patel, PV Dinesh Kumar, Prasanta Kumar Majhi, Deepak Kumar Mahanta, and Tanmaya Kumar Bhoi. 2023. "Unveiling the genetic symphony: Harnessing CRISPR-Cas genome editing for effective insect pest management." *Plants* 12 (23):3961.

Kumar, Pradeep, Abhishek Singh, Vishnu D Rajput, Ajit Kumar Singh Yadav, Pravin Kumar, Anil Kumar Singh, and Tatiana Minkina. 2022. "Role of artificial intelligence, sensor technology, big data in agriculture: Next-generation farming." In *Bioinformatics in agriculture*, 625–639. Elsevier.

Kyrou, Kyros, Andrew M Hammond, Roberto Galizi, Nace Kranjc, Austin Burt, Andrea K Beaghton, Tony Nolan, and Andrea Crisanti. 2018. "A CRISPR – Cas9 gene drive targeting doublesex causes complete population suppression in caged Anopheles gambiae mosquitoes." *Nature Biotechnology* 36 (11):1062–1066.

Leung, Kelley, Erica Ras, Kim Ferguson, Simone Ariëns, Dirk Babendreier, Piter Bijma, Kostas Bourtzis, Jacques Brodeur, Margreet Bruins, and Alejandra Centurión. 2019. "Next generation biological control: The need for integrating genetics and evolution." *Biological Reviews*, 95 (6): 1838–1854.

Li, Ting, Bo Liu, Martin H Spalding, Donald P Weeks, and Bing Yang. 2012. "High-efficiency TALEN-based gene editing produces disease-resistant rice." *Nature Biotechnology* 30 (5):390–392.

Lundgren, Jonathan G, and Jian J Duan. 2013. "RNAi-based insecticidal crops: Potential effects on nontarget species." *Bioscience* 63 (8):657–665.

McDougall, S., Creek, A., & Napier, T., & Duff, J. 2005. Improving lettuce pest management – NSW and SE Queensland: Final Report, Project VG01028. Yanco, NSW: National Vegetable Industry Centre, NSW Department of Primary Industries. ISBN 0 7347 1691 5.

Messelink, Gerben J, Jérôme Lambion, Arne Janssen, and Paul CJ van Rijn. 2021. "Biodiversity in and around greenhouses: Benefits and potential risks for pest management." *Insects* 12 (10):933.

Meyling, Nicolai V, and Jørgen Eilenberg. 2007. "Ecology of the entomopathogenic fungi Beauveria bassiana and Metarhizium anisopliae in temperate agroecosystems: Potential for conservation biological control." *Biological Control* 43 (2):145–155.

Midega, Charles AO, Jimmy O Pittchar, John A Pickett, Girma W Hailu, and Zeyaur R Khan. 2018. "A climate-adapted push-pull system effectively controls fall armyworm, Spodoptera frugiperda (JE Smith), in maize in East Africa." *Crop Protection* 105:10–15.

Monica, Kansiime K, Makale Fernadis, Chacha Duncan, Nunda Winnie, Karanja Daniel, and Durocher-Granger Léna. 2024. "Area-wide pest management and prospects for fall armyworm control on smallholder farms in Africa: A review." *Sustainable Environment* 10 (1):2345464.

Moses-Gonzales, Nathan, and Michael J Brewer. 2021. *A special collection: Drones to improve insect pest management*. Oxford University Press US.

Naranjo, Steven E, and Peter C Ellsworth. 2009. "Fifty years of the integrated control concept: Moving the model and implementation forward in Arizona." *Pest Management Science: Formerly Pesticide Science* 65 (12):1267–1286.

Norton, George W, KL Heong, David Johnson, and Serge Savary. 2010. "Rice pest management: Issues and opportunities." In *Rice in the global economy: Strategic research and policy issues for food security*. IRRI, Los Banos.

Oye, Kenneth A, Kevin Esvelt, Evan Appleton, Flaminia Catteruccia, George Church, Todd Kuiken, Shlomiya Bar-Yam Lightfoot, Julie McNamara, Andrea Smidler, and James P Collins. 2014. "Regulating gene drives." *Science* 345 (6197):626–628.

Parra, José Roberto Postali. 2014. "Biological control in Brazil: An overview." *Scientia Agricola* 71:420–429.

Pereira, H. C., Pereira, F. F., Fernandes, W. C., Carneiro, Z. F., Lucchetta, J. T., Andrade, G. S., & Zanuncio, J. C. (2022). Parasitism rate of Plutella xylostella (Lepidoptera: Plutellidae) larvae in greenhouse by Tetrastichus howardi (Hymenoptera: Eulophidae) females at different densities. *Brazilian Journal of Biology*, 82: e263443.

Pijnakker, Juliette, Dominiek Vangansbeke, Marcus Duarte, Rob Moerkens, and Felix L Wäckers. 2020. "Predators and parasitoids-in-first: From inundative releases to preventative biological control in greenhouse crops." *Frontiers in Sustainable Food Systems* 4:595630.

Plata-Rueda, Angelica, Luis Carlos Martínez, José Cola Zanuncio, and José Eduardo Serrão. 2022. "Advances zoophytophagous stinkbugs (Pentatomidae) use in agroecosystems: Biology, feeding behavior and biological control." *Journal of Pest Science* 95 (4):1485–1500.

Pretty, Jules, Tim G Benton, Zareen Pervez Bharucha, Lynn V Dicks, Cornelia Butler Flora, H Charles J Godfray, Dave Goulson, Sue Hartley, Nic Lampkin, and Carol Morris. 2018. "Global assessment of agricultural system redesign for sustainable intensification." *Nature Sustainability* 1 (8): 441–446.

Reid, Steven, Henry De Malmanche, Leslie Chan, Holly Popham, and Monique M Van Oers. 2023. "Production of entomopathogenic viruses." In *Mass production of beneficial organisms*, 375–406. Elsevier.

Rosenheim, Jay A, Lawrence R Wilhoit, and Christine A Armer. 1993. "Influence of intraguild predation among generalist insect predators on the suppression of an herbivore population." *Oecologia* 96:439–449.

Sanchez, Juan Antonio, E López-Gallego, M Pérez-Marcos, and L Perera-Fernández. 2021. "The effect of banker plants and pre-plant release on the establishment and pest control of Macrolophus pygmaeus in tomato greenhouses." *Journal of Pest Science* 94 (2):297–307.

Shaurub, El-Sayed H. 2024. "Remote sensing and geographic information system applications as early-warning tools in monitoring fall armyworm, Spodoptera frugiperda: A review." *International Journal of Tropical Insect Science*:1–18.

Singh, Abhaya Pal, Amol Yerudkar, Valerio Mariani, Luigi Iannelli, and Luigi Glielmo. 2022. "A bibliometric review of the use of unmanned aerial vehicles in precision agriculture and precision viticulture for sensing applications." *Remote Sensing* 14 (7):1604.

Sishodia, R. P., Ray, R. L., & Singh, S. K. 2020. Applications of remote sensing in precision agriculture: A review. *Remote Sensing*, 12 (19): 3136. https://doi.org/10.3390/rs12193136

Snyder, William E, and Anthony R Ives. 2009. "Population dynamics and species interactions." In *Integrated pest management*, 62–74. Cambridge University Press, Cambridge, UK.

Stenberg, Johan A. 2017. "A conceptual framework for integrated pest management." *Trends in Plant Science* 22 (9):759–769.

Su, Wen-Hao. 2020. "Advanced machine learning in point spectroscopy, RGB-and hyperspectral-imaging for automatic discriminations of crops and weeds: A review." *Smart Cities* 3 (3):767–792.

Sunderland, KD. 1996. "Progress in quantifying predation using antibody techniques." *Systematics Association Special Volume* 53:419–456.

van Lenteren, Joop C. 2012. "The state of commercial augmentative biological control: Plenty of natural enemies, but a frustrating lack of uptake." *BioControl* 57 (1):1–20.

van Lenteren, Joop C, Karel Bolckmans, Jürgen Köhl, Willem J Ravensberg, and Alberto Urbaneja. 2018. "Biological control using invertebrates and microorganisms: Plenty of new opportunities." *BioControl* 63:39–59.

Vangansbeke, Dominiek, Duc Tung Nguyen, Joachim Audenaert, Bruno Gobin, Luc Tirry, and Patrick De Clercq. 2016. "Establishment of Amblyseius swirskii in greenhouse crops using food supplements." *Systematic and Applied Acarology* 21 (9):1174–1184.

Wäckers, Felix Leopold, Paul Cornelis Jacobus van Rijn, and Jan Bruin. 2005. *Plant-provided food for carnivorous insects: A protective mutualism and its applications*. Cambridge University Press.

Whyard, Steven, Aditi D Singh, and Sylvia Wong. 2009. "Ingested double-stranded RNAs can act as species-specific insecticides." *Insect Biochemistry and Molecular Biology* 39 (11):824–832.

Zhang, Hao, Hai-Chao Li, and Xue-Xia Miao. 2013. "Feasibility, limitation and possible solutions of RNAi-based technology for insect pest control." *Insect Science* 20 (1):15–30.

Zhu, Fang, Jingjing Xu, Roshan Palli, Joseph Ferguson, and Subba R Palli. 2011. "Ingested RNA interference for managing the populations of the Colorado potato beetle, Leptinotarsa decemlineata." *Pest Management Science* 67 (2):175–182.

Zhu, Yanping, Jieru Pan, Junzhi Qiu, and Xiong Guan. 2008. "Optimization of nutritional requirements for mycelial growth and sporulation of entomogenous fungus Aschersonia aleyrodis Webber." *Brazilian Journal of Microbiology* 39: 770–775.

18 Metamorphosis of IPM

Integrating Pest Management for Sustainable Agriculture

A. Suganthi, Bimal Kumar Sahoo, Mohammad Ikram, Iqra Ali, A.R. Ramya, M. Alagar, and S. Sivasankari*
Department of Agricultural Entomology, Tamil Nadu Agricultural University, Coimbatore 641003, India
*Corresponding author

18.1 INTRODUCTION

In the ancient human civilization, though human beings lived alongside crop pests, they never attempted to control them (Palumbi, 2001). However, for the growing human population, when the demand for food production increased, humans began to protect their crops from these notorious insect pests through various methods, like mechanical, cultural, physical, and biological means (Abrol and Shankar, 2012). The world's population is increasing quickly and is expected to reach nearly 9.7 billion by 2050 and 10.9 billion by 2100 (Gu et al., 2021). This escalating population growth urges a strong need to produce more food to meet the hunger of the growing population. However, during farming practices, the crop faces several challenges, with insect pests being a major threat and key limiting factor to productivity (Kumar et al., 2024). Several factors are responsible for the varied global loss of crop due to insect pests, such as pest biology, geographical location, host plant characteristics, migration of new invasive pests, introduction of new host crops, crop intensification, as well as pest management tactics. According to NIFA (2023), insect pests can cause an estimated crop loss of 20–40% annually, while plant pathogens and invasive insects are expected to cause a global economic loss of $220 and $70 billion, respectively. In the 1990s, the green revolution set a significant benchmark by increasing food production beyond the population growth using high-yielding cultivars as well as synthetic chemical pesticides and fertilizers. In the mid-20th century, some of the new synthetic insecticides were introduced that showed much effectiveness in managing insect pests, resulting to the sole tactics for insect pest control (Metcalf, 1980; Dhaliwal et al., 1998). Overreliance on these chemical pesticides resulted in serious concerns, such as severe ecological imbalances, resistance development in insect pests, environmental contamination, biodiversity loss, groundwater contamination, pest resurgence, as well as impaired animal and human health (Rundlof et al., 2015). It also impacted agricultural trade, as some products were rejected due to pesticide residues. In 1962, the book *Silent Spring* published by Rachel Carson set a milestone for reducing the use of toxic pesticides in controlling insect pests, and that set the tone for modifying the concept of pest management.

IPM concept was developed in the mid-1970s, with a focus on reducing the overuse of chemical pesticides, as a holistic approach designed to combine biological, cultural, and chemical methods for controlling pest populations more sustainably (Kogan, 1998). It uses detailed knowledge about the biology and life cycles of insect pests, as well as their interaction with the environment. In addition to new technologies, traditional pest control methods were also incorporated into IPM to manage pests ecologically while minimizing harm to both the environment and humans (Karlsson et al., 2020). By the late 20th century, IPM became widely accepted as a standard pest management approach. According to Dent (1995), IPM is a pest control strategy that takes into account both pest populations and their environment, while also considering the economic aspects of farming. It is also based on the ecological principles that integrate a variety of complementary methods to keep the pest population below injury level and without compromising the natural enemy population (Barzman et al., 2015).

Even though pesticides are required in the worst-case situation when dealing with a pest incidence problem, they are frequently unreliable. The term *wicked problem* has also been used to describe the emergence of pesticide resistance (Gould et al., 2018). Therefore, alternate approaches of managing these pests are desperately needed. The significance of using an evolutionary perspective to pest management has been highlighted by the studies of Hicks et al. (2018) and Thrall et al. (2011). According to Peterson et al. (2018), this approach would make it easier to assess the efficacy of particular pest control strategy and forecast their long-term effects on agro-ecosystems. According to Barzman et al. (2015), IPM can be thought of as a hierarchical structure with preventative measures at the base and

DOI: 10.1201/9781003516538-18

curative methods usually involving chemical control at the top. Chemical control should be used as a last choice, when other integrated strategies are unable to keep pests from reaching economically damaging levels.

Today, IPM stands at the forefront of sustainable agriculture, incorporating cutting-edge technology and ecological principles (Zia et al., 2024). Advances in biotechnology, remote sensing, genetic engineering, and artificial intelligence are reshaping pest management by enabling more precise and targeted interventions (Adetunji et al., 2023). The future of IPM looks promising, with ongoing research focusing on further enhancing its adaptability, effectiveness, and sustainability in the face of new challenges, such as climate change and the emergence of new pest species.

Over the past 60 years, IPM has supported sustainable farming, reduced the use of synthetic pesticides, and helped address many socioeconomic, environmental, and health challenges (Deguine et al., 2021). However, despite its clear benefits, IPM is still not widely used by farmers. Policymakers, researchers, and agricultural support agencies need to identify the obstacles preventing its adoption and develop strategies to encourage more widespread use of IPM. In this dynamic journey from ancient practices to the advanced methodologies of today and tomorrow, the metamorphosis of IPM illustrates humanity's growing understanding of the complex relationships between agriculture, pests, and the environment, offering a path toward a more sustainable and balanced approach to pest management.

18.2 HISTORICAL EVOLUTION OF IPM

18.2.1 Before World War II

Before World War II, pest control mainly relied on experience and understanding of pest biology (Gaines, 1957), and traditional methods, like crop rotation, mechanical practices, and cultural techniques, such as deep ploughing, field sanitation, manual removal of insects, and destroying infested plants (Koppenhofer et al., 2013; Flint and van den Bosch, 2012), were passed down from generation to generation (Smith et al., 1976). For example, the Chinese used plant-based treatments for seeds, ants for natural pest control, and wood ash to manage insects in small spaces. In the Middle East, chrysanthemum flower powder was used as an insecticide. By the early 20th century, synthetic inorganic insecticides, including those made from arsenic, mercury, tin, and copper, were developed (Perkins, 1980).

18.2.2 Post–World War II

DDT was the most widely used pesticide during World War II to prevent yellow fever outbreaks and control disease-transmitting insect vectors. DDT and other organic compounds gained popularity and started to be used as insecticides due to their outstanding success in World War II (Roberts et al., 2016). Following the success of DDT, the use of chemical insecticides became more common in controlling insect pests as well as disease-spreading vectors, and it reached its peak during the 1950s and the 1960s (Dhawan and Peshin, 2009). This period is referred to as the "dark ages" of pest control, as researchers focused mainly on chemical solutions, ignoring non-chemical methods and pest biology (Pimentel and Perkins, 2019; Koppenhofer et al., 2013). As a result, from 1950 to 1970, farmers became overreliant on chemical pesticides as the primary method of pest control, without considering other options. Pesticides were used routinely, irrespective of the presence or absence of pests in the field, leading to serious ecological issues, like phytotoxicity, resistance, resurgence, and environmental contamination (Smith et al., 1976; Flint and van den Bosch, 2012; Kogan, 1998). By the late 1950s, concerns about the excessive use of insecticides began to emerge, especially from areas known for biological pest control, such as California (Ripper, 1956) and the cotton-producing regions in North and South America, as well as in Canada and Europe (MacPhee and MacLellan, 1971). Meanwhile, severe economic crisis was encountered by farmers due to the high cost of chemical pesticides. For example, cotton farmers in Peru, Egypt, Central America, and Texas applied insecticides up to 60 times, making cotton production unprofitable and causing the collapse of the cotton industry in these areas (Dhawan and Peshin, 2009).

The publication of *Silent Spring* by Rachel Carson in 1962 was a major turning point, which criticized existing pest control methods by defining the chemical pesticides as biocides due to their toxic nature. Some of the negative implications from these biocides, such as environmental pollution, resistance development, biodiversity loss, and harm to beneficial non-target organisms, were described in this book. To combat the development of resistance in insects, farmers followed the practice of spraying larger quantities of chemical pesticides more frequently, leading to a "pesticide treadmill." This was followed by a 1963 report from the United States President's Science Advisory Committee called "Use of Pesticides," which warned about the harmful effects of pesticides on beneficial plants and animals (PSAC, 1963). These concerns played a significant role in the development of integrated pest management (IPM), which aimed to reduce pesticide use and its harmful effects on the environment while controlling crop pests (FIGURE 18.1)

18.3 PHASES OF IPM DEVELOPMENT

The concept of integrated pest management evolved in three phases, viz., integrated control, pest management, and lastly, integrated pest management.

18.3.1 Phase I: Integrated Control

The term *integrated pest management* (IPM) has evolved over time and now includes various meanings and interpretations. Different approaches have been attached to its

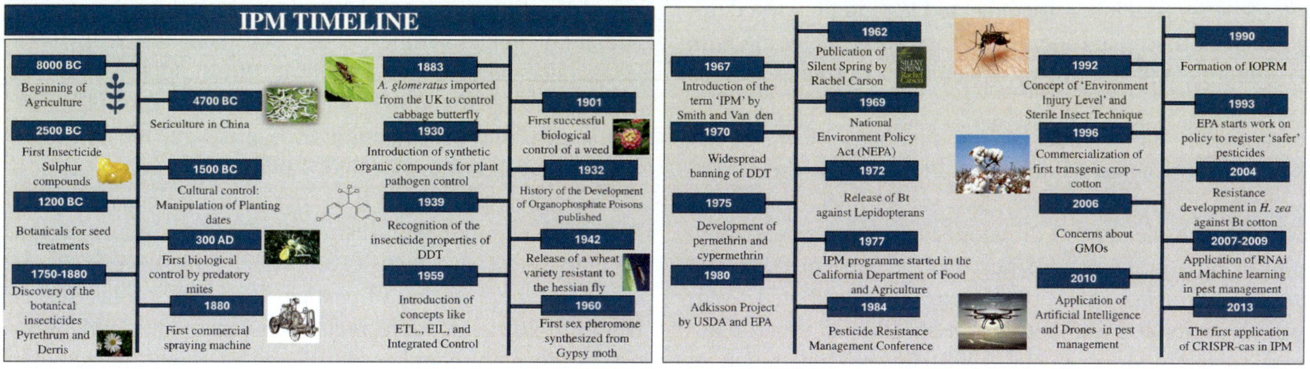

FIGURE 18.1 Key historical events that contributed to the development of IPM. It highlights major scientific discoveries, policy changes, and technological advancements that shaped IPM strategies over time. The progression from early pest control methods to modern, ecologically based approaches is depicted, emphasizing the integration of biological, cultural, mechanical, and chemical control measures to achieve sustainable pest management.

use over the past 40 years (Lewis et al., 1997). In the year 1956, Bartlett introduced the concept of integrated control (IC) in the United States, which was further expanded by Stern et al. (1959) and laid the foundation for IPM (Ehi-Eromosele et al., 2013). IC was originally proposed by Smith and Allen (1954), with the idea that the timing and conditions of pesticide application could reduce the negative impact of chemicals on natural enemies. Stern et al. (1959) described it as a pest management approach that combines both chemical and biological control methods. This concept was later expanded to include the coordinated use of cultural, biological, and artificial techniques (van den Bosch and Stern, 1962).

18.3.2 Phase II: Pest Management

During the 12th International Congress of Entomology in London, the concept of "protective population management" (Geier and Clark, 1960), which was later shortened to "pest management" (Geier, 1966), gained significant attention. Pest management focused on understanding pest biology and argued that the complete eradication of pests was not necessary. Geier (1966) minimized the semantic controversy surrounding the term, claiming that *pest management* was simply a pragmatic term used to represent the idea of strategically manipulating nature for human benefit, much like *wildlife management*. This was true even though the concept of pest management quickly gained the attention of the scientific community.

18.3.3 Phase III: Integrated Pest Management

Although the concepts of integrated control and pest management eventually merged, the term *integrated pest management* (IPM) was not widely accepted by the scientific community until 1972. During the 1970s, organizations like the Biological and Integrated Control of Noxious Animals and Plants (IOBC) in Europe and the Food and Agriculture Organization (FAO), helped promote IPM. In 1972, IPM became an important milestone in the field of plant protection, gaining recognition from academic societies. That same year, IPM was also included in the guidelines for Environmental Protection (Nixon, 1975). In 1973, the IOBC defined *IPM* as "[a] pest control system that uses a combination of methods to meet economic, ecological, and toxicological requirements, with a focus on natural control while respecting tolerance thresholds" (Ferron, 1999).

IPM was first introduced in the United States and later spread globally, supported by the Huffaker IPM project (1972–1978), a major and influential interdisciplinary and interagency initiative. This project initially focused on IPM practices for six selected crops: *Medicago sativa*, *Citrus* spp., *Glycine max*, *Gossypium hirsutum*, and *Pinus* spp. Following this, another large-scale project, called the Consortium for Integrated Pest Management, or the Atkinson IPM Project (1979–1985), focused on IPM for crops like alfalfa, apple, cotton, and soybean was implemented. While the Huffaker project primarily concentrated on the entomological aspect of IPM, the Atkinson Project took a more comprehensive approach, integrating economic evaluation of IPM with plant pathology, weed science, and entomology (Kogan, 1998).

The Report of the United States Council on Environmental Quality (1972) included a section on IPM following the "Special Message to Congress Outlining the 1972 Environmental Program" (Nixon, 1975). These efforts contributed to a national movement in United States which focused on protecting human health and the environment through pest management programs (Smith et al., 1976). In the early 1990s, the International Organization for Biological and Integrated Control (IOBC) and the LEAF (Linking Agriculture and Farming) group advocated for expanding IPM principles to integrated agricultural production. This led to the introduction of terms like *integrated farming* (El Titi et al., 1993) and *integrated crop management* (Leake, 2000). This shift emphasized that crop protection should align with broader agroecosystem goals. In integrated production, resource conservation, such as soil, water, energy, and labor, is connected to reducing pest damage and improving agronomic performance, including both the quantity and quality of agricultural output (Deguine et al., 2021).

In 1985, the United States National IPM Center (NIPMC) redirected some IPM funding from the USDA Cooperative State Research Service (CSRS) under the supervision of deans and directors at the experimental stations of land-grant universities. Unlike the Huffaker and Atkinson Projects, these IPM programs distributed funds through regional coordinators, which allowed more states to access the funding (Kogan, 1998). Several national symposiums and workshops on IPM were organized with support from CSRS and NIPMC. These events led to a call for a national commitment to IPM, which eventually resulted in the Clinton administration's national IPM initiative in 1993 by US Congress.

In 1991, the concept of "biologically intensive IPM" (BIIPM) or "bio-intensive IPM" (BIPM) was introduced by Frisbie and Smith, focused on biological control, cultural control, and host plant resistance (Dhaliwal et al., 2004). Dhaliwal et al. (2013) expanded on BIIPM by proposing a new term, *biopesticide-based IPM* (BBIPM), which is defined as the manipulation of natural enemies (such as predators, parasitoids, and microbes), biorational compounds (including semiochemicals, growth regulators, plant-derived materials, microbe-derived materials, etc.), and transgenic crops in a way that keeps pest populations below damaging levels while maintaining ecosystem balance. In 1996, a special committee of the Natural Research Council's Board of Agriculture proposed *ecological pest management* (EPM). Later, Koul and Cuperus (2007) emphasized the importance of deploying ecologically based IPM (EBIPM), which involves understanding the ecological relationships between host plants and management practices (such as cultural control, biological control, and host plant resistance).

18.4 CONCEPT OF ECONOMIC INJURY, THRESHOLD, AND IPM PYRAMIDS

Stone and Pedigo (1972) proposed a research approach to bridge the gap between theory and practice, presenting the economic and biological formula for determining the economic injury level (EIL) 10 years after the pioneering work of Stern et al. (1959). They defined *EIL* as the lowest pest population density that can cause economic damage. In the year 1991, Pedigo described EIL as the smallest number of insects that can cause economic harm. Injury equivalents are the standard measures used in pest control to indicate the total damage a single insect can cause over its life cycle (Higley and Pedigo, 1996). This is also known as the total injury equivalent (for a population) at a specific moment, assuming that management actions are taken swiftly and losses are completely prevented. In 1992, Pedigo and Huxley developed the environmental economic injury level (EEIL) to assess management strategies by considering both their direct costs and benefits, as well as their environmental impact (Higley et al., 2009). In the year 1991, Pedigo also defined the economic threshold level (ETL), which is similar to the EIL but incorporates a time element. In this case, the pest population is measured as an insect equivalent (Higley and Pedigo, 1996).

The concept of the "IPM pyramid" emphasizes that IPM involves more than just combining pest management technologies; it also focuses on establishing a hierarchy of practices, where the use of pesticides is seen as a last resort (Naranjo, 2001). A more recent version of the IPM pyramid, proposed by Stenberg (2017), pointed out the lack of a holistic approach to IPM that incorporates both traditional and modern tools. However, his framework mainly concentrated on the ecological aspects of pest management, highlighting the importance of an interdisciplinary research approach. Several reports have indicated that the successful implementation of IPM depends on multiple factors, such as education levels, economic and social conditions, environmental awareness, rational thinking, moral values, regulatory frameworks, government policies, availability of IPM tools, extension education, consumer preferences, and retail marketing (Lefebvre et al., 2015; Jayasooriya and Aheeyar, 2016; Rezaei et al., 2019).

18.5 DEFINITIONS OF IPM

IPM can be interpreted in several ways by individuals who create, advocate for, or use IPM techniques. Despite being a simple three-word term, it has as many definitions as there are authors, and each definition highlights one aspect of IPM based on their perspective. According to the FAO (1967), *integrated pest management* (IPM) is defined as a pest management system that uses all suitable techniques in an environmentally compatible way to keep pest populations below the economic injury level (National Academy of Sciences, 1969). The term *integrated pest management* was formally adopted by the Council on Environmental Quality in 1972, with a detailed explanation of each component: *I* stands for "integration," which refers to the harmonious application of various techniques to reduce the impact of one or more pests; *P* stands for "pest," which includes any organism harmful to humans, such as vertebrates, invertebrates, weeds, or pathogens; and *M* stands for "management," which involves a set of decisions or regulations based on ecological, economic, and social factors. The United States Department of Agriculture – Agricultural Research Service (USDA-ARS, 2018) defines *IPM* as a science-based, sustainable decision-making process that integrates biological, cultural, physical, and chemical tools to identify, manage, and reduce pest risks (Dara, 2019).

In addition to the definitions provided by the FAO and USDA-ARS, several other authors have also offered their perspectives on IPM. According to Apple et al. (1979), IPM is the optimization of pest control in a way that is both economically and environmentally sound – meaning, that it should achieve both goals. Similarly, Higley et al. (1989)

described pest management as a comprehensive technology that employs various techniques to maintain pest populations at manageable levels while preserving a high-quality environment. Luckmann and Metcalf (1975) defined *IPM* as the intelligent selection and application of pest control strategies that ensure positive economic, ecological, and sociological outcomes. Kogan (1998) emphasized that IPM involves selecting and using pest control tactics, either individually or in coordination, based on cost–benefit analyses that consider the interests and impacts on producers, society, and the environment (Peterson et al., 2018).

18.6 CORE PRINCIPLES OF IPM

The concept of integrated pest management is built upon eight key principles (Barzman et al., 2015), outlined as follows:

1. *Prevention and suppression.* The goal is not to completely eliminate pests but to prevent any single pest from becoming dominant or causing significant damage within a cropping system.
2. *Monitoring.* Applied once the crop is established, this principle emphasizes the importance of decision-making based on monitoring pest populations and the thresholds that indicate when control measures are needed. It involves tracking actual or predicted pest occurrences during the growing season.
3. *Decision-making.* Pest management decisions should be based on the biology and ecology of major pest species and their natural enemies, ensuring an informed approach to pest control.
4. *Prioritization of non-chemical methods.* The use of non-chemical methods is prioritized to reduce the risks of resistance and pesticide residues. These methods include practices such as soil solarization, trap cropping, mechanical control, biological control, and various biotechnological approaches. The availability and effectiveness of these practices can vary.
5. *Pesticide selection.* When pesticides are necessary, the choice should be based on targeting the specific pest while minimizing the impact on human health, non-target organisms, and the environment.
6. *Reduced pesticide use.* This principle encourages reducing pesticide usage by lowering dosages, limiting the frequency of applications, and employing partial applications. The goal is to minimize risks to human health and the environment while maintaining effective pest control.
7. *Anti-resistance strategies.* This principle aims to prevent the development of pest resistance to control methods, particularly pesticides. Repeated exposure to the same control methods can lead to resistance, reducing their effectiveness over time.
8. *Evaluation.* Regular assessment of pest control strategies is essential to ensure they are effectively managing pest populations, minimizing environmental impact, and protecting non-target species. This evaluation helps refine and improve pest management practices.

18.7 SEMIOCHEMICAL-BASED IPM

Semiochemical-based strategies became widely used in the 1970s, promoting environmentally friendly solutions for pest control (Cui and Zhu, 2016). Pheromones have been identified in over 1,000 insect species, with annual pheromone lure production exceeding 20 million units for monitoring and mass trapping (El-Sayed, 2011; Witzgall et al., 2010). Semiochemical-based pest management strategies are widely used to control insect pests through various approaches. Monitoring involves the use of pheromones and kairomones to track pest populations and predict outbreaks. For example, pheromone-baited traps have been employed to monitor *Choristoneura fumiferana* in Canada (Abd El-Ghany et al., 2016). Mass trapping relies on trapping pests using large numbers of pheromone-baited traps to reduce pest populations, with Mosquito Magnet® Pro traps being effective in reducing *Ochlerotatus taeniorhynchus* annoyance (Kline, 2007). Mating disruption utilizes synthetic sex pheromones to interfere with mating behaviors, such as Isonet CGX111 dispensers, which have been used to control the honeydew moth, *Cryptoblabes gnidiella*, in Mediterranean vineyards (Ricciardi et al., 2021). The attract-and-kill method uses semiochemicals to lure pests and then kill them, as seen with SPLAT® combined with Torac® + *Acttra plutella*, which reduced larval activity of the diamondback moth in cabbage (Muthomi et al., 2025). Lastly, repellents use chemicals to deter pests from approaching crops, such as the synthetic repellent urethane, which showed 67% repellence against *Chrysomya megacephala* (Bhaskaran et al., 2018). Allelochemicals and pheromones used in pest management have been classified into several types (Table 18.1 and 18.2).

18.8 PHEROMONE AND SEMIOCHEMICAL-BASED PEST DETECTION IN IPM

Semiochemicals are organic compounds that mediate chemical communication among insects. They are classified into pheromones, which facilitate intraspecific (same species) communication, and allelochemicals, which influence interspecific (different species) interactions. Pheromones have been of significant value to insect communication for millions of years, regulating mating, foraging, and territorial marking behavior. The initial scientific isolation and synthesis of insect pheromones began in the late 1950s with Adolf Butenandt's isolation and synthesis of bombykol, the silkworm moth (*Bombyx mori*) sex pheromone (Butenandt, 1959). The use of pheromones has been

TABLE 18.1
Types of Allelochemicals and Their Interaction

Allelochemicals	Interaction Type	Function	Source and Pest	Principal Components	References
Allomones	Beneficial to emitter, harmful to receiver	Act as a defense mechanism by deterring predators and pests	*Aromia bungii*	Rose oxide	(Chen et al., 2024)
			Pinus sylvestris against *Blastophagus piniperda*	α-Pinene, 3-Carene	(Senthil-Nathan et al., 2013)
Kairomones	Beneficial to receiver, harmful to emitter	Attract predators or parasites to the emitter's location	Honeydew volatiles of *Trialeurodes vaporariorum* attracted *Encarsia formosa*	(Z)-3-hexenol, δ-3-carene, 3-octanone, α-phellandrene, methyl salicylate, β-ocimene, β-myrcene, and (E)-β-caryophyllene	(Ayelo et al., 2022)
Synomones	Beneficial to both emitter and receiver	Facilitate mutualistic interactions, such as pollination or natural enemy recruitment	*Lantana montevidensis* attracting the omnivorous predator *Campylomma chinensis*	Farnesol	(Ma et al., 2017)
Antimones	Harmful to both emitter and receiver	Cause repellent or harmful responses in both interacting species	–	–	(El-Ghany, 2019)
Apneumones	Emitted by non-living sources; beneficial to receiver, harmful to other species	Cause favorable responses in one species while harming others	Rabbit stools attract sandfly females for egg-laying	Hexanal and 2-methyl-2-butanol	(Dougherty et al., 1995)

TABLE 18.2
Classification of Pheromones, Their Functions, and Key Examples

Pheromones	Category	Function	Pest	Principal Components	References
Aggregation pheromones	Releaser	Attract individuals to a common location	*Rhinostomus barbirostris*	diastereoisomers of 5-hydroxy-4-methyl-3-heptanone	(Reis et al., 2018)
			Cosmopolites sordidus	(2R,5S)-theaspirane	(Abagale et al., 2019)
			Holotrichia parallela	Pentadecyl acetate, *cis*-13-docosenol, and behenic acid	(Qu et al., 2025)
Alarm pheromones	Releaser	Trigger escape or defensive behavior	*Formica argentea*	Undecane	(Lenz et al., 2013)
			Ooceraea biroi	4-methyl-3-heptanone and 4-methyl-3-heptanol	(Lopes et al., 2023)
Sex pheromones	Releaser	Stimulate mating behavior	*Conopomorpha cramerella*	*E, Z, Z-* and *E, E, Z*-4,6,10-hexadecatrienyl acetates	(Amalin et al., 2023)
			Rhagoletis batava	(-)-δ-heptalactone	(Buda et al., 2020)
Trail pheromones	Releaser	Guide others to food sources or nesting sites	*Camponotus modoc*	(2S,4R,5S)-2,4-dimethyl-5-hexanolide	(Renyard et al., 2019)
			Crematogaster scutellaris	Tridecan-2-ol	(Scarano et al., 2024)
Royal pheromones	Primer	Regulate reproductive roles and caste development in social insects.	*Reticulitermes flavipes*	Heneicosane	(Funaro et al., 2018)
Oviposition-deterrent pheromones	Primer	Discourage egg-laying by influencing reproductive behavior	*Ceratitis capitata*	Linalool	(Papanastasiou et al., 2020)
			Phthorimaea operculella	Hexanal, octanal, nonanal and 1-octen-3-ol	(Anfora et al., 2014)
Home recognition pheromones	Primer	Help individuals recognize colony members.	wasp	δ-n-hexadecalactone	(El-Ghany et al., 2019)

particularly effective against major agricultural pests, such as *Cydia pomonella* (codling moth), *Pectinophora gossypiella* (pink bollworm), and *Tuta absoluta* (tomato leaf miner) (Evenden and Silk, 2016). Unlike conventional insecticides, pheromone-based control methods minimize non-target effects, reduce chemical residues, and help delay the development of insecticide resistance (Trematerra, 2012). Continued advances in molecular biology and chemical analysis have further developed knowledge of semiochemicals, which have been used in pest control. In early research during the 1960s, gas chromatography, mass spectrometry, and electroantennogram recordings were employed to identify the response of insects to semiochemicals, and these were the precursors to pheromone-based pest control today (Reddy and Guerrero, 2004).

18.8.1 Advances in Pheromone Identification and Application

To enhance the effectiveness of pheromone-based pest control, researchers have integrated electroantennogram (EAG) recordings and gas chromatography mass spectrometry (GC-MS) to identify and evaluate semiochemicals with precision. GC-MS can separate and structurally identify volatile compounds, while EAG confirms their biological activity by measuring the neural responses of insect antennae (El-Sayed et al., 2013). The techniques have enabled the discovery of new pheromones and the design of their optimization for use in the field (Smart et al., 2014). However, challenges such as pheromone degradation, high production costs, and the need for improved controlled-release technologies remain. Continued research is crucial to overcoming these limitations and expanding the role of pheromones in sustainable pest management (Mafra-Neto et al., 2014).

18.8.2 Innovations in Semiochemical-Based Pest Detection

Recent advances in pheromone-based pest detection have transformed its effectiveness and practicality. The lure-and-infect strategy combines pheromones with entomopathogenic fungi to attract and infect pests, making it an efficient method for performing biological control. Autodissemination strategies utilize semiochemicals to motivate pests to disperse microbial pathogens across their populations and have successfully targeted pests like the fall armyworm. Additionally, polymer-based controlled-released formulations could extend the frequency of pheromone-based traps by providing a slow and continuous release profile of the active compounds (Reddy et al., 2020).

Innovative technologies such as specialized pheromone and lure application technology (SPLAT) employ wax-based formulations for extended pheromone release, minimizing the need for frequent reapplications (Kumar et al., 2023). Meanwhile, pheromonereceptor-based biosensors provide real-time pest monitoring by the detection of particular pheromone signals to reduce manual inspection of insect bait traps (Zaki et al., 2024). The researchers are also investigating whether genetic engineering can enable plants to produce pheromones; this would offer a form of low-cost pest control to crops as they work to synthesize their own pheromones (Reddy et al., 2020). These developments highlight the growing role of semiochemicals in eco-friendly pest management strategies.

18.9 BIOLOGICAL CONTROL IN IPM

With growing emphasis on sustainable agriculture, biological control has emerged as a cornerstone of IPM. This method uses living organisms, such as predators, parasitoids, and microbial agents, to regulate pest populations while reducing the dependence on chemical pesticides. Advances in biotechnology, artificial intelligence, and formulation techniques have significantly increased the efficacy of biocontrol strategies, making them a viable alternative to conventional pest management techniques (Galli et al., 2024). The concept of biological control has been practiced for centuries, with early records showing the use of predatory organisms to manage pest populations. A noteworthy historical example is the introduction of the vedalia beetle (*Rodolia cardinalis*) in the late 1800s to control cottony cushion scale in citrus orchards. The formalization of biological control as a scientific discipline began in the mid-20th century, leading to its integration within IPM frameworks. Biocontrol strategies are now classified into two major categories: macrobial and microbial control (Sabbahi et al., 2022).

18.9.1 Macrobial and Microbial Biocontrol in IPM

Microbial biocontrol relies on beneficial insects and nematodes to naturally suppress pest population, while *microbial biocontrol* utilizes bacteria, fungi, and viruses to suppress pest populations (Table 18.3). Advancements in microbial biocontrol have led to the development of RNA interference (RNAi) technology, which enhances microbial pesticide efficacy by silencing essential pest genes (Weiberg et al., 2013). Additionally, combination products integrating microbial biopesticides with plant extracts and semiochemicals have been formulated to improve pest suppression. Biopolymer encapsulation technologies now protect microbial agents from environmental degradation, extending their viability and effectiveness in the field (Galli et al., 2024).

The advent of genetic engineering has further expanded the capabilities of biological control. Genetically modified agents such as *Agrobacterium radiobacter* strain K1026 have been developed to prevent the horizontal transfer of resistance genes, improving their safety and efficacy (Kerr and Bullard, 2020). Similarly, Bt-based recombinant products, such as MVPTM and M-TrackrTM, utilize *Pseudomonas fluorescens* as a carrier for insecticidal proteins, offering improved environmental stability. Genetically engineered baculoviruses have also been modified to produce additional insecticidal proteins, increasing their potency against resistant pest populations (Mawcha et al., 2024).

TABLE 18.3
Macrobial and Microbial Biocontrol Strategies in IPM

Category	Biocontrol Agent	Target Pests/Diseases	Mechanism of Action	Recent Advances	References
		Macrobial Biocontrol			
Predatory insects	Coccinellidae (Ladybird beetles)	Aphids, whiteflies	Consumption of soft-bodied pests	Genetic enhancement for environmental resilience	(Mawcha et al., 2024)
	Chrysopidae (Lacewings)	Thrips, aphids	Larval predation	Mass-rearing techniques	(Chaudhary et al., 2024)
Parasitoids	*Encarsia formosa*	Whiteflies	Parasitism of host	Greenhouse-specific applications	(Sabbahi et al., 2022)
	Trichogramma spp.	Lepidopteran pests (corn borers, armyworms)	Egg parasitism	Genetic improvement for field stability	(Galli et al., 2024)
Entomopathogenic nematodes	*Steinernema* spp., *Heterorhabditis* spp.	Soil-dwelling pests	Release of symbiotic bacteria, causing mortality	Formulation improvements for enhanced persistence	(Sabbahi et al., 2022)
Microbial Biocontrol					
Bacterial biopesticides	*Bacillus thuringiensis* (Bt)	Lepidopteran larvae	Insecticidal protein production	Bt crops and recombinant Bt products	(Chaudhary et al., 2024)
	Pseudomonas fluorescens	Soilborne pathogens	Antifungal metabolite production	Integration with plant growth–promoting traits	(Sabbahi et al., 2022)
Fungal biopesticides	*Beauveria bassiana*, *Metarhizium anisopliae*	Aphids, beetles, thrips	Cuticle penetration, fungal colonization	UV-resistant formulations for longevity	(Mawcha et al., 2024)
	Trichoderma spp.	Soilborne fungal pathogens	Mycoparasitism and ISR induction	Engineered strains for pathogen specificity	(Galli et al., 2024)
Viral biopesticides	Nucleopolyhedroviruses (NPVs), granuloviruses (GVs)	Lepidopteran pests (*Spodoptera frugiperda*)	Viral infection leading to pest mortality	Recombinant baculoviruses with enhanced virulence	(Sabbahi et al., 2022)
Genetically modified biocontrol agents	*Agrobacterium radiobacter* K1026	Crown gall disease (*Agrobacterium tumefaciens*)	Biocontrol through competitive exclusion	Engineered resistance to prevent genetic transfer	(Mawcha et al., 2024)
	Bt-based recombinant products	Multiple insect pests	Expression of delta-endotoxin	Microbial encapsulation for field stability	(Mawcha et al., 2024)
	Genetically modified baculoviruses	Resistant pest populations	Enhanced viral pathogenicity	Molecular biology–driven strain improvement	(Chaudhary et al., 2024)

18.9.2 Market Potential and Economic Impact

The global market for biocontrol agents has been expanding, driven by increasing restrictions on chemical insecticides and growing consumer concerns over food safety. Projections indicate that the biocontrol market will grow from $6.51 billion in 2022 to $18.15 billion by 2029, with the EPA having registered 299 biopesticide active ingredients for commercial use. The adoption of biopesticides is particularly strong in organic farming and high-value crops, such as fruits, vegetables, and pulses (Chaudhary et al., 2024).

18.9.3 Challenges and Limitations in Biocontrol Adoption

Despite its advantages, biological control faces several challenges. Regulatory barriers remain a significant hurdle in many parts of the world, delaying product commercialization. Field effectiveness can also be inconsistent due to environmental variability, while short shelf life and high production costs hinder widespread adoption. Additionally, slow action and potential pest resistance pose concerns, as biopesticides often take longer to suppress pest populations compared to synthetic alternatives (Galli et al., 2024).

18.10 CULTURAL CONTROL IN IPM

Cultural control is one of the oldest and most fundamental approaches within IPM. It involves modifying agricultural practices to create conditions unfavorable for pests while enhancing crop resilience. It also reduces reliance on chemical pesticides, promoting sustainability and environmental conservation. These methods date back thousands of years, with early agricultural societies using techniques such as crop rotation and intercropping to minimize pest outbreaks. The 20th century saw a decline in these practices due to the widespread adoption of synthetic pesticides. However, growing concerns over pesticide resistance, environmental degradation, and human health risks have led to a resurgence in cultural control as a core component of IPM (Bashyala et al., 2022). Recent advances, including precision agriculture and AI-driven decision-making, have significantly improved the efficiency of cultural control methods (Zhou et al., 2024).

18.10.1 Key Cultural Control Strategies in IPM

Cultural control breaks pest life cycles and induces plant resilience. Crop rotation reduces host plant density, lowering soilborne disease, and intercropping introduces diversity and discourages pests through masking effects (*Ocimum basilicum*, *Mentha* spp.). Trap cropping, like push–pull technology, disperses pests from primary crops. Sanitation practices, such as removal of infested plant debris, and sanitation of tools, limit pest colonization. Disposal of fallen fruits reduces *Amyelois transitella* (navel orangeworm) populations in orchards. Resistant crop varieties, like *Bt cotton*, lower pesticide use, though resistance management remains critical. Adjusting planting and harvesting schedules helps crops evade peak pest activity, as seen with early cowpea planting reducing aphid infestations in Uganda. Soil and fertilizer management enhances plant health through improvement of microbial diversity and pest suppression. However, excessive nitrogen increases crop susceptibility. Tillage exposes soil pests to predators, while mulching (with *Azadirachta indica*) repels insects and conserves moisture (Bashyala et al., 2022; Zhou et al., 2024).

18.10.2 Recent Advances in Cultural Control and Their Impact on IPM

18.10.2.1 Systemic Acquired Resistance (SAR) and Induced Systemic Resistance (ISR)

SAR and ISR are plant-mediated defense mechanisms that provide long-term resistance against pests and diseases. SAR is primarily regulated by salicylic acid, which triggers systemic protection and enhances the production of pathogenesis-related proteins (Kamle et al., 2020). ISR, on the other hand, is activated by beneficial microbes, such as plant growth–promoting rhizobacteria, and functions through jasmonic acid and ethylene pathways (Berendsen et al., 2012). SAR and ISR have emerged as powerful tools in integrated pest management (IPM), enhancing plant defense against various insect pests. Beneficial microbes like *Pseudomonas putida* and *Serratia marcescens* have been shown to trigger ISR in cucumber and tomato, reducing infestations by sap-sucking pests, such as aphids and whiteflies (Raupach et al., 1996). Similarly, *Bacillus subtilis* and *Bacillus amyloliquefaciens* produce volatile organic compounds that disrupt the feeding behavior of root knot nematodes and leaf miners, lowering pest populations without the need for chemical pesticides (Ryu et al., 2004). Fungal biocontrol agents like *Trichoderma harzianum* and *Trichoderma virens* have been found to bolster plant resistance against chewing insects, including caterpillars and beetles in crops such as maize and legumes (Lamdan et al., 2015). Additionally, *Micromonospora* species, naturally present in plant rhizospheres, have been reported to induce systemic resistance in tomato, significantly reducing thrips infestations (Martínez-Hidalgo et al., 2015). Beyond pest control, SAR and ISR have also demonstrated efficacy in managing plant diseases, such as *Fusarium wilt* and *zucchini yellow mosaic virus* (Kamle et al., 2020), making them invaluable components of sustainable agricultural practices.

18.10.2.2 Integration of Microbial Biocontrol and Cultural Control

Beneficial rhizobacteria (*Pseudomonas fluorescens*, *Bacillus subtilis*) enhance plant defenses and suppress soilborne diseases. Advances in microbiome engineering are improving soil microbial diversity to maximize ISR activity (Berendsen et al., 2012). Additionally, biochar and microbial amendments are being explored to boost soil health and reduce chemical pesticide dependence (Bashyala et al., 2022).

18.10.2.3 Challenges and Limitations in Cultural Control Adoption

Despite the long-term benefits, cultural control is faced with several adoption challenges. Since the methods are labor-intensive and require technical know-how, the feasibility for resource-constrained farmers is limited. Additionally, the effectiveness of strategies like intercropping and altered planting dates is highly climate-dependent, reducing their reliability in unpredictable weather patterns (Zhou et al., 2024). Farmers often prefer synthetic pesticides due to their immediate results, making education and financial incentives crucial for encouraging cultural control adoption (Bashyala et al., 2022).

18.11 HOST PLANT RESISTANCE–BASED IPM

Over the past 350 million years, plants and insects have interacted in various ways, including pollination benefits and plant defenses against insect herbivory. This defense mechanism in plants against insect arthropods is referred to as host plant resistance (HPR) (Gatehouse, 2002). Developing and deploying crop varieties resistant to arthropod herbivores is the focus of host plant resistance (Stout et al., 2024). Painter (1951) introduced the concept of host plant resistance, which has since evolved and remains a significant area of research. Plant resistance strategies involve screening of resistant traits, categorization of resistance types, breeding to introduce resistance genes, and implementation of resistant varieties as pest management programs (Dwivedi and Pandit, 2024). Host plant resistance is classified into three mechanisms, namely, antibiosis, antixenosis, and tolerance (Smith, 2005). *Antibiosis* refers to the plant traits that affect pest survival, growth, and reproduction. *Antixenosis* refers to the plant traits that affect the herbivore behavior from feeding or laying eggs (Pamidi et al., 2025). *Tolerance* refers to the plant's ability to withstand the stress caused by the pest and recover from the stress without any loss in the yield (Stout, 2019). Fraenkel (1959) and Ehrlich and Raven (1964) explored the interaction between plants and insects in terms of ecology and evolution focusing on plant defense and insect adaptions.

Secondary plant metabolites play a key role in the evolutionary relationships between plants and herbivory arthropods (Karinho-Betancourt, 2020). Plant resistance can be classified as constitutive resistance (always present in the plant), induced resistance (activated only after pest attack), direct resistance (directly affects herbivores), and indirect plant resistance (attracts natural enemies to control pests) (Stout and Davis, 2009). With understanding of plant–insect interaction, analytical techniques such as biochemical analysis, proteomics, metabolomics, molecular breeding, microscopy imaging, and electrophysiological techniques are used to identify the host plant resistant traits and the defense mechanism and such studies have significantly expanded over time (Zebelo et al., 2015; Oates et al., 2016; Hall et al., 2017; Haldhar et al., 2018; Keszthelyi et al., 2020; Smith, 2021). After the advancement of genetic engineering, Bt crops, such as modified maize, cotton, and soybean varieties, were introduced in the 1990s. These are genetically modified to produce insecticidal proteins from *Bacillus thuringiensis*, providing protection against insect pests (Tabashnik *et al.*, 2023). Faria et al. (2021) reported maize genotypes, such as 90XB06, IAC 8390, XB 9003, and DKB 310 PRO3, that exhibited antixenosis by reducing leaf hopper oviposition, while genotypes 60XB14, 90XB06, IAC 8046, XB 8018, and IAC 8390 showed antibiosis and/or antixenosis by decreasing nymphal viability. The use of RNAi and the newest and most sophisticated genome-editing technology using CRISPR-Cas 9 mechanism in developing insecticide-resistant plant with a distinct mode of action, particularly for insect resistance (Kumari et al., 2022), appears to be a highly effective and novel component of IPM. CRISPR-Cas gene editing is emerging as an effective tool for the development of insect-resistant plants to promote sustainable agriculture. This technique includes changing the effect or target of interactions, knocking out host-susceptible genes, and uncoupling the antagonistic action of defense hormones. Zhang et al. (2022) reported that CRISPR/Cas9-mediated mutation of *GmUGT* enhanced soybean resistance by modifying flavonoid levels and regulating defense-related gene expression. CRISPR-Cas technology mediated the deletion of *Orco* gene from *Spodopthera litura*; the knockout of odorant receptor gene 16 *(OR16)* from *Helicoverpa armigera* altered their response to host plants, while mutation of the α6 subunit of nicotinic acetylcholine receptor limited insecticide resistance (Koutroumpa et al., 2016; Zimmer et al., 2016; Chang et al., 2017). Similarly, Ashok et al. (2024) edited two spermatogenesis-related genes, *viz.*, *testis-specific zinc finger protein* (*topi*) and *testis-specific serine protein kinase 1 (Tssk1)*, of *B. dorsalis* to impart male sterility and its impact on further progeny, which resulted in a decreased number of viable offsprings.

18.12 CHEMICAL CONTROL–BASED IPM

Botanicals were among the earliest insecticides as the first insecticides with the neem tree's use mentioned in Veda over 4,000 years ago (El-Wakeil, 2013). Pyrethrin from *Dalmatian pyrethrum* flowers was used against cockroaches, bedbugs, flies, and mosquitoes (Davies et al., 2007), while aqueous tobacco extract was effectively used for the control of aphids (Sohail et al., 2012). In the 19th century, parapesticides, such as arsenics, lead arsenate, mercury, cryolite, and borax, were used as insect baits before the development of modern synthetic pesticides (Hassan, 2019). Paris green was used against *Colorado potato beetles* and *Anopheles* mosquito from 1871 until the 20th century (Majori, 2012; Oberemok et al., 2015). DDT was the first modern insecticide used against agricultural pest but was later banned due to its harmful effects on non-target organisms, such as mammals, birds, and reptiles (Jarman and Ballschmiter, 2012). Later, carbamates and organophosphates replaced DDT, and Ops still constitutes 19% of the global insecticide market (Casida and Durkin, 2013). Pyrethroid, such as permethrin, cypermethrin, and deltamethrin, derived from pyrethrin, gained popularity for their low toxicity. By the 1970s, pyrethroid accounted for 17% of the global insecticide market and remains widely used (Davies et al., 2007). Following the success of pyrethroids, neonicotinoids were introduced, which now occupy 24% of the global market but is reported to have negative impact on non-target organism (Jeschke et al., 2011). Since many of the key insects have developed resistance to old insecticide molecules, there is a need to develop newer molecules with different modes of action. Phenylpyrazoles, avermectins, diamides, spinosyns, insect growth regulators, pyrazole insecticides, macrocyclic lactone insecticides, formamidine, and metadiamide insecticides are all considered newer insecticidal molecules (Oberemok et al., 2015) and are now widely included in the IPM programs.

The goal of insecticide resistance management (IRM) is to delay resistance development or restore insect susceptibility to insecticides. Insects and mites might develop metabolic resistance to multiple insecticides due to enhanced enzyme activity or changes in behavior and pesticide penetration, making it essential to rotate insecticides with different modes of action to manage resistance effectively. The Insecticide Resistance Action Committee (IRAC) has classified insecticides into 37 groups, each targeting specific processes/receptors in the insect body (IRAC, 2025). Biorational pesticides, such as bacteria *Bacillus thuringiensis*, fungi *Trichoderma*, and Baculoviruses, are considered environmentally friendly pesticides due to low toxicity (Mahawer et al., 2024).

The harmful effects of conventional pesticides drove the need for innovative, eco-friendly nanopesticides since the early 2000s. These advanced nanoformulations enhance solubility, stability, and controlled release for prolonged pest control. Nanoparticles act as targeted carriers, minimizing pesticide use while maximizing efficacy (Dangi and Verma, 2021). Pesticide nanoformulations include microemulsions, nanoemulsions, nanodispersions, and nanoencapsulation, each with distinct characteristics. Chitosan, copper, zinc oxide, titanium dioxide, and silver nanoparticles are widely used for their antimicrobial properties, controlled release capabilities, and role as carrier materials in various applications (Priyanka et al., 2019). One such example of a nanopesticide is graphene, which exhibits enhanced insecticidal activity against *Sitophilus oryzae*, *Rhyzopertha dominica*, and *Prostephanus truncatus* (Batool et al., 2020).

18.13 DRONE (UNMANNED AERIAL VEHICLES [UAVS])–BASED IMAGING FOR PEST HOTSPOT MAPPING IN IPM

The introduction of drones was first done in Japan in the early 1980s for the application of pesticides on rice. Further, it was commercially introduced into the market in 1997 by the Japanese government in collaboration with Yamaha, a multinational company. In the early 2000s, drones were employed in pest survey, monitoring, and precision spraying of pesticides, which could reduce the number of sprays. For instance, fixed-wing drones were first employed in monitoring locust swarms in France in 2001 (Matthews, 2021). Then arrived the modern concept of sensing drones and actuation drones. Drones used for the detection of pest hotspots are referred to as *sensing drones*, while drones used for precision distribution of solutions are called *actuation drones*. In particular, actuation drones could reduce the area where pesticide applications are necessary and reduce the costs of dispensing natural enemies (Iost Filho et al., 2020).

Recent advances in drone technology may fulfill many aspects of IPM, like biological control, reduced pesticide usage, pest monitoring, sterile insect technique, and so on. Using drones, sterile insect technique was for the first time employed by releasing sterile insects of *Cydia pomonella* in apple orchards of Canada by Judd and Gardiner (2005). This technique was considered cheap, time-saving, and less laborious than ground release (Iost Filho et al., 2020). Meanwhile, the concept of spraying natural enemies with the use of aircraft was introduced in the early 1980s (Pickett et al., 1987). But it was not able to cover larger areas earlier. Gradually, the concept again flared back in 2017, by the introduction of actuation drones, which makes augmentative biocontrol seem easy. A large-scale effort has been done by a California-based company by offering services to distribute predatory mites *Phytoseiulus persimilis* Athias-Henriot using actuation drones on crops such as strawberry to combat two-spotted spider mite (Parabug, 2019). Even drones were employed in the application of egg parasitoid, *Trichogramma* sp., in many countries, such as Germany, France, and Austria, from 2018 (Airborne Robotics, 2018). Further, drones were also deployed in semiochemical-based pest management. For example, mating disruptors like SPLAT (specialized pheromone and lure application technology) are being followed in cranberry pest management at the University of Wisconsin (Seely, 2018).

18.14 PEST MONITORING AND SURVEY MADE EASY BY ARTIFICIAL INTELLIGENCE AND REMOTE SENSING

Remote sensing technologies allow a non-contact, large-scale monitoring of plant diseases and pests, which has an important advantage over conventional scouting methods. These technologies serve as a sort of "radiodiagnosis" of plants, recording infection symptoms, physiological responses, and structural alterations via spectral imaging techniques. Initial studies demonstrated the feasibility of aerial and satellite imagery in identifying pest-infested areas (Riley, 1989; Nilsson, 1995). This is especially true with the recent use of multispectral (MS), thermal, fluorescence, synthetic aperture radar (SAR), and light detection and ranging (lidar) imaging for pest monitoring applications (Zhang et al., 2019). These technologies enable precision agriculture by facilitating automated insect pest and disease identification and precision spraying, thereby improving the efficiency of pest control. Selecting the most appropriate spectral bands and thermal signatures that correspond to specific pest damage is critical for efficacious pest detection (Mahlein, 2016).

Advances in AI and remote sensing have enhanced pest monitoring and cultural control implementation. AI-driven models can predict aphid outbreaks with over 90% accuracy, allowing for directed interventions. Artificial intelligence and machine learning–integrated imaging technology–based high-resolution images captured by drones, called aerial imagery, are utilized in monitoring pest and beneficial insect population and observing crop health through the detection of symptoms caused by pest attack (Sneha et al., 2024; Aziz et al., 2025). The physical and physiological symptoms caused by insect attack, such as chlorosis, insect mining, reduction in photosynthesis, leaf rolling, etc., are recorded and used in the identification of plant stress levels to pest attack (Franklin, 2001).

Drones equipped with multispectral cameras could identify pest hotspots, enabling site-specific control measures (Zhou et al., 2024). The Indian Council of Agricultural Research has developed an AI-based smart pheromone trap that provides real-time monitoring of pink bollworm infestations in cotton, sending accurate insect data to alert farmers directly. This technique helps optimize pest management and reduce pesticide usage while ensuring timely interventions for crop protection (ICAR, 2025).

By 2010, RS was combined with the GIS technique, which was able to identify insect migration, breeding sites, and habitat detection. With this technique, locust migration and habitat detection were recorded by Latchininsky and Sivanpillai (2010). UAVs attached with various sensors, like multispectral scanners, have been used for large-scale pest surveillance (Figure 18.2). For example, leafhopper infestation in cotton was monitored with the help of hyperspectral

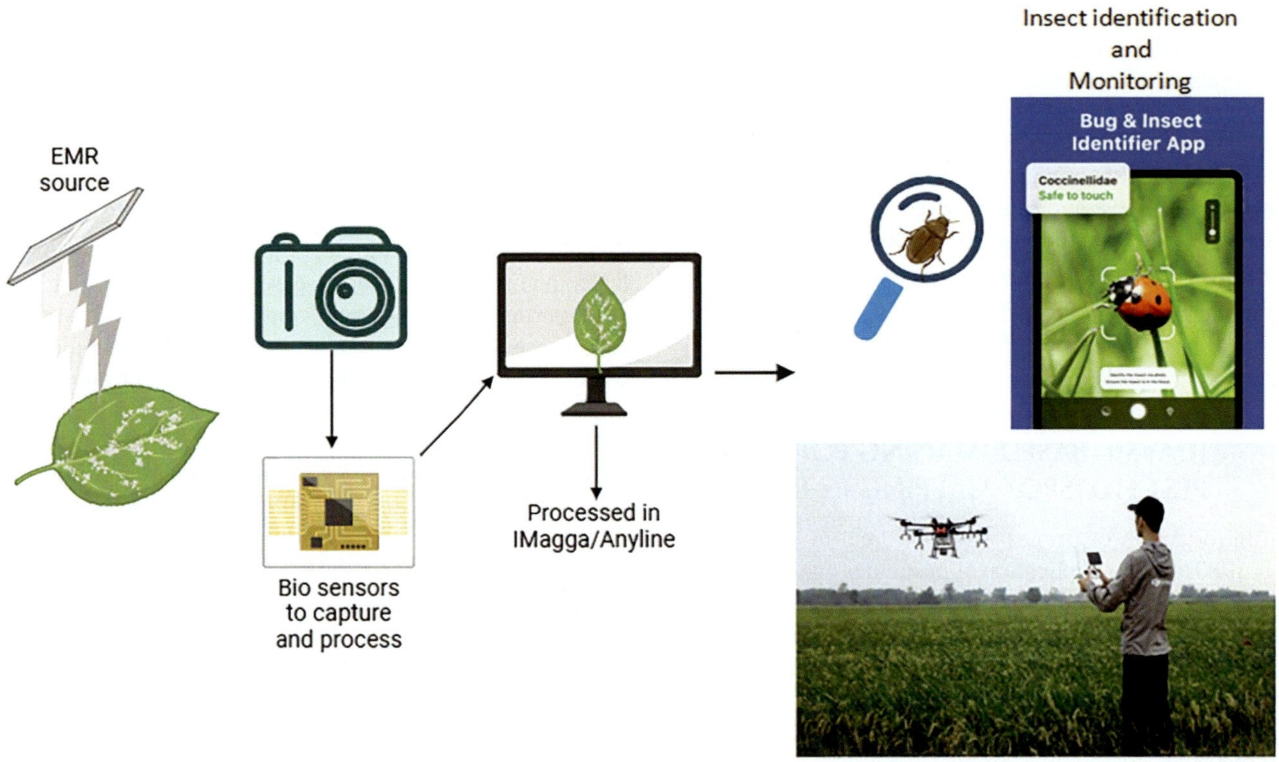

FIGURE 18.2 Sequential process of image capturing and processing in remote sensing for insect identification and monitoring. It begins with an electromagnetic radiation (EMR) source which interacts with plant surfaces, influencing reflectance properties. The reflected signals are captured by biosensors integrated with cameras, which process the data for further analysis. The captured images undergo processing using software like IMagga or Anyline, where features are analyzed and enhanced. The processed data is then used for insect identification through specialized applications, such as a bug and insect identifier app. The final step involves monitoring and decision-making, often facilitated by UAVs (drones) deployed in agricultural fields for real-time surveillance and pest management.

RS, which recorded the severity of damage and stress caused by the pest (Prabhakar et al., 2012). In addition to symptoms recorded, a new technology, called E-nose electronic sensor with fluorescence imaging, is being used to detect insect attack based on change in volatile organic compounds released in response to pest infestation (Fuentes et al., 2018). The real-time data collection and analysis possible with AI and IoT integration has revolutionized detection accuracy. Sensors and imaging tools are used as IoT devices to capture pest-related information about agriculture and processed using AI algorithms.

18.15 AI AND ML MODELS FOR AUTOMATED PEST IDENTIFICATION

Machine learning (ML) models play a significant role in automating pest identification and improving classification accuracy (Zhang et al., 2012). These models are developed to conduct remote sensing analyses to differentiate pest species, estimate infection severity, and identify early-stage infestations (Zhang et al., 2019). Spectral angle mapping, maximum likelihood classification, random forest, and support vector machines (SVMs) are the commonly used techniques (Bauriegel et al., 2011; Römer et al., 2011). Although data scarcity continues to pose a challenge, deep learning models, like convolutional neural networks (CNNs), continue to enhance classification accuracy when trained on large datasets (Ferentinos, 2018). The integration of ML with remote sensing enables precise and timely insect pest detection, which lowers crop losses and improves the overall agricultural productivity.

Using datasets, including images, sounds, and environmental data, machine learning models can detect and classify pests with a high precision (Kasinathan et al., 2021). A study employing a CNN-Bi-LSTM model achieved an accuracy of 98.91% in detecting sound analytics for pests surpassing traditional methods (Ali et al., 2023). Furthermore, AI-based monitoring systems enable the early detection of potential pest threats, facilitating timely and targeted interventions. The advanced insect detection network (AIDN) is a novel deep learning–based system designed to improve the accuracy of UAV-based insect detection. AIDN improves IPM strategies by enhancing the precision of pest identification and lowering the need for broad-spectrum pesticide applications. It tackles the challenges of identifying tiny and swift-moving insects in complex natural environments, where conventional image processing methods often fail due to dynamic backgrounds and scale of the insects (Safarov et al., 2025). AIDN makes use of multi-scale feature fusion, sophisticated activation techniques, and a custom-tailored loss function to enhance the efficacy (Khujamatov et al., 2025). This advanced system provides a strong solution for ecological monitoring and agricultural pest control, ensuring timely and precise pest detection (Santoso et al., 2025).

18.16 INTEGRATION OF TRADITIONAL TAXONOMY WITH AI FOR INSECT IDENTIFICATION

Integrating taxonomy with AI will enable the rapid and primary identification of pest. This makes us further decide the management strategies at a very early phase, before the pest reaches ETL. Initially, reflectance spectra with sensors were used for identifying pests. The device with multispectral reflectance spectra was able to detect mite infestation in cotton (Fitzgerald, 2000). This further evolved to hyperspectral imaging with modern sensors, such as multispectral scanner, for species identification. An ML-based algorithm was designed based on hyperspectral imaging to identify pests occurring in coffee by Lee and Tardaguila (2023). A Python-based software was developed with Keras and Tensflow framework by Saikumar et al. (2023) for detecting pests of brinjal crop. Many smartphone-based apps, like Plantix, Agrobase, Pestoz, CropDoctor, and Bioleaf, are also being used in insect pest identification.

18.17 GEOSPATIAL PEST MODELLING IN PEST MANAGEMENT

Using geographic information systems (GIS) to analyze the spatial distribution of pest populations and forecast possible outbreaks is known as geospatial pest modelling.

These models evaluate pest risks in various agroecological zones by combining geographical, ecological, and climatic data. Researchers can better assess risk and plan for management by mapping pest patterns and connecting phenology models to environmental conditions through the use of geostatistical algorithms. Geospatial pest modelling reduces pesticide use and enhances agricultural interventions by facilitating targeted pest control measures with actionable insights (Fand et al., 2014).

18.18 RECENT ADVANCES IN INTEGRATED PEST MANAGEMENT

IPM strategies are advancing with pheromone-based innovations, including meso- and mega-pheromone dispensers, which release larger quantities of pheromones compared to traditional hand-applied dispensers. Mega-dispensers can release up to 336 mg of pheromone per unit, significantly reducing labor and application costs, especially in large-scale agricultural settings (McGhee, 2014). Sustainability has also come to the forefront, with additional research on biodegradable pheromone lure-and-trap materials, such as polycoated recycled paper, lacto-polymers, and sawdust (Samet, 2018). Attract-and-kill technology combining pheromone lures with selective insecticides and auto-dissemination methods using entomopathogenic fungi further increase the effectiveness of pest control in inaccessible areas (Lacey et al., 2001). The integration of computer-assisted monitoring has further enhanced IPM strategies. These systems use AI and data analytics to estimate insect population dynamics, spatial distribution, and pest bionomics, allowing for precise, data-driven decision-making (Trematerra, 2012). Advances in chemical ecology have also contributed to IPM, with research focusing on reverse chemical ecology – a technique that identifies novel semiochemicals by studying odorant-binding proteins (OBPs) in insects (Evenden and Silk, 2016). This approach has the potential to uncover new pheromone-based solutions for pest management (Table 18.4).

Recent advances in IPM have also significantly benefited from genetic engineering, CRISPR technology, and RNA interference (RNAi), offering innovative and sustainable pest control solutions (Laichattiwar et al., 2024). Genetic engineering has enabled the development of genetically modified crops, such as Bt cotton and Bt maize, which

TABLE 18.4
Evolution of RS and AI in Aspects of IPM (Pest Surveillance, Monitoring, and Identification)

Year of Development	Technology Used	References
2005	Fluorescence imaging	Jia et al., 2007
2007–2016	Reflectance spectra (RS), multispectral RS, hyperspectral RS, proximal RS, satellite RS, leaf reflectance spectroscopy combined RS with GIS, RGB camera	Wang et al., 2016
2019	Harmonic RADAR	Maggiora et al., 2019
2021	AI with Python and Imagga cloud	Bhoi et al., 2021
2022	Convolutional recurrent neural network classifier (CRNN),	Aziz et al., 2025
2023	AI tools with ML algorithms	Aziz et al., 2025
2024	IoT-based UAVs with AI and python with RGB, multispectral, and infrared cameras	Yadav et al., 2024

express insecticidal proteins from *Bacillus thuringiensis*, providing built-in resistance to specific pests (Tilgam et al., 2021). CRISPR/Cas9 gene editing technology has shown promise in creating pest-resistant crops by knocking out genes essential for pest survival or reproduction, offering a precise, targeted approach to pest control (Komal et al., 2023). RNAi, which silences specific genes in pests, is also gaining traction as a biotechnological tool for pest management. By introducing RNAi constructs into crops, scientists can target essential genes in pests, like the corn rootworm, effectively reducing pest damage without harming beneficial organisms (Darlington et al., 2022). These technologies represent a shift toward more sustainable, environmentally friendly pest management, minimizing the reliance on chemical pesticides and reducing the impact on non-target species. Together, these advances in genetic engineering, CRISPR, and RNAi are revolutionizing the way pests are managed, providing novel solutions for sustainable agricultural practices (FIGURE 18.3).

18.19 CONSTRAINTS FACED IN IMPLEMENTING IPM

As IPM has been in existence for three decades, still it has not been practiced by not even one-third of the world's population. In particular, small-scale farmers face various types of challenges while implementing IPM. These constraints can be categorized into technical, economic, social, and institutional.

18.19.1 Technical Constraints

Small-scale farmers often had less accessibility regarding biocontrol approaches because of the limited reach of extension services. There was lack of involvement of technical expertise and absence of community participation. So information regarding pest monitoring and pest surveillance was not reached at farmers' level (Kumar et al., 2020). There was a saying: "What farmers don't understand, they won't follow and adopt." Many farmers are still unaware of the existence of few biocontrol agents, like predatory mites (Wyckhuys et al. 2019). Now, interventions are being made through farmers field schools, and farmers are guided to adopt community-level IPM programs (Bakker et al., 2021). So policies that encourage farmers toward greater levels of engagement and more proactive information seeking, such as through advisory professionals, more experienced peers through crop walks, open days and discussion groups, should be encouraged.

18.19.2 Economic Constraints

The high initial investment needed for pheromone-based pest control, or any other bio-based inputs, and the lack of subsidy to inputs, and also the lack of remuneration if crop faces any loss by pest attack (Gupta et al., 2020), are all the constraining factors. In case of drone usage, it was practically so difficult to be adopted by small-area holders because of the rental costs and equipment handling (Parsa

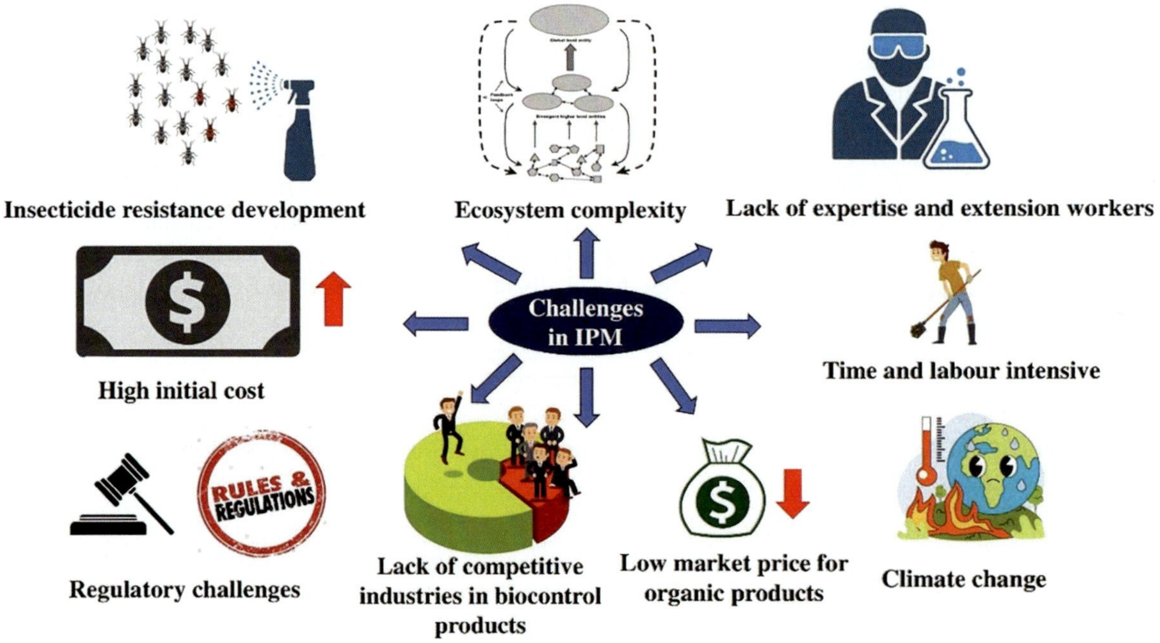

FIGURE 18.3 Major challenges that hinder the effective implementation of integrated pest management (IPM). These challenges include ecosystem complexity, lack of expertise and extension workers, resistance development in pests, regulatory and policy limitations, high initial cost, climate change, and complexity of integrating multiple control methods. Understanding these barriers is crucial for developing strategies to enhance the adoption and success of IPM program

et al., 2014). Also, drones, which are particularly designed in a way to cover larger area holdings, served as a disadvantage for marginal and small-scale farmers (Matthews, 2021).

18.19.3 Social Constraints

The availability of chemical pesticides was easier in the market than the availability of other biopesticides. The reason for this is, farmers are greatly influenced by agrochemical companies to buy their products, where farmers are also in the mindset of immediate control. They are much reluctant to give up on other aspects, as chemical control is cost-effective and effective within a short period of time (Deguine et al., 2021). Meanwhile, neither are there proper guidelines for the sale of biocontrol agents in developing countries nor a premium price for organically cultivated produces (Kumar et al., 2020).

18.19.4 Institutional Constraint

As far as research is concerned, there are many research gaps among the scientific community, such as not considering ecology while speaking about IPM. IPM programs are all pest-centric and not holistic/interdisciplinary. This was more pest-oriented than crop- and farmer-oriented (Barrera, 2020). Despite IPM's existence for over three decades, its adoption has been limited due to technical, economic, social, and institutional constraints. To enhance IPM adoption, strategies such as advisory services, improved access to technical expertise, and the promotion of community-driven IPM programs through FFS must be prioritized. Additionally, encouraging proactive information-seeking behaviors among farmers can significantly boost IPM adoption rates. Addressing these constraints with a more holistic, farmer-centric approach that integrates ecological considerations will be vital in achieving sustainable pest management practices globally.

18.20 CONCLUSION AND FUTURE DIRECTIONS

Over the last 60 years, integrated pest management has significantly contributed to sustainable agriculture, though several challenges have hindered its complete success. These challenges arise from the varying interpretations of IPM, uneven regional practices, and reliance on chemical pesticides. There are many programs which focuses on the usage of chemicals rather than supporting IPM's aim of reducing the use of pesticides, leading to a decline in studies on ecological dynamics in agroecosystems. It is now evident that IPM alone is inadequate to solve modern-day agricultural issues, such as biodiversity loss, climate change, and the need for food security.

Agroecological crop protection (ACP)/agroecosystem analysis (AESA)–based pest management serves as a promising alternative offering a holistic, systems-based approach. ACP focuses on health and the resilience of entire agroecosystems rather than focusing on controlling pest by IPM. Even without depending on synthetic inputs, ACP emphasizes biodiversity, soil health, and ecosystem services. Rather than focusing on individual pests, ACP follows a crop-oriented approach, considering the broader system and promoting preventive measures, such as crop diversification, soil management, and conservation of beneficial organisms. Transitioning from IPM to ACP requires a basic design of regional agricultural practices, shifting the focus from pest control to agroecosystem health. This shift requires interdisciplinary collaboration and participatory governance, involving farmers, researchers, and policymakers. Biotechnological advancements, such as genetic engineering and CRISPR gene editing technologies, could enable the development of pest-resistant crops, reducing the need for chemical interventions and improving resilience in ACP. Biopesticides and genetically engineered microorganisms can enhance agroecological practices by offering sustainable and effective pest control solutions. ACP, along with biotechnological innovations, promises great potential to transform a sustainable agriculture. By incorporating ecological principles with cutting-edge technologies, we need to develop a sustainable farming system that addresses the global food security challenges of the future.

REFERENCES

Abagale SA, Woodcock CM, Chamberlain K, Osafo-Acquaah S, van Emden H, Birkett MA, Pickett JA, Braimah H. 2019. Attractiveness of host banana leaf materials to the banana weevil, Cosmopolites sordidus in Ghana for development of field management strategies. *Pest Management Science* 75(2), 549–555.

AbdEl-Ghany NM, Abdel-Wahab ES, Ibrahim SS. 2016. Population fluctuation and evaluation the efficacy of pheromone-based traps with different color on tomato leafminer moth, Tuta absoluta (Lepidoptera: Gelechiidae) in Egypt. *Research Journal of Pharmaceutical Biological and Chemical Sciences* 7(4), 1533–1539.

Abrol DP, Shankar U. 2012. History, overview and principles of ecologically-based pest management. In *Integrated pest management: Principles and practice*. CABI, Wallingford, UK, 1–26.

Adetunji CO, Olaniyan OT, Anani OA, Inobeme A, Osemwegie OO, Hefft D, Akinbo O. 2023. Artificial intelligence and automation for precision pest management. In *Sensing and artificial intelligence solutions for food manufacturing*. CRC Press, 49–70.

Airborne Robotics. 2018. *Agriculture & forestry*. Available from www.air6systems.com/portfolio/agriculture-forestry/

Ali MA, Dhanaraj RK, Nayyar A. 2023. A high performance-oriented AI-enabled IoT-based pest detection system using sound analytics in large agricultural field. *Microprocessors and Microsystems* 103, 104946.

Amalin DM, Arcelo M, Almarinez BJM, Castillo RC, Legaspi JC, Santos KLT, Zhang A. 2023. Field evaluation of the sex pheromone of the cacao pod borer (Conopomorpha cramerella Snellen) in the Philippines. *Frontiers in Agronomy* 5, 1165299.

Anfora G, Vitagliano S, Larsson MC, Witzgall P, Tasin M, Germinara GS, De Cristofaro A. 2014. Disruption of

Phthorimaea operculella (Lepidoptera: Gelechiidae) oviposition by the application of host plant volatiles. *Pest Management Science* 70(4), 628–635.

Apple JL, Benepal PS, Berger R, Bird GW, Gutierrez AP, Maxwell F, Ruesnik WG, Santelmann P, White GB. 1979. *Integrated pest management: A program of research for the state agricultural experiment stations and colleges of 1890.* A Study conducted by the Intersociety Consortium for Plant Protection.

Ashok K, Bhargava CN, Pradeep C, Pradhan SK, Jha GK, Maligeppagol M, Asokan R. 2024. Toward the genetic suppression of *Bactrocera dorsalis* (Diptera: Tephritidae) through CRISPR/Cas9-mediated editing of spermatogenesis-related genes, Tssk1 and topi for imparting male sterility. *Annals of the Entomological Society of America* 117 (5), 270–279.

Ayelo PM, Yusuf AA, Chailleux A, Mohamed SA, Pirk CW, Deletre E. 2022. Chemical cues from honeydew and cuticular extracts of Trialeurodes vaporariorum serve as kairomones for the parasitoid Encarsia formosa. *Journal of Chemical Ecology* 48(4), 370–383.

Aziz D, Rafiq S, Saini P, Ahad I, Gonal B, Rehman SA, Rashid S, Saini P, Rohela GK, Aalum K, Singh G. 2025. Remote sensing and artificial intelligence: Revolutionizing pest management in agriculture. *Frontiers in Sustainable Food Systems* 9, 1551460.

Bakker T, Dugué P, De Tourdonnet S. 2021. Assessing the effects of Farmer Field Schools on farmers' trajectories of change in practices. *Agronomy for Sustainable Development* 41(2), 18.

Barrera JF. 2020 Holistic pest management. In *Aera-wide management of fruit fly pests.* CRC Press, Boca Raton, 275–293.

Barzman M, Bàrberi P, Birch ANE, Boonekamp P, Dachbrodt-Saaydeh S, Graf B, Hommel B, Jensen JE, Kiss J, Kudsk P, Lamichhane JR. 2015. Eight principles of integrated pest management. *Agronomy for Sustainable Development* 35, 1199–1215.

Bashyala S, Poudela D, Gautamb B. 2022. A review on cultural practice as an effective pest management approach under integrated pest management. *Tropical Agroecosystems (TAEC)* 3, 34–40.

Batool M, Hussain D, Akrem A, Najam-ul-Haq M, Saeed S, Zaka SM, Nawaz MS, Buck F, Saeed Q. 2020. Graphene quantum dots as cysteine protease nanocarriers against stored grain insect pests. *Scientific Reports* 10(1), 3444.

Bauriegel E, Giebel A, Geyer M, Schmidt U, Herppich WB. 2011. Early detection of Fusarium infection in wheat using hyperspectral imaging. *Computers and Electronics in Agriculture* 75(2), 304–312.

Berendsen RL, Pieterse CM, Bakker PA. 2012. The rhizosphere microbiome and plant health. *Trends in Plant Science* 17, 478–486.

Bhaskaran TV, Nithin CT, Bindu J, Gopal TKS, Kumar KA, Ravishankar CN. 2018. Drying fish preference assessment and efficacy of semiochemicals as repellents to blow fly Chrysomya megacephala (F.)(Diptera: Calliphoridae) during sun drying of fish. *ENTOMON* 43(3), 157–164.

Bhoi SK, Jena KK, Panda SK, Long HV, Kumar R, Subbulakshmi P, Jebreen HB. 2021. An Internet of Things assisted unmanned aerial vehicle based artificial intelligence model for rice pest detection. *Microprocessors and Microsystems* 80, 103607.

Buda V, Blazyte-Cereskiene L, Radziute S, Apsegaite V, Stamm P, Schulz S, Aleknavicius D, Mozuraitis R. 2020. Male-produced (−)-δ-heptalactone, pheromone of fruit fly Rhagoletis batava (Diptera: Tephritidae), a sea buckthorn berries pest. *Insects* 11(2), 138.

Butenandt A. 1959. Wirkstoffe des insektenreiches. *Naturw issenschaften* 46(15), 461–471.

Casida JE, Durkin KA. 2013. Anticholinesterase insecticide retrospective. *Chemico-biological Interactions*203(1), 221–225.

Chang H, Liu Y, AiD. 2017. A pheromone antagonist regulates optimal mating time in the moth *Helicoverpa armigera. Current Biology* 27(11), 1610–1615.

Chaudhary R, Nawaz A, Khattak Z, Butt MA, Fouillaud M, Dufossé L, Munir M, ul Haq I, Mukhtar H. 2024. Microbial bio-control agents: A comprehensive analysis on sustainable pest management in agriculture. *Journal of Agriculture and Food Research*, 101421.

Chen R, Chen C, Zhao X, Chen L, Xu T, Hao D. 2024. Identification and potential application of a putative allomone component in the defensive secretion of a destructive invasive species: The red-necked longhorn beetle, Aromia bungii. *Journal of Pest Science* 97(2), 993–1004.

CuiG Z, Zhu JJ. 2016. Pheromone-based pest management in China: Past, present, and future prospects. *Journal of Chemical Ecology* 42, 557–570.

Dangi K, Verma AK. 2021. Efficient & eco-friendly smart nanopesticides: Emerging prospects for agriculture. *Materials Today: Proceedings* 45, 3819–3824.

Dara SK. 2019. The new integrated pest management paradigm for the modern age. *Journal of Integrated Pest Management* 10(1), 12.

Darlington M, Reinders JD, Sethi A, Lu AL, Ramaseshadri P, Fischer JR, Boeckman CJ, Petrick JS, Roper JM, Narva KE, Vélez AM. 2022. RNAi for western corn rootworm management: Lessons learned, challenges, and future directions. *Insects*13(1), 57.

Davies TGE, Field LM, Usherwood PNR, Williamson MS. 2007. DDT, pyrethrins, pyrethroids and insect sodium channels. *IUBMB Life* 59(3), 151–162.

Deguine JP, Aubertot JN, Flor RJ, Lescourret F, Wyckhuys KA, Ratnadass A. 2021. Integrated pest management: Good intentions, hard realities. A review. *Agronomy for Sustainable Development* 41(3), 38.

Dent DR. 1995. Principles of integrated pest. *Integrated Pest Management* 8.

Dhaliwal GS, Koul O, Arora R. 2004. Integrated pest management: Retrospect and prospect. In Koul O, Dhaliwal GS (Eds.), *Integrated pest management: Potential, constraints and challenges.* CABI Publishing, Wallingford, UK, 1–20.

Dhaliwal GS, Koul O, Khokhar S, Ram S. 2013. Biopesticide-based integrated pest management for sustainable agriculture. In Koul O, Dhaliwal GS (Eds.), *Biopesticides in sustainable agriculture: Progress and potential.* Scientific Publishers (India), Jodhpur, 1–21.

Dhaliwal GS, Walia US, Brar LS. 1998. Response of Phalaris minor Retz. biotypes to various herbicides. *Indian Journal of Weed Science* 30, 116–120.

Dhawan AK, Peshin R. 2009. Integrated pest management: Concept, opportunities and challenges. *Integrated Pest Management: Innovation-Development Process* 1, 51–81.

Dougherty MJ, Guerin PM, Ward RD. 1995. Identification of oviposition attractants for the sandfly Lutzomyia longipalpis (Diptera: Psychodidae) in volatiles of faeces from vertebrates. *Physiological Entomology* 20(1), 23–32.

Dwivedi SA, Pandit TR. 2024. Host plant resistance and sustainable management of insect pests. In *Antimicrobial resistance in agriculture and its consequences.* CRC Press, 95–110.

Ehi-Eromosele CO, Nwinyi OC, Ajani OO. 2013. Integrated pest management. In *Weed and pest control-conventional and new challenges.* IntechOpen.

Ehrlich PR, Raven PH. 1964. Butterflies and plants: A study in coevolution. *Evolution*, 586–608.

El Titi A, Boller EF, Gendrier JP. 1993. *Integrated production: Principles and technical guidelines*. 3rd ed. IOBC/WPRS Bull 16.

El-Ghany NMA. 2019. Semiochemicals for controlling insect pests. *Journal of Plant Protection Research* 59(1).

El-Sayed AM. 2011. The pherobase: Database of insect pheromones and semiochemicals. The Horticulture and Food Research Institute of New Zealand, HortResearch. Available from http://www. pherobase.com

El-Sayed AM, Gibb AR, Mitchell VJ, Manning LAM, Revell J, Thistleton B, Suckling DM. 2013. Identification of the sex pheromone of *Conogethes pluto*: A pest of Alpinia. *Chemoecology* 23 (2), 93–101.

El-Wakeil NE. 2013. Botanical pesticides and their mode of action. *Gesunde Pflanzen* 65(4).

Evenden ML, Silk PJ. 2016. The influence of Canadian research on semiochemical-based management of forest insect pests in Canada. *The Canadian Entomologist* 148(S1), S170-S209.

Fand BB, Choudhary JS, Kumar M, Bal SK. 2014. Phenology modelling and GIS applications in pest management: A tool for studying and understanding insect-pest dynamics in the context of global climate change. *Approaches to Plant Stress and their Management*, 107–124.

FAO. 1967. *Report of the first session of the FAO panel of experts on integrated pest control*. Food and Agriculture Organization, Rome, Italy.

Faria RD, Baldin EL, Takaku VS, Canassa VF. 2021. Variable levels of antibiosis and/or antixenosis of Bt and non-Bt maize genotypes on Dalbulus maidis (Hemiptera: Cicadellidae). *Arthropod-Plant Interactions* 15(4), 457–465.

Ferentinos KP. 2018. Deep learning models for plant disease detection and diagnosis. *Computers and Electronics in Agriculture* 145, 311–318.

Ferron P. 1999. Protection intégrée des cultures: Évolution du concept et de son application. *Cahiers d'Etudes et de Recherches Francophones Agricultures (France)* 8(5).

Fitzgerald G. 2000. Bug checking for mites – from the sky. *Australian Cotton Grow* 21, 29–31.

Flint ML, Van den Bosch R. 2012. *Introduction to integrated pest management*. Springer Science & Business Media.

Fraenkel GS. 1959. The raison dètre of secondary plant substances: These odd chemicals arose as a means of protecting plants from insects and now guide insects to food. *Science* 129(3361), 1466–1470.

Franklin SE. 2001. *Remote sensing for sustainable forest management*. CRC Press.

Fuentes MT, Lenardis A, De la Fuente EB. 2018. nsect assemblies related to volatile signals emitted by different soybean – weeds – herbivory combinations. *Agriculture, Ecosystems & Environment* 255, 20–26.

Funaro CF, Böröczky K, Vargo EL, Schal C. 2018. Identification of a queen and king recognition pheromone in the subterranean termite Reticulitermes flavipes. *Proceedings of the National Academy of Sciences* 115(15), 3888–3893.

Gaines JC. 1957. Cotton insects and their control in the United States. *Annual Review of Entomology* 2(1), 319–338.

Galli M, Feldmann F, Vogler UK, Kogel KH. 2024. Can biocontrol be the game-changer in integrated pest management? A review of definitions, methods and strategies. *Journal of Plant Diseases and Protection* 131(2), 265–291.

Gatehouse JA. 2002. Plant resistance towards insect herbivores: A dynamic interaction. *New Phytologist* 156(2), 145–169.

Geier PW. 1966. Management of insect pests. *Annual Review of Entomology* 11(1), 471–490.

Geier PW, Clark LR. 1960. An ecological approach to pest control. In *Proceedings of the eighth technical meeting*. International Union of Conservation of Nature and Natural Resources, Warsaw, 10–18.

Gould F, Brown ZS, Kuzma J. 2018. Wicked evolution: Can we address the sociobiological dilemma of pesticide resistance? *Science* 360(6390), 728–732.

Gu D, Andreev K, Dupre ME 2021. Major trends in population growth around the world. *China CDC Weekly* 3(28), 604.

Gupta BK, Mishra BP, Singh V, Patel D, Singh MP. 2020. Constraints faced by vegetable growers in adoption of IPM in Bundelkhand region of Uttar Pradesh. *Indian Journal of Extension Education* 56(4), 92–97.

Haldhar SM, Berwal MK, Samadia DK, Kumar R, Gora JS, Choudhary S. 2018. Biochemical basis of plant-insect interaction in arid horticulture crops: A scientific review. *Journal of Agriculture and Ecology* 6, 1–16.

Hall CR, Carroll AR, Kitching RL. 2017. A meta-analysis of the effects of galling insects on host plant secondary metabolites. *Arthropod-Plant Interactions* 11, 463–473.

Hassan AS. 2019. Inorganic-based pesticides: A review article. *Egyptian Scientific Journal of Pesticides* 5, 39–52.

Hicks HL, Comont D, Coutts SR, Crook L, Hull R, Norris K, Neve P, Childs DZ, Freckleton RP. 2018. The factors driving evolved herbicide resistance at a national scale. *Nature Ecology & Evolution* 2(3), 529–536.

Higley LG, Karr LL, Pedigo LP. 1989. *Manual of entomology and pest management*. MacMillan Publishing Company.

Higley LG, Pedigo LP. 1996. The EIL concept. In *Economic thresholds for integrated pest management*. University of Nebraska Press, Lincoln, NE, 9–21.

Higley LG, Peterson RKD. 2009. Economic decision rules for IPM. In Radcliffe EB, Hutchison WD, Cancelado RE (Eds.), *Integrated pest management: Concepts, tactics, strategies and case studies*. Cambridge University Press, Cambridge, UK, 25–32.

ICAR. 2025. *CAR-CICR demonstrates potential of AI technology for real-time field monitoring of cotton pink bollworms in Punjab*. Available from www.icar.org.in/icar-cicr-demonstrates-potential-ai-technology-real-time-field-monitoring-cotton-pink-bollworms

Iost Filho FH, Heldens WB, Kong Z, de Lange ES. 2020. Drones: Innovative technology for use in precision pest management. *Journal of Economic Entomology* 113(1), 1–25.

IRAC. 2025. *Mode of action classification scheme version 11.1, January 2024*. Available from https://irac-online.org/documents/moa-classification/ (accessed January 2025).

Jarman WM, Ballschmiter K. 2012. From coal to DDT: The history of the development of the pesticide DDT from synthetic dyes till Silent Spring. *Endeavour* 36(4), 131–142.

Jayasooriya HJC, Aheeyar MMM. 2016. Adoption and factors affecting on adoption of integrated pest management among vegetable farmers in Sri Lanka. *Procedia Food Science* 6, 208–212.

Jeschke P, Nauen R, Schindler M, Elbert A. 2011. Overview of the status and global strategy for neonicotinoids. *Journal of Agricultural and Food Chemistry* 59(7), 2897–2908.

Jia F, Maghirang E, Dowell F, Abel C, Ramaswamy S. 2007. Differentiating tobacco budworm and corn earworm using near-infrared spectroscopy. *Journal of Economic Entomology* 100(3), 759–764.

Judd GJ, Gardiner MG. 2005. Towards eradication of codling moth in British Columbia by complimentary actions of mating disruption, tree banding and sterile insect technique: Five-year study in organic orchards. *Crop Protection* 24 (8), 718–733.

Kamle M, Borah R, Bora H, Jaiswal AK, Singh RK, Kumar P. 2020. Systemic acquired resistance (SAR) and induced systemic resistance (ISR): Role and mechanism of action against phytopathogens. *Fungal Biotechnology and Bioengineering*, 457–470.

Karinho-Betancourt E. 2020. Coevolution: Plant-herbivore interactions and secondary metabolites of plants. In *Co-evolution of secondary metabolites*, Springer International Publishing, Cham, 47–76.

Karlsson GK, Stenberg JA, Lankinen A. 2020. Making sense of Integrated Pest Management (IPM) in the light of evolution. *Evolutionary Applications* 13(8), 1791–1805.

Kasinathan T, Singaraju D, Uyyala SR. 2021. Insect classification and detection in field crops using modern machine learning techniques. *Information Processing in Agriculture* 8(3), 446–457.

Kerr A, Bullard G. 2020. Biocontrol of crown gall by Rhizobium rhizogenes: Challenges in biopesticide commercialisation. *Agronomy* 10(8), 1126.

Keszthelyi S, Pónya Z, Csóka Á, Bázár G, Morschhauser T, Donkó T. 2020. Non-destructive imaging and spectroscopic techniques to investigate the hidden-lifestyle arthropod pests: A review. *Journal of Plant Diseases and Protection* 127(3), 283–295.

Khujamatov H, Muksimova S, Abdullaev M, Cho J, Jeon HS. 2025. Advanced insect detection network for UAV-based biodiversity monitoring. *Remote Sensing* 17(6), 962.

Kline DL. 2007. Semiochemicals, traps/targets and mass trapping technology for mosquito management. *Journal of the American Mosquito Control Association* 23(sp2), 241–251.

Kogan M. 1998. Integrated pest management: Historical perspectives and contemporary developments. *Annual Review of Entomology* 43(1), 243–270.

Komal J, Desai HR, Samal I, Mastinu A, Patel RD, Kumar PD, Majhi PK, Mahanta DK, Bhoi TK. 2023. Unveiling the genetic symphony: Harnessing CRISPR-Cas genome editing for effective insect pest management. *Plants* 12(23), 3961.

Koppenhofer AM, Latin R, McGraw BA, Brosnan JT, Crow WT. 2013. Integrated pest management. *Turfgrass: Biology, Use, and Management* 56, 933–1006.

Koul O, Cuperus GW (Eds.). 2007. *Ecologically based integrated pest management*. CABI.

Koutroumpa FA, Monsempes C, François MC, de Cian A, Royer C, Concordet JP, Jacquin-Joly E. 2016. Heritable genome editing with CRISPR/Cas9 induces anosmia in a crop pest moth. *Scientific Reports* 6(1), 29620.

Kumar H, Sagar D, Nebapure SM. 2023. SPLAT: A green technology for insect pest management. *Insect Environment* 26(3), 369–374.

Kumar S, Sharma A, Samal RR. 2024. Management of agricultural pests: Strategies and constraints. IntechOpen. 1–17 doi: 10.5772/intechopen.1006075

Kumar S, Singh, A, Shankar R. 2020. Challenges in adoption of integrated pest management (IPM) practices in developing countries: A critical review. *Journal of Environmental Management* 280, 111692.

Kumari P, Jasrotia P, Kumar D, Kashyap PL, Kumar S, Mishra CN, Kumar S, Singh GP. 2022. Biotechnological approaches for host plant resistance to insect pests. *Frontiers in Genetics* 13, 914029.

Lacey LA, Frutos R, Kaya HK, Vail P. 2001. Insect pathogens as biological control agents: Do they have a future? *Biological Control* 21(3), 230–248.

Laichattiwar MA, Ubale NB. 2024. Innovations in integrated pest control strategies. In *Entomology redefined*, CRC Press, 120–133.

Lamdan NL, Shalaby S, Ziv T, Kenerley CM, Horwitz BA. 2015. Secretome of Trichoderma interacting with maize roots: Role in induced systemic resistance*[S]. *Molecular & Cellular Proteomics* 14(4), 1054–1063.

Latchininsky A., Sivanpillai R. 2010. Locust habitat monitoring and risk assessment using remote sensing and GIS technologies. In *Integrated management of arthropod pests and insect borne diseases*. Springer Netherlands, Dordrecht, 163–188.

Leake A. 2000. The development of integrated crop management in agricultural crops: Comparisons with conventional methods. *Pest Management Science: Formerly Pesticide Science* 56(11), 950–953.

Lee WS, Tardaguila J. 2023. Pest and disease management. In *Advanced automation for tree fruit orchards and vineyards*. Springer International Publishing, Cham, 93–118.

Lefebvre M, Langrell SRH, Gomez-y-Paloma S. 2015. Incentives and policies for integrated pest management in Europe: A review. *Agronomy for Sustainable Development* 35, 27–45.

Lenz EL, Krasnec MO, Breed MD. 2013. Identification of undecane as an alarm pheromone of the ant Formica argentea. *Journal of Insect Behavior* 26, 101–108.

Lewis WJ, Van Lenteren JC, Phatak SC, Tumlinson LII JH. 1997. A total system approach to sustainable pest management. *Proceedings of the National Academy of Sciences* 94(23), 12243–12248.

Lopes LE, Frank ET, Kárpáti Z, Schmitt T, Kronauer DJ. 2023. The alarm pheromone and alarm response of the clonal raider ant. *Journal of Chemical Ecology* 49(1), 1–10.

Luckmann WH, Metcalf RL. 1975. *Introduction to insect pest management*. John Wiley & Sons.

Ma FM, Zheng LX, Gao ZZ, Wu WJ. 2017. Farnesol, a synomone component between lantana (Lamiales: Verbenaceae) and the omnivorous predator, Campylomma chinensis Schuh (Hemiptera: Miridae). *Arthropod-Plant Interactions* 11, 703–708.

MacPhee AW, MacLellan CR. 1971. Cases of naturally-occurring biological control in Canada. In *Biological control: Proceedings of an AAAS symposium on biological control, held at Boston, Massachusetts December 30–31, 1969*. Springer US, Boston, MA, 312–328.

Mafra-Neto A, Fettig CJ, Munson AS, Rodriguez-Saona C, Holdcraft R, Faleiro JR, Villagran KM. 2014. Development of specialized pheromone and lure application technologies (SPLAT®) for management of coleopteran pests in agricultural and forest systems. In *Biopesticides: State of the art and future opportunities*. American Chemical Society, 211–242

Maggiora R, Saccani M, Milanesio D, Porporato M. 2019. An innovative harmonic radar to track flying insects: The case of Vespa velutina. *Scientific Reports* 9(1), 11964.

Mahawer SK, Karakoti H, Bargali P, Kumar R, Koli P, Prakash O, Chitara MK. 2024. An introduction to biorational pest control agents. *Biorationals and Biopesticides: Pest Management*, 1.

Mahlein AK. 2016. Plant disease detection by imaging sensors – parallels and specific demands for precision agriculture and plant phenotyping. *Plant Disease* 100(2), 241–251.

Majori G. 2012. Short history of malaria and its eradication in Italy with short notes on the fight against the infection in the Mediterranean basin. *Mediterranean Journal of Hematology and Infectious Diseases* 4(1), e2012016.

Martínez-Hidalgo P, García JM, Pozo MJ 2015. Induced systemic resistance against Botrytis cinerea by Micromonospora strains isolated from root nodules. *Frontiers in Microbiology* 6, 922.

Matthews GA. 2021. New technology for desert locust control. *Agronomy* 11 (6), 1052.

Mawcha KT, Malinga L, Muir D, Ge J, Ndolo D. 2024. Recent advances in biopesticide research and development with a focus on microbials. *F1000Research* 13, 1071.

McGhee PS. 2014. *Impact of high releasing mating disruption formulations on (male) codling moth, Cydia pomonella L., behavior*. Michigan State University.

Metcalf RL. 1980. Changing role of insecticides in crop protection. *Annual Review of Entomology* 25, 215–250.

Muthomi PK, Rhodes E, Liburd OE. 2025. Evaluation of a SPLAT®-based semiochemical with insecticides for diamondback moth management in cabbage. *Arthropod Management Tests* 50(1), tsaf014.

Naranjo SE. 2001. Conservation and evaluation of natural enemies in IPM systems for Bemisia tabaci. *Crop Protection* 20(9), 835–852.

National Academy of Sciences. 1969. Insect-pest management and control. Principles of plant and animal pest control. *National Academy of Sciences Publication* 1695 (3), 448–449.

NIFA, USDA. 2023. Available from www.nifa.usda.gov/about-nifa/blogs/researchers-helping-protect-crops-pests.

Nilsson HE. 1995. Remote sensing and image analysis in plant pathology. *Canadian Journal of Plant Pathology* 17(2), 154–166.

Nixon RM. 1975. *Public papers of the presidents of the United States: Richard M. Nixon, 1974*. Best Books On.

Oates CN, Denby KJ, Myburg AA, Slippers B, Naidoo S. 2016. Insect gallers and their plant hosts: From omics data to systems biology. *International Journal of Molecular Sciences* 17(11), 1891.

Oberemok VV, Laikova KV, Gninenko YI, Zaitsev AS, Nyadar PM, Adeyemi TA. 2015. A short history of insecticides. *Journal of Plant Protection Research* 55(3).

Painter RH. 1951. *Insect resistance in crop plants*. The University of Kansas Press, Lawrence, KS.

Palumbi, SR. 2001. Humans as the world's greatest evolutionary force. *Science* 293(5536), 1786–1790.

Pamidi LS, Sekhar JC, Yathish KR, Karjagi CG, Rao KS, Suby SB, Rakshit S. 2025. Characterization of antixenosis and antibiosis resistance to the fall armyworm, Spodoptera frugiperda (JE Smith) in maize. *Phytoparasitica* 53(1), 4.

Papanastasiou SA, Ioannou CS, Papadopoulos NT. 2020. Oviposition-deterrent effect of linalool – a compound of citrus essential oils – on female Mediterranean fruit flies, Ceratitis capitata (Diptera: Tephritidae). *Pest Management Science* 76(9), 3066–3077.

Parabug. 2019. *Parabug, biocontrol by drone*. Available from www. parabug.solutions/

Parsa S, Morse S, Bonifacio A, Chancellor TCB, Condori B, Crespo-Pérez V, Hobbs SL. 2014. Obstacles to integrated pest management adoption in developing countries. *Proceedings of the National Academy of Sciences* 111(10), 3889–3894.

Perkins JH. 1980. The quest for innovation in agricultural entomology, 1945–1978. In *Pest control: Cultural and environmental aspects*. Westview Press. Boulder, CO, 23–80.

Peterson RKD, Higley LG, Pedigo LP. 2018. Whatever happened to IPM? *American Entomologist* 64(3), 146–150.

Pickett CH, Gilstrap FE, Morrison RK, Bouse LF. 1987. Release of predatory mites (Acari: Phytoseiidae) by aircraft for the biological control of spider mites (Acari: Tetranychidae) infesting corn. *Journal of Economic Entomology* 80(4), 906–910.

Pimentel D, Perkins JH. 2019. *Pest control: Cultural and environmental aspects*. CRC Press.

Prabhakar M, Prasad YG, Rao MN. 2012. Remote sensing of biotic stress in crop plants and its applications for pest management. In *Crop stress and its management: Perspectives and strategies*. Springer, New York, 517–549.

President's Science Advisory Committee (PSAC). 1963. *Use of pesticides*. U.S. Gov. Print. Office, Washington, DC. Available from www.jlibrary.org/Asset-Viewer/Archives/JFKPOF-087-003.aspx (accessed 17 May 2012).

Priyanka P, Kumar D, Yadav K, Yadav A. 2019. Nanopesticides: Synthesis, formulation and application in agriculture. *Nanobiotechnology Applications in Plant Protection* 2, 129–143.

Qu YF, Li X, Dong HH, Wang WJ, Li XF, Feng H, Li K.2025. Identification and field verification of the aggregation pheromone components produced by male Holotrichia parallela Motschulsky (Coleoptera: Scarabaeidae). *Pest Management Science* 81(1), 427–437.

Raupach GS, Liu L, Murphy JF, Tuzun S, Kloepper JW. 1996. Induced systemic resistance in cucumber and tomato against cucumber mosaic cucumovirus using plant growth-promoting rhizobacteria (PGPR). *Plant Disease* 80(8), 891–894.

Reddy GV, Guerrero A. 2004. Interactions of insect pheromones and plant semiochemicals. *Trends in Plant Science* 9(5), 253–261.

Reddy GV, Sharma A, Guerrero A. 2020. Advances in the use of semiochemicals in integrated pest management: Pheromones. In *Biopesticides for sustainable agriculture*. Burleigh Dodds Science Publishing, 251–282.

Reis AC, Neta PL, Jordão JP, Moura JIL, Vidal DM, Zarbin PH, Fávaro CF. 2018. Aggregation pheromone of the bearded weevil, Rhinostomus barbirostris (Coleoptera: Curculionidae): Identification, synthesis, absolute configuration and bioactivity. *Journal of Chemical Ecology* 44, 463–470.

Renyard A, Alamsetti SK, Gries R, Munoz A, Gries G. 2019. Identification of the trail pheromone of the carpenter ant Camponotus modoc. *Journal of Chemical Ecology* 45, 901–913.

Rezaei R, Safa L, Damalas CA, Ganjkhanloo MM. 2019. Drivers of farmers' intention to use integrated pest management: Integrating theory of planned behavior and norm activation model. *Journal of Environmental Management* 236, 328–339.

Ricciardi R, Di Giovanni F, Cosci F, Ladurner E, Savino F, Iodice A, Benelli G, Lucchi A. 2021. Mating disruption for managing the honeydew moth, Cryptoblabes gnidiella (Millière), in mediterranean vineyards. *Insects* 12(5), 390.

Riley JR. 1989. Remote sensing in entomology. Annual Review of Entomology 34, 247–271.

Ripper WE.1956. Effect of pesticides on balance of arthropod populations. *Annual Review of Entomology* 1(1), 403–438.

Roberts D, Tren R, Bate R, Zambone J. 2016. *The excellent powder: DDT's political and scientific history*. Dog Ear Publishing.

Römer C, Bürling K, Hunsche M, Rumpf T, Noga G, Plümer L. 2011. Robust fitting of fluorescence spectra for pre-symptomatic wheat leaf rust detection with support vector machines. *Computers and Electronics in Agriculture* 79(2), 180–188.

Rundlof M, Andersson GK, Bommarco R, Fries I, Hederström V, Herbertsson L, Jonsson O, Klatt BK, Pedersen TR, Yourstone J, Smith HG. 2015. Seed coating with a neonicotinoid insecticide negatively affects wild bees. *Nature* 521(7550), 77–80.

Ryu CM, Farag MA, Hu CH, Reddy MS, Kloepper JW, Paré PW. 2004. Bacterial volatiles induce systemic resistance in Arabidopsis. *Plant Physiology* 134(3), 1017–1026.

Sabbahi R, Hock V, Azzaoui K, Saoiabi S, Hammouti B. 2022. A global perspective of entomopathogens as microbial biocontrol agents of insect pests. *Journal of Agriculture and Food Research* 10, 100376.

Safarov F, Khojamuratova U, Komoliddin M, Bolikulov F, Muksimova S, Cho YI. 2025. MBGPIN: Multi-branch generative prior integration network for super-resolution satellite imagery. *Remote Sensing* 17(5), 805.

Saikumar N, Emmanuel N, Krishna KSP, Chinnabbai C, Krishna KU. 2023. Artificial intelligence for classification and detection of major insect pests of Brinjal. *Indian Journal of Entomology* 85(3).

Samet C. 2018. U.S. Patent Application No. 15/549,757.

Santoso CB, Singadji M, Purnama DG, Abdel S, Kharismawardani A. 2025. Enhancing apple leaf disease detection with deep learning: From model training to android app integration. *Journal of Applied Data Sciences* 6(1), 377–390.

Scarano F, Giannetti D, Trenti F, Giacomazzi F, Vigna J, Guella G, Grasso DA, Haase A. 2024. Trail pheromone identification in the ant Crematogaster scutellaris. *Scientific Reports* 14(1), 7883.

Seely R. 2018. Drones, joysticks, and data-driven farming. In *Grow*. University of Wisconsin-Madison College of Agricultural and Life Sciences, 16–21. Available from https://grow.cals.wisc.edu/wp-content/uploads/sites/14/2018/06/Grow-Summer2018-web.pdf

Senthil-Nathan S. 2013. Physiological and biochemical effect of neem and other Meliaceae plants secondary metabolites against Lepidopteran insects. *Frontiers in Physiology* 4, 359.

Smart LE, Aradottir GI, Bruce TJA. 2014. Role of semiochemicals in integrated pest management. *Integrated Pest Management*, 93–109.

Smith CM. 2005. *Plant resistance to arthropods: Molecular and conventional approaches*. Springer, Dordrecht, The Netherlands, 423.

Smith CM. 2021. Conventional breeding of insect-resi0stant crop plants: Still the best way to feed the world population. *Current Opinion in Insect Science* 45, 7–13.

Smith RF, Allen WW. 1954. Insect control and the balance of nature. *Scientific American* 190(6), 38–43.

Smith RF, Apple JL, Bottrell DG. 1976. The origins of integrated pest management concepts for agricultural crops. *Integrated Pest Management*, 1–16.

Sneha P, Nagarathna N. 2024. Traffic sign recognition for automatic vehicles. In *2024 IEEE international conference for women in innovation, technology & entrepreneurship (ICWITE)*. IEEE, 534–538.

Sohail A, Hamid FS, Waheed A, Ahmed N, Aslam N, Zaman Q, Ahmed F, Islam S. 2012. Efficacy of different botanical materials against aphid Toxoptera aurantii on tea (Camellia sinensis L.) cuttings under high shade nursery. *Journal of Materials and Environmental Science* 3(6), 1065–1070.

Stenberg JA. 2017. A conceptual framework for integrated pest management. *Trends in Plant Science* 22(9), 759–769.

Stern VM, Smith RF, Van den Bosch R, Hagen KS. 1959. The integration of chemical and biological control of the spotted alfalfa aphid. *Hilgardia* 29(2), 81–101.

Stone JD, Pedigo LP. 1972. Development and economic-injury level of the green cloverworm on soybean in Iowa. *Journal of Economic Entomology* 65(1), 197–201.

Stout M, Davis J. 2009. Keys to the increased use of host plant resistance in integrated pest management. *Integrated Pest Management: Innovation-development Process* 1, 163–181.

Stout MJ. 2019. Plant-insect interactions, host-plant resistance, and integrated pest management. In *Integrated management of insect pests*. Burleigh Dodds Science Publishing, 191–224.

Stout MJ, Bernaola L, Acevedo F. 2024. Recent history and future trends in host – plant resistance. *Annals of the Entomological Society of America* 117(3), 139–149.

Tabashnik BE, Fabrick JA, Carrière Y. 2023. Global patterns of insect resistance to transgenic Bt crops: The first 25 years. *Journal of Economic Entomology* 116(2), 297–309.

Thrall PH, Oakeshott JG, Fitt G, Southerton S, Burdon JJ, Sheppard A, Russell RJ, Zalucki M, Heino M, Ford Denison R. 2011. Evolution in agriculture: The application of evolutionary approaches to the management of biotic interactions in agro-ecosystems. *Evolutionary Applications* 4(2), 200–215.

Tilgam J, Kumar K, Jayaswal D, Choudhury S, Kumar A, Jayaswall K, Saxena AK. 2021. Success of microbial genes based transgenic crops: Bt and beyond Bt. *Molecular Biology Reports* 48(12), 8111–8122.

Trematerra P. 2012. Advances in the use of pheromones for stored-product protection. *Journal of Pest Science* 85, 285–299.

USDA-ARS (United States Department of Agriculture-Agricultural Research Service). 2018. A national road map for integrated pest management. https://www.ars.usda.gov/ARSUserFiles/OPMP/IPM%20Road%20Map%20FINAL.pdf

Van den Bosch R, Stern VM. 1962. The integration of chemical and biological control of arthropod pests. *Annual Review of Entomology* 7(1), 367–386.

Wang Y, Nansen C, Zhang Y. 2016. Integrative insect taxonomy based on morphology, mitochondrial DNA, and hyperspectral reflectance profiling. *Zoological Journal of the Linnean Society* 177(2), 378–394.

Weiberg A, Wang M, Lin FM, Zhao H, Zhang Z, Kaloshian I, Huang HD, Jin H. 2013. Fungal small RNAs suppress plant immunity by hijacking host RNA interference pathways. *Science* 342(6154), 118–123.

Witzgall P, Kirsch P, Cork A. 2010. Sex pheromones and their impact on pest management. *Journal of Chemical Ecology* 36, 80–100.

Wyckhuys KAG, Pozsgai G, Lovei GL, Vasseur L, Wratten SD, Gurr GM, Reynolds OL, Goettel M. 2019. Global disparity in public awareness of the biological control potential of invertebrates. *Science of the Total Environment* 660, 799–806.

Yadav M, Vashisht BB, Vullaganti N, Jalota SK, Yadav SL, Singh G, Kumar A, Kothiyal S. 2024. IoT-enabled unmanned aerial vehicle: An emerging trend in precision farming. In *Artificial intelligence and smart agriculture: Technology and applications*, Springer, Singapore, 271–292.

Zaki O, Scorsone E, Lissorgues G, Français O, Rousseau L. 2024. Development of pheromone-receptor-based biosensors for the early detection of pest insects. *Proceedings* 97, 220.

Zebelo SA, Maffei ME. 2015. Role of early signalling events in plant – insect interactions. *Journal of Experimental Botany* 66(2), 435–448.

Zhang J, Huang Y, Pu R, Gonzalez-Moreno P, Yuan L, Wu K, Huang W. 2019. Monitoring plant diseases and pests through remote sensing technology: A review. *Computers and Electronics in Agriculture* 165, 104943.

Zhang J, Pu R, Huang W, Yuan L, Luo J, Wang J. 2012. Using in-situ hyperspectral data for detecting and discriminating yellow rust disease from nutrient stresses. *Field Crops Research* 134, 165–174.

Zhang Y, Guo W, Chen L, Shen X, Yang H, Fang Y, Ouyang W, Mai S, Chen H, Chen S, Hao Q. 2022. CRISPR/Cas9-mediated targeted mutagenesis of GmUGT enhanced soybean resistance against leaf-chewing insects through flavonoids biosynthesis. *Frontiers in Plant Science* 13, 802716.

Zhou W, Arcot Y, Medina RF, Bernal J, Cisneros-Zevallos L, Akbulut ME. 2024. Integrated pest management: An update on the sustainability approach to crop protection. *ACS Omega* 9(40), 41130–41147.

Zia Ul Haq M, Ali I (Eds.). 2024. *Revolutionizing pest management for sustainable agriculture*. IGI Global.

Zimmer CT, Garrood WT, Puinean AM, Eckel-Zimmer M, Williamson MS, Davies TE, Bass C. 2016. A CRISPR/Cas9 mediated point mutation in the alpha 6 subunit of the nicotinic acetylcholine receptor confers resistance to spinosad in Drosophila melanogaster. *Insect Biochemistry and Molecular Biology* 73, 62–69.

Index

P

Q

R

S

T

W

X

For Product Safety Concerns and Information please contact our EU representative GPSR@taylorandfrancis.com Taylor & Francis Verlag GmbH, Kaufingerstraße 24, 80331 München, Germany